# NORMAL FORMS AND BIFURCATION OF PLANAR VECTOR FIELDS

# NORMAL FORMS AND BIFURCATION OF PLANAR VECTOR FIELDS

**SHUI-NEE CHOW**
*Georgia Institute of Technology*

**CHENGZHI LI**
*Peking University*

**DUO WANG**
*Tsinghua University*

CAMBRIDGE UNIVERSITY PRESS
Cambridge, New York, Melbourne, Madrid, Cape Town, Singapore, São Paulo, Delhi

Cambridge University Press
The Edinburgh Building, Cambridge CB2 8RU, UK

Published in the United States of America by Cambridge University Press, New York

www.cambridge.org
Information on this title: www.cambridge.org/9780521372268

First published 1994
This digitally printed version 2008

*A catalogue record for this publication is available from the British Library*

*Library of Congress Cataloguing in Publication data*

Chow, Shui-Nee.
Normal forms and bifurcation of planar vector fields / Shui-Nee
Chow, Chengzhi Li, Duo Wang.
p. cm.
Includes bibliographical references and index.
ISBN 0-521-37226-7
1. Bifurcation theory. 2. Vector fields. 3. Normal forms
(Mathematics) I. Li, Chengzhi. II. Wang, Duo. III. Title.
QA372.C548 1994
515′.35—dc20 93-26510 CIP

ISBN 978-0-521-37226-8 hardback
ISBN 978-0-521-10223-0 paperback

# Contents

# Preface

The theory of bifurcation of vector fields is the study of a family of equations that are close to a given equation. For example, the family of equations could be a system of vector fields depending on several parameters. An important problem is to understand how the topological structure of the flow generated by the family of vector fields changes qualitatively as parameters are varied. The main purpose of this book is to present some methods and results of the theory of bifurcations of planar vector fields.

Since simplifying equations is often a necessary first step in many bifurcation problems, we introduce the theory of center manifolds and the theory of normal forms. Center manifold theory is important for the reduction of equations to ones of lower dimension, and normal-form theory gives a tool for simplifying the forms of equations to the ones with the simplest possible higher-order terms near their equilibria. We introduce versal deformations of vector fields and define the codimension of a bifurcation of vector fields. This is illustrated by saddle-node and Hopf bifurcations. We discuss in detail all known codimension-two bifurcations of planar vector fields. Some special cases of higher-codimension bifurcations are also considered.

In Chapter 1, we introduce briefly the basic concepts of center manifolds. We show the existence, uniqueness, and smoothness of global center manifolds. The existence, asymptotic behavior, and foliation of local center manifolds are also discussed.

In Chapter 2, we present the theory of normal forms. We first discuss in detail normal forms of vector fields near their equilibria. We introduce two methods for computing normal forms: the matrix representation method and the method of adjoints. We also introduce normal forms of equations with periodic coefficients or with symmetries. Normal forms of diffeomorphisms and Hamiltonian systems are discussed.

Complete proofs of Poincaré and Siegel linearization theorems are presented. Takens's Theorem gives a relation between diffeomorphisms near fixed points and the time-one maps of flows of vector fields near equilibria. We introduce also versal deformations of matrices and of infinitesimally symplectic matrices and normal forms of vector fields of codimension one and two.

In Chapters 3, 4, and 5, we discuss bifurcation problems of vector fields with some degeneracies. We assume that the problems to be considered are restricted to local center manifolds and are in their normal forms up to some order. In Chapter 3, we introduce the concepts of versal deformations and the codimension of a bifurcation of vector fields. Bifurcations of codimension one near singularities and homoclinic orbits are considered. In Chapter 4, we deal with bifurcations of codimension two. For vector fields whose linear parts have double zero eigenvalues, we consider a nonsymmetrical case and the cases with $1:q$ symmetries ($q = 2, 3, 4$ and $q \geq 5$). The case of $1:4$ symmetry is the most difficult and is far from being solved completely. For the cases in which the linear parts have one zero and one pair of purely imaginary eigenvalues, or two pairs of purely imaginary eigenvalues, we reduce them to planar systems and then give complete bifurcation diagrams. In Chapter 5, we discuss higher-codimension bifurcation problems, including Hopf and homoclinic bifurcations with any codimension and cusp bifurcations with codimension three and four.

In the last section of each chapter we give briefly the history and literature of material covered in the chapter. We have tried to make our references as complete as possible. However, we are sure that many are missing.

We would like to express our special acknowledgment to Max Ashkenazi, Freddy Dumortier, Jibin Li, Kening Lu, Robert Roussarie, Christiane Rousseau, Lan Wen, and Henryk Żoładek. They read all or part of the original manuscript and made many helpful suggestions which enabled us to correct some mistakes and make improvements.

The second and the third authors would also like to thank Professor Zhifen Zhang and Professor Tongren Ding for many helpful discussions. They would also like to thank the Department of Mathematics at Michigan State University and the Center for Dynamical Systems and Nonlinear Studies and the School of Mathematics at the Georgia Institute of Technology for their kind hospitality, since most of the book was written while they were visiting there.

This work was partially supported by grants from DARPA and NSF (USA) and from the National Natural Science Foundation of China.

# 1

# Center Manifolds

The main goal of this book is to study some bifurcation phenomena of vector fields. This is, in general, a complicated problem. As a preliminary step, it is necessary to simplify the problem as much as possible without changing the dynamic behavior of the original vector field. There are two steps for this purpose: to reduce the dimension of the bifurcation problem by using the center-manifold theory, which will be introduced in this chapter, and to make the equation as simple as possible by using normal-form theory which will be discussed in the next chapter.

We first give some rough ideas about center manifolds. Consider a differential equation

$$\dot{x} = Ax + f(x), \qquad \qquad (A)_f$$

where $x \in \mathbb{R}^n$, $A \in \mathscr{L}(\mathbb{R}^n, \mathbb{R}^n)$, $f \in C^k(\mathbb{R}^n, \mathbb{R}^n)$ for some $k \geq 1$, $f(0) = 0$, and $Df(0) = 0$.

We write the spectrum $\sigma(A)$ of $A$ as

$$\sigma(A) = \sigma_s \cup \sigma_c \cup \sigma_u,$$

where

$$\sigma_s = \{\lambda \in \sigma(A) \,|\, \mathrm{Re}\, \lambda < 0\},$$

$$\sigma_c = \{\lambda \in \sigma(A) \,|\, \mathrm{Re}\, \lambda = 0\},$$

$$\sigma_u = \{\lambda \in \sigma(A) \,|\, \mathrm{Re}\, \lambda > 0\}.$$

Let $E_s$, $E_c$, and $E_u$ be the generalized eigenspaces corresponding to $\sigma_s$,

1

$\sigma_c$, and $\sigma_u$, respectively. Then we have

$$\mathbb{R}^n = E_s \oplus E_c \oplus E_u,$$

with corresponding projections

$$\pi_s\colon \mathbb{R}^n \to E_s, \qquad \pi_c\colon \mathbb{R}^n \to E_c, \qquad \pi_u\colon \mathbb{R}^n \to E_u.$$

It is well known that if $A$ is hyperbolic, that is, $\sigma_c = \varnothing$, then the flow of $(A)_f$ in a small neighborhood $\Omega$ of the equilibrium point $x = 0$ is topologically equivalent to the flow of the linearized equation at $x = 0$

$$\dot{x} = Ax. \tag{$A)_0$}$$

Since $x(t) = e^{At}x(0)$ is the solution of $(A)_0$ and $\sigma_c = \varnothing$, any nonzero solution in $E_s$ (or $E_u$) tends to the equilibrium $x = 0$ exponentially as $t \to +\infty$ (or $t \to -\infty$). Therefore, the structure of flow in $\Omega$ is simple; it is also stable with respect to any small perturbation on the right-hand side of equation $(A)_0$. See Hartman [1], for example.

However, if $\sigma_c \neq \varnothing$, then the situation will be different from the above in two aspects. First, the topological structure for $(A)_f$ is not, in general, the same as for $(A)_0$ any more; this will be shown in a lot of examples in Chapters 3–5. Second, more complicated structure of the flow for $(A)_f$ may exist on an invariant manifold $W^c(f)$, and the dimension of $W^c(f)$ is equal to the dimension of $E_c$.

In fact, if $f \equiv 0$, then all bounded solutions of $(A)_0$, including all equilibria and periodic orbits, are contained in the subspace $E_c$, which is invariant under $(A)_0$. So we take $W^c(0) = E_c$. We will prove that the aforementioned $W^c(f)$ exists for $f \not\equiv 0$, it is tangent to $E_c$ at $x = 0$, and $W^c(f)$ contains all solutions of $(A)_f$ that stay in $\Omega$ for all $t \in \mathbb{R}^1$. In particular, $W^c(f)$ contains all sufficiently small equilibria, periodic orbits, and homoclinic and heteroclinic orbits. And if $\sigma_u = \varnothing$, then all solutions of $(A)_f$ (in $\Omega$) will converge exponentially to some solutions on $W^c(f)$ as $t \to +\infty$. Therefore, instead of the $n$-dimensional equation $(A)_f$, we can consider a lower-dimensional equation on $W^c(f)$ for a bifurcation problem, and $W^c(f)$ is called a center manifold. The precise definition will be given subsequently in Section 1.1.

We will prove the existence, uniqueness, and smoothness of global center manifolds in Sections 1.1–1.2 under a quite strong condition which says the Lipschitz constant of $f$ is globally small. In Section 1.3 the cut-off technique is used to get the local center manifolds from the global theory, and the above Lipschitz condition will be satisfied automatically since $f(0) = 0$ and $Df(0) = 0$. But a new problem arises: The local center manifold is not unique. In fact, different cut-off functions can give different local center manifolds. Hence, it is needed to show the equivalence (in some sense) between different local center manifolds concerning the bifurcation problems. Finally, in Section 1.4 we discuss the center-stable and center-unstable manifolds, give the asymptotic behavior of any solution of (1.1) in $\mathbb{R}^n$, and describe the invariant foliation structure.

### 1.1  Existence and Uniqueness of Global Center Manifolds

Consider the equation

$$\dot{x} = Ax + f(x), \tag{1.1}$$

where $x \in \mathbb{R}^n$, $A \in \mathcal{L}(\mathbb{R}^n, \mathbb{R}^n)$, $f \in C^k(\mathbb{R}^n, \mathbb{R}^n)$ for some $k \geq 1$, $f(0) = 0$, and $Df(0) = 0$.

We keep the notations $E_s, E_c, E_u$ and $\pi_s, \pi_c, \pi_u$ throughout this chapter, and let

$$E_h = E_s \oplus E_u, \qquad \pi_h = \pi_s + \pi_u.$$

As usual, we denote by $|y|$ the norm of $y$ in some Banach space. Let $X, Y$ be Banach spaces and $C^k(X, Y)$ be the set of all $C^k$ mappings from $X$ into $Y$. We define the Banach space

$$C_b^k(X, Y) = \left\{ w \in C^k(X, Y) \,\middle|\, \|w\|_{C^k} := \max_{0 \leq j \leq k} \sup_{x \in X} |D^j w(x)| < \infty \right\}.$$

If $X = Y$, we write $C_b^k(X, X)$ as $C_b^k(X)$. We let

$$\|Dw\| = \sup_{x \in X} |Dw(x)|.$$

Similarly, we define

$$C_b^{k,1}(X, Y) = \left\{ w \in C_b^k(X, Y) \,\middle|\, \sup \frac{\|D^k w(x) - D^k w(y)\|}{\|x - y\|_X} < \infty, \right.$$

$$\left. \vphantom{\frac{1}{1}} x, y \in X, x \neq y \right\}$$

with norm

$$\|w\|_{C^{k,1}} = \|w\|_{C^k} + \sup \frac{\|D^k w(x) - D^k w(x)\|}{\|x - y\|_X}, \qquad x, y \in X, x \neq y.$$

Finally, we denote by $\bar{x}(t, x)$ the solution of (1.1) with the initial condition $\bar{x}(0, x) = x$.

Now we state the main result of this section, and will prove it by using several lemmas.

**Theorem 1.1.** (i) *There is a positive number $\delta_0$ which depends only on $A$ in (1.1) such that if $f \in C_b^{0,1}(\mathbb{R}^n)$ and $\mathrm{Lip}(f) < \delta_0$, then the set*

$$W^c := \left\{ x \in \mathbb{R}^n \,\middle|\, \sup_{t \in \mathbb{R}} |\pi_h \bar{x}(t, x)| < \infty \right\} \tag{1.2}$$

*is invariant under (1.1) and is a Lipschitz submanifold of $\mathbb{R}^n$; more precisely, there exists a unique Lipschitz function $\psi \in C_b^0(E_c, E_h)$ such that*

$$W^c = \{ x_c + \psi(x_c) \mid x_c \in E_c \}. \tag{1.3}$$

(ii) *If $\phi \in C_b^0(E_c, E_h)$, and the set*

$$M_\phi := \{ x_c + \phi(x_c) \mid x_c \in E_c \} \tag{1.4}$$

*is invariant under (1.1), then $M_\phi = W^c$ and $\phi = \psi$.*

**Definition 1.2.** $W^c$ is called the global center manifold of (1.1).

**Remark 1.3.** If $f \in C_b^1(\mathbb{R}^n)$, then we will usually replace the condition $\mathrm{Lip}(f) < \delta_0$ by $\|Df\| < \delta_0$.

**Remark 1.4.** The uniqueness conclusion (ii) should be understood in the following sense: If $M_\phi$ is invariant under (1.1), then $\phi \in C_b^0(E_c, E_h)$ is determined uniquely. This is not true if we replace the condition

$\phi \in C_b^0(E_c, E_h)$ by $\phi \in C^0(E_c, E_h)$ unless $A|_{E_c}$ is semisimple and $f$ has compact support (see Sijbrand [1] and Vanderbauwhede [3] for more details).

**Lemma 1.5.** *For any integer $k > 0$, there are constants $K \geq 1$, $\alpha > 0$, and $\beta > 0$ such that $k\alpha < \beta$, and*

$$|e^{At}\pi_c| \leq Ke^{\alpha|t|}, \qquad t \in \mathbb{R},$$

$$|e^{At}\pi_s| \leq Ke^{-\beta t}, \qquad t \geq 0, \tag{1.5}$$

$$|e^{At}\pi_u| \leq Ke^{\beta t}, \qquad t \leq 0.$$

*Proof.* Let

$$\beta = \min\{|\text{Re }\lambda| \,\big|\, \lambda \in \sigma_u \cup \sigma_s|\} - \epsilon,$$

$$0 < \epsilon < \alpha < \frac{\beta}{k},$$

where $\epsilon$ and $\alpha$ are sufficiently small. Thus, the existence of $K$ is obvious by the properties of $e^{At}$. $\qquad\square$

Let $\gamma$ satisfy

$$\alpha < \gamma < \beta. \tag{1.6}$$

Define a Banach space by

$$C_\gamma := \left\{x \in C^0(\mathbb{R}, \mathbb{R}^n) \,\Big|\, \|x\|_\gamma := \sup_{t \in \mathbb{R}} e^{-\gamma|t|}|x(t)| < \infty\right\}.$$

The following lemma gives a different criterion for $W^c$.

**Lemma 1.6.** *Suppose $f \in C_b^{0,1}(\mathbb{R}^n)$ and (1.6) is satisfied. Then*
(i)

$$W^c = \left\{x \in \mathbb{R}^n \mid \bar{x}(\cdot, x) \in C_\gamma\right\}. \tag{1.7}$$

(ii) *Consider the integral equation*

$$y(t) = e^{At}\pi_c x + \int_0^t e^{A(t-\tau)}\pi_c f(y(\tau))\,d\tau$$

$$+ \int_\infty^t e^{A(t-\tau)}\pi_u f(y(\tau))\,d\tau + \int_{-\infty}^t e^{A(t-\tau)}\pi_s f(y(\tau))\,d\tau. \quad (1.8)$$

*We have*

$$W^c = \{y(0) \in \mathbb{R}^n \mid y(\cdot) \in C_\gamma \text{ and satisfies (1.8) for some } x \in \mathbb{R}^n\}.$$

$$(1.9)$$

*Proof.* By the variation of constants formula, for $t_0, t \in \mathbb{R}$ we have

$$\tilde{x}(t,x) = e^{A(t-t_0)}\tilde{x}(t_0,x) + \int_{t_0}^t e^{A(t-\tau)}f(\tilde{x}(\tau,x))\,d\tau. \quad (1.10)$$

Denote by $\tilde{W}^c$ the right-hand side of (1.7), and by $\widetilde{\widetilde{W^c}}$ the right-hand side of (1.9). We will show that $W^c \subset \tilde{W}^c \subset \widetilde{\widetilde{W^c}} \subset W^c$.

(a) Suppose $x \in W^c$; then by (1.2)

$$\sup_{t\in\mathbb{R}} e^{-\gamma|t|}|\pi_h\tilde{x}(t,x)| \le \sup_{t\in\mathbb{R}}|\pi_h\tilde{x}(t,x)| < \infty. \quad (1.11)$$

Taking $t_0 = 0$ in (1.10) we obtain

$$\pi_c\tilde{x}(t,x) = e^{At}\pi_c x + \int_0^t e^{A(t-\tau)}\pi_c f(\tilde{x}(\tau,x))\,d\tau. \quad (1.12)$$

Using Lemma 1.5 and (1.6), we have from (1.12) that

$$|\pi_c\tilde{x}(t,x)| \le Ke^{\gamma|t|}|x| + K\|f\|_{C^0}\left|\int_0^t e^{\gamma(t-\tau)}d\tau\right| \le Ke^{\gamma|t|}\left(|x| + \frac{\|f\|_{C^0}}{\gamma}\right),$$

whence

$$\sup_{t\in\mathbb{R}} e^{-\gamma|t|}|\pi_c\tilde{x}(t,x)| < \infty. \quad (1.13)$$

It follows from (1.11) and (1.13) that $x \in \tilde{W}^c$, and this implies $W^c \subset \tilde{W}^c$.

(b) Suppose now $x \in \bar{W}^c$, that is, $\tilde{x}(\cdot, x) \in C_\gamma$. From (1.10) we have

$$\pi_u \tilde{x}(t, x) = e^{A(t-t_0)}\pi_u \tilde{x}(t_0, x) + \int_{t_0}^t e^{A(t-\tau)}\pi_u f(\tilde{x}(\tau, x))d\tau. \quad (1.14)$$

Fixing $t \in \mathbb{R}$ and $t_0 \geq \max(t, 0)$, we obtain from (1.5)

$$|e^{A(t-t_0)}\pi_u \tilde{x}(t_0, x)| \leq Ke^{\beta(t-t_0)}|\tilde{x}(t_0, x)|$$

$$\leq Ke^{\beta t - (\beta - \gamma)t_0}\|\tilde{x}(\cdot, x)\|_\gamma \to 0 \quad \text{as } t_0 \to +\infty,$$

since $\tilde{x}(\cdot, x) \in C_\gamma$ and $\gamma < \beta$. Therefore, taking the limit in (1.14) as $t_0 \to +\infty$, we have

$$\pi_u \tilde{x}(t, x) = \int_{-\infty}^t e^{A(t-\tau)}\pi_u f(\tilde{x}(\tau, x))d\tau. \quad (1.15)$$

Similarly, we can obtain

$$\pi_s \tilde{x}(t, x) = \int_{-\infty}^t e^{A(t-\tau)}\pi_s f(\tilde{x}(\tau, x))d\tau. \quad (1.16)$$

Combining (1.12), (1.15), and (1.16), we see that $\tilde{x}(\cdot, x)$ satisfies (1.8). Thus $x = \tilde{x}(0, x) \in \widetilde{W^c}$. Therefore $\bar{W}^c \subset \widetilde{W^c}$.

(c) Suppose $y_0 \in \widetilde{W^c}$, that is, there is a function $y(\cdot) \in C_\gamma$, which satisfies (1.8) for some $x \in \mathbb{R}^n$ and $y(0) = y_0$. Then from (1.8)

$$y(t) = e^{At}\left\{\pi_c x + \int_{-\infty}^0 e^{-A\tau}\pi_s f(y(\tau))d\tau + \int_\infty^0 e^{-A\tau}\pi_u f(y(\tau))d\tau\right\}$$

$$+ \int_0^t e^{A(t-\tau)}f(y(\tau))d\tau$$

$$= e^{At}y_0 + \int_0^t e^{A(t-\tau)}f(y(\tau))d\tau.$$

Hence $y(t)$ is the solution of (1.1) with initial value $y(0) = y_0$. Using (1.5) and (1.8), it follows that

$$|\pi_u y(t)| \leq K\|f\|_{C^0}\int_t^\infty e^{\beta(t-\tau)}d\tau = \frac{K}{\beta}\|f\|_{C^0} < \infty,$$

and

$$|\pi_s y(t)| \le \frac{K}{\beta} \|f\|_{C^0} < \infty,$$

since $f \in C_b^0(\mathbb{R}^n)$. Hence $|\pi_h y(t)| < \infty$. Thus $y_0 \in W^c$. This implies $\widetilde{W^c} \subset W^c$.                                                                □

Now we consider the integral equation defined by (1.8). Let $F$: $E_c \to C_\gamma$ be defined by

$$F(\xi)(t) = e^{At}\xi, \qquad \xi \in E_c, \tag{1.17}$$

and $G: C_\gamma \to C_\gamma$ be defined by

$$G(y(\cdot))(t) = \int_0^t e^{A(t-\tau)}\pi_c y(\tau)d\tau + \int_\infty^t e^{A(t-\tau)}\pi_u y(\tau)d\tau$$

$$+ \int_{-\infty}^t e^{A(t-\tau)}\pi_s y(\tau)d\tau. \tag{1.18}$$

We denote the previous three integrals by $G_c(y(\cdot))(t)$, $G_u(y(\cdot))(t)$ and $G_s(y(\cdot))(t)$, respectively. We will use these notations repeatedly in this chapter.

Define $J: E_c \times C_\gamma \to C_\gamma$ by

$$J(\xi, y) = F(\xi) + G(f(y(\cdot))). \tag{1.19}$$

Obviously, if $\xi \in E_c$ then $y = y^*(\cdot)$ is a fixed point of $J(\xi, \cdot)$ if and only if $y^*(t)$ is a solution of (1.8) with $x = \xi$.

**Lemma 1.7.** *There is a number $\delta_0 > 0$, which depends only on $A$, such that, if $\mathrm{Lip}(f) < \delta_0$, then for any $\xi \in E_c$, $J(\xi, y)$, defined by (1.19), has a unique fixed point $y = x^*(\cdot, \xi)$.*

*Proof.* Note that

$$J(\xi, y_1) - J(\xi, y_2) = G(f(y_1(\cdot))) - G(f(y_2(\cdot)))$$

$$= G(f(y_1(\cdot)) - f(y_2(\cdot))), \tag{1.20}$$

and by (1.5) we have

$$\left| G_c(f(y_1(\cdot)) - f(y_2(\cdot)))(t) \right|$$

$$= \left| \int_0^t e^{A(t-\tau)} \pi_c(f(y_1(\tau)) - f(y_2(\tau))) d\tau \right|$$

$$\leq K \operatorname{Lip}(f) \left| \int_0^t e^{\alpha|t-\tau|} |y_1(\tau) - y_2(\tau)| d\tau \right|$$

$$\leq K \operatorname{Lip}(f) \left| \int_0^t e^{\alpha|t-\tau|} e^{\gamma|\tau|} \left( \sup_{\tau \in \mathbb{R}} e^{-\gamma|\tau|} |y_1(\tau) - y_2(\tau)| \right) d\tau \right|$$

$$\leq \frac{e^{\gamma|t|}}{\gamma - \alpha} K \operatorname{Lip}(f) \|y_1 - y_2\|_\gamma. \qquad (1.21)$$

Similarly, we have

$$|G_u(f(y_1(\cdot)))(t) - G_u(f(y_2(\cdot)))(t)| \leq \frac{e^{\gamma|t|}}{\beta - \gamma} K \operatorname{Lip}(f) \|y_1 - y_2\|_\gamma,$$

$$(1.22)$$

$$|G_s(f(y_1(\cdot)))(t) - G_s(f(y_2(\cdot)))(t)| \leq \frac{e^{\gamma|t|}}{\beta - \gamma} K \operatorname{Lip}(f) \|y_1 - y_2\|_\gamma.$$

$$(1.23)$$

These estimates give

$$\sup_{t \in \mathbb{R}} e^{-\gamma|t|} |G(f(y_1(\cdot)))(t) - G(f(y_2(\cdot)))(t)|$$

$$\leq K \left( \frac{1}{\gamma - \alpha} + \frac{2}{\beta - \gamma} \right) \operatorname{Lip}(f) \|y_1 - y_2\|_\gamma.$$

We choose

$$\delta_0 = \frac{1}{3} \left[ K \left( \frac{1}{\gamma - \alpha} + \frac{2}{\beta - \gamma} \right) \right]^{-1}.$$

If $\mathrm{Lip}(f) < \delta_0$ then

$$K\left(\frac{1}{\gamma - \alpha} + \frac{2}{\beta - \gamma}\right)\mathrm{Lip}(f) < \frac{1}{3}, \tag{1.24}$$

and

$$\|G(f(y_1(\cdot))) - G(f(y_2(\cdot)))\|_\gamma \leq \frac{1}{3}\|y_1 - y_2\|_\gamma. \tag{1.25}$$

Thus, for any $\xi \in E_c$, by (1.20), we have

$$\|J(\xi, y_1) - J(\xi, y_2)\|_\gamma \leq \frac{1}{3}\|y_1 - y_2\|_\gamma, \tag{1.26}$$

as long as $\mathrm{Lip}(f) < \delta_0$.

By the Uniform Contraction Mapping Theorem, $J(\xi, \cdot)$ has a unique fixed point $y = x^*(t, \xi)$ for each $\xi \in E_c$. $\square$

**Lemma 1.8.** *If* $\mathrm{Lip}(f) < \delta_0$, *then there exists a unique Lipschitz function* $\psi \in C_b^0(E_c, E_h)$ *such that*

$$W^c = \{x_c + \psi(x_c) \mid x_c \in E_c\}.$$

*Proof.* By Lemmas 1.7 and 1.6, (1.8) has a unique solution $x^*(t, \xi) = \tilde{x}(t, x^*(0, \xi))$, for any $\xi \in E_c$. By Lemma 1.6,

$$W^c = \{x^*(0, \xi) \mid \xi \in E_c\}.$$

Note that

$$x^*(0, \xi) = J(\xi, x^*(\cdot, \xi))(0) = \xi + \psi(\xi), \qquad \xi \in E_c,$$

where

$$\psi(\xi) = \int_\infty^t e^{-A\tau}\pi_u f(x^*(\tau, \xi))d\tau + \int_{-\infty}^0 e^{-A\tau}\pi_s f(x^*(\tau, \xi))d\tau. \tag{1.27}$$

We need to prove the boundedness and Lipschitz continuity of $\psi$.

From (1.5) it follows that

$$\left| \int_\infty^0 e^{-A\tau} \pi_u f(x^*(\tau,\xi))d\tau \right| = \left| \int_{-\infty}^0 e^{A\tau} \pi_u f(x^*(-\tau,\xi))d\tau \right|$$

$$\leq K\|f\|_{C^0} \int_{-\infty}^0 e^{\beta\tau}d\tau = \frac{K}{\beta}\|f\|_{C^0} < \infty,$$

$$(1.28)$$

since $f \in C_b^0(\mathbb{R}^n)$. Similarly,

$$\left| \int_{-\infty}^0 e^{-A\tau} \pi_s f(x^*(\tau,\xi))d\tau \right| < \infty. \qquad (1.29)$$

Hence $\psi$ is bounded.

In (1.22) and (1.23), we take $y_1(t) = x^*(t,\xi)$ and $y_2(t) = x^*(t,\hat{\xi})$, $\xi, \hat{\xi} \in E_c$, and then using the condition (1.24) and letting $t = 0$, we obtain

$$\left| \psi(\xi) - \psi(\hat{\xi}) \right| \leq \frac{1}{3} \left\| x^*(\cdot,\xi) - x^*(\cdot,\hat{\xi}) \right\|_\gamma. \qquad (1.30)$$

On the other hand, for $\xi, \hat{\xi} \in E_c$

$$\left| x^*(t,\xi) - x^*(t,\hat{\xi}) \right|$$

$$= \left| J(\xi, x^*(\cdot,\xi))(t) - J(\hat{\xi}, x^*(\cdot,\hat{\xi}))(t) \right|$$

$$\leq \left| F(\xi - \hat{\xi})(t) \right| + \left| G(f(x^*(\cdot,\xi)))(t) - G(f(x^*(\cdot,\hat{\xi})))(t) \right|.$$

Using (1.5) and (1.25), respectively, we have

$$\left\| x^*(\cdot,\xi) - x^*(\cdot,\hat{\xi}) \right\|_\gamma$$

$$\leq Ke^{-(\gamma-\alpha)|t|}|\xi - \hat{\xi}| + \frac{1}{3} \left\| x^*(\cdot,\xi) - x^*(\cdot,\hat{\xi}) \right\|_\gamma,$$

whence,

$$\left\| x^*(\cdot,\xi) - x^*(\cdot,\hat{\xi}) \right\|_\gamma \leq \frac{3K}{2}|\xi - \hat{\xi}|. \qquad (1.31)$$

From (1.30) and (1.31) we have finally that

$$|\psi(\xi) - \psi(\hat{\xi})| \le \frac{K}{2} |\xi - \hat{\xi}| \qquad \text{for } \xi, \hat{\xi} \in E_c. \qquad (1.32)$$

Thus the lemma is proved.                                            □

**Remark 1.9.** By the definition of $C_\gamma$, we can rewrite (1.31) as

$$\left| x^*(t, \xi) - x^*(t, \hat{\xi}) \right| \le e^{\gamma |t|} \frac{3K}{2} |\xi - \hat{\xi}| \qquad (1.33)$$

for any $\gamma \in (\alpha, \beta)$, $\xi, \hat{\xi} \in E_c$ and all $t \in \mathbb{R}$.

*Proof of Theorem 1.1.* Since $\tilde{x}(t_1, \tilde{x}(t_2, x)) = \tilde{x}(t_1 + t_2, x)$, the set $W^c$ defined by (1.2) is invariant under (1.1). The remaining conclusions in (i) are proved in Lemma 1.8.

Now we prove the uniqueness of $\phi$ in (ii). Suppose $\phi \in C_b^0(E_c, E_h)$ and $M_\phi$ defined by (1.4) is invariant under (1.1). Then $\tilde{x}(t, x_c + \phi(x_c)) \in M_\phi$ for all $t \in \mathbb{R}$ and any $x_c \in E_c$. By the definition of $M_\phi$, it follows that

$$\pi_h \tilde{x}(t, x_c + \phi(x_c)) = \phi(\pi_c \tilde{x}(t, x_c + \phi(x_c))).$$

Since $\phi \in C_b^0(E_c, E_h)$, the boundedness of $\phi$ implies the boundedness of $\pi_h \tilde{x}(t, x_c + \phi(x_c))$, and hence, by the definition (1.2), $x_c + \phi(x_c) \in W^c$ for any $x_c \in E_c$. In Lemma 1.8 we have proved that such a $\phi$ is unique. Hence $\phi = \psi$, and $M_\phi = W^c$.                □

## 1.2 Smoothness of the Global Center Manifolds

We have proved the existence and uniqueness of the global center manifolds $W^c$ under the conditions $f \in C_b^{0,1}(\mathbb{R}^n)$ and $\text{Lip}(f)$ is sufficiently small. If, in addition, $f \in C_b^k(\mathbb{R}^n)$ for some $k \ge 1$, then we will show that $W^c$ is smooth. The main result of this section is the following theorem.

**Theorem 2.1.** *Suppose* $f \in C_b^k(\mathbb{R}^n)$ *for some* $k \geq 1$, $f(0) = 0$ *and* $Df(0)$ $= 0$. *Then there is a number* $\delta_k > 0$ *such that if* $\|Df\| < \delta_k$, *the unique global center manifold* $W^c$ *is of class* $C^k$, *that is,* $\psi \in C_b^k(E_c, E_h)$, *where* $\psi$ *is given by (1.27) and is related to* $W^c$ *by (1.3). Moreover,* $\mathrm{Lip}(\psi) <$ 1, $\psi(0) = 0$ *and* $D\psi(0) = 0$. *Furthermore, if* $\hat{x} \in W^c$ *and* $\bar{x}_c(t) :=$ $\pi_c \bar{x}(t, \hat{x})$, *then* $\bar{x}_c(t)$ *satisfies the following equation*

$$\dot{x}_c = Ax_c + \pi_c f(x_c + \psi(x_c)), \qquad x_c \in E_c. \tag{2.1}$$

We will prove this theorem by induction on $k$, and consider first the case $k = 1$.

We remark here that if $\sigma > 0$, then $C_\gamma \subset C_{\gamma+\sigma}$ and $\|x\|_{\gamma+\sigma} \leq \|x\|_\gamma$. Hence, there exists a continuous inclusion from $C_\gamma$ into $C_{\gamma+\sigma}$. The choices of spaces $\{C_\eta\}$ for different $\eta$ in the following discussion are very important.

To prove $\psi \in C^1$, by (1.27), we need to prove first that $x^*(t, \xi) \in C^1$ with respect to $\xi \in E_c$. Since $x^*(t, \xi)$ is the unique solution of (1.8), we have

$$x^*(t, \xi) = e^{At}\xi + G(f(x^*(\cdot, \xi)))(t), \tag{2.2}$$

where $G$ is defined in (1.18).

**Lemma 2.2.** *Suppose that* $f \in C_b^1(\mathbb{R}^n)$, $\|Df\| < \delta_0$. *Then there exists a number* $\sigma > 0$ *such that the map* $\xi \mapsto x^*(\cdot, \xi)$ *from* $E_c$ *to* $C_{\gamma+\sigma}$, *is differentiable.*

*Proof.* Let

$$u(t, \xi, \hat{\xi}) = x^*(t, \xi) - x^*(t, \hat{\xi}), \qquad \xi, \hat{\xi} \in E_c, \tag{2.3}$$

and

$$f^*(t, \xi, \hat{\xi}) = f(x^*(t, \xi)) - f(x^*(t, \hat{\xi})) - Df(x^*(t, \hat{\xi}))u(t, \xi, \hat{\xi}). \tag{2.4}$$

Define

$$L\big(u(\,\cdot\,,\xi,\hat{\xi})\big) = G\big(Df(x^*(\,\cdot\,,\hat{\xi}))u(\,\cdot\,,\xi,\hat{\xi})\big), \tag{2.5}$$

and

$$N\big(u(\,\cdot\,,\xi,\hat{\xi})\big) = G\big(f^*(\,\cdot\,,\xi,\hat{\xi})\big). \tag{2.6}$$

Then we obtain from (2.2) that

$$(I - L)\big(u(\,\cdot\,,\xi,\hat{\xi})\big) = F(\xi - \hat{\xi}) + N\big(u(\,\cdot\,,\xi,\hat{\xi})\big), \tag{2.7}$$

where $I$ is the identity operator and $F$ is defined in (1.17). Obviously, $F$ is a bounded linear operator. If we replace $(f(y_1) - f(y_2))$ by $Df(x^*(t,\hat{\xi}))u(t,\xi,\hat{\xi})$ and replace $\mathrm{Lip}(f)$ by $\|Df\|$ in (1.21)–(1.24), then instead of (1.25) we can obtain

$$\big\| L\big(u(\,\cdot\,,\xi,\hat{\xi})\big)\big\|_\gamma \le \frac{1}{3}\big\| u(\,\cdot\,,\xi,\hat{\xi})\big\|_\gamma. \tag{2.8}$$

This implies the norm of $L$, as an operator from $C_\gamma$ to $C_\gamma$, satisfies

$$\|L\| \le \frac{1}{3}.$$

Hence $(I - L)^{-1}$ exists and is bounded, and (2.7) can be written as

$$u(\,\cdot\,,\xi,\hat{\xi}) = (I - L)^{-1}F(\xi - \hat{\xi}) + (I - L)^{-1}N\big(u(\,\cdot\,,\xi,\hat{\xi})\big). \tag{2.9}$$

We will prove that there exists a $\sigma > 0$ such that

$$\big\| N\big(u(\,\cdot\,,\xi,\hat{\xi})\big)\big\|_{\gamma+\sigma} = o(|\xi - \hat{\xi}|) \qquad \text{as } \xi \to \hat{\xi}. \tag{2.10}$$

Hence, by the definition of a derivative, the map $\xi \mapsto x^*(\,\cdot\,,\xi)\colon E_c \to C_{\gamma+\sigma}$ is differentiable.

It is obvious that if $\sigma > 0$ is sufficiently small, and we replace $\gamma$ by $\gamma + \sigma$ (in some cases later, we need to replace $\gamma$ by $\gamma + k\sigma$ for some integer $k > 0$), then (1.6) and (1.24) still hold. We fix such a number $\sigma$.

We will prove that for every small $\epsilon > 0$, there exists a $\mu > 0$ such that if $|\xi - \hat{\xi}| \le \mu$, $\xi, \hat{\xi} \in E_c$, then

$$\sup_{t\in\mathbb{R}} e^{-(\gamma+\sigma)|t|}\big| N\big(u(\,\cdot\,,\xi,\hat{\xi})\big)(t)\big| \le \epsilon|\xi - \hat{\xi}|. \tag{2.11}$$

From (2.6) and (2.4) we have

$$N\big(u(\,\cdot\,,\xi,\hat{\xi})\big)(t) = N_c + N_u + N_s, \tag{2.12}$$

where

$$N_c = G_c\big(f^*(\cdot, \xi, \hat{\xi})\big)(t),$$

$$N_u = G_u\big(f^*(\cdot, \xi, \hat{\xi})\big)(t), \tag{2.13}$$

$$N_s = G_s\big(f^*(\cdot, \xi, \hat{\xi})\big)(t).$$

We will find an estimate only for $N_c$, since it is similar for $N_u$ and $N_s$. Choose $T > 0$ so large that

$$\frac{3K^2}{\gamma - \alpha} \|Df\| e^{-\sigma T} < \frac{\epsilon}{6}, \tag{2.14}$$

where the constants $\alpha$ and $K$ are the same as in (1.5).

We consider two cases:

(i) $|t| \le T$.

Without loss of generality, we assume $0 \le t \le T$. By (2.13), (2.4) and (1.31), we have

$$|N_c| = \left| \int_0^t e^{A(t-\tau)} \pi_c \big[ f(x^*(\tau, \xi)) - f(x^*(\tau, \hat{\xi})) \right.$$

$$\left. - Df(x^*(\tau, \hat{\xi}))(x^*(\tau, \xi) - x^*(\tau, \hat{\xi})) \big] d\tau \right|$$

$$\le \left| \int_0^t e^{A(t-\tau)} \pi_c \left( \int_0^1 \big[ Df(\lambda x^*(\tau, \xi) + (1-\lambda) x^*(\tau, \hat{\xi})) \right.\right.$$

$$\left.\left. - Df(x^*(\tau, \hat{\xi})) \big] d\lambda \right)(x^*(\tau, \xi) - x^*(\tau, \hat{\xi})) d\tau \right|$$

$$\le \frac{3K^2}{2} e^{\alpha|t|} |\xi - \hat{\xi}| \int_0^{|t|} e^{(\gamma - \alpha)\tau}$$

$$\times \left( \int_0^1 |Df(\lambda x^*(\tau, \xi) + (1-\lambda) x^*(\tau, \hat{\xi})) - Df(x^*(\tau, \hat{\xi}))| d\lambda \right) d\tau$$

$$\le \frac{3K^2}{2} e^{\gamma|t|} |\xi - \hat{\xi}| \int_0^T \int_0^1 |Df(\lambda x^*(\tau, \xi) + (1-\lambda) x^*(\tau, \hat{\xi}))$$

$$- Df(x^*(\tau, \hat{\xi}))| d\lambda d\tau. \tag{2.15}$$

Since $f \in C_b^1(\mathbb{R}^n)$ and the last integral is taken over a compact region $[0, T] \times [0, 1] \subset \mathbb{R}^2$, there exists a $\mu_1 > 0$ such that if $|\xi - \hat{\xi}| \le \mu_1$ then

$$\sup_{|t| \le T} e^{-(\gamma+\sigma)|t|}|N_c| \le \frac{\epsilon}{3}|\xi - \hat{\xi}|. \qquad (2.16)$$

(ii) $|t| > T$.

Without loss of generality, we assume $t > T$, and let $N_c = N_c^{(1)} + N_c^{(2)}$, where

$$N_c^{(1)} = \int_0^T e^{A(t-\tau)}\pi_c f^*(\tau, \xi, \hat{\xi})d\tau, \qquad N_c^{(2)} = \int_T^t e^{A(t-\tau)}\pi_c f^*(\tau, \xi, \hat{\xi})d\tau.$$

Similarly to (2.15) and (2.16), there exists a $\mu_2 > 0$ such that if $|\xi - \hat{\xi}| < \mu_2$, then

$$\sup_{|t| > T} e^{-(\gamma+\sigma)|t|}|N_c^{(1)}| \le \frac{\epsilon}{6}|\xi - \hat{\xi}|. \qquad (2.17)$$

Using (2.4) and (1.5) we have

$$|N_c^{(2)}| = \left|\int_T^t e^{A(t-\tau)}\pi_c f^*(\tau, \xi, \hat{\xi})d\tau\right|$$

$$\le K\int_T^t e^{\alpha(t-\tau)}\big(2\|Df\|\big|x^*(\tau, \xi) - x^*(\tau, \hat{\xi})\big|\big)d\tau.$$

From (1.33) it follows that

$$\left|x^*(\tau, \xi) - x^*(\tau, \hat{\xi})\right| \le \frac{3K}{2}e^{\gamma\tau}|\xi - \hat{\xi}|. \qquad (2.18)$$

Hence

$$|N_c^{(2)}| \le 3K^2\|Df\||\xi - \hat{\xi}|\int_T^t e^{\alpha t}e^{(\gamma-\alpha)\tau}d\tau \le 3K^2\|Df\||\xi - \hat{\xi}|\frac{1}{\gamma-\alpha}e^{\gamma t}.$$

From the above estimate and condition (2.14) we obtain

$$\sup_{|t| > T} e^{-(\gamma+\sigma)|t|}|N_c^{(2)}| \le \frac{3K^2}{\gamma-\alpha}\|Df\|e^{-\sigma T}|\xi - \hat{\xi}| < \frac{\epsilon}{6}|\xi - \hat{\xi}|. \quad (2.19)$$

We choose $\mu_c = \min(\mu_1, \mu_2)$. If $|\xi - \hat{\xi}| < \mu_c$ then (2.16), (2.17) and (2.19) give

$$\|N_c\|_{\gamma+\sigma} \le \frac{\epsilon}{3}|\xi - \hat{\xi}|.$$

Similarly, we can find $\mu_u$ and $\mu_s$ such that

$$\|N_u\|_{\gamma+\sigma} \le \frac{\epsilon}{3}|\xi - \hat{\xi}| \quad \text{when } |\xi - \hat{\xi}| \le \mu_u,$$

and

$$\|N_s\|_{\gamma+\sigma} \le \frac{\epsilon}{3}|\xi - \hat{\xi}| \quad \text{when } |\xi - \hat{\xi}| \le \mu_s.$$

Let $\mu = \min(\mu_c, \mu_u, \mu_s)$. If $|\xi - \hat{\xi}| \le \mu$ and $\xi, \hat{\xi} \in E_c$ then (2.11) holds, and hence (2.10) holds. $\qquad\square$

The following lemma gives a more general result which will be used repeatedly in the rest of this section.

**Lemma 2.3.** *Suppose that $E$ is a Euclidean space with norm $\|\cdot\|_E$ and, for each $y \in E$, the map $y \mapsto g(\cdot, y)$ from $E$ to $C_\rho$ for some $\rho \in (\alpha, \beta)$ satisfies*
(i) *$g(t, y)$ is continuous in $(t, y) \in \mathbb{R} \times E$;*
(ii) *$\|g(\cdot, y)\|_\rho \le M$ for some constant $M > 0$, where $M$ is independent of $y$.*
*Then for any $\zeta \in (\rho, \beta)$ and $y_0 \in E$, we have*

$$\lim_{\|y - y_0\|_E \to 0} \|G(g(\cdot, y)) - G(g(\cdot, y_0))\|_\zeta = 0,$$

*where $G: C_\rho \to C_\zeta$ is defined in (1.18).*

*Proof.* This lemma can be proved by using the same arguments as in the proof of Lemma 2.2. For a given small $\epsilon > 0$, we find a $T > 0$ such that $Me^{-(\zeta - \rho)T} < \epsilon/6$. Then divide the integrals in $G$ into two parts $G^{(1)}$ and $G^{(2)}$. For the noncompact part $|t| \ge T$, we use condition (ii) and the continuous inclusion from $C_\rho$ to $C_\zeta$ to get $\|G^{(1)}(g(\cdot, y) - $

$g(\cdot, y_0))\|_\zeta \le \epsilon/2$; for the compact part $|t| \le T$, we use the uniform continuity of $g(t, y)$ (condition (i)) to find a $\mu > 0$ such that if $|y - y_0| < \mu$ then $\|G^{(2)}(g(\cdot, y) - g(\cdot, y_0))\|_\zeta \le \epsilon/2$. □

We have proved that under the hypothesis of Lemma 2.2, $x^*(\cdot, \xi)$, as a mapping from $E_c$ to $C_{\gamma+\sigma}$, is differentiable.

It is known that

$$D_\xi x^*(t, \xi)\eta = \lim_{\lambda \to 0} \frac{x^*(t, \xi + \lambda\eta) - x^*(t, \xi)}{\lambda}. \qquad (2.20)$$

Note that for each $\xi \in E_c$, $D_\xi x^*(\cdot, \xi)$ is a linear mapping from $T_\xi E_c = E_c$ to $T_{x^*(\cdot, \xi)}C_{\gamma+\sigma} = C_{\gamma+\sigma}$, where $T_\xi E_c$ and $T_{x^*(\cdot, \xi)}C_{\gamma+\sigma}$ are the tangent spaces of $E_c$ and $C_{\gamma+\sigma}$ at $\xi$ and $x^*(\cdot, \xi)$, respectively. For all $\xi \in E_c$, we consider $D_\xi x^*(\cdot, \xi)$ as a mapping from $E_c$ to $\mathscr{L}(E_c, C_{\gamma+\sigma})$. We will show in the next two lemmas that $D_\xi x^*(\cdot, \xi)$, as a mapping from $E_c$ to $\mathscr{L}(E_c, C_{\gamma+3\sigma})$, is continuous in $\xi \in E_c$.

**Lemma 2.4.** *Suppose that $f \in C_b^1(\mathbb{R}^n)$ and $\|Df\| < \delta_0$. Then the map $\xi \mapsto D_\xi x^*(\cdot, \xi)$: $E_c \to \mathscr{L}(E_c, C_{\gamma+2\sigma})$ satisfies the following integral equation*

$$v(t)\eta = e^{At}\eta + G(Df(x^*(\cdot, \xi))v(\cdot)\eta)(t), \qquad \forall \eta \in E_c, \quad t \in \mathbb{R}, \tag{2.21}$$

*where $G: C_{\gamma+\sigma} \to C_{\gamma+2\sigma}$ is defined in (1.18).*

*Proof.* Let $\xi, \eta \in E_c$ be fixed, $\lambda \ne 0$, and

$$g(t, \lambda) = \frac{x^*(t, \xi + \lambda\eta) - x^*(t, \xi)}{\lambda}. \tag{2.22}$$

Since $\lim_{\lambda \to 0}(g(t, \lambda))_{\gamma+\sigma} = D_\xi x^*(t, \xi)\eta$ exists (Lemma 2.2) and $x^*(\cdot, \xi)$ is Lipschitz continuous in $\xi$ in $C_\gamma$-norm (see (1.31)), $g(t, \lambda)$ is continuous in $(t, \lambda) \in \mathbb{R} \times \mathbb{R}$. From (2.2) we have for $\lambda \ne 0$,

$$g(t, \lambda) = e^{At}\eta + G(h(\cdot, \lambda))(t), \tag{2.23}$$

where

$$h(t,\lambda) = \frac{f(x^*(t,\xi+\lambda\eta)) - f(x^*(t,\xi))}{\lambda}$$

$$= \left(\int_0^1 Df(\theta x^*(t,\xi+\lambda\eta) + (1-\theta)x^*(t,\xi))d\theta\right)g(t,\lambda).$$

To prove Lemma 2.4, we need to take limits on both sides of (2.23) in $C_{\gamma+\sigma}$, and show that as $\lambda \to 0$,

$$G(h(\cdot,\lambda)) \to G(h(\cdot,0)) \quad \text{in } C_{\gamma+2\sigma}.$$

In fact, the continuity of $h(t,\lambda)$ comes from the continuity of $g(t,\lambda)$ and $f \in C_b^1(\mathbb{R}^n)$. By (2.22) and (1.31), we have

$$\|h(\cdot,\lambda)\|_{\gamma+\sigma} \le \|h(\cdot,\lambda)\|_{\gamma} \le \delta_0 \frac{3K}{2}|\eta|.$$

Thus the hypotheses of Lemma 2.3 are satisfied. $\qquad\qquad\square$

**Lemma 2.5.** *Suppose $f \in C_b^1(\mathbb{R}^n)$. Then there exists a number $\delta_1 \le \delta_0$ such that if $\|Df\| < \delta_1$, then the map $\xi \mapsto D_\xi x^*(\cdot,\xi)$ from $E_c$ to $\mathscr{L}(E_c, C_{\gamma+3\sigma})$ is continuous in $\xi \in E_c$.*

*Proof.* We consider $D_\xi x^*(\cdot,\xi)$ as a solution of (2.21), and $G$ as a mapping from $C_{\gamma+2\sigma}$ to $C_{\gamma+3\sigma}$. Then for any $\eta \in E_c$ we have

$$\left\|\left(D_\xi x^*(\cdot,\xi) - D_\xi x^*(\cdot,\hat{\xi})\right)\eta\right\|_{\gamma+3\sigma}$$

$$\le \left\|G\left[Df(x^*(\cdot,\xi))\left(D_\xi x^*(\cdot,\xi) - D_\xi x^*(\cdot,\hat{\xi})\right)\eta\right]\right\|_{\gamma+3\sigma}$$

$$+ \left\|G\left[\left(Df(x^*(\cdot,\xi)) - Df(x^*(\cdot,\hat{\xi}))\right)D_\xi x^*(\cdot,\hat{\xi})\eta\right]\right\|_{\gamma+3\sigma}.$$

$$(2.24)$$

Similarly to obtaining the estimate (1.25) (it comes from (1.21), (1.22), and (1.23)), we can find $\delta_1 > 0$, $\delta_1 \leq \delta_0$, such that if $\|Df\| < \delta_1$, then

$$\left\| G\Big[ Df(x^*(\cdot,\xi))\big(D_\xi x^*(\cdot,\xi) - D_\xi x^*(\cdot,\hat\xi)\big)\eta \Big] \right\|_{\gamma+3\sigma}$$

$$\leq \frac{1}{3} \left\| \big(D_\xi x^*(\cdot,\xi) - D_\xi x^*(\cdot,\hat\xi)\big)\eta \right\|_{\gamma+3\sigma}. \qquad (2.25)$$

Noting $Df(x^*(t,\xi))$ is continuous in $(t,\xi) \in \mathbb{R} \times E_c$, and

$$\|Df\| < \delta_1, \qquad \left\| D_\xi x^*(\cdot,\hat\xi)\eta \right\|_{\gamma+2\sigma} \leq \frac{3K}{2}|\eta|.$$

we can use the same argument as in Lemmas 2.2 and 2.3 to find $\mu > 0$ such that if $|\xi - \hat\xi| \leq \mu$ then

$$\left\| G\Big[ Df(x^*(\cdot,\xi)) - Df(x^*(\cdot,\hat\xi)) D_\xi x^*(\cdot,\hat\xi)\eta \Big] \right\|_{\gamma+3\sigma} \leq \epsilon|\eta|. \qquad (2.26)$$

From (2.24), (2.25), and (2.26) we obtain finally

$$\left\| \big(D_\xi x^*(\cdot,\xi) - D_\xi x^*(\cdot,\hat\xi)\big)\eta \right\|_{\gamma+3\sigma} \leq \frac{3}{2}\epsilon|\eta|,$$

which implies

$$\left\| D_\xi x^*(\cdot,\xi) - D_\xi x^*(\cdot,\hat\xi) \right\|_{\mathscr{L}(E_c, C_{\gamma+3\sigma})} \leq \frac{3}{2}\epsilon$$

if $|\xi - \hat\xi| < \mu$.                                                           $\square$

**Lemma 2.6.** *Suppose* $f \in C_b^1(\mathbb{R}^n)$, $f(0) = 0$, *and* $Df(0) = 0$. *Suppose* $\|Df\| < \delta_1$, *where* $\delta_1$ *is given by Lemma 2.4. Then* $\psi \in C_b^1(E_c, E_h)$, $\psi(0) = 0$, $D\psi(0) = 0$, $\mathrm{Lip}(\psi) < 1$, *where* $\psi$ *is given by (1.27). Furthermore, if* $\hat x \in W^c$ *and* $\tilde x_c(t) := \pi_c \tilde x(t, \hat x)$, *then* $\tilde x_c(t)$ *satisfies equation (2.1).*

*Proof.* From Lemma 2.5 we have $D_\xi x^*(\cdot,\xi) \in C^0(E_c, \mathscr{L}(E_c, C_{\gamma+3\sigma}))$, and from (1.33) we have

$$|D_\xi x^*(\tau,\xi)\eta| \leq \frac{3K}{2} e^{\gamma|\tau|}|\eta| \leq \frac{3K}{2} e^{(\gamma+3\sigma)|\tau|}|\eta|, \quad \forall \xi, \eta \in E_c, \quad \tau \in \mathbb{R}.$$

This inequality and (1.5) imply that for each $\eta \in E_c$ and all $\xi \in E_c$

$$\left| \int_{\infty}^0 e^{-A\tau} \pi_u Df(x^*(\tau, \xi)) D_\xi x^*(\tau, \xi) \eta \, d\tau \right|$$

$$\leq \frac{3K^2 \delta_1}{2} |\eta| \int_{-\infty}^0 e^{[\beta - (\gamma + 3\sigma)]\tau} d\tau$$

$$= \frac{3K^2 \delta_1 |\eta|}{2(\beta - (\gamma + 3\sigma))}. \qquad (2.27)$$

Similarly, for each $\eta \in E_c$ and all $\xi \in E_c$ we have

$$\left| \int_{-\infty}^0 e^{-A\tau} \pi_s Df(x^*(\tau, \xi)) D_\xi x^*(\tau, \xi) \eta \, d\tau \right| \leq \frac{3K^2 \delta_1 |\eta|}{2(\beta - (\gamma + 3\sigma))}.$$

$$(2.28)$$

Hence, we can take the derivative under the integral sign with respect to $\xi \in E_c$ on the right-hand side of (1.27). This gives for each $\eta \in E_c$

$$D\psi(\xi)\eta = \int_{\infty}^0 e^{-A\tau} \pi_u Df(x^*(\tau, \xi)) D_\xi x^*(\tau, \xi) \eta \, d\tau$$

$$+ \int_{-\infty}^0 e^{-A\tau} \pi_s Df(x^*(\tau, \xi)) D_\xi x^*(\tau, \xi) \eta \, d\tau. \quad (2.29)$$

Moreover, the uniform convergence of the above integrals with respect to $\xi \in E_c$ (see (2.27) and (2.28)), and the continuity of $Df$, $x^*(\cdot, \xi)$, and $D_\xi x^*(\cdot, \xi)$ imply the continuity of $D\psi(\xi)$. From equations (2.27) and (2.28), we obtain the boundedness of $D\psi$. Therefore, $\psi \in C_b^1(E_c, E_h)$.

Since $f(0) = 0$, $Df(0) = 0$, by using the uniqueness of solutions of (1.1) with initial conditions, we have

$$x^*(t, 0) = \tilde{x}(t, x^*(0,0)) = \tilde{x}(t, 0) = 0,$$

and by (1.27) and (2.29)

$$\psi(0) = 0 \qquad \text{and} \qquad D\psi(0) = 0.$$

It is obvious from (2.27) and (2.28) that if we choose $\delta_1$ small enough, then

$$\mathrm{Lip}(\psi) < 1.$$

Finally for $\hat{x} \in W^c$, if we take $\tilde{x}(t, \hat{x}) = \xi(t) + \psi(\xi(t))$, where $\xi(t) \in E_c$, then

$$(I + D\psi)\dot{\xi} = A\xi + A\psi(\xi) + f(\xi + \psi(\xi)).$$

Projecting both sides of the above equality onto $E_c$, and noting that $\psi$: $E_c \rightarrow E_h$, $D\psi$: $TE_c = E_c \rightarrow TE_h = E_h$, we have

$$\dot{\xi} = A\xi + \pi_c f(\xi + \psi(\xi)), \qquad \xi \in E_c,$$

and this is equation (2.1). $\qquad\qquad\qquad\qquad\qquad\qquad\qquad\qquad\square$

*Proof of Theorem 2.1.* The conclusions for the case $k = 1$ have been proved in Lemma 2.6. The case $k \geq 2$ is slightly different from the case $k = 1$, although the basic arguments are eventually the same. We will prove the case $k = 2$; the general case can be obtained by an induction on $k$. In the following we assume $\alpha < \gamma < k\gamma < \beta$ and $\sigma > 0$ is sufficiently small.

Suppose that $f \in C_b^2(\mathbb{R}^n)$. Then by Lemma 2.5, $D_\xi x^*(t, \xi) \in C^0(E_c, \mathscr{L}(E_c, C_{\gamma+3\sigma}))$. We prove first that $\xi \mapsto D_\xi x^*(\cdot, \xi)$, as a mapping from $E_c$ to $\mathscr{L}(E_c, C_{2(\gamma+3\sigma)+\sigma})$, is differentiable. We will use the same idea as in the proof of Lemma 2.2 and consider equation (2.21) instead of (2.2). Let

$$v = v(t, \xi, \hat{\xi}, \eta) = D_\xi x^*(t, \xi)\eta - D_\xi x^*(t, \hat{\xi})\eta,$$

$$\xi, \hat{\xi}, \eta \in E_c, \quad t \in \mathbb{R},$$

and

$$L(v) = G\big(Df\big(x^*(\cdot, \hat{\xi})\big)v(\cdot, \xi, \hat{\xi}, \eta)\big).$$

Then from (2.21) we obtain

$$(I - L)v = G\big(g(\cdot, \xi, \hat{\xi}, \eta)\big), \qquad (2.30)$$

where

$$g(t, \xi, \hat{\xi}, \eta)$$

$$= Df(x^*(t, \xi)) D_\xi x^*(t, \xi) \eta$$

$$- Df\left(x^*(t, \hat{\xi})\right) D_\xi x^*(t, \hat{\xi}) \eta - Df\left(x^*(t, \hat{\xi})\right) v$$

$$= \left(Df(x^*(t, \xi)) - Df\left(x^*(t, \hat{\xi})\right)\right) D_\xi x^*(t, \xi) \eta$$

$$= \left[\int_0^1 D^2 f\left(\theta x^*(t, \xi) + (1 - \theta) x^*(t, \hat{\xi})\right) d\theta\right]$$

$$\times \left\langle \left(x^*(t, \xi) - x^*(t, \hat{\xi})\right), D_\xi x^*(t, \xi) \eta \right\rangle$$

$$= \left[\int_0^1 D^2 f\left(\theta x^*(t, \xi) + (1 - \theta) x^*(t, \hat{\xi})\right) d\theta\right]$$

$$\times \left\langle D_\xi x^*(t, \xi) \eta, \left[\int_0^1 D_\xi x^*(t, \theta \xi + (1 - \theta)\hat{\xi}) d\theta\right] (\xi - \hat{\xi}) \right\rangle$$

and $\langle \cdot, \cdot \rangle$ denotes the action of the bilinear map $D^2 f$. We define the following bilinear form:

$$B(t, \xi, \hat{\xi})\langle a, b \rangle = B(t, \xi, \hat{\xi}) ab$$

$$= \left[\int_0^1 D^2 f(\theta x^*(t, \xi) + (1 - \theta) x^*(t, \hat{\xi})) d\theta\right]$$

$$\times \left\langle D_\xi x^*(t, \xi) \cdot a, \int_0^1 D_\xi x^*(t, \theta \xi + (1 - \theta)\hat{\xi}) d\theta \cdot b \right\rangle.$$

Thus, (2.30) becomes

$$(I - L)(v) = G\left[\left(B(\cdot, \hat{\xi}, \hat{\xi})\right) \eta (\xi - \hat{\xi})\right]$$

$$+ G\left[\left(B(\cdot, \xi, \hat{\xi}) - B(\cdot, \hat{\xi}, \hat{\xi})\right) \eta (\xi - \hat{\xi})\right]. \quad (2.31)$$

Hence, in order to prove the differentiability of $D_\xi x^*(\cdot, \xi)$ (as a mapping from $E_c$ to $\mathcal{L}(E_c, C_{2(\gamma+3\sigma)+\sigma}))$, we only need to verify the following two facts:

(i) $G(B(\cdot, \hat{\xi}, \hat{\xi})\langle \cdot, \cdot \rangle)$ is a bounded bilinear operator from $E_c^2$ to $C_{2(\gamma+3\sigma)+\sigma}$, where $E_c^2 = E_c \times E_c$;

(ii) $\|G[(B(\cdot, \xi, \hat{\xi}) - B(\cdot, \hat{\xi}, \hat{\xi}))\langle \cdot, \cdot \rangle]\|_{\mathcal{L}(E_c^2, C_{2(\gamma+3\sigma)+\sigma})} \to 0$ as $|\xi - \hat{\xi}| \to 0$.

Note that

$$B(t, \hat{\xi}, \hat{\xi})\eta_1 \eta_2 = D^2 f(x^*(t, \hat{\xi})) D_\xi x^*(t, \hat{\xi})\eta_1 \cdot D_\xi x^*(t, \hat{\xi})\eta_2.$$

For any $\eta_1, \eta_2 \in E_c$,

$$\left\| G\left( B(\cdot, \hat{\xi}, \hat{\xi})\eta_1 \eta_2 \right) \right\|_{2(\gamma+3\sigma)+\sigma}$$

$$\leq \sup_{t \in \mathbb{R}} e^{-(2(\gamma+3\sigma)+\sigma)|t|} \|f\|_{C^2} \left( \frac{1}{2(\gamma+3\sigma) - \alpha} + \frac{1}{\beta - (2\gamma+6\sigma)} \right)$$

$$\cdot \left( \frac{3K}{2} \right)^2 e^{2(\gamma+3\sigma)t} |\eta_1| |\eta_2|$$

$$\leq M|\eta_1| |\eta_2|, \quad \text{for some } M > 0.$$

Thus, (i) holds. On the other hand, for any $\eta_1, \eta_2 \in E_c$, we consider $G((h(\cdot, \xi, \hat{\xi}) - h(\cdot, \hat{\xi}, \hat{\xi}))\eta_1\eta_2)$ as a mapping from $C_{2(\gamma+3\sigma)}$ to $C_{2(\gamma+3\sigma)+\sigma}$, and can prove that

$$\left\| G\left( h(\cdot, \xi, \hat{\xi}) - h(\cdot, \hat{\xi}, \hat{\xi}) \right)\eta_1\eta_2 \right\|_{2(\gamma+3\sigma)+\sigma} \to 0 \quad \text{as } |\xi - \hat{\xi}| \to 0$$

by completely the same way as in the proof of Lemma 2.3. Hence (ii) holds.

Similarly to Lemma 2.4, we can obtain an equation satisfied by $D_\xi^2(x^*(\cdot, \xi))$: $E_c \to \mathcal{L}(E_c^2, C_{2(\gamma+3\sigma)+2\sigma})$.

In fact, if we let

$$\bar{g}(t, \lambda) = \frac{D_\xi x^*(t, \xi + \lambda\eta_2)\eta_1 - D_\xi x^*(t, \xi)\eta_1}{\lambda}, \quad \eta_1, \eta_2 \in E_c,$$

then

$$\lim_{\lambda \to 0} \tilde{g}(t, \lambda) = D_\xi^2 x^*(t, \xi) \eta_1 \eta_2,$$

and by (2.21)

$$\tilde{g}(t, \lambda) = G(\tilde{h}(\cdot, \lambda))(t), \tag{2.32}$$

where

$$\tilde{h}(t, \lambda) = \frac{1}{\lambda} \left[ Df(x^*(t, \xi + \lambda \eta_2)) D_\xi x^*(t, \xi + \lambda \eta_2) \eta_1 \right.$$

$$\left. - Df(x^*(t, \xi)) D_\xi x^*(t, \xi) \eta_1 \right]$$

$$= \frac{1}{\lambda} \left[ Df(x^*(t, \xi + \lambda \eta_2)) \right.$$

$$\times \left( D_\xi x^*(t, \xi + \lambda \eta_2) \eta_1 - D_\xi x^*(t, \xi) \eta_1 \right)$$

$$\left. + \left( Df(x^*(t, \xi + \lambda \eta_2)) - Df(x^*(t, \xi)) \right) D_\xi x^*(t, \xi) \eta_1 \right]$$

$$= Df(x^*(t, \xi + \lambda \eta_2)) \tilde{g}(t, \lambda)$$

$$+ \left( \int_0^1 D^2 f(\theta x^*(t, \xi + \lambda \eta_2) + (1 - \theta) x^*(t, \xi)) d\theta \right)$$

$$\cdot D_\xi x^*(t, \xi) \eta_1 \left( \int_0^1 D_\xi x^*(t, \theta(\xi + \lambda \eta_2) + (1 - \theta)\xi) d\theta \right) \eta_2.$$

Using Lemma 2.3 and taking limits on both sides of (2.32), we obtain

$$D_\xi^2 x^*(t, \xi)(\eta_1 \eta_2) = G\left( Df(x^*(\cdot, \xi)) D_\xi^2 x^*(\cdot, \xi) \eta_1 \eta_2 \right.$$

$$+ D^2 f(x^*(\cdot, \xi)) D_\xi x^*(\cdot, \xi) \eta_1 D_\xi x^*(\cdot, \xi) \eta_2 \right)$$

$$:= G\left( Df(x^*(\cdot, \xi)) D_\xi^2 x^*(\cdot, \xi) \eta_1 \eta_2 + H_1(\cdot, \xi) \right). \tag{2.33}$$

As in the proof of Lemma 2.5, we obtain that there exists a number $\delta_2 < \delta_1$ such that if $\|Df\| < \delta_2$, then $D_\xi^2 x^*(\cdot, \xi)$ as a mapping from $E_c$

to $\mathscr{L}(E_c^2, C_{2(\gamma+3\sigma)+3\sigma})$ is continuous in $\xi \in E_c$. In fact, the term $e^{At}\eta$ in (2.21) has no influence in the proof. The only difference is the existence of the additional term $H_1(t, \xi)$ in (2.33). But $H_1(t, \xi)$ is continuous in $(t, \xi) \in \mathbb{R} \times E_c$. Hence, by Lemma 2.3,

$$\left\| G\big(H_1(\cdot, \xi) - H_1(\cdot, \hat{\xi})\big) \right\|_{C_{2(\gamma+3\sigma)+3\sigma}} \to 0 \quad \text{as } |\xi - \hat{\xi}| \to 0.$$

Here we consider $G$ as a mapping from $C_{2(\gamma+3\sigma)+2\sigma}$ to $C_{2(\gamma+3\sigma)+3\sigma}$.

Finally, by the same reasoning as in Lemma 2.6, we can take the derivative under the integral sign with respect to $\xi$ in (2.29), and obtain

$$D^2\psi(\xi)\eta_1\eta_2$$

$$= \int_\infty^0 e^{-A\tau} \pi_u\big(Df(x^*(\tau, \xi)) D_\xi^2 x^*(\tau, \xi)\eta_1\eta_2 + H_1(\tau, \xi)\big) d\tau$$

$$+ \int_{-\infty}^0 e^{-A\tau} \pi_s\big(Df(x^*(\tau, \xi)) D_\xi^2 x^*(\tau, \xi)\eta_1\eta_2 + H_1(\tau, \xi)\big) d\tau,$$

$$(2.34)$$

where $H_1(\tau, \xi)$ is defined in (2.33). Besides, the continuity and boundedness of $D^2f$, $D_\xi^2 x^*(t, \xi)$, and $H_1(t, \xi)$ imply $\psi \in C_b^2(E_c, E_h)$.

We have just proved Theorem 2.1 for $k = 2$. Suppose now the conclusions are true for $k = j \geq 2$, that is:

(1) $D_\xi^j x^*(\cdot, \xi)$ exists as a mapping from $E_c$ to $\mathscr{L}(E_c^j, C_{j(\gamma+3\sigma)+(j-2)3\sigma+\sigma})$, where $E_c^j = E_c \times E_c \times \cdots \times E_c$ ($j$ times), and satisfies the equation (as a mapping from $E_c$ to $\mathscr{L}(E_c^j, C_{j(\gamma+3\sigma)+(j-2)3\sigma+2\sigma})$)

$$D_\xi^j x^*(t, \xi)(\eta_1 \cdots \eta_j)$$

$$= G\big(Df(x^*(\cdot, \xi)) D_\xi^j x^*(\cdot, \xi)(\eta_1 \cdots \eta_j) + H_{j-1}(\cdot, \xi)\big)(t),$$

$$\eta_i \in E_c, \quad (2.35)$$

where $H_{j-1}(t, \xi)$ is a finite sum of terms involving $Df(x^*(t, \xi)), \ldots, D^j f(x^*(t, \xi))$; $D_\xi x^*(t, \xi)\eta_i$ ($i = 1, \ldots, j - 1$), $\ldots, D_\xi^{j-1} x^*(t, \xi)(\eta_1 \cdots \eta_{j-1})$.

(2) There exists a number $\delta_j < \delta_{j-1}$ such that if $\|Df\| < \delta_j$, then $D^j_\xi x^*(t, \xi)$ as a mapping from $E_c$ to $\mathscr{L}(E^j_c, C_{j(\gamma+3\sigma)+(j-1)3\sigma})$ is continuous in $\xi \in E_c$, and hence
(3) $\psi \in C^j_b(E_c, E_h)$.

If $f \in C^{j+1}_b(\mathbb{R}^n)$, we need to prove the above conclusions are true for $k = j + 1$. But the procedure is completely the same as that done for $k = 2$. Therefore, Theorem 2.1 is proved.                                    □

## 1.3 Local Center Manifolds

In the previous two sections we established the existence, uniqueness, and smoothness of the global center manifolds for equation (1.1). The condition $f \in C^k_b(\mathbb{R}^n)$ is natural. But the hypothesis $\text{Lip}(f) < \delta_0$ (or $\|Df\| < \delta_0$) for a small $\delta_0 > 0$ is quite strong. If we consider a bifurcation problem only near an equilibrium point of (1.1), then we need a local center manifold. This can be obtained from the global center manifold of a modified equation by using the cut-off technique, and the hypothesis $\|Df\| < \delta_0$ will be satisfied automatically since $f(0) = 0$ and $Df(0) = 0$. Let us discuss this in detail.

We consider a cut-off function $\chi: \mathbb{R}^n \to \mathbb{R}$ with the following properties:
(i) $\chi(x) \in C^\infty$;
(ii) $0 \le \chi(x) \le 1, \forall x \in \mathbb{R}^n$;
(iii) $\chi(x) = 1$ if $|x| \le 1$ and $\chi(x) = 0$ if $|x| \ge 2$.

Related to $f(x)$ in (1.1) and for a given $\rho > 0$, we define

$$f_\rho(x) = f(x)\chi\left(\frac{x}{\rho}\right), \qquad \forall x \in \mathbb{R}^n. \tag{3.1}$$

Thus, as a modification of equation (1.1), we consider

$$\dot{x} = Ax + f_\rho(x). \tag{3.2}$$

Obviously, if we restrict $x$ to the domain $|x| < \rho$, then equations (1.1) and (3.2) are the same. The following lemma shows that if we choose $\rho$ sufficiently small, then $\|Df_\rho\|$ can be very small. Hence we can apply the global center manifold theory to (3.2), and get some local results for (1.1).

**Lemma 3.1.** *If $f \in C^k(\mathbb{R}^n)$ for some $k \geq 1$, and $f(0) = 0$ and $Df(0) = 0$, then $f_\rho(x) \in C_b^k(\mathbb{R}^n)$ for a given $\rho > 0$, and*

$$\lim_{\rho \to 0} \|Df_\rho\| = 0. \qquad (3.3)$$

*Proof.* Since $f \in C^k$ and $\chi \in C^\infty$, $f_\rho \in C^k$. For a given $\rho > 0$, $f_\rho(x) = 0$ if $|x| \geq 2\rho$, whence $f_\rho \in C_b^k(\mathbb{R}^n)$. From (3.1) we have that

$$Df_\rho(x) = Df(x) \cdot \chi\left(\frac{x}{\rho}\right) + \frac{1}{\rho} f(x) \cdot D\chi\left(\frac{x}{\rho}\right).$$

Hence,

$$\|Df_\rho\| \leq \sup_{|x| \leq 2\rho} |Df(x)| + \frac{1}{\rho} \|D\chi\| \sup_{|x| \leq 2\rho} |f(x)|. \qquad (3.4)$$

The condition $f(0) = 0$ implies $f(x) = \int_0^1 Df((1 - \lambda)x)x d\lambda$. This gives

$$\sup_{|x| \leq 2\rho} |f(x)| \leq \sup_{|x| \leq 2\rho} |x| \left( \sup_{|x| \leq 2\rho} |Df(x)| \right) \leq 2\rho \sup_{|x| \leq 2\rho} |Df(x)|.$$

Substituting the above inequality into (3.4), we obtain that

$$\|Df_\rho\| \leq (1 + 2\|D\chi\|) \sup_{|x| \leq 2\rho} |Df(x)|.$$

The desired result (3.3) follows from the above estimate and the condition $Df(0) = 0$. $\qquad \square$

**Theorem 3.2.** *Suppose that $f \in C^k(\mathbb{R}^n)$ for some $k \geq 1$, and $f(0) = 0$, $Df(0) = 0$. Then there exists $\psi \in C_b^k(E_c, E_h)$ and an open neighborhood $\Omega$ of $x = 0$ in $\mathbb{R}^n$ such that*
(i) *the manifold*

$$M_\psi := \{x_c + \psi(x_c) \mid x_c \in E_c\} \qquad (3.5)$$

*is locally invariant under (1.1). More precisely,*

$$\tilde{x}(t, x) \in M_\psi, \qquad \forall x \in M_\psi \cap \Omega, \quad \forall t \in J_\Omega(x),$$

*where $\tilde{x}(t, x)$ is the flow of (1.1) with $\tilde{x}(0, x) = x$, and $J_\Omega(x)$ is the maximal interval of existence of the solution $\tilde{x}(\cdot, x)$ with respect to $\Omega$;*
*(ii) $\psi(0) = 0$ and $D\psi(0) = 0$;*
*(iii) if $x \in \Omega$ and $J_\Omega(x) = \mathbb{R}$, then $x \in M_\psi$.*

*Proof.* Let $\delta_k$ be given as in Theorem 2.1. By Lemma 3.1, we can find a $\rho > 0$ such that $f_\rho \in C_b^k(\mathbb{R}^n)$ and $\|Df_\rho\| < \delta_k$, where $f_\rho$ is defined in (3.1). By Theorem 1.1 and Theorem 2.1, we can obtain the global center manifold $M_\psi$ of equation (3.2), $M_\psi$ is defined by (3.5), $\psi \in C_b^k(\mathbb{R}^n)$, and $\psi(0) = 0$, $D\psi(0) = 0$.

On the other hand, from the properties of the cut-off function $\chi$, we know that the equations (3.2) and (1.1) are the same if $x \in \Omega :=$ $\{x \in \mathbb{R}^n \mid \|x\| < \rho\}$. Thus, the conclusions (i) and (ii) are proved.

We suppose now that $x \in \Omega$ and $J_\Omega(x) = \mathbb{R}$. Then $\tilde{x}(t, x) \equiv \tilde{x}_\rho(t, x) \subset \Omega, \forall t \in \mathbb{R}$, whence $\sup_{t \in \mathbb{R}} |\pi_h \tilde{x}(t, x)| < \infty$. By (1.2), $x \in M_\psi$, and conclusion (iii) follows. $\square$

**Definition 3.3.** If $\phi \in C^k(E_c, E_h)$, $k \geq 1$, $\phi(0) = 0$, $D\phi(0) = 0$, and $M_\phi := \{x_c + \phi(x_c) \mid x_c \in E_c\}$ is locally invariant for the flow of (1.1), then $M_\phi$ is called a $C^k$ local center manifold of (1.1).

**Lemma 3.4.** *Suppose that $f \in C^1(\mathbb{R}^n)$, $f(0) = 0$, and $Df(0) = 0$; and $\phi \in C^1(E_c, E_h)$, $\phi(0) = 0$, and $D\phi(0) = 0$. Then $M_\phi := \{x_c + \phi(x_c) \mid x_c \in E_c\}$ is a local center manifold of (1.1) if and only if there is a neighborhood $\Omega_c$ of the origin in $E_c$ such that for all $x_c \in \Omega_c$,*

$$D\phi(x_c)\pi_c(Ax_c + f(x_c + \phi(x_c))) = \pi_h(A\phi(x_c) + f(x_c + \phi(x_c))).$$

$$(3.6)$$

*Proof.* Suppose that such an $\Omega_c$ exists and (3.6) holds. For each $x_c \in \Omega_c$, let $\tilde{x}_c(t)$ be the solution of the following initial value problem

$$\dot{\tilde{x}}_c = \pi_c(A\tilde{x}_c + f(\tilde{x}_c + \phi(\tilde{x}_c))), \qquad \tilde{x}_c(0) = x_c,$$

and let $\tilde{x}(t) = \tilde{x}_c(t) + \phi(\tilde{x}_c(t))$. It is obvious that $x_c(t) \in \Omega_c$ if $|t|$ is sufficiently small. Then by (3.6) we have

$$\dot{\bar{x}}(t) = (I + D\phi(\bar{x}_c(t)))\pi_c(A\bar{x}_c(t) + f(\bar{x}_c(t) + \phi(\bar{x}_c(t))))$$

$$= (\pi_c + \pi_h)(A\bar{x}(t) + f(\bar{x}(t))), \qquad |t| \text{ is sufficiently small.}$$

This means $\bar{x}(t) = \bar{x}_c(t) + \phi(\bar{x}_c(t))$ is a solution of (1.1) if $|t|$ is sufficiently small. Hence, $M_\phi$ is locally invariant under (1.1), and, by Definition 3.3, it is a local center manifold of (1.1).

Suppose that $M_\phi$ is locally invariant under (1.1) in a neighborhood $\Omega$ of the origin in $\mathbb{R}^n$. Let $\Omega_c$ be an open neighborhood of the origin in $E_c$ such that $x_c + \phi(x_c) \in \Omega$ if $x_c \in \Omega_c$. For any $x_c \in \Omega_c$, let $\bar{x}(t) = \bar{x}(t, x_c + \phi(x_c))$, which is the solution of (1.1) with the initial condition $\bar{x}(0) = x_c + \phi(x_c)$. Then the local invariance of $M_\phi$ under (1.1) implies that for $|t|$ sufficiently small we have

$$\pi_h \bar{x}(t) = \phi(\pi_c \bar{x}(t)).$$

Differentiating the above equality with respect to $t$, and using (1.1), we obtain

$$\pi_h(A\bar{x}(t) + f(\bar{x}(t))) = D\phi(\pi_c \bar{x}(t))\pi_c(A\bar{x}(t) + f(\bar{x}(t))).$$

Taking $t = 0$ and noting $\phi: E_c \to E_h$, we get (3.6).                     □

Theorem 3.2 gives the existence of a local center manifold. But, in general, it is not unique.

**Example 3.5.** Consider the planar system

$$\dot{x} = x^2, \qquad \dot{y} = -y.$$

It is easy to see that $E_c = \{(x, 0) \mid x \in \mathbb{R}\}$ and $E_h = \{(0, y) \mid y \in \mathbb{R}\}$. Suppose $\phi \in C^1(E_c, E_h)$ gives the local center manifold $M_\phi = \{x + \phi(x) \mid x \in \mathbb{R}\}$. Then by Lemma 3.4, we have

$$\phi'(x)x^2 = -\phi(x), \qquad \phi(0) = \phi'(0) = 0.$$

Hence

$$\phi(x) = \begin{cases} \alpha e^{1/x} & \text{for } x < 0; \\ 0 & \text{for } x \geq 0, \end{cases}$$

where $\alpha \in \mathbb{R}$ is a constant. Each different $\alpha$ gives a different $M_\phi$. This means that local center manifolds are not unique, even in a sufficiently small neighborhood of the origin.

Fortunately, the nonuniqueness of local center manifolds is not a serious problem when we consider bifurcation phenomena of vector fields. In fact, every local center manifold of (1.1) contains all bounded solutions of (1.1), for example, equilibrium points, periodic orbits, or homoclinic or heteroclinic orbits, provided they stay in a sufficiently small neighborhood of the origin. To show this, we need the following result which says that each local center manifold of (1.1) can be obtained from the global center manifold of a related vector field.

**Theorem 3.6.** *Suppose $f \in C^k(\mathbb{R}^n)$ for some $k \geq 1$, $f(0) = 0$ and $Df(0) = 0$, and $\phi \in C^{k+1}(E_c, E_h)$ defines a local center manifold $M_\phi$ of (1.1). Let $\delta \in (0, \delta_0]$, where $\delta_0$ is defined in Theorem 1.1. Then there exists a neighborhood $\Omega$ of the origin in $\mathbb{R}^n$ and mappings $\tilde{f} \in C_b^k(\mathbb{R}^n)$ and $\psi \in C_b^{k+1}(E_c, E_h)$ such that*
*(i) $f(x) = \tilde{f}(x)$, $\forall x \in \Omega$;*
*(ii) $\|D\tilde{f}\| < \delta$;*
*(iii) $M_\phi \cap \Omega = M_\psi \cap \Omega$,*
*where $M_\psi$ is the unique global center manifold of the following equation*

$$\dot{x} = Ax + \tilde{f}(x). \tag{3.7}$$

*Proof. Part (I): a special case.* We suppose that $M_\phi = E_c$ is a local center manifold of (1.1), that is, there exists some neighborhood $\Omega_1$ of the origin in $\mathbb{R}^n$, such that

$$\phi(x_c) \equiv 0, \qquad \forall x_c \in E_c \cap \Omega_1.$$

By Lemma 3.4, there exists a $d > 0$ such that

$$\pi_h f(x_c) \equiv 0, \qquad \forall x_c \in E_c \quad \text{and} \quad \|x_c\| < d. \tag{3.8}$$

Let $0 < \rho < d/2$, and $\chi \in C^\infty(\mathbb{R}^n, \mathbb{R})$ be a cut-off function. We define

$$\tilde{f}(x) = f(x)\chi\left(\frac{x}{\rho}\right), \qquad \forall\, x \in \mathbb{R}^n;$$

then

$$\pi_h \tilde{f}(x_c) = (\pi_h f(x_c))\chi\left(\frac{x_c}{\rho}\right) = 0, \qquad \forall\, x_c \in E_c. \qquad (3.9)$$

In fact, if $|x_c| < 2\rho < d$, then $\pi_h f(x_c) \equiv 0$ by (3.8); if $|x_c| \geq 2\rho$, then $\chi(x_c/\rho) \equiv 0$. Lemma 3.4 and (3.9) give the invariance of $E_c$ under (3.7). By Lemma 3.1, we can choose $\rho$ so small that $\|D\tilde{f}\| < \delta$. On the other hand, since $\tilde{f} \in C_b^k(\mathbb{R}^n)$ and $\|D\tilde{f}\| < \delta \leq \delta_0$, it follows from Theorem 1.1 that (3.7) has a unique global center manifold $M_0$. Hence, $M_0 = E_c$. We define $\psi(x_c) \equiv 0$ and $\Omega = \{x \in \mathbb{R}^n \mid |x| < \rho\}$. Hence, $\psi \in C_b^{k+1}(E_c, E_h)$ and the conclusions (i)–(iii) are satisfied.

*Part (II): the general case.* Let $\phi \in C^{k+1}(E_c, E_h)$, and $M_\phi$ is a local center manifold of (1.1). By Lemma 3.4, we can find some $d_0 > 0$ such that (3.6) holds for $|x_c| < d_0$ and $x_c \in E_c$. Let $d \in (0, d_0)$ and define $\psi \in C_b^{k+1}(E_c, E_h)$ by

$$\psi(x_c) = \phi(x_c)\chi\left(\frac{x_c}{d}\right), \qquad \forall\, x_c \in E_c. \qquad (3.10)$$

We make a transformation

$$y = x - \psi(\pi_c x) := \Psi(x), \qquad \forall\, x \in \mathbb{R}^n. \qquad (3.11)$$

Then it is easy to verify that
(a) $\Psi(M_\psi) = E_c$;
(b) $\Psi^{-1}(y) = y + \psi(\pi_c y), \forall\, y \in \mathbb{R}^n$.
Under the transformation (3.11), (1.1) becomes

$$\dot{y} = Ay + g(y), \qquad (3.12)$$

where

$$g(y) = A\psi(\pi_c y) + f(y + \psi(\pi_c y))$$

$$- D\psi(\pi_c y)\pi_c(Ay + f(y + \psi(\pi_c y))). \qquad (3.13)$$

Since $E_c$ is a local center manifold of (3.12) (see the property (a)), the results in Part (I) imply that there exist some $\bar{g} \in C_b^k(\mathbb{R}^n)$ and some neighborhood $\tilde{\Omega}$ of the origin in $\mathbb{R}^n$ such that

$$g(y) = \bar{g}(y), \qquad \forall\, y \in \tilde{\Omega}, \tag{3.14}$$

and $E_c$ is the unique global center manifold of the equation

$$\dot{y} = Ay + \bar{g}(y). \tag{3.15}$$

Furthermore, $\|D\bar{g}\|$ can be smaller than any given positive number. Now if we take the inverse transformation $x = \Psi^{-1}(y) = y + \psi(\pi_c y)$, (3.15) becomes an equation of the form (3.7), where

$$\tilde{f}(x) = D\psi(\pi_c x)A\pi_c x - A\psi(\pi_c x) + \bar{g}(x - \psi(\pi_c x))$$

$$+ D\psi(\pi_c x)\pi_c \bar{g}(x - \psi(\pi_c x)). \tag{3.16}$$

Since $\bar{g} \in C_b^k(\mathbb{R}^n)$, $\psi \in C_b^{k+1}(E_c, E_h)$, and $\psi$ has bounded support, we have $\tilde{f} \in C_b^k(\mathbb{R}^n)$. We claim that if we take $\Omega = \Psi^{-1}(\tilde{\Omega})$ and let $\tilde{\Omega}$ be sufficiently small, then the conclusions (i)–(iii) are satisfied.

In fact, $\forall\, x \in \Omega = \Psi^{-1}(\tilde{\Omega})$, $y = x - \psi(\pi_c x) \in \tilde{\Omega}$, and hence (3.14) holds. Noting $\psi, D\psi\colon E_c \to E_h$, and substituting (3.13) into (3.16), we obtain $\tilde{f}(x) = f(x)$ for $x \in \Omega$.

Next, by using the following equalities

$$\psi(\pi_c x) = \left(\int_0^1 D\psi((1 - \theta)\pi_c x)d\theta\right)(-\pi_c x),$$

$$\bar{g}(x - \psi(\pi_c x)) = \left(\int_0^1 D\bar{g}((1 - \theta)(x - \psi(\pi_c x)))d\theta\right)(\psi(\pi_c x) - x),$$

we can obtain from (3.16) that

$$\|D\tilde{f}\| \le (2|A|\|\pi_c\|)\|D\psi\| + (1 + \|\pi_c\|\|D\psi\|)^2\|D\bar{g}\|.$$

By (3.10) and Lemma 3.1, $\|D\psi\| \to 0$ as $d \to 0$. Hence we can choose $d$ so small that $(2|A|\|\pi_c\|)\|D\psi\| < \delta/2$. Fix such a $d > 0$; then $\psi$ and $\bar{g}$ are well defined. By Part (I), we can choose $\tilde{\Omega}$ properly such that $(1 + \|\pi_c\|\|D\psi\|)^2\|D\bar{g}\| < \delta/2$. Thus, we have $\|D\tilde{f}\| < \delta$.

Finally, from (3.10) it is obvious that $\phi(x_c) = \psi(x_c)$ if $|x_c| \le d$. Let $\tilde{\Omega}$ be sufficiently small so that $|\pi_c x| \le d$ if $x \in \Omega$. Hence, $M_\phi \cap \Omega = M_\psi \cap \Omega$. Since $E_c$ is the unique global center manifold of (3.15) and $\Psi^{-1}(E_c) = M_\psi$, $M_\psi$ is invariant under (3.7). But $\|D\tilde{f}\| < \delta \le \delta_0$, so by Theorem 1.1, $M_\psi$ is the unique global center manifold of (3.7). $\square$

**Theorem 3.7.** *Under the assumptions of Theorem 3.6, there exists a bounded neighborhood $\Omega$ of the origin in $\mathbb{R}^n$ such that if $x \in \Omega$ and $J_\Omega(x) = \mathbb{R}$, then $x \in M_\phi$. (For the definition of $J_\Omega(x)$, see Theorem 3.2).*

*Proof.* We use Theorem 3.6 with $\delta = \delta_0$, and assume that $\Omega$ found by Theorem 3.6 is bounded (otherwise, just shrink it to a bounded one). Suppose that $x \in \Omega$ and $J_\Omega(x) = \mathbb{R}$, which means $\tilde{x}(t, x) \in \Omega$ for all $t \in \mathbb{R}$, where $\tilde{x}(t, x)$ is the solution of (1.1) with $\tilde{x}(0, x) = x$. By the conclusion (i) of Theorem 3.6, $\tilde{x}(t, x)$ is a solution of (3.7), and it is globally bounded. By the conclusion (ii), we can use Theorem 1.1, and hence $x \in M_\psi$. By the conclusion (iii), $x \in M_\phi$. $\square$

**Remark 3.8.** Theorem 3.7 says that if $M_\phi$ is a $C^k$ ($k \ge 1$) local center manifold of (1.1), then it must contain all small bounded solutions of (1.1). In particular, $M_\phi$ must contain all sufficiently small equilibria, periodic orbits, and homoclinic and heteroclinic orbits.

**Theorem 3.9.** *Suppose that $f \in C^k(\mathbb{R}^n)$ for some $k \ge 1$, $f(0) = 0$ and $Df(0) = 0$, and $M_{\phi_1}$ and $M_{\phi_2}$ are two $C^{k+1}$ local center manifolds of (1.1). Then we have*

$$D^j\phi_1(0) = D^j\phi_2(0), \qquad 1 \le j \le k. \tag{3.17}$$

*Proof.* We use Theorem 3.6 for $\phi_1$ and $\phi_2$ with $\delta = \delta_k \le \delta_0$, where $\delta_0$ and $\delta_k$ are defined in Theorems 1.1 and 2.1, respectively, and they depend only on $A \in \mathscr{L}(\mathbb{R}^n, \mathbb{R}^n)$ in (1.1). Then there exist corresponding $\Omega_i$, $\tilde{f}_i \in C_b^k(\mathbb{R}^n)$, and $\psi_i \in C_b^{k+1}(E_c, E_h)$ satisfying the conclusions (i)–(iii) for $i = 1$ and 2, respectively. Let $\Omega = \Omega_1 \cap \Omega_2$; then we have

the following conclusions:
(a) $f(x) = \tilde{f}_1(x) = \tilde{f}_2(x)$ for $x \in \Omega$;
(b) $\|D\tilde{f}_i\| < \delta = \delta_k \le \delta_0$ for $i = 1, 2$;
(c) $M_{\phi_i} \cap \Omega = M_{\psi_i} \cap \Omega$ for $i = 1, 2$,
where $M_{\psi_i}$ is the unique global center manifold of the equation

$$\dot{x} = Ax + \tilde{f}_i(x).$$

From (1.27) and the proof of Theorem 2.1 we know that $D^j\psi_i(0)$ $(1 \le j \le k)$ is completely determined by $A$ and $\tilde{f}_i(x)$ for $x$ in a sufficiently small neighborhood of the origin. Hence, by (a) and (b), $D^j\psi_1(0) = D^j\psi_2(0)$, $1 \le j \le k$, and then by (c), $D^j\phi_1(0) = D^j\phi_2(0)$, $1 \le j \le k$. $\qquad\square$

**Remark 3.10.** The conclusions in Theorem 3.7 and Theorem 3.9 give partial uniqueness of local center manifolds. At the end of the next section we will introduce a new result by Burchard, Deng, and Lu [1] which says that the flows on any two $C^{k+1}$ local center manifolds of (1.1) are locally $C^k$ conjugate. Hence, we can choose any local center manifold to study the bifurcation phenomena.

### 1.4 Asymptotic Behavior and Invariant Foliations

In this section we will generalize the results on global center manifolds in Sections 1.1–1.2 to the cases of global center-stable and center-unstable manifolds. Then we will discuss asymptotic behavior of solutions outside these invariant manifolds. This is related to foliations of $\mathbb{R}^n$. We will go back to the local situation by using the cut-off technique, as in Section 1.3, and study stability properties of local center manifolds.

Denote

$$\pi_{cu} = \pi_c + \pi_u, \qquad \pi_{cs} = \pi_c + \pi_s,$$

$$E_{cu} = E_c \oplus E_u, \qquad E_{cs} = E_c \oplus E_s.$$

We now introduce the center-unstable manifold for which the proof of results is completely similar to that in Sections 1.1 and 1.2. Suppose that the positive numbers $\alpha, \beta$ are determined in Lemma 1.5, and $\alpha < \gamma < \beta$.

**Theorem 4.1.** (i) *There is a positive number $\delta_{cu}$ such that if $f \in C_b^{0,1}(\mathbb{R}^n)$ and $\mathrm{Lip}(f) < \delta_{cu}$, then the set*

$$W^{cu} := \left\{ x \in \mathbb{R}^n \mid \sup_{t \le 0} |\pi_s \tilde{x}(t, x)| < \infty \right\}$$

$$= \left\{ x \in \mathbb{R}^n \mid \sup_{t \le 0} e^{\gamma t} |\tilde{x}(t, x)| < \infty \right\} \tag{4.1}$$

*is invariant under (1.1), and is a Lipschitz submanifold of $\mathbb{R}^n$, that is, there exists a unique Lipschitz function $\psi \in C_b^0(E_{cu}, E_s)$ such that*

$$W^{cu} = \{ x_{cu} + \psi(x_{cu}) \mid x_{cu} \in E_{cu} \}. \tag{4.2}$$

(ii) *If $\phi \in C_b^0(E_{cu}, E_s)$ and the manifold*

$$M_\phi := \{ x_{cu} + \phi(x_{cu}) \mid x_{cu} \in E_{cu} \} \tag{4.3}$$

*is invariant under (1.1), then $M_\phi = W^{cu}$, $\phi = \psi$.*

We say that $W^{cu}$ is the unique global center-unstable manifold of (1.1). Using the same approach of Section 1.2, for any $k \ge 1$, we have the following:

**Theorem 4.2.** *Suppose $f \in C_b^k(\mathbb{R}^n)$ for some $k \ge 1$. Then there is a number $\delta_{cu}^k > 0$ such that when $\|Df\| < \delta_{cu}^k$, the conclusion of Theorem 4.1 holds with $\psi \in C_b^k(E_{cu}, E_s)$.*

We next consider the existence of global invariant foliations. For $\gamma \in \mathbb{R}$, a Banach space is defined by

$$C_\gamma^+ = \left\{ z \in C^0(\mathbb{R}, \mathbb{R}^n) \mid \|z\|_\gamma^+ = \sup_{t \in \mathbb{R}^+} e^{\gamma t} |z(t)| < \infty \right\}.$$

We fix $\gamma$ such that $\alpha < \gamma \le k\gamma < \beta$, where $k$ is the positive integer in Theorem 4.2. Suppose that $\tilde{x}(t, \hat{x})$, $\tilde{x}(t, x)$ are the two solutions of (1.1) satisfying the initial conditions $\tilde{x}(0, \hat{x}) = \hat{x}$ and $\tilde{x}(0, x) = x$, $\hat{x}, x \in \mathbb{R}^n$.

We define the set

$$M_s(x) = \left\{ \hat{x} \in \mathbb{R}^n \mid z(t) = \tilde{x}(t, \hat{x}) - \tilde{x}(t, x) \in C_\gamma^+ \right\}. \quad (4.4)$$

We call $M_s(x)$ the stable leaf of $x \in \mathbb{R}^n$.

**Theorem 4.3.** *Suppose $f \in C_b^{0,1}(\mathbb{R}^n)$ and $\mathrm{Lip}(f) < \delta_{cu}$. Then there exists a uniformly continuous mapping $J_s: \mathbb{R}^n \times E_s \to E_{cu}$ such that for each $x \in \mathbb{R}^n$ the following conclusions hold:*
(i) *$M_s(x) = \{x_s + J_s(x, x_s) \mid x_s \in E_s\}$ is a Lipschitz manifold;*
(ii) *$M_s(x)$ has a unique intersection point with $W^{cu}$;*
(iii) *there is a stable foliation of $\mathbb{R}^n$:*

$$\mathbb{R}^n = \bigcup_{x \in W^{cu}} M_s(x). \quad (4.5)$$

(iv) *The foliation is invariant under (1.1), that is,*

$$M_s(\tilde{x}(t, x)) = \tilde{x}(t, M_s(x)), \qquad \forall\, t > 0, \qquad \forall\, x \in \mathbb{R}^n.$$

The proof of Theorem 4.3 needs several lemmas.

**Lemma 4.4.** *Suppose $f \in C_b^{0,1}(\mathbb{R}^n)$, $\mathrm{Lip}(f) < \delta_{cu}$, and $z(t) \in C_\gamma^+$. For $x \in \mathbb{R}^n$, $\tilde{x}(\cdot, x) + z$ is a solution of (1.1) if and only if there exists some $x_s \in E_s$ such that*

$$z(t) = e^{At} x_s + \int_0^t e^{A(t-\tau)} \pi_s \tilde{f}(\tau; x, z(\tau))\, d\tau$$

$$+ \int_\infty^t e^{A(t-\tau)} \pi_{cu} \tilde{f}(\tau; x, z(\tau))\, d\tau, \quad (4.6)$$

*where $\tilde{f}: \mathbb{R} \times \mathbb{R}^n \times \mathbb{R}^n \to \mathbb{R}^n$ is defined by*

$$\tilde{f}(t; x, z(t)) = f(\tilde{x}(t, x) + z(t)) - f(\tilde{x}(t, x)). \quad (4.7)$$

*Proof.* If $\tilde{x}(\cdot, x) + z$ is a solution of (1.1), then $z$ satisfies the equation

$$\dot{z} = Az + \tilde{f}(t; x, z). \quad (4.8)$$

The variation of constants formula gives

$$z(t) = e^{A(t-t_0)}z(t_0) + \int_{t_0}^{t} e^{A(t-\tau)}\tilde{f}(\tau; x, z(\tau))d\tau. \qquad (4.9)$$

Taking $t_0 = 0$ in (4.9) and applying $\pi_s$, we have

$$\pi_s z(t) = e^{At}\pi_s z(0) + \int_0^t e^{A(t-\tau)}\pi_s \tilde{f}(\tau; x, z(\tau))d\tau. \qquad (4.10)$$

Since $z(t) \in C_\gamma^+$ we obtain $\lim_{t_0 \to \infty} e^{A(t-t_0)}\pi_{cu}z(t_0) = 0$. Writing $x_s = \pi_s z(0)$ and applying $\pi_{cu}$ in (4.9), and then combining it with (4.10), we obtain (4.6).

Conversely, we suppose (4.6) holds. Noting

$$\tilde{f}(t; x, z(t)) + f(\bar{x}(t, x)) = f(\bar{x}(t, x) + z(t)), \qquad (4.11)$$

we then have

$$\bar{x}(t, x) + z(t) = e^{At}(x + z(0)) + \int_0^t e^{A(t-\tau)}f(\bar{x}(\tau, x) + z(\tau))d\tau.$$

This means that $\bar{x}(t, x) + z(t)$ is a solution of (1.1).     □

**Lemma 4.5.** *Under the conditions of Lemma 4.4, for each $(x, x_s) \in \mathbb{R}^n \times E_s$, there exists a unique solution $z = z^*(x, x_s) \in C_\gamma^+$ of (4.6) which is uniformly continuous in $(x, x_s) \in \mathbb{R}^n \times E_s$ such that*

$$M_s(x) = \{x + z^*(x, x_s)(0) \mid x_s \in E_s\}. \qquad (4.12)$$

*Proof.* Let $J^+: \mathbb{R}^n \times E_s \times C_\gamma^+ \to C_\gamma^+$ be defined by

$$J^+(x, x_s, z(\cdot))(t) = e^{At}x_s + \int_0^t e^{A(t-\tau)}\pi_s \tilde{f}(\tau; x, z(\tau))d\tau$$

$$+ \int_\infty^t e^{A(t-\tau)}\pi_{cu}\tilde{f}(\tau; x, z(\tau))d\tau. \qquad (4.13)$$

By using a similar estimate as in the proof of Lemma 1.7, for each $z_1, z_2 \in C_\gamma^+$, we have

$$\|J^+(x, x_s, z_1(\cdot)) - J^+(x, x_s, z_2(\cdot))\|_\gamma^+$$

$$\leq K\left(\frac{1}{\gamma - \alpha} + \frac{2}{\beta - \gamma}\right)\text{Lip}(f)\|z_1 - z_2\|_\gamma^+.$$

Hence, there is a $\delta_{cu} > 0$ such that if $\text{Lip}(f) < \delta_{cu}$, then

$$\|J^+(x, x_s, z_1(\cdot)) - J^+(x, x_s, z_2(\cdot))\|_\gamma^+ < \frac{1}{3}\|z_1 - z_2\|_\gamma^+.$$

Therefore, it follows from the Uniform Contraction Mapping Theorem that for each $(x, x_s) \in \mathbb{R}^n \times E_s$, $J^+(x, x_s, z(\cdot))$ has a unique fixed point $z = z^*(x, x_s)(\cdot) \in C_r^+$, which is the unique solution of (4.6). Since for fixed $x$, $J^+(x, \cdot, \cdot)$ is Lipschitz in $x_s$ and $z(\cdot)$, $z^*(x, x_s)$ is Lipschitz in $x_s$. By Lemma 4.4, $\tilde{x}(t, x) + z^*(x, x_s)(t)$ is a solution of (1.1). Hence there exists a $\hat{x} \in \mathbb{R}^n$ such that

$$\tilde{x}(t, \hat{x}) = \tilde{x}(t, x) + z^*(x, x_s)(t),$$

where $\hat{x} = x + z^*(x, x_s)(0)$. We obtain (4.12) from (4.4).

It remains to show the continuity of $z^*(x, x_s)$ in $(x, x_s) \in \mathbb{R}^n \times E_s$. We note that $C_{\gamma+\sigma}^+ \subset C_\gamma^+$ and $\|z\|_\gamma^+ \leq \|z\|_{\gamma+\sigma}^+$ for $\sigma > 0$, hence there is a continuous inclusion from $C_{\gamma+\sigma}^+$ into $C_\gamma^+$. If $\sigma > 0$ is sufficiently small, then we can apply the Contraction Mapping Theorem to $J^+$: $\mathbb{R}^n \times E_s \times C_{\gamma+\sigma}^+ \to C_{\gamma+\sigma}^+$ to obtain the fixed point $z = z_\sigma^*(x, x_s) \in C_{\gamma+\sigma}^+ \subset C_\gamma^+$ for given $(x, x_s) \in \mathbb{R}^n \times E_s$. By uniqueness, we have $z_\sigma^*(x, x_s) = z^*(x, x_s), \forall x \in \mathbb{R}^n, x_s \in E_s$.

Now we consider $J^+$ as a mapping from $\mathbb{R}^n \times E_s \times C_{\gamma+\sigma}^+$ to $C_\gamma^+$. We will prove that for any given $\epsilon > 0$, there exists a $\mu > 0$, such that if $x, \bar{x} \in \mathbb{R}^n$, $x_s, \bar{x}_s \in E_s$, and $|x - \bar{x}| + |x_s - \bar{x}_s| < \mu$, then

$$\Delta J^+ := \|J^+(x, x_s, z^*(x, x_s)(\cdot)) - J^+(\bar{x}, \bar{x}_s, z^*(\bar{x}, \bar{x}_s)(\cdot))\|_\gamma^+ < \epsilon.$$

$$(4.14)$$

Note that $\Delta J^+ = \|z^*(x, x_s)(\cdot) - z^*(\bar{x}, \bar{x}_s)(\cdot)\|_\gamma^+$. This implies that the map $(x, x_s) \mapsto z^*(x, x_s)(\cdot)$ from $\mathbb{R}^n \times E_s$ to $C_\gamma^+$ is uniformly continuous.

From (4.13) we have

$$|J^+(x, x_s, z^*(x, x_s)(t)) - J^+(\bar{x}, \bar{x}_s, z^*(\bar{x}, \bar{x}_s)(t))|$$

$$\leq |e^{At}\pi_s(x_s - \bar{x}_s)| + |\tilde{G}(g(\cdot, x, x_s, \bar{x}, \bar{x}_s))(t)|, \quad (4.15)$$

where

$$\tilde{G}(g(\cdot, x, x_s, \bar{x}, \bar{x}_s))(t) := \int_0^t e^{A(t-\tau)}\pi_s g(\tau, x, x_s, \bar{x}, \bar{x}_s)d\tau$$

$$+ \int_\infty^t e^{A(t-\tau)}\pi_{cu}g(\tau, x, x_s, \bar{x}, \bar{x}_s)d\tau,$$

and

$$g(\tau, x, x_s, \bar{x}, \bar{x}_s) := \tilde{f}(\tau, x, z^*(x, x_s)(\tau)) - \tilde{f}(\tau, \bar{x}, z^*(\bar{x}, \bar{x}_s)(\tau))$$

$$= f(\tilde{x}(\tau, x) + z^*(x, x_s)(\tau)) - f(\tilde{x}(\tau, x))$$

$$-f(\tilde{x}(\tau, \bar{x}) + z^*(\bar{x}, \bar{x}_s)(\tau)) + f(\tilde{x}(\tau, \bar{x})).$$

By using (1.5) and the definition of $C_\gamma^+$, we obtain from (4.15) that

$$\Delta J^+ \leq \sup_{t \geq 0} e^{-(\beta-\gamma)t}|x_s - \bar{x}_s| + \|\tilde{G}(g(\cdot, x, x_s, \bar{x}, \bar{x}_s))\|_\gamma^+$$

$$\leq |x_s - \bar{x}_s| + \|\tilde{G}(g(\cdot, x, x_s, \bar{x}, \bar{x}_s))\|_\gamma^+. \quad (4.16)$$

We can use the same technique as in Lemma 2.3 to prove that
$\|\tilde{G}(g(\cdot, x, x_s, z^*(\cdot)))\|_\gamma^+ < 2\epsilon/3$ provided $|(x, x_s) - (\bar{x}, \bar{x}_s)| < \mu$. In
fact, since

$$|g(\tau, x, x_s, \bar{x}, \bar{x}_s)| \leq \text{Lip}(f)(|z^*(x, x_s)(\tau)| + |z^*(\bar{x}, \bar{x}_s)(\tau)|),$$

and $z^* \in C_{\gamma+\sigma}^+$, we can find $T > 0$ such that the integral over the
noncompact part ($t \geq T$) is smaller than $\epsilon/3$. Then, using the uniform
continuity of $g$ in the compact part, we can find $\mu > 0$ ($\mu < \epsilon/3$) such
that if $|x - \bar{x}| + |x_s - \bar{x}_s| < \mu$, then the integral over the compact part
($0 \leq t \leq T$) is smaller than $\epsilon/3$. Therefore, (4.14) follows from (4.16).

$\square$

**Lemma 4.6.** *Assume the conditions of Theorem 4.1 are satisfied. Then for each $x \in \mathbb{R}^n$ there exists a unique point $\hat{x} \in W^{cu}$ such that*

$$W^{cu} \cap M_s(x) = \{\hat{x}\}.$$

*Proof.* From (4.1) and (4.4) we see that $\hat{x} \in W^{cu} \cap M_s(x)$ if and only if

$$\sup_{t \leq 0} e^{\gamma t}|\tilde{x}(t, \hat{x})| < \infty \qquad \text{and} \qquad \sup_{t \geq 0} e^{\gamma t}|\tilde{x}(t, \hat{x}) - \tilde{x}(t, x)| < \infty.$$

Denote

$$w(t, x) = \begin{cases} \tilde{x}(t, x), & t \geq 0, \\ x_{cu}(t, x), & t \leq 0, \end{cases}$$

where $x_{cu}(t, x)$ satisfies the equation

$$\dot{x} = A\pi_{cu}x + \pi_{cu}f(x),$$

with $x_{cu}(0, x) = x$.

Suppose $\hat{x} \in W^{cu} \cap M_s(x)$. Let $z(t) = \tilde{x}(t, \hat{x}) - w(t, x)$. It is easy to see that $z(t)$ belongs to the Banach space

$$C_\gamma^{+-} := \left\{ z \in C^0(\mathbb{R}, \mathbb{R}^n) \,\middle|\, \|z\|_\gamma^{+-} = \sup_{t \in \mathbb{R}} e^{\gamma t}|z(t)| < \infty \right\}.$$

Conversely, it is also obvious that if $z(t) \in C_\gamma^{+-}$ and $w(t, x) + z(t)$ is a solution of (1.1) then $\hat{x} = x + z(0) \in W^{cu} \cap M_s(x)$.

Using the same method as in the proof of Lemma 4.4, we can show that if $z(\cdot) \in C_\gamma^{+-}$, then $w(t, x) + z(t)$ is a solution of (1.1) if and only if

$$z(t) = -\pi_s w(t, x) + \int_{-\infty}^t e^{A(t-\tau)}\pi_s f(w(\tau, x) + z(\tau))d\tau$$

$$+ \int_\infty^t e^{A(t-\tau)}\pi_{cu}[f(w(\tau, x) + z(t)) - f(w(\tau, x))]d\tau. \quad (4.17)$$

By the condition of the lemma and the uniform contraction principle, it follows that equation (4.17) has for each $x \in \mathbb{R}^n$ a unique continuous solution $z = z^*(x) \in C_\gamma^{+-}$.

We define a mapping $H_{cu}: \mathbb{R}^n \to W^{cu}$ by

$$H_{cu}(x) := x + z^*(x)(0) \qquad \text{for all } x \in \mathbb{R}^n. \qquad (4.18)$$

From the uniqueness of the solution $z = z^*(x)$ it follows that $\hat{x} = H_{cu}(x) \in W^{cu} \cap M_s(x)$ is unique for each $x \in \mathbb{R}^n$. $\qquad\qquad\square$

*Proof of Theorem 4.3.* By Lemma 4.5, $M_s(x) = \{x + z^*(x, \bar{x}_s)(0) \mid \bar{x}_s \in E_s\}$, where $z^*(x, \bar{x}_s)(t)$ is the unique solution of (4.6) in $C_\gamma^+$ with $x_s = \bar{x}_s$. From (4.6) we have

$$\pi_s(x + z^*(x, \bar{x}_s)(0)) = \pi_s x + \bar{x}_s.$$

Hence

$$x + z^*(x, \bar{x}_s)(0) = \pi_s x + \bar{x}_s + \pi_{cu} x + \pi_{cu} z^*(x, \bar{x}_s)(0)$$

$$= y_s + \pi_{cu} x + \pi_{cu} z^*(x, y_s - \pi_s x)(0),$$

where $y_s := \bar{x}_s + \pi_s x \in E_s$. Now, replace $y_s$ by $\bar{x}_s$ in the above expression and let

$$J_s(x, \bar{x}_s) = \pi_{cu} x + \pi_{cu} z^*(x, \bar{x}_s - \pi_s x)(0).$$

Since $z^*(x, \bar{x}_s)$ is Lipschitz in $\bar{x}_s$, conclusion (i) follows. Conclusion (ii) follows from Lemma 4.6. Conclusion (iii) follows from (ii) and the fact that if $y \in M_s(x)$, then $M_s(y) = M_s(x)$. To prove conclusion (iv), we assume $\hat{x} \in M_s(x)$. By (4.4), $z(t) := \bar{x}(t, \hat{x}) - \bar{x}(t, x) \in C_\gamma^+$. Hence for any $t_1 > 0$,

$$z(t + t_1) = \bar{x}(t, \bar{x}(t_1, \hat{x})) - \bar{x}(t, \bar{x}(t_1, x)) \in C_\gamma^+.$$

This means

$$M_s(\bar{x}(t, x)) = \{\bar{x}(t, \hat{x}) \mid \hat{x} \in M_s(x)\} = \bar{x}(t, M_s(x)). \qquad\square$$

From Theorem 4.6 we have the following two corollaries. Corollary 4.7 will be needed in the proof of Theorem 4.13.

**Corollary 4.7.** *Assume the conditions of Theorem* 4.1 *are satisfied. Then for each* $y \in W^{cu}$ *and* $\epsilon > 0$ *there exists a* $\delta > 0$ *such that if* $|x - y| < \delta$, $x \in \mathbb{R}^n$, *then*

$$|\tilde{x}(t, x) - \tilde{x}(t, H_{cu}(x))| \leq \epsilon e^{-\gamma t}, \qquad \forall\, t \geq 0. \qquad (4.19)$$

*Proof.* From the proof of Lemma 4.6 we have

$$z^*(x)(t) = \tilde{x}(t, \hat{x}) - w(t, x) \in C_\gamma^{+-},$$

where $\hat{x} = H_{cu}(x)$, and $w(t, x) = \tilde{x}(t, x)$ if $t \geq 0$. Hence

$$z^*(x)(t) = \tilde{x}(t, H_{cu}(x)) - \tilde{x}(t, x), \qquad \text{for all } t \geq 0, \qquad x \in \mathbb{R}^n. \tag{4.20}$$

If $y \in W^{cu}$, then $H_{cu}(y) = y$, which gives

$$z^*(y)(t) = 0, \quad \text{for } t \geq 0, \qquad y \in W^{cu}. \tag{4.21}$$

Finally, by the continuity of $z^*$ we can find $\delta > 0$ such that if $|x - y| < \delta$, then $\|z^*(x) - z^*(y)\|_\gamma^{+-} \leq \epsilon$. By (4.21) we obtain

$$\sup_{t \geq 0} e^{\gamma t}|z^*(x)(t)| = \|z^*(x) - z^*(y)\|_\gamma^+ \leq \|z^*(x) - z^*(y)\|_\gamma^{+-} \leq \epsilon. \tag{4.22}$$

Thus, (4.19) follows from (4.22) and (4.20). $\qquad\qquad\qquad\qquad\square$

**Corollary 4.8.** *Assume the conditions of Theorem* 4.1 *are satisfied. Then*

$$\sup_{t \geq 0} e^{\gamma t}|\tilde{x}(t, \hat{x}) - \tilde{x}(t, x)| < \infty, \qquad x \in \mathbb{R}^n, \qquad \hat{x} \in W^{cu},$$

*if and only if* $\hat{x} = H_{cu}(x)$. $\qquad\qquad\qquad\qquad\qquad\qquad\square$

**Remark 4.9.** The above result gives the asymptotic behavior of solutions of (1.1) that do not lie on the center-unstable manifold. It says that any solution $\tilde{x}(t, x)$, $x \in \mathbb{R}^n$, converges exponentially for $t \to +\infty$ to a uniquely determined solution $\tilde{x}(t, H_{cu}(x))$ which is on the center-unstable manifold. In particular, if $\sigma_u = \varnothing$, that is, $W^{cu} = W^c$, then any solution of (1.1) converges exponentially as $t \to +\infty$ to a uniquely determined solution on the center manifold. This gives the stability property of center manifolds. We will give a local version of this property in Theorem 4.12.

**Theorem 4.10.** *Suppose, in addition to the conditions of Theorem 4.1, that $f \in C_b^k(\mathbb{R}^n)$ for some $k \geq 1$. Then for each $x \in \mathbb{R}^n$ the stable leaf $M_s(x)$ is $C^k$, that is, the mapping $x_s \to J_s(x, x_s)$ given by Theorem 4.3 is $C^k$ from $E_s$ into $E_{cu}$.*

**Remark 4.11.** By reversing time in Theorems 4.1, 4.3, and 4.10, we can obtain analogous results for the center-stable manifold $W^{cs}$, the unstable leaf $M_u(x)$, and the unstable foliation

$$\mathbb{R}^n = \bigcup_{x \in W^{cs}} M_u(x).$$

As in Section 2, we can get local results from the above global results.

**Theorem 4.12.** (*Asymptotic Phase*) *Suppose that $f \in C^1(\mathbb{R}^n)$, $f(0) = 0$ and $Df(0) = 0$, and $\sigma_u = \varnothing$. Let $M_\phi$ be a $C^2$ local center manifold of (1.1). Then we can find a neighborhood $\Omega$ of the origin in $\mathbb{R}^n$ and some constants $\gamma > 0$ ($\alpha < \gamma < \beta$) such that if $x \in \Omega$ and $\mathrm{cl}\{\tilde{x}(t, x) \mid t \geq 0\} \subset \Omega$, then there exist some $t_0 \geq 0$, $M > 0$ and $y \in M_\phi \cap \Omega$ such that*

$$|\tilde{x}(t, x) - \tilde{x}(t - t_0, y)| \leq M e^{-\gamma t} \qquad \text{for all } t \geq t_0. \quad (4.23)$$

*Proof.* We use Theorem 3.6 for (1.1) with $\delta = \delta_0$. Then there exist a neighborhood $\Omega$ of the origin in $\mathbb{R}^n$, and mappings $\tilde{f} \in C_b^1(\mathbb{R}^n)$, $\psi \in C_b^2(E_c, E_h)$ such that (3.7) has the global center manifold $M_\psi$ and $M_\psi \cap \Omega = M_\phi \cap \Omega$.

Since $\sigma_u = \varnothing$, we can take $\delta_{cu} = \delta_0$, and the global center-unstable manifold $W^{cu}$ of (3.7) coincides with $M_\psi$. Letting $x \in \Omega$, by Lemma 4.6 and Remark 4.9, we can find $\hat{x} = H_{cu}(x) \in M_\psi$ such that

$$\sup_{t \geq 0} e^{\gamma t} |\bar{x}(t, x) - \bar{x}(t, \hat{x})| < \infty. \tag{4.24}$$

Note that $\hat{x}$ may not belong to $\Omega$. The condition $\mathrm{cl}\{\bar{x}(t, x) \mid x \geq 0\} \subset \Omega$ and (4.24) imply that there exits a $t_0 \geq 0$ such that $\bar{x}(t, \hat{x}) \in \Omega$ for all $t \geq t_0$. Let $y = \bar{x}(t_0, \hat{x})$, then $y \in \Omega$, and $\bar{x}(t - t_0, y) = \bar{x}(t, \hat{x}) \in \Omega$ for $t \geq t_0$. Therefore (4.23) follows from (4.24). □

Suppose that $f \in C^1(\mathbb{R}^n)$, $M_\phi$ is a local center manifold of (1.1), and $y \in M_\phi$. Let $x_c(t) = \pi_c \bar{x}(t, y)$, then by Lemma 2.6 and Theorem 3.6 $x_c(t)$ satisfies the equation

$$\dot{x}_c = Ax_c + \pi_c f(x_c + \phi(x_c)), \qquad x_c \in E_c. \tag{4.25}$$

The following theorem gives the relationship between $x_c(t)$, as a solution of (4.25), and $\bar{x}(t, y)$, as a solution of (1.1).

**Theorem 4.13.** (*Pliss Reduction Principle*) *Assume the conditions of Theorem 4.12 are satisfied. Suppose that $y \in M_\phi \cap \Omega$ and $\mathrm{cl}\{\bar{x}(t, y) \mid t \geq 0\} \subset \Omega$. Then $\bar{x}(t, y)$ is stable (asymptotically stable, unstable) if and only if $x_c(t)$ is stable (asymptotically stable, unstable).*

*Proof.* We use the same method and notations as in the proof of Theorem 4.12, that is, extend the local center manifold $M_\phi$ of (1.1) to the global center manifold $M_\psi$ of (3.7) in order to use some global results, and then restrict to $\Omega$ to get corresponding local results.

Suppose $x_c(t) = \pi_c \bar{x}(t, y)$ is stable as a solution of (4.25). Then $x_c(t)$ is also stable as a solution of

$$\dot{x}_c = Ax_c + \pi_c \tilde{f}(x_c + \psi(x_c)), \tag{4.26}$$

where $\tilde{f}$ is given in Theorem 3.6, and the equation is globally defined.

We consider a special case first. Suppose $\hat{x} \in M_\psi$, then $\tilde{x}(t, \hat{x}) = \pi_c \tilde{x}(t, \hat{x}) + \psi(\pi_c \tilde{x}(t, \hat{x}))$, $\pi_c \tilde{x}(t, \hat{x})$ is a solution of (4.26). Thus

$$|\tilde{x}(t, \hat{x}) - \tilde{x}(t, y)|$$

$$\leq |\pi_c \tilde{x}(t, \hat{x}) + \psi(\pi_c \tilde{x}(t, \hat{x})) - (x_c(t) + \psi(x_c(t)))|$$

$$\leq (1 + \|D\psi\|)|\pi_c \tilde{x}(t, \hat{x}) - x_c(t)|.$$

On the other hand, $|\pi_c \tilde{x}(0, \hat{x}) - x_c(0)| \leq \|\pi_c\||\hat{x} - y|$. The stability property of $x_c(t)$ implies that for $\epsilon > 0$ there exists a $\delta_1 > 0$ such that

$$|\tilde{x}(t, \hat{x}) - \tilde{x}(t, y)| \leq \epsilon/2 \quad \text{for } t \geq 0, \quad \hat{x} \in M_\psi, \quad \text{and } |\hat{x} - y| < \delta_1.$$

$$(4.27)$$

Now we consider the general case. Suppose $x \in \mathbb{R}^n$, then we can use the continuous mapping $H_{cu}: \mathbb{R}^n \to M_\psi = W^{cu}$ (since $\sigma_u = \varnothing$) to find $\hat{x} = H_{cu}(x) \in M_\psi$. Noting $H_{cu}(y) = y$, we can find a $\delta > 0$ such that $|\hat{x} - y| < \delta_1$ if $|x - y| < \delta$. Let $\delta$ be sufficiently small so that Corollary 4.7 holds. By using (4.19) and (4.27), we have

$$|\tilde{x}(t, x) - \tilde{x}(t, y)| \leq |\tilde{x}(t, x) - \tilde{x}(t, \hat{x})| + |\tilde{x}(t, \hat{x}) - \tilde{x}(t, y)|$$

$$\leq \frac{\epsilon}{2} e^{-\gamma t} + \frac{\epsilon}{2} \leq \epsilon, \quad \text{for } t \geq 0, \quad (4.28)$$

as long as $|x - y| < \delta$. This gives the stability property of $\tilde{x}(t, y)$ as a solution of (3.7). Restricting to $\Omega$, we obtain that $\tilde{x}(t, y)$ is stable as a solution of (1.1).

If $x_c(t)$ is asymptotically stable, then instead of (4.27) we can obtain

$$|\tilde{x}(t, \hat{x}) - \tilde{x}(t, y)| \to 0$$

$$\text{as } t \to +\infty \quad \text{for } \hat{x} \in M_\psi, |\hat{x} - y| < \delta_1. \quad (4.29)$$

Therefore, (4.29) and (4.28) imply that $\tilde{x}(t, y)$ is asymptotically stable as a solution of (1.1). On the other hand, if $\tilde{x}(t, y)$ is stable (or asymptotically stable), then $x_c(t) = \pi_c \tilde{x}(t, y)$ is obviously stable (or asymptotically stable).

The above results mean that $\tilde{x}(t, y)$ is stable (asymptotically stable) if and only if $x_c(t)$ is also. Hence, $\tilde{x}(t, y)$ must be unstable if $x_c(t)$ is unstable.                                                                                          □

By using the foliation structure, Burchard, Deng and Lu [1] proved the following theorem.

**Theorem 4.14.** *Suppose that $U$ is a neighborhood of the origin in $\mathbb{R}^n$, $f \in C^{k+1,1} (U, \mathbb{R}^n)$ for some $k \geq 0$, $f(0) = 0$, and $Df(0) = 0$. Then the flows on two arbitrary $C^{k+1,1}$ local center manifolds $W_1^c$ and $W_2^c$ of (1.1) in $U$ are locally conjugate. More precisely, there is a neighborhood $V \subset U$ of the origin in $\mathbb{R}^n$ and a $C^k$ diffeomorphism $\phi: W_1^c \cap V \to W_2^c \cap V$ such that*

$$\tilde{x}(t, \phi(x)) = \phi(\tilde{x}(t, x))$$

*for all $x \in W_1^c \cap V$ and all $t \in \mathbb{R}^1$ as long as $\tilde{x}(t, x) \in W_1^c \cap V$.*

**Outline of the Proof.** There is a $C^{k,1}$ local center-stable manifold $W^{cs}$ containing $W_1^c$, and there is a $C^{k,1}$ local center-unstable manifold $W^{cu}$ containing $W_2^c$. The intersection $W^{cs} \cap W^{cu}$ must be a $C^{k,1}$ center manifold of (1.1) which is denoted by $W_3^c$. Since $W_1^c$ and $W_3^c$ are contained in the same center-stable manifold $W^{cs}$, the unstable foliation on $W^{cs}$ gives a $C^k$ conjugacy between $W_1^c$ and $W_3^c$. Similarly, the stable foliation on $W^{cu}$ gives a $C^k$ conjugacy between $W_2^c$ and $W_3^c$ which are contained in $W^{cu}$. Hence, $W_1^c$ and $W_2^c$ are $C^k$ conjugate.

## 1.5 Bibliographical Notes

The invariant manifold theory has a long history, and the center-manifold theory for the finite-dimensional case has been developed by Carr [1], Chow and Hale [1], Chow and Lu [3], Chow and Yi [1], Fenichel [1–4], Guckenheimer and Holmes [1], Hirsch and Pugh [1], Hirch, Pugh and Shub [1], Kelley [1], Marsden [1], Palmer [1], Pliss [1], Sijbrand [1], Vanderbauwhede [2–3], Vanderbauwhede and van Gils [1], Wan [2], Wells [1], Yi [1], and others. The center manifold theory for the infinite-dimensional case has been studied by Bates and Jones [1],

Burchard, Deng, and Lu [1], Chow, Lin and Lu [1], Chow and Lu [2–3], Chow, Lu, and Sell [1], Hale and Lin [1], Hale, Magalhães, and Oliva [1], Henry [1], Mielke [1], Sell [1], Sell and You [1], Temam [1–2], Vanderbauwhede and Iooss [1], and many others.

In this chapter we present a short introduction to the basic concepts and results in center-manifold theory for the finite-dimensional case. From this point of view, a clear description has been given by Vanderbauwhede [3]. We follow some of his approaches and notations, as well as some of his proofs (Lemma 3.1, Lemma 3.4, Theorem 3.6, Corollary 4.7, and Theorems 4.12 and 4.13). The proofs in Sections1.1–1.4 are essentially due to Chow and Lu [1]. The proof of the existence of center manifolds in Section 1.1 appears in many of the above references. The proof of the smoothness of center manifolds might be the most difficult part in this theory. We would like to mention some other works. Vanderbauwhede and van Gils [1] use the contractions on embedded Banach spaces, Vanderbauwhede [2] uses an approximation argument, and the fiber contraction theorem has been used in Vanderbauwhede [3]. Our approach in Section 1.2 uses only the definition of the derivative and a specific estimate described in Lemma 2.2.

Van Strien [1] gives an example to show that there may not exist any $C^\infty$ local center manifold of (1.1) for $f \in C^\infty$ (even if $f$ is analytic). Similar examples can be found in Carr [1], Guckenheimer and Holmes [1], Sijbrand [1], and Vanderbauwhede [3]. However, under certain conditions, $C^\infty$ center manifolds do exist, see Sijbrand [1], for example.

In Sections 1.3 and 1.4, we follow the standard approach to deal with local center manifolds from the global theory by using Theorem 3.6. Palmer [1] gives a direct proof of the local results (Theorems 4.12 and 4.13) by using Gronwall's inequality. We note that in Palmer [1], we only need to assume that $M_\phi$ is $C^1$. Theorem 4.14 belongs to Burchard, Deng, and Lu [1].

# 2

# Normal Forms

It is well known that a linear change of coordinates

$$x = Ty$$

transforms a linear differential equation

$$\dot{x} = Ay$$

to the form

$$\dot{y} = (T^{-1}AT)y,$$

where $x, y \in \mathbb{R}^n$, $A$ and $T$ are $n \times n$ matrices, and $T$ is nondegenerate. Therefore without changing the topological structure of the orbits, we can study the case that $A$ is in its Jordan form.

One may ask if it is possible to do a similar procedure for a nonlinear differential equation, that is, to obtain the simplest possible form by a suitable (nonlinear) change of coordinates? The answer is positive, and this is just the subject of this chapter.

We remark here that in contrast to the linear case, the results of normal-form theory will be local, and the normal-form equation is not unique. Nevertheless, the normal-form theory is useful for the study of bifurcation problems.

## 2.1 Normal Forms for Differential Equations near a Critical Point

In this section we will consider a vector field near a critical point which we will take to be the origin. Consider a $C^{r+1}$ differential equation,

49

$r \geq 2$:

$$\dot{x} = Ax + h(x), \qquad x \in \mathbb{C}^n, \tag{1.1}$$

where $A \in \mathbb{C}^{n \times n}$, the $n \times n$ matrices with complex entries, and $h(x) = O(|x|^2)$ as $|x| \to 0$.

Consider a $C^r$ transformation in a neighborhood $\Omega$ of the origin:

$$x = \xi(y), \qquad y \in \Omega, \tag{1.2}$$

where $\xi(0) = 0$. By substituting (1.2) into (1.1), we get:

$$\dot{y} = \xi_y^{-1}(y)A\xi(y) + \xi_y^{-1}(y)h(\xi(y)), \qquad y \in \Omega, \tag{1.3}$$

where $\xi_y(y)$ denotes the derivative of $\xi(y)$ with respect to $y$ and $\xi_y^{-1}(y)$ is the inverse of $\xi_y(y)$ in $\Omega$. Note that the linear part of (1.3) is $\xi_y^{-1}(0)A\xi_y(0)y$. Thus, if $A$ is already in a canonical form, we may assume that the diffeomorphism $\xi(y)$ in (1.2) takes the form

$$\xi(y) = y + O(|y|^2) \quad \text{as } y \to 0. \tag{1.4}$$

Therefore we may write (1.3) as

$$\dot{y} = Ay + g(y), \qquad y \in \Omega, \tag{1.5}$$

where $g(y) = O(|y|^2)$ as $|y| \to 0$.

Our goal is to determine a change of coordinates (1.2) such that the transformed equation (1.5) will be in the simplest possible form, so that the essential features of the flow of (1.1) near the critical point $x = 0$ become more evident. The desired simplification of (1.1) will be obtained, up to terms of a specified order, by performing inductively a sequence of near identity change of coordinates of the form

$$\xi(y) = y + \xi^k(y), \qquad y \in \Omega_k, \tag{1.6}$$

where $\xi^k \colon \mathbb{C}^n \to \mathbb{C}^n$ is a homogeneous polynomial of order $k \geq 2$ and $\Omega_k$ is a neighborhood of the origin in $\mathbb{C}^n$. Notice that any map of the form (1.6) is a diffeomorphism in some neighborhood of the origin. To see how far $g$ can be simplified, we write $h(x)$ as a formal power series using superscripts to denote the order of the homogeneous terms

$$h(x) = h^2(x) + h^3(x) + \cdots, \tag{1.7}$$

where for each $k \geq 2$, $h^k \in H_n^k$, the vector space of homogeneous polynomials of order $k$ in $n$ variables with values in $\mathbb{C}^n$. From (1.6) we get

$$\xi_y(y) = I + \xi_y^k(y), \tag{1.8}$$

and then

$$\left(\xi_y(y)\right)^{-1} = I - \xi_y^k(y) + O\left(|y|^{2k-2}\right), \qquad y \in \Omega_k, \tag{1.9}$$

where $\Omega_k$ is a small neighborhood of the origin in $\mathbb{C}^n$. Substituting (1.6)–(1.9) into (1.3) we obtain

$$\dot{y} = Ay + h^2(y) + \cdots + h^{k-1}(y)$$
$$+ \left\{ h^k(y) - \left[ \xi_y^k(y) Ay - A\xi^k(y) \right] \right\} + O\left(|y|^{k+1}\right), \qquad y \in \Omega_k.$$

$$\tag{1.10}$$

To simplify the term $h^k(y)$ we have to choose a suitable $\xi^k(y)$ before we make transformation (1.6). In order to see clearly the dependence of $\xi^k$ on $h^k$ (1.10) suggests introducing for each $k \geq 2$ a linear operator $L_A^k: H_n^k \to H_n^k$ defined by

$$\left( L_A^k \xi^k \right)(y) = \xi_y^k(y) Ay - A\xi^k(y), \qquad \xi^k \in H_n^k. \tag{1.11}$$

Then (1.10) can be expressed as

$$\dot{y} = Ay + h^2(y) + \cdots + h^{k-1}(y)$$
$$+ \left( h^k(y) - L_A^k \xi^k(y) \right) + O\left(|y|^{k+1}\right), \qquad y \in \Omega_k. \tag{1.12}$$

Let $\mathcal{R}^k$ be the range of $L_A^k$ in $H_n^k$ and $\mathcal{C}^k$ be any complementary subspace to $\mathcal{R}^k$ in $H_n^k$. We have

$$H_n^k = \mathcal{R}^k \oplus \mathcal{C}^k, \qquad k \geq 2. \tag{1.13}$$

The following theorem gives the desired simplification of (1.1).

**Theorem 1.1.** *Let* $X: \mathbb{C}^n \to \mathbb{C}^n$ *be a* $C^{r+1}$ *vector field with* $X(0) = 0$ *and* $DX(0) = A$. *Let the decomposition (1.13) of* $H_n^k$ *be given for* $k = 2, \ldots, r$. *Then there exists a sequence of near identity transformations* $x = y + \xi^k(y)$, $y \in \Omega_k$, *where* $\xi^k \in H_n^k$ *and* $\Omega_k$ *is a neighborhood of the origin,* $\Omega_{k+1} \subseteq \Omega_k$, $k = 2, \ldots, r$, *such that equation (1.1) is transformed into:*

$$\dot{y} = Ay + g^2(y) + \cdots + g^r(y) + O(|y|^{r+1}), \qquad y \in \Omega_r, \quad (1.14)$$

*where* $g^k \in \mathscr{C}^k$ *for* $k = 2, \ldots, r$.

*Proof.* Let $X(x) = Ax + h^2(x) + \cdots + h^r(x) + O(|x|^{r+1})$, as $x \to 0$. For $k = 2$, (1.12) becomes

$$\dot{y} = Ay + \left(h^2(y) - L_A^2 \xi^2(y)\right) + O(|y|^3), \qquad y \in \Omega_2, \quad (1.15)$$

where $\Omega_2$ is so small that $I + \xi_y^2(y)$ is invertible on it. Since for each $h^2 \in H_n^2$ there exist $f^2 \in \mathscr{R}^2$ and $g^2 \in \mathscr{C}^2$ such that $h^2 = f^2 + g^2$, we can find a $\xi^2 \in H_n^2$ with $L_A^2 \xi^2 = f^2$, and then (1.15) becomes

$$\dot{y} = Ay + g^2(y) + O(|y|^3), \qquad y \in \Omega_2.$$

Next we proceed by induction. Assume that Theorem 1.1 is true for $2 \le k \le s - 1 < r$. By a change of variables, we may assume that (1.1) becomes

$$\dot{x} = Ax + g^2(x) + \cdots + g^{s-1}(x) + h^s(x) + O(|x|^{s+1}),$$

$$x \in \Omega_{s-1},$$

where $g^k \in \mathscr{C}^k$ for $k = 2, \ldots, s - 1$, $h^s \in H_n^s$ (we remark that $h^s$ here may be different from the one in (1.1)), and $\Omega_{s-1}$ is a neighborhood of the origin. Let $x = y + \xi^s(y)$, $y \in \Omega_s$, where $\xi^s \in H_n^s$ is chosen according to (1.13) so that $h^s = L_A^s \xi^s + g^s$, $g^s \in \mathscr{C}^s$, and $\Omega_s \subseteq \Omega_{s-1}$ is a neighborhood of the origin on which $y + \xi^s(y)$ is invertible. Then from (1.12) with $k = s$ we obtain:

$$\dot{y} = Ay + g^2(y) + \cdots + g^{s-1}(y) + \left(h^s(y) - L_A^s \xi^s(y)\right) + O(|y|^{s+1})$$

$$= Ay + g^2(y) + \cdots + g^{s-1}(y) + g^s(y) + O(|y|^{s+1}), \qquad y \in \Omega_s.$$

This completes the proof.                                                    □

**Definition 1.2.** Suppose that the decompositions (1.13) are given. The following truncated equation of (1.14)

$$\dot{y} = Ay + g^2(y) + \cdots + g^r(y) \tag{1.16}$$

where $g^k \in \mathscr{C}^k$, $k = 2, \ldots, r$, is called an $A$-normal form of equation (1.1) up to order $r$.

We note that an $A$-normal form is not unique for the fixed $A$. In fact, it depends on the choices of the complementary subspaces $\mathscr{C}^k$ $(k = 2, \ldots, r)$.

**Remark 1.3.** Let $K = \{k \in \mathbb{N} \mid \mathscr{C}^k \neq \varnothing\}$. Suppose $\dim \mathscr{C}^k = n_k \geq 1$ and $\{v_1^k, \ldots, v_{n_k}^k\}$ is a basis of $\mathscr{C}^k$ for $k \in K$. Then (1.16) can be written as

$$\dot{y} = Ay + \sum_{\substack{k=2 \\ k \in K}}^{r} \sum_{j=1}^{n_k} a_{kj} v_j^*, \tag{1.17}$$

where $a_{kj} \in \mathbb{C}$ for all $j = 1, \ldots, n_k$, $k = 2, \ldots, r$. Then an $A$-normal form of (1.1) up to order $r$ is of the form (1.17). Generally it is not easy to determine the coefficients of (1.17) for a particular equation (1.1) and it is not easy to find the transformation which transforms (1.1) into (1.17). Numerical and symbolic computational methods are available for users to find such transformations and to determine the coefficients in $A$-normal form equations. Equation (1.17) with arbitrary coefficients $\{a_{kj}\}$ is called a general form of an $A$-normal form up to order $r$.

From the above discussion the $A$-normal forms are determined by the choices of the complementary subspaces $\mathscr{C}^k$ $(k = 2, \ldots, r)$ and these subspaces are determined by the matrix $A$ only. In general, it is not easy to find complementary subspaces $\mathscr{C}^k$ for $k = 2, \ldots, r$. However in the case when the matrix $A$ is diagonal, it is very easy. We will consider this case first.

A monomial in $H_n^k$ is an expression of the form $x^\alpha e_j$, where $\alpha = (\alpha_1, \alpha_2, \ldots, \alpha_n)$ with nonnegative integers $\alpha_i$ is a multi-index and $|\alpha| \equiv \alpha_1 + \alpha_2 + \cdots + \alpha_n = k$, $x^\alpha = x_1^{\alpha_1} x_2^{\alpha_2} \cdots x_n^{\alpha_n}$, $e_j = (0, \ldots, 1, \ldots, 0)^T$ is the standard basis element of $\mathbb{C}^n$ with only the $j$th component being 1 and all other components zero. A basis for the vector space $H_n^k$, $n \geq 1$, is given by

$$\{x^\alpha e_j \mid |\alpha| = k, 1 \leq j \leq n\}.$$

The dimension of $H_n^k$ is

$$\dim H_n^k = n \cdot \binom{n + k - 1}{k}. \tag{1.18}$$

**Definition 1.4.** If $\sigma(A) = \{\lambda_1, \ldots, \lambda_n\}$ is the spectrum of $A$, then the following relations are called resonant conditions:

$$\lambda \cdot \alpha - \lambda_j = 0, \tag{1.19}$$

where $\lambda = (\lambda_1, \ldots, \lambda_n)^T$, $\alpha = (\alpha_1, \ldots, \alpha_n)^T$, $|\alpha| \geq 2$, $1 \leq j \leq n$, and $\lambda \cdot \alpha = \lambda_1 \alpha_1 + \cdots + \lambda_n \alpha_n$. Let $(x_1, \ldots, x_n)$ be coordinates with respect to the standard basis $\{e_1, \ldots, e_n\}$ of $\mathbb{C}^n$ in which the matrix $A$ is in Jordan normal form with diagonal elements $(\lambda_1, \ldots, \lambda_n)$. Then a monomial $x^\alpha e_j(|\alpha| = k \geq 2$ and $1 \leq j \leq n)$ is called a resonant monomial of order $k$ if and only if (1.19) holds for $\alpha$ and $j$.

**Theorem 1.5.** *Let $A = \mathrm{diag}(\lambda_1, \ldots, \lambda_n)$. Then an A-normal form up to order $r \geq 2$ can be chosen so that its nonlinear part consists of all resonant monomials up to order $r$.*

*Proof.* A direct calculation shows that for any monomial $x^\alpha e_j$ with $|\alpha| = k \geq 2$ and $1 \leq j \leq n$,

$$L_A^k(x^\alpha e_j) = (\lambda \cdot \alpha - \lambda_j) x^\alpha e_j. \tag{1.20}$$

Hence $\mathrm{Ker}(L_A^k)$ is obviously a complementary subspace to the range of $L_A^k$ in $H_n^k$ and $\mathrm{Ker}(L_A^k)$ is spanned by all resonant monomials of order $k$ for each $k \geq 2$. Then the desired result follows. $\qquad \square$

In the following, we will present two methods for finding the complementary subspaces $\mathscr{C}^k$ for a given matrix $A$. The first is the adjoint operator method. Since an inner product can be introduced in $H_n^k$, a possible choice for $\mathscr{C}^k$ is the orthogonal complement of $\mathscr{R}^k$, which will be characterized as $\text{Ker}((L_A^k)^*)$, the null space of the adjoint operator $(L_A^k)^*$ of $L_A^k$. Other choices will be obtained from $\text{Ker}((L_A^k)^*)$ by using linear algebraic techniques. The second is the matrix representation method. Each $L_A^k$ is a linear operator defined on a finite-dimensional linear space $H_n^k$. If $\tilde{L}_A^k$ is the matrix representation of $L_A^k$ with respect to a basis of $H_n^k$, then our problem is reduced to finding a complementary subspace to the range of $\tilde{L}_A^k$ in $\mathbb{C}^{d_k}$, where $d_k = \dim H_n^k$.

If $p(x) = \sum_{j=1}^n \sum_{|\alpha|=k} p_{\alpha j} x^\alpha e_j$ and $q(x) = \sum_{j=1}^n \sum_{|\alpha|=k} q_{\alpha j} x^\alpha e_j$, where $p_{\alpha j}$ and $q_{\alpha j}$ are complex constants, then we define

$$\langle p, q \rangle = \sum_{j=1}^n \sum_{|\alpha|=k} p_{\alpha j} \bar{q}_{\alpha j} \alpha!, \tag{1.21}$$

where $\alpha! = \alpha_1! \alpha_2! \ldots \alpha_n!$.

**Example 1.6.** Let $|\alpha| = |\beta| = k$ and $1 \le i, j \le n$. Then

$$\langle x^\alpha e_i, x^\beta e_j \rangle = \delta_{ij} \delta_{\alpha\beta} \alpha!,$$

where $\delta_{ij}$ and $\delta_{\alpha\beta}$ are the Kronecker symbols.

It is easy to see that $\langle \cdot, \cdot \rangle$ is an inner product in $H_n^k$.

**Theorem 1.7.** *Operator $L_{A^*}^k$ is the adjoint operator of $L_A^k$ with respect to the inner product $\langle \cdot, \cdot \rangle$ in $H_n^k$ for each $k \ge 2$, where $A^*$ is the adjoint operator of $A$ with respect to the usual product $(\cdot, \cdot)$ in $\mathbb{C}^n$.*

*Proof.* Let $p, q \in H_n^k$, $p(x) = \sum_{i=1}^n \sum_{|\alpha|=k} p_{\alpha i} x^\alpha e_i$, $q(x) = \sum_{j=1}^n \sum_{|\beta|=k} q_{\beta j} x^\beta e_j$. Then, using the linearity of $L_A^k$, $L_{A^*}^k$ and properties of the inner product, we get

$$\langle L_A^k p, q \rangle = \sum_{i=1}^n \sum_{j=1}^n \sum_{|\alpha|=k} \sum_{|\beta|=k} p_{\alpha i} \bar{q}_{\beta j} \langle L_A^k(x^\alpha e_i), x^\beta e_j \rangle,$$

$$\langle p, L_{A^*}^k q \rangle = \sum_{i=1}^n \sum_{j=1}^n \sum_{|\alpha|=k} \sum_{|\beta|=k} p_{\alpha i} \bar{q}_{\beta j} \langle x^\alpha e_i, L_{A^*}^k(x^\beta e_j) \rangle.$$

Therefore it is enough to prove that

$$\langle L_A^k(x^\alpha e_i), x^\beta e_j \rangle = \langle x^\alpha e_i, L_{A^*}^k(x^\beta e_j) \rangle$$

for any $\alpha, \beta, i, j$ with $|\alpha| = |\beta| = k$ and $1 \le i, j \le n$. An easy computation shows:

$$L_A^k(x^\alpha e_i) = D(x^\alpha e_i) Ax - A(x^\alpha e_i)$$

$$= \sum_{l=1}^n \sum_{m=1}^n \alpha_l a_{lm} \frac{x^\alpha x_m}{x_l} e_i - \sum_{l=1}^n a_{li} x^\alpha e_l,$$

$$L_{A^*}^k(x^\beta e_j) = D(x^\beta e_j) A^* x - A^*(x^\beta e_j)$$

$$= \sum_{m=1}^n \sum_{l=1}^n \beta_m \bar{a}_{lm} \frac{x^\beta x_l}{x_m} e_j - \sum_{m=1}^n \bar{a}_{jm} x^\beta e_m,$$

where $D$ is the differential operator. Therefore

$$\langle L_A^k(x^\alpha e_i), x^\beta e_j \rangle$$

$$= \begin{cases} \left( \sum_{l=1}^n \alpha_l a_{ll} - a_{ii} \right) \alpha! & \text{if } i = j \text{ and } \alpha = \beta, \\ \alpha_l a_{lm} \beta! & \text{if } i = j; \beta_l = \alpha_l - 1, \beta_m = \alpha_m + 1 \\ & \text{for some } l \ne m; \\ & \beta_s = \alpha_s \text{ for any } s \ne l, m, \\ -a_{ji} \alpha! & \text{if } i \ne j \text{ but } \alpha = \beta, \\ 0 & \text{otherwise;} \end{cases}$$

$$\langle x^\alpha e_i, L_{A^*}^k(x^\beta e_j) \rangle$$

$$= \begin{cases} \left( \sum_{l=1}^n \beta_l a_{ll} - a_{ii} \right) \beta! & \text{if } i = j \text{ and } \beta = \alpha, \\ \beta_m a_{lm} \alpha! & \text{if } i = j; \alpha_l = \beta_l + 1, \alpha_m = \beta_m - 1 \\ & \text{for some } l \ne m; \\ & \alpha_s = \beta_s \text{ for any } s \ne l, m, \\ -a_{ji} \beta! & \text{if } i \ne j \text{ but } \alpha = \beta, \\ 0 & \text{otherwise,} \end{cases}$$

The two expressions are equal in each case. Thus the theorem is proved.                                                        □

**Corollary 1.8.** Ker($L^k_{A*}$) is the orthogonal complementary subspace to $\mathcal{R}^k$ with respect to the inner product $\langle\ \cdot\ ,\ \cdot\ \rangle$ in $H^k_n$ for each $k \geq 2$.

**Remark 1.9.** If we define $L_A\colon C^1(\mathbb{C},\mathbb{C}^n) \to C^0(\mathbb{C},\mathbb{C}^n)$ by

$$(L_A\xi)(x) = \xi_x(x)Ax - A\xi(x),$$

then $L^k_A = L_A|_{H^k_n}$, $L^k_{A*} = L_{A*}|_{H^k_n}$. Thus a polynomial of order $r$, $g(x) = g^2(x) + \cdots + g^r(x)$, where $g^k \in H^k_n$, $k = 2, \ldots, r$, belongs to Ker($L_{A*}$) if and only if $g^k \in$ Ker $L^k_{A*}$, $k = 2, \ldots, r$. Therefore to find $A$-normal form equations up to order $r$, it is sufficient to solve the partial differential equation $L_{A*}\xi = 0$ for $r$th-order polynomial solutions with no constant and linear terms.

**Example 1.10.** Let

$$A = \begin{bmatrix} 0 & 1 \\ 0 & 0 \end{bmatrix} \quad \text{and} \quad \xi(x) = \begin{bmatrix} \xi_1(x_1, x_2) \\ \xi_2(x_1, x_2) \end{bmatrix},$$

where $\xi_1(x_1, x_2)$ and $\xi_2(x_1, x_2)$ are scalar polynomials of degree $r \geq 2$. Then

$$L_{A*}\xi(x) = \begin{bmatrix} \dfrac{\partial \xi_1}{\partial x_1} & \dfrac{\partial \xi_1}{\partial x_2} \\[2ex] \dfrac{\partial \xi_2}{\partial x_1} & \dfrac{\partial \xi_2}{\partial x_2} \end{bmatrix} \begin{bmatrix} 0 & 0 \\ 1 & 0 \end{bmatrix} \begin{bmatrix} x_1 \\ x_2 \end{bmatrix} - \begin{bmatrix} 0 & 0 \\ 1 & 0 \end{bmatrix} \begin{bmatrix} \xi_1(x_1, x_2) \\ \xi_2(x_1, x_2) \end{bmatrix}$$

$$= \begin{bmatrix} x_1 \dfrac{\partial \xi_1}{\partial x_2} \\[2ex] x_1 \dfrac{\partial \xi_2}{\partial x_2} - \xi_1 \end{bmatrix}.$$

It is easy to see that the $r$th-order polynomial solutions (without constant and the first-order terms) of the equation $L_{A*}\xi(x) = 0$, that is,

$$\begin{cases} x_1\dfrac{\partial\xi_1}{\partial x_2} = 0, \\[2mm] x_1\dfrac{\partial\xi_2}{\partial x_2} - \xi_1 = 0, \end{cases}$$

are

$$\xi_1(x_1, x_2) = x_1^2\phi_1(x_1), \quad \xi_2(x_1, x_2) = x_1 x_2\phi_1(x_1) + x_1^2\phi_2(x_1),$$

where $\phi_1(x_1)$ and $\phi_2(x_1)$ are arbitrary scalar polynomials of order $r - 2$. Thus an $A$-normal form equation up to order $r$ is

$$\begin{cases} \dot{x}_1 = x_2 + x_1^2\phi_1(x_1), \\ \dot{x}_2 = x_1^2\phi_2(x_1) + x_1 x_2\phi_1(x_1). \end{cases}$$

We note that $r \geq 2$ can be any integer. An $A$-normal form up to order 2 is

$$\begin{cases} \dot{x}_1 = x_2 + bx_1^2, \\ \dot{x}_2 = ax_1^2 + bx_1 x_2, \end{cases} \tag{1.22}$$

where $a, b$ are complex constants.

We can also choose span$\{x_1^2 e_2, x_1 x_2 e_2\}$ as a complementary subspace to $\mathscr{R}^2$. In fact, if $v_1 = x_1^2 e_2$, $v_2 = x_1^2 e_1 + x_1 x_2 e_2$, $w_1 = \frac{1}{2}x_1^2 e_2$, and $w_2 = x_1 x_2 e_2$, then $\{v_1, v_2\}$ is a basis of Ker$(L_{A*}^2)$ and it is easy to see that $\langle v_i, w_j \rangle = \delta_{ij}$ for $i, j = 1, 2$. Hence span$\{w_1, w_2\}$ is another complementary subspace to $\mathscr{R}^2$. Thus a different $A$-normal form up to order 2 is given by

$$\begin{cases} \dot{x}_1 = x_2, \\ \dot{x}_2 = ax_1^2 + bx_1 x_2, \end{cases} \qquad a, b \in \mathbb{C}.$$

**Example 1.11.** Let

$$A = \begin{bmatrix} 0 & 1 & 0 \\ 0 & 0 & 1 \\ 0 & 0 & 0 \end{bmatrix} \quad \text{and} \quad \xi(x) = \begin{bmatrix} \xi_1(x_1, x_2, x_3) \\ \xi_2(x_1, x_2, x_3) \\ \xi_3(x_1, x_2, x_3) \end{bmatrix},$$

where $\xi_i(x_1, x_2, x_3)$ are scalar polynomials of order $r \geq 2$, $i = 1, 2, 3$. Suppose that $\xi(x)$ is a solution of equation $L_{A^*}\xi = 0$, that is,

$$x_1 \frac{\partial \xi_1}{\partial x_2} + x_2 \frac{\partial \xi_1}{\partial x_3} = 0,$$

$$x_1 \frac{\partial \xi_2}{\partial x_2} + x_2 \frac{\partial \xi_2}{\partial x_3} = \xi_1, \qquad (1.23)$$

$$x_1 \frac{\partial \xi_3}{\partial x_2} + x_2 \frac{\partial \xi_3}{\partial x_3} = \xi_2.$$

The two independent first integrals of the first equation of (1.23) are $p_1 = x_1$ and $p_2 = 2x_1x_3 - x_2^2$.

Suppose $\xi_1(x_1, x_2, x_3) = \sum_{i+j+k=m} c_{ijk} x_1^i x_2^j x_3^k$ and $\xi_1(x_1, x_2, x_3) = \Phi(p_1, p_2)$. Let $p_1 \neq 0$. Then $x_3 = (p_2 + x_2^2)/(2p_1)$. Hence

$$\Phi(p_1, p_2) = \sum_{i+j+k=m} \frac{c_{ijk}}{2^k} p_1^{i-k} x_2^j (p_2 + x_2^2)^k.$$

Let $x_2 = 0$. Then

$$\Phi(p_1, p_2) = \sum_{i+k=m} \frac{c_{i0k}}{2^k} p_1^{i-k} p_2^k.$$

Since $\Phi$ is differentiable in $p_1$, $i \geq k$. Thus $\Phi$ is a polynomial of $p_1$ and $p_2$ and hence so is $\xi_1$. Let

$$\xi_1(x_1, x_2, x_3) = x_1 \phi_1(p_1, p_2) + \psi_1(p_2),$$

where $\phi_1$ and $\psi_1$ are polynomials in their arguments. Then $\xi_2$ can be expressed as

$$\xi_2(x_1, x_2, x_3) = x_2\phi_1(p_1, p_2) + \frac{x_2}{x_1}\psi_1(p_2) + \phi(p_1, p_2),$$

where $\phi$ is a differentiable function in $p_1$ and $p_2$. By multiplying $\xi_2$ by $x_1$ and then taking $x_1 = 0$, we can show that $\psi_1 \equiv 0$ and $\phi$ is a polynomial in $p_1$ and $p_2$. Let $\phi(p_1, p_2) = x_1\phi_2(p_1, p_2) + \psi_2(p_2)$, where $\phi_2$ and $\psi_2$ are polynomials in their arguments. Then it can be shown in a similar way that $\psi_2 \equiv 0$ and

$$\xi_3(x_1, x_2, x_3) = x_3\phi_1(p_1, p_2) + x_2\phi_2(p_1, p_2) + \phi_3(p_1, p_2),$$

where $\phi_3$ is a polynomial in $p_1$ and $p_2$. Hence an $A$-normal form up to order $r$ is

$$\begin{cases} \dot{x}_1 = x_2 + x_1\phi_1(x_1, 2x_1x_3 - x_2^2), \\ \dot{x}_2 = x_3 + x_2\phi_1(x_1, 2x_1x_3 - x_2^2) + x_1\phi_2(x_1, 2x_1x_3 - x_2^2), \\ \dot{x}_3 = x_3\phi_1(x_1, 2x_1x_3 - x_2^2) + x_2\phi_2(x_1, 2x_1x_3 - x_2^2) \\ \qquad + \phi_3(x_1, 2x_1x_3 - x_2^2), \end{cases}$$

where $\phi_i$ are polynomials in their arguments such that $x_3\phi_1$, $x_2\phi_2$, and $\phi_3$ are polynomials in $x_1$, $x_2$, and $x_3$ of order $r$ without constant and linear terms.

A different method for finding $A$-normal forms is to use the matrix representation of the linear operator $L_A^k$ with respect to a given basis of $H_n^k$. First, we give an ordering of the elements of the basis of $H_n^k$, $\{x^\alpha e_j \mid |\alpha| = k, 1 \le j \le n\}$. This is taken to be the reverse lexicographic ordering, that is,

$$x^\alpha e_i < x^\beta e_j \quad \text{if and only if} \quad (i, \alpha_1, \ldots, \alpha_n) > (j, \beta_1, \ldots, \beta_n),$$

where $(i, \alpha_1, \ldots, \alpha_n) > (j, \beta_1, \ldots, \beta_n)$ if and only if $i > j$ or $i = j$ and the first unequal components, say, $\alpha_s \neq \beta_s$, satisfy $\alpha_s > \beta_s$, $1 \le s \le n$. We shall write $i \sim (j, \alpha)$ if $x^\alpha e_j$ is the $i$th basis element with respect to reverse lexicographic ordering.

Let $d_k = \dim H_n^k$ and $U_k = \{u_1, \ldots, u_{d_k}\}$ be an orthogonal basis of $H_n^k$. We will use hereafter only basis $U_k$ of the form

$$u_i(x) = x^\alpha e_j, \qquad |\alpha| = k, \quad j = 1, \ldots, n, \tag{1.24}$$

where $i \sim (j, \alpha)$ is in the reverse lexicographic ordering for $i = 1, \ldots, d_k$. We denote by $\tilde{L}_A^k$ the matrix representation of $L_A^k$ with respect to the basis $U_k$ of $H_n^k$. Then $\tilde{L}_A^k$ is a $d_k \times d_k$ matrix which can be viewed also as a linear operator on $\mathbb{C}^{d_k}$. Let $\tilde{\mathscr{R}}^k$ be the range of $\tilde{L}_A^k$ in $\mathbb{C}^{d_k}$ and $\tilde{\mathscr{C}}^k$ any complementary subspace, that is, $\mathbb{C}^{d_k} = \tilde{\mathscr{R}}^k \oplus \tilde{\mathscr{C}}^k$. If we define:

$$\begin{cases} \mathscr{R}^k = \left\{ \xi^k = \sum_{i=1}^{d_k} a_i u_i \in H_n^k \,\middle|\, (a_1, \ldots, a_{d_k}) \in \tilde{\mathscr{R}}^k \right\}, \\[2mm] \mathscr{C}^k = \left\{ \xi^k = \sum_{i=1}^{d_k} a_i u_i \in H_n^k \,\middle|\, (a_1, \ldots, a_{d_k}) \in \tilde{\mathscr{C}}^k \right\}, \end{cases} \tag{1.25}$$

then $\mathscr{R}^k$ is the range of $L_A^k$ in $H_n^k$ and $\mathscr{C}^k$ is a complementary subspace to $\mathscr{R}^k$ in $H_n^k$. Therefore, finding a complementary subspace $\mathscr{C}^k$ to $\mathscr{R}^k$ in $H_n^k$ is equivalent by (1.25) to finding a complementary subspace $\tilde{\mathscr{C}}^k$ to $\tilde{\mathscr{R}}^k$ in $\mathbb{C}^{d_k}$. Such a complementary subspace is provided by $\tilde{\mathscr{C}}^k = \text{Ker}((\tilde{L}_A^k)^*)$, which is the orthogonal complementary subspace of $\tilde{\mathscr{R}}^k$ in $\mathbb{C}^{d_k}$ with respect to the inner product $(\cdot, \cdot)$ in $\mathbb{C}^{d_k}$. Other complementary subspaces to $\tilde{\mathscr{R}}^k$ can be obtained from $\text{Ker}((\tilde{L}_A^k)^*)$ by performing elementary algebraic calculations.

**Remark 1.12.** Since the size of the matrix $\tilde{L}_A^k$ is $d_k \times d_k = n\binom{n+k-1}{n-1} \times n\binom{n+k-1}{n-1}$, which increases rapidly as $k$ increases, to calculate matrices $\tilde{L}_A^k$ and to find bases of complementary subspaces $\tilde{\mathscr{C}}^k$ become generally more and more difficult. However, the matrix $\tilde{L}_A^k$ depends only on the matrix $A$ and $k$, so it involves only computing coefficients of the basis of $H_n^k$ in the expansion:

$$L_A^k(x^\alpha e_j) = \left( \sum_{i=1}^n \sum_{l=1}^n \frac{\alpha_i}{x_i} a_{il} x_l \right) x^\alpha e_j - \sum_{i=1}^n a_{ij} x^\alpha e_i. \tag{1.26}$$

Even though the computations might be tedious they can be performed in principle, especially when $n$ and $k$ are small, and one can use a computer to do it. Once $\tilde{L}_A^k$ is known, to find a basis of $\mathrm{Ker}((\tilde{L}_A^k)^*)$ or some other complementary subspace $\tilde{\mathscr{C}}^k$, and hence $\mathscr{C}^k$ by identification (1.25), becomes an algebraic problem.

In some cases, as we shall see, $\tilde{L}_A^k$ is easy to compute and so is a complementary subspace $\tilde{\mathscr{C}}^k$ to $\tilde{\mathscr{R}}^k$ in $\mathbb{C}^{d_k}$.

**Lemma 1.13.** *If $A = \mathrm{diag}(\lambda_1, \ldots, \lambda_n)$, then $\tilde{L}_A^k$ is also diagonal; if $A$ is upper (lower) triangular with diagonal elements $(\lambda_1, \ldots, \lambda_n)$, then $\tilde{L}_A^k$ is lower (upper) triangular. Furthermore, for both cases, the ith element of the diagonal of $\tilde{L}_A^k$ is $\lambda \cdot \alpha - \lambda_j$, where $i \sim (j, \alpha)$.*

*Proof.* The conclusion follows from (1.26) and the definition of the reverse lexicographic ordering of the basis $U_k$ of $H_n^k$. $\qquad\square$

**Remark 1.14.** When $A$ is diagonal or upper triangular it is fairly easy to obtain a basis of a complementary subspace since in those cases the matrix $\tilde{L}_A^k$ is diagonal or lower triangular and its range is spanned by the columns of the matrix. From the structure of the matrix it is easy to read off a basis of a complementary subspace, or to obtain the column echelon form from which this basis can be read off. This is the advantage of giving the reverse lexicographical ordering to the basis. When $A$ is not upper triangular we can make a linear transformation so that the linear part of the resulting equation is upper triangular.

**Example 1.15.** Consider the following equation in $\mathbb{C}^2$:

$$\dot{x} = Ax + O(|x|^2), \qquad x \in \mathbb{C}^2,$$

where

$$A = \begin{bmatrix} 0 & 1 \\ 0 & 0 \end{bmatrix}.$$

For any $k \geq 2$, we shall determine a complemntary subspace $\mathscr{C}^k$ to the range $\mathscr{R}^k$ of $L_A^k$ in $H_2^k$. We have $\dim H_2^k = 2(k+1)$ and the basis of

$H_2^k$ in the reverse lexicographic ordering is

$$\{u_i(x)\}_{i=1}^{2(k+1)} = \{x_1^k x_2^0 e_2, x_1^{k-1} x_2 e_2, \ldots,$$

$$x_1^0 x_2^k e_2, x_1^k x_2^0 e_1, x_1^{k-1} x_2 e_1, \ldots, x_1^0 x_2^k e_1\},$$

that is, $u_i(x) = x_1^{k-i+1} x_2^{i-1} e_2$, $u_{k+i+1}(x) = x_1^{k-i+1} x_2^{i-1} e_1$ for $1 \le i \le k + 1$, where $\{e_1, e_2\}$ is the standard basis of $\mathbb{C}^2$. For $\xi = (\xi_1, \xi_2)^T \in H_2^k$ we have

$$L_A^k \begin{bmatrix} \xi_1 \\ \xi_2 \end{bmatrix} = \begin{bmatrix} \dfrac{\partial \xi_1}{\partial x_1} & \dfrac{\partial \xi_1}{\partial x_2} \\ \dfrac{\partial \xi_2}{\partial x_1} & \dfrac{\partial \xi_2}{\partial x_2} \end{bmatrix} \begin{bmatrix} 0 & 1 \\ 0 & 0 \end{bmatrix} \begin{bmatrix} x_1 \\ x_2 \end{bmatrix} - \begin{bmatrix} 0 & 1 \\ 0 & 0 \end{bmatrix} \begin{bmatrix} \xi_1 \\ \xi_2 \end{bmatrix}$$

$$= \begin{bmatrix} x_2 \dfrac{\partial \xi_1}{\partial x_1} - \xi_2 \\ x_2 \dfrac{\partial \xi_2}{\partial x_1} \end{bmatrix}.$$

Then, applying $L_A^k$ to the elements of our basis we get

$$\begin{cases} L_A^k u_i = (k - i + 1)u_{i+1} - u_{i+k+1}, \\ L_A^k u_{i+k+1} = (k - i + 1)u_{k+i+2}, \end{cases} \quad 1 \le i \le k + 1,$$

with the convention $u_{2k+3} = 0$. So we have the following $2(k + 1) \times 2(k + 1)$ matrix for $\tilde{L}_A^k$.

$$\tilde{L}_A^k = \left[\begin{array}{cccc:cccc}
0 & 0 & \cdots & 0 & 0 & 0 & \cdots & 0 & 0 \\
k & 0 & \cdots & 0 & 0 & 0 & \cdots & 0 & 0 \\
0 & k-1 & \cdots & 0 & 0 & 0 & \cdots & 0 & 0 \\
\vdots & \vdots & \ddots & \vdots & \vdots & \vdots & & \vdots & \vdots \\
0 & 0 & \cdots & 1 & 0 & 0 & \cdots & 0 & 0 \\
\hdashline
-1 & 0 & \cdots & 0 & 0 & 0 & \cdots & 0 & 0 \\
0 & -1 & \cdots & 0 & 0 & k & \cdots & 0 & 0 \\
0 & 0 & \ddots & & 0 & 0 & k-1 & \cdots & 0 & 0 \\
\vdots & \vdots & & -1 & \vdots & \vdots & & \ddots & \vdots & \vdots \\
0 & 0 & \cdots & 0 & -1 & 0 & 0 & \cdots & 1 & 0
\end{array}\right]_{2(k+1) \times 2(k+1)}$$

Let $\{\tilde{e}_i \,|\, i = 1, \ldots, 2k + 2\}$ be the standard basis of $\mathbb{C}^{2k+2}$. Then a basis of $\mathrm{Ker}((\tilde{L}_A^k)^*)$ is given by

$$\tilde{v}_1 = \tilde{e}_1, \qquad \tilde{v}_2 = \tilde{e}_2 + k\tilde{e}_{k+2}.$$

We can choose also $\tilde{\mathscr{C}}^k = \mathrm{span}\{\tilde{w}_1, \tilde{w}_2\}$, where $\tilde{w}_1 = \tilde{e}_1$, $\tilde{w}_2 = \tilde{e}_{k+2}$, as a complementary subspace to $\tilde{R}^k$ in $\mathbb{C}^{2k+2}$. By the correspondence between $H_2^k$ and $\mathbb{C}^{2k+2}$

$$\mathscr{C}^k = \mathrm{span}\left\{\begin{bmatrix} 0 \\ x_1^k \end{bmatrix}, \begin{bmatrix} x_1^k \\ 0 \end{bmatrix}\right\}$$

is a complementary subspace to $\mathscr{R}^k$ in $H_2^k$ and an $A$-normal form equation up to order $r$ is

$$\dot{x} = \begin{bmatrix} x_2 \\ 0 \end{bmatrix} + \sum_{k=2}^{r} \begin{bmatrix} a_k x_1^k \\ b_k x_1^k \end{bmatrix}, \tag{1.27}$$

where $a_k, b_k \in \mathbb{C}$, $k = 2, \ldots, r$. We can rewrite (1.27) as

$$\begin{cases} \dot{x}_1 = x_2 + x_1^2 \phi_1(x_1), \\ \dot{x}_2 = x_1^2 \phi_2(x_1), \end{cases}$$

where $\phi_1, \phi_2$ are polynomials of degree $r - 2$, for $r \geq 2$.

Since we can also choose $\tilde{\mathscr{C}}^k = \mathrm{span}\{\tilde{w}_1, \tilde{w}_2\}$, where $\tilde{w}_1 = \tilde{e}_1$, $\tilde{w}_2 = \tilde{e}_2$, as a basis for a complementary subspace to $\tilde{\mathscr{R}}^k$ in $\mathbb{C}^{2k+2}$, the corresponding $A$-normal form is

$$\dot{x} = \begin{bmatrix} x_2 \\ 0 \end{bmatrix} + \sum_{k=2}^{r} \begin{bmatrix} 0 \\ a_k x_1^k + b_k x_1^{k-1} x_2 \end{bmatrix},$$

where $a_k, b_k \in \mathbb{C}$, $k = 2, \ldots, r$, or

$$\begin{cases} \dot{x}_1 = x_2, \\ \dot{x}_2 = x_1^2 \phi_1(x_1) + x_1 x_2 \phi_2(x_1), \end{cases}$$

where $\phi_1, \phi_2$ are polynomials of degree $r - 2$, $r \geq 2$. In particular,

$$\begin{cases} \dot{x}_1 = x_2, \\ \dot{x}_2 = ax_1^2 + bx_1x_2, \end{cases}$$

where $a, b \in \mathbb{C}$ is an $A$-normal form equation up to order 2.

To deal with the case when $A$ is not diagonalizable we can apply the $S - N$ decompositions of linear operators in finite-dimensional vector spaces.

**Definition 1.16.** Let $L$ be a linear operator in a finite-dimensional vector space $V$ over $\mathbb{C}$. $L = S + N$ is called the $S - N$ decomposition of $L$ if $S$ is semisimple, that is, the matrix representation of $S$ with respect to a basis of $V$ is diagonal, $N$ is nilpotent, and $SN = NS$.

It is well known that for any linear operator in finite-dimensional vector spaces there exists a unique $S - N$ decomposition.

**Theorem 1.17.** *Let $L$ be a linear operator in a finite-dimensional vector space $V$. If $L = S + N$ is its $S - N$ decomposition, then*

$$\mathrm{Ker}(L) = \mathrm{Ker}(S) \cap \mathrm{Ker}(N).$$

*Proof.* It is not hard to see that $V = \mathrm{Im}(S) \oplus \mathrm{Ker}(S)$ since $S$ is semisimple. Then it follows that $S$ is invertible on $\mathrm{Im}(S)$. $\mathrm{Ker}(S) \cap \mathrm{Ker}(N) \subseteq \mathrm{Ker}(L)$ is also apparent. Let $v \in \mathrm{Ker}(L)$. Then $Sv = -Nv$. Since $SN = NS$, $S^m v = (-1)^m N^m v = 0$ for some positive integer $m$. If $m = 1$, then $Sv = -Nv = 0$. Thus $v \in \mathrm{Ker}(S) \cap \mathrm{Ker}(N)$. If $m > 1$, then $S^{m-1}(Sv) = 0$ and then we still have $Sv = 0$ since $S$ is invertible on $\mathrm{Im}(S)$. Consequently $Nv = 0$. The proof is completed. $\square$

**Theorem 1.18.** *If $A = S + N$ is the $S - N$ decomposition of $A$, then $L_A^k = L_S^k + L_N^k$ is the $S - N$ decomposition of $L_A^k$ for each $k \geq 2$.*

*Proof.* Let $\{\eta_1, \ldots, \eta_n\}$ be such a basis of $\mathbb{C}^n$ that the matrix of $A$ with respect to the basis is in Jordan form denoted by

$$A = \begin{bmatrix} \lambda_1 I_1 + N_1 & & \\ & \ddots & \\ & & \lambda_s I_s + N_s \end{bmatrix},$$

where $\lambda_i$ is an eigenvalue of $A$, $I_i$ is the $n_i \times n_i$ identity matrix,

$$N_i = \begin{bmatrix} 0 & 1 & & \\ & \ddots & \ddots & \\ & & \ddots & 1 \\ & & & 0 \end{bmatrix}_{n_i \times n_i},$$

$i = 1, \ldots, s$, and $n_1 + \cdots + n_2 = n$. By the uniqueness of the $S - N$ decomposition of $A$,

$$S = \begin{bmatrix} \lambda_1 I_1 & & \\ & \ddots & \\ & & \lambda_s I_s \end{bmatrix} \quad \text{and} \quad N = \begin{bmatrix} N_1 & & \\ & \ddots & \\ & & N_s \end{bmatrix}.$$

The matrix representations of $L_S^k$ and $L_N^k$ with respect to the basis $U_k$ are diagonal and strictly lower triangular, respectively, by Lemma 1.13. Hence $L_S^k$ is semisimple and $L_N^k$ is nilpotent. Furthermore, for any monomial $x^\alpha e_j$, if $n_1 + \cdots + n_{i-1} < j \le n_1 + \cdots + n_i$ for some $1 \le i \le s$, then by (1.26)

$$L_S^k(x^\alpha e_j) = (\lambda \cdot \alpha - \lambda_i) x^\alpha e_j, \tag{1.28}$$

$$L_N^k(x^\alpha e_j) = \left( \sum_{l=1}^{n-1} a_{l,l+1} \frac{\alpha_l}{x_l} x_{l+1} \right) x^\alpha e_j - a_{j-1,j} x^\alpha e_{j-1}, \tag{1.29}$$

where $a_{l,l+1} = 1$ if $l \ne n_1 + \cdots + n_t$ for any $1 \le t < s$ or $a_{l,l+1} = 0$ if $l = n_1 + \cdots + n_t$ for $t = 1, \ldots, s - 1$, and $a_{0,1} = 0$. A simple calculation shows that if $l \ne n_1 + \cdots + n_t$ for any $1 \le t < s$ and if $n_1 + \cdots + n_{i-1} < j \le n_1 + \cdots + n_i$ for some $1 \le i \le s$, then

$$L_S^k \left( \frac{x_{l+1}}{x_l} x^\alpha e_j \right) = (\lambda \cdot \alpha - \lambda_i) \frac{x_{l+1}}{x_l} x^\alpha e_j.$$

Therefore if $n_1 + \cdots + n_{i-1} < j \le n_1 + \cdots + n_i$ for some $1 \le i \le s$,

$$L_S^k L_N^k (x^\alpha e_j) = (\lambda \cdot \alpha - \lambda_i) L_N^k (x^\alpha e_j),$$

and

$$L_N^k L_S^k (x^\alpha e_j) = (\lambda \cdot \alpha - \lambda_i) L_N^k (x^\alpha e_j).$$

Since $x^\alpha e_j$ is arbitrary it follows that $L_N^k$ commutes with $L_S^k$. The theorem is then proved. $\qquad\square$

**Corollary 1.19.** *Let $A = S + N$ be the $S - N$ decomposition of $A$ and $F(x)$ be an $n$-vector valued polynomial in $x \in \mathbb{C}^n$. Then $F(x) \in \mathrm{Ker}(L_{A*})$ if and only if $F(x)$ is a solution of the following system of partial differential equations*:

$$\begin{cases} L_{S*} F(x) = 0, \\ L_{N*} F(x) = 0. \end{cases}$$

We note that when $S$ is diagonal $\mathrm{Ker}(L_S^k)$ and $\mathrm{Ker}(L_{S*}^k)$ are the same for any $k \ge 2$. Then we have the following corollary.

**Corollary 1.20.** *Let $A = S + N$ be the $S - N$ decomposition of $A$. If $S = \mathrm{diag}(\lambda_1, \ldots, \lambda_n)$, then an $A$-normal form up to order $r \ge 2$ can be chosen so that its nonlinear part is spanned by resonant monomials up to order $r$.*

It is not hard to see that if $A = S + N$ is the $S - N$ decomposition of $A$ then $\mathrm{Ker}(L_S^k) \cap \mathrm{Ker}(L_{N*}^k)$ is also a complementary subspace to the range of $L_A^k$ for any $k \ge 2$. Hence we can also get $A$-normal forms by finding polynomial solutions of the system of linear partial differential equations

$$\begin{cases} L_S F(x) = 0, \\ L_{N*} F(x) = 0. \end{cases}$$

We note that if $S = \mathrm{diag}(\lambda_1, \ldots, \lambda_n)$ then any resonant monomial commutes with $\exp(S)$. In general we have the following corollary.

**Corollary 1.21.** *Let $A = S + N$ be the $S - N$ decomposition of $A$. Then an $A$-normal form can be chosen so that it is invariant under the linear transformation $x \rightarrow e^S x$.*

Now we discuss $A$-normal forms of the real equation

$$\dot{x} = Ax + f(x), \qquad x \in \Omega \subseteq \mathbb{R}^n, \tag{1.30}$$

where $\Omega$ is a neighborhood of the origin of $\mathbb{R}^n$ and $f(x) = O(|x|^2)$ as $x \rightarrow 0$.

Let $\bar{H}_n^k$ be the linear space of all homogeneous polynomials of order $k$ in $n$ real variables with values in $\mathbb{R}^n$ and the operator $\bar{L}_A^k: \bar{H}_n^k \rightarrow \bar{H}_n^k$ be defined by the same formula (1.11). We can get a real $A$-normal form of (1.30) by solving the partial differential equation $\bar{L}_{A^*} F(x) = 0$ for its real polynomial solutions or by the matrix representation method.

If $A$ is diagonalizable over $\mathbb{C}$, we cannot apply resonant conditions to find its real $A$-normal form directly. Let $B = P^{-1}AP$ be diagonal, where $P$ is a nonsingular complex matrix. We change variables in (1.30) by $x = Pz$, where $z \in \mathbb{C}^n$ and $\overline{Pz} = Pz$. Then (1.30) becomes

$$\dot{z} = Bz + h(z), \tag{1.31}$$

where $h(z) = O(|z|^2)$ as $z \rightarrow 0$. Let

$$E = \{z | \overline{Pz} = Pz, z \in \mathbb{C}^n\}.$$

It can be shown that $E$ is a real $n$-dimensional linear space. Thus (1.31) is an equation in the real linear space $E$. We note that the $k$th-order homogeneous part of $h(z)$ in (1.31) belongs to the real linear space

$$\tilde{H}_n^k = \{g(z) \in H_n^k | \overline{Pg(z)} = Pg(z) \text{ for any } z \in E\}$$

and $\tilde{H}_n^k$ is $L_B^k$-invariant. Since $B$ is diagonal, $\text{Ker}(L_B^k) \cap \tilde{H}_n^k$ is a complementary subspace to the range of $L_B^k|_{\tilde{H}_n^k}$ in $\tilde{H}_n^k$. Therefore we can find a near identity transformation in $E$,

$$z = v + \xi(v), \qquad v \in \Omega_r, \tag{1.32}$$

where $\xi(v) = O(|v|^2)$ as $v \rightarrow 0$ and $\Omega_r$ is a neighborhood of the origin in $E$, to obtain a $B$-normal form of (1.31)

$$\dot{v} = Bv + G^2(v) + \cdots + G^r(v), \qquad v \in \Omega_r, \tag{1.33}$$

where $G^k$ is a linear combination of all resonant monomials of order $k$ with suitable complex coefficients such that $\overline{PG^k(v)} = PG^k(\bar{v})$ for $v \in E$, $k = 2, \ldots, r$. We change variables by $y = Pv$ in (1.33). Then the resulting equation

$$\dot{y} = Ay + PG^2(P^{-1}y) + \cdots + PG^r(P^{-1}y) \tag{1.34}$$

is an $A$-normal form of (1.30), which is real. From (1.32), we get the required real transformation $x = y + P\xi(P^{-1}y)$. We illustrate this method by the next example.

**Example 1.22.** Consider the equation:

$$\dot{x} = Ax + O(|x|^2), \qquad x = (x_1, x_2)^T \in \mathbb{R}^2, \tag{1.35}$$

where

$$A = \begin{bmatrix} 0 & -1 \\ 1 & 0 \end{bmatrix}.$$

We change variables by $z_1 = x_1 + ix_2$ and $z_2 = x_1 - ix_2$. We note that $z_2 = \bar{z}_1$. Then (1.35) becomes

$$\begin{cases} \dot{z}_1 = iz_1 + O(|z_1|^2 + |z_2|^2), \\ \dot{z}_2 = -iz_2 + O(|z_1|^2 + |z_2|^2), \end{cases} \tag{1.36}$$

where the second equation of (1.36) is conjugate to the first one. Since the matrix of the linear part of (1.36) $\Lambda = \text{diag}\{i, -i\}$, the resonant conditions are

$$\begin{cases} \alpha_1 - \alpha_2 - 1 = 0, & j = 1, \\ \alpha_1 - \alpha_2 + 1 = 0, & j = 2, \end{cases}$$

where $\alpha_1 + \alpha_2 \geq 2$. Therefore, $\alpha_1 + \alpha_2$ must be odd and $\alpha_1 - \alpha_2 = (-1)^{j-1}$. Thus if $k = 2m + 1$, $m \geq 1$, then

$$\text{Ker}(L^k_{\Lambda^*}) = \text{span}\{z_1^{m+1}z_2^m e_1, z_1^m z_2^{m+1} e_2\}.$$

A $\Lambda$-normal form up to order $r$ will be of the form

$$\begin{cases} \dot{z}_1 = iz_1 + c_1 z_1^2 z_2 + \cdots + c_s z_1^{s+1} z_2^s, \\ \dot{z}_2 = -iz_2 + d_1 z_1 z_2^2 + \cdots + d_s z_1^s z_2^{s+1}, \end{cases} \tag{1.37}$$

where $r - 1 \le 2s + 1 \le r$, $c_k$ and $d_k$ are complex constants, and $d_k = \bar{c}_k$, $k = 1, \ldots, s$. Applying the change of coordinates $z_1 = x_1 + ix_2$ and $z_2 = x_1 - ix_2$ to (1.37), we get a real $A$-normal form up to order $r$ as follows:

$$\begin{cases} \dot{x}_1 = -x_2 + \sum_{k=1}^{s} \left(x_1^2 + x_2^2\right)^k (a_k x_1 - b_k x_2), \\ \dot{x}_2 = x_1 + \sum_{k=1}^{s} \left(x_1^2 + x_2^2\right)^k (a_k x_2 + b_k x_1), \end{cases}$$

where $a_k$ and $b_k$ are $\text{Re}(c_k)$ and $\text{Im}(c_k)$, respectively, $k = 1, \ldots, s$. A different type of normal form can be obtained by making a change to polar coordinates, $z = z_1 = re^{i\theta}$, in the first equation of (1.37) (from the second equation of (1.37), we can change variables by $z = z_2 = re^{-i\theta}$, but we will get the same result). Then, we get

$$\begin{cases} \dot{r} = a_1 r^3 + \cdots + a_s r^{2s+1}, \\ \dot{\theta} = 1 + b_1 r^2 + \cdots + b_s r^{2s}, \end{cases}$$

where $a_i$ and $b_i$ are real constants, $i = 1, \ldots, s$.

We note that in (1.37) the second equation is conjugate to the first one. Let $z = z_1$. Then $\bar{z} = z_2$. We may say that

$$\dot{z} = iz + c_1 |z|^2 z + \cdots + c_s |z|^{2s} z, \qquad z \in \mathbb{C},$$

where $c_1, \ldots, c_s$ are complex constants, $r - 1 \le 2s + 1 \le r$, is an $A$-normal form (in $\mathbb{C}$) of (1.35).

## 2.2 Poincaré's Theorem and Siegel's Theorem

Consider the differential equation

$$\dot{x} = Ax + f(x), \qquad x \in \mathbb{C}^n, \tag{2.1}$$

where $A$ is an $n \times n$ complex matrix, $f(x) = O(|x|^2)$ as $x \to 0$, and $f(x)$ is analytic in $x$. If the resonant conditions for $A$ do not hold for any $\alpha$ and $j$ with $|\alpha| \geq 2$, $1 \leq j \leq n$, then it is clear that (2.1) can be formally transformed into a linear equation

$$\dot{x} = Ax. \tag{2.2}$$

In this section we will give sufficient conditions on $A$ for (2.1) to be transformed to (2.2) by an analytic transformation.

**Definition 2.1.** If the convex hull of the spectrum $\sigma(A)$ of $A$ in the complex plane does not contain the origin of $\mathbb{C}$, then $\sigma(A)$ is said to be in the Poincaré domain. If the origin of $\mathbb{C}$ lies inside the convex hull of $\sigma(A)$, then we say $\sigma(A)$ is in the Siegel domain.

**Lemma 2.2.** *If $\sigma(A) = \{\lambda_1, \ldots, \lambda_n\}$ is in the Poincaré domain, then there are at most finitely many resonant monomials.*

*Proof.* Let $D$ be the convex hull of $\sigma(A)$ in the complex plane. By the assumption, $d = \mathrm{dist}(0, D) > 0$. Then for any $\alpha$ with $|\alpha| \geq 2$, $\frac{\alpha \cdot \lambda}{|\alpha|} = \frac{\alpha}{|\alpha|} \cdot \lambda \in D$ and hence $\frac{|\alpha \cdot \lambda|}{|\alpha|} \geq d$. Let $M = \max_{1 \leq j \leq n}\{|\lambda_j|\}$. Thus for any $j = 1, \ldots, n$, if $|\alpha| \geq \frac{2M}{d}$, then

$$\frac{|\alpha \cdot \lambda - \lambda_j|}{|\alpha|} \geq \frac{|\alpha \cdot \lambda|}{|\alpha|} - \frac{|\lambda_j|}{|\alpha|} \geq \frac{d}{2} > 0.$$

This proves the lemma.                                                    □

**Corollary 2.3.** *If $\sigma(A)$ is in the Poincaré domain and the resonant conditions for $A$ do not hold for any $\alpha$ and $j$ with $|\alpha| \geq 2$ and $1 \leq j \leq n$, then there exists a constant $C_0 > 0$ such that*

$$|\alpha \cdot \lambda - \lambda_j| \geq C_0|\alpha|, \qquad |\alpha| \geq 2, \quad 1 \leq j \leq n.$$

**Theorem 2.4.** *(Poincaré) Let $A = \mathrm{diag}(\lambda_1, \ldots, \lambda_n)$. If $\sigma(A)$ is in the Poincaré domain and the resonant conditions for $A$ do not hold for any $\alpha$ and $j$ with $|\alpha| \geq 2$ and $1 \leq j \leq n$, then there exists an analytic change of* *variables* $x = y + \xi(y)$, $y \in \Omega$, *where* $\xi(y) = O(|y|^2)$ *as* $y \to 0$ *and* $\Omega$ *is a neighborhood of the origin in* $\mathbb{C}^n$, *which transforms (2.1) into (2.2).*

To prove Poincaré's Theorem 2.4 it is sufficient to show that the following equation

$$D\xi(y)Ay - A\xi(y) - f(y + \xi(y)) = 0, \tag{2.3}$$

where $D\xi(y)$ is the derivative of $\xi$ with respect to $y$, has a solution $\xi(y)$ which is analytic in $y \in \Omega$ (where $\Omega$ is a neighborhood of the origin in $\mathbb{C}^n$) and $\xi(y) = O(|y|^2)$ as $y \to 0$. We will apply the Implicit Function Theorem to solve (2.3) for any given analytic function $f$ with $f(x) = O(|x|^2)$ as $x \to 0$. To do so, we first introduce some Banach spaces of analytic functions.

Let $\{X, |\cdot|_X\}$ be a Banach space with norm $|\cdot|_X$ and $L_S^k(\mathbb{C}^n, X)$ be the linear space of all bounded symmetric $k$-linear maps from $\mathbb{C}^n$ into $X$ ($k \geq 1$). In $\mathbb{C}^n$ we use the norm $|\cdot|$ defined by $|x| = \max_{1 \leq i \leq n}|x_i|$ for $x = (x_1, \ldots, x_n)^T \in \mathbb{C}^n$. Let $e_1, \ldots, e_n$ be the usual basis of $\mathbb{C}^n$. A $k$-linear map $m$ from $\mathbb{C}^n$ to $X$ has the form

$$m(x_1, x_2, \ldots, x_k) = \sum_{i_1=1}^{n} \cdots \sum_{i_k=1}^{n} \alpha_{i_1 i_2 \cdots i_k} x_{1i_1} \cdots x_{ki_k},$$

where $x_j = (x_{j1}, \ldots, x_{jn})$ and $\alpha_{i_1 i_2 \ldots i_k} = m(e_{i_1}, \ldots, e_{i_k}) \in X$.

The map $m$ is symmetric if and only if a permutation of the subscripts of the $\alpha$s leaves them unchanged. We define

$$|m|_k = \sum_{i_1=1}^{n} \cdots \sum_{i_k=1}^{n} |\alpha_{i_1 \cdots i_k}|_x, \quad \text{for } m \in L_S^k(\mathbb{C}^n, X).$$

From this definition we can show that

(i)   $\{L_S^k(\mathbb{C}^n, X), |\cdot|_k\}$ is a Banach space;
(ii)  $|m(x_1, \ldots, x_k)| \leq |m|_k |x_1| |x_2| \cdots |x_k|$ for all $x_i \in \mathbb{C}^n$;
(iii) the usual isomorphism of $L_S^{k+h}(\mathbb{C}^n, X)$ into $L_S^h(\mathbb{C}^n, L_S^k(\mathbb{C}^n, X))$ is a norm-preserving isomorphism.

If $f: \mathbb{C}^n \to X$ is an analytic function, then $f$ can be represented as a power series of the form

$$f(x) = \sum_{k=0}^{\infty} f_k(x^k),$$

where $f_k \in L_S^k(\mathbb{C}^n, X)$, $x^k = (x, \ldots, x) \in \mathbb{C}^n \times \mathbb{C}^n \times \ldots \times \mathbb{C}^n$ ($k$ times), $k = 1, 2, \ldots$, and $f_0(x^0) = f(0)$ is an element in $X$. If we denote $f_k(x^k) = \sum_{|\alpha|=k} C_\alpha x_1^{\alpha_1} \cdots x_n^{\alpha_n}$, where $x = (x_1, x_2, \ldots, x_n) \in X$, $\alpha = (\alpha_1, \alpha_2, \ldots, \alpha_n)$ is a multi-index, and $C_\alpha \in X$, then it is easy to see that $|f_k|_k = \sum_{|\alpha|=k} |C_\alpha|_x$, $k \geq 1$. We define $|f_0|_0 = |f(0)|_x$.

Now we define:

$$\mathscr{D}_r = \{x \in \mathbb{C}^n \mid |x| < r\}, \qquad r > 0.$$

$$C^\omega(\mathscr{D}_r, X) = \{f \mid f\colon \mathscr{D}_r \to X \text{ is analytic}\}.$$

$$A_{0,r}(\mathbb{C}^n, X) = \left\{f \in C^\omega(\mathscr{D}_r, X) \mid f(x) = \sum_{k=0}^{\infty} f_k(x^k),\right.$$

$$\left. f_k \in L_S^k(\mathbb{C}^n, X), \sum_{k=0}^{\infty} |f_k|_k r^k < \infty\right\}.$$

$$A_{1,r}(\mathbb{C}^n, X) = \{f \in A_{0,r}(\mathbb{C}^n, X) \mid Df \in A_{0,r}(\mathbb{C}^n, L(\mathbb{C}^n, X))\}.$$

For $f \in A_{0,r}(\mathbb{C}^n, X)$ we define $|f|_{0,r} = \sum_{k=0}^\infty |f_k|_k r^k$. For $f \in A_{1,r}(\mathbb{C}^n, X)$ we define $|f|_{1,r} = |f|_{0,r} + |Df|_{0,r}$. Then $\{A_{0,r}(\mathbb{C}^n, X), |\cdot|_{0,r}\}$ and $\{A_{1,r}(\mathbb{C}^n, X), |\cdot|_{1,r}\}$ are Banach spaces. Let $B_{i,r}(\delta) = \{g \in A_{i,r}(\mathbb{C}^n, \mathbb{C}^n) \mid |g|_{i,r} < \delta\}$, $i = 0, 1$.

**Lemma 2.5.** *If* $f \in A_{0,r}(\mathbb{C}^n, X)$ *and* $0 < \delta < r$, *then* $D^k f \in A_{0,\delta}(\mathbb{C}^n, L_S^k(\mathbb{C}^n, X))$ *and*

$$D^k f(x) = k! \sum_{j=k}^\infty \binom{j}{k} f_j(x^{j-k}, \cdot), \quad |D^k f|_{0,\delta} \leq k! \frac{|f|_{0,r}}{(r-\delta)^k},$$

*for* $k = 1, 2, \ldots$, *where* $f_k(x^0, \cdot) = f_k(\cdot)$.

*Proof.* Let $f = \sum_{k=0}^\infty f_k(x^k)$. Then

$$|f|_{0,r} = \sum_{k=0}^\infty |f_k|_k r^k = \sum_{k=0}^\infty |f_k|_k (\delta + (r-\delta))^k.$$

Hence

$$|f|_{0,r} = \sum_{k=0}^{\infty} \sum_{i=0}^{k} \binom{k}{i} |f_k|_k \delta^{k-i} (r - \delta)^i. \tag{2.4}$$

Since $|f|_{0,r} < \infty$ and the terms in the above series are all nonnegative, we can rearrange terms. Thus

$$|f|_{0,r} = \sum_{i=0}^{\infty} \left\{ \sum_{k=i}^{\infty} \binom{k}{i} |f_k|_k \delta^{k-i} \right\} (r - \delta)^i. \tag{2.5}$$

Let

$$g(z) = \sum_{i=0}^{\infty} \left\{ \sum_{k=i}^{\infty} \binom{k}{i} |f_k|_k \delta^{k-i} \right\} z^i, \qquad z \in \mathbb{C} \text{ and } |z| \le r - \delta.$$

Then $g(z)$ is analytic in $\Omega_{r-\delta} = \{z \in \mathbb{C} \,|\, |z| < r - \delta\}$ since (2.5) majorizes the series of $g(z)$. Applying Cauchy's inequality to $g(z)$, we have

$$\sum_{k=i}^{\infty} \binom{k}{i} |f_k|_k \delta^{k-i} \le \frac{M}{(r - \delta)^i},$$

where $M = \max_{|z|=r-\delta} \{|g(z)|\}$. From (2.5), $M \le |f|_{0,r}$. Therefore

$$\sum_{k=i}^{\infty} \binom{k}{i} |f_k|_k \delta^{k-i} \le \frac{|f|_{0,r}}{(r - \delta)^i}. \tag{2.6}$$

Let $|x| \le \delta$ and $|y| < r - \delta$. Then $|x + y| < r$ and then

$$f(x + y) = \sum_{k=0}^{\infty} f_k((x + y)^k) = \sum_{k=0}^{\infty} \left\{ \sum_{i=0}^{k} \binom{k}{i} f_k(x^{k-i}, y^i) \right\}.$$

We note that the last series is majorized by series (2.4) and hence is absolutely convergent. Thus we can rearrange the terms such that

$$f(x + y) = \sum_{i=0}^{\infty} \left\{ \sum_{k=i}^{\infty} \binom{k}{i} f_k(x^{k-i}, y^i) \right\}$$

$$= \sum_{i=0}^{\infty} \left\{ \sum_{k=i}^{\infty} \binom{k}{i} f_k(x^{k-i}, \cdot) \right\} (y^i).$$

From (2.5) and by the uniqueness of the Taylor's series of $f$ at $x$, $D^i f(x) = i! \sum_{k=i}^{\infty} \binom{k}{i} f_k(x^{k-i}, \cdot)$, $i = 0, 1, 2, \ldots$. Let $g_j(x^j) = f_k(x^j, \cdot)$ for each $j \le k$. Then

$$|D^i f|_{0,\delta} = i! \sum_{k=i}^{\infty} \binom{k}{i} |g_{k-i}|_{k-i} \delta^{k-i} = i! \sum_{k=i}^{\infty} \binom{k}{i} |f_k|_k \delta^{k-i}.$$

The desired conclusion follows from (2.6). □

**Lemma 2.6.** *Let $f \in A_{0,r}(\mathbb{C}^n, X)$, $g \in B_{0,\delta}(r)$. Then $f \circ g \in A_{0,\delta}(\mathbb{C}^n, X)$ and $|f \circ g|_{0,\delta} \le |f|_{0,r}$.*

*Proof.* Since $|g(x)| \le |g|_{0,\delta} < r$ for $|x| \le \delta$, $f \circ g \in C^\omega(\mathscr{D}_\delta, X)$. Let $f(x) = \sum_{k=0}^{\infty} f_k(x^k)$ and $g(x) = \sum_{l=0}^{\infty} g_l(x^l)$, where $f_k \in L_S^k(\mathbb{C}^n, X)$ and $g_l \in L_S^l(\mathbb{C}^n, \mathbb{C}^n)$ for $k \ge 0$ and $l \ge 0$. Then for $|x| \le \delta$

$$f \circ g(x) = \sum_{k=0}^{\infty} f_k\left(\left(\sum_{l=0}^{\infty} g_l(x^l)\right)^k\right)$$

$$= \sum_{i=0}^{\infty} \sum_{k=0}^{\infty} \sum_{|l|=i} f_k\big(g_{l_1}(x^{l_1}), \ldots, g_{l_k}(x^{l_k})\big),$$

where $|l| = l_1 + \cdots + l_k$ and each $l_i$ is a nonnegative integer.

$$|f \circ g|_{0,\delta} \le \sum_{i=0}^{\infty} \left(\sum_{k=0}^{\infty} |f_k|_k \left(\sum_{|l|=i} |g_{l_1}|_{l_1} \cdots |g_{l_k}|_{l_k}\right)\right) \delta^i$$

$$= \sum_{k=0}^{\infty} |f_k|_k \sum_{i=0}^{\infty} \left(\sum_{|l|=i} |g_{l_1}|_{l_1} \cdots |g_{l_k}|_{l_k}\right) \delta^i$$

$$= \sum_{k=0}^{\infty} |f_k|_k \left(\sum_{l=0}^{\infty} |g_l|_l \delta^l\right)^k < \sum_{k=0}^{\infty} |f_k|_k r^k = |f|_{0,r}. \quad \square$$

**Lemma 2.7.** *Let $E: A_{0,r}(\mathbb{C}^n, X) \times B_{1,\delta}(r) \to A_{0,\delta}(\mathbb{C}^n, X)$ be defined by $E(f, g) = f \circ g$, where $f \in A_{0,r}(\mathbb{C}^n, X)$ and $g \in B_{1,\delta}(r)$. Then $E$ is continuous.*

*Proof.* Let $f$ and $f_1 \in A_{0,r}(\mathbb{C}^n, X)$, $g \in B_{1,\delta}(r)$. By Lemma 2.6

$$|E(f + f_1, g) - E(f, g)|_{0,\delta} = |f_1 \circ g|_{0,\delta} \leq |f_1|_{0,r}.$$

Hence $E$ is uniformly continuous in its first argument.

Let $|g|_{1,\delta} = \alpha < r$, $\beta = (r - \alpha)/3$, and $h \in B_{1,\delta}(\beta)$. Then $E(f, g + h)(x) = f(g(x) + h(x)) \in C^\omega(\mathscr{D}_\delta, X)$. Let $g(x) = \sum_{l=0}^\infty g_l(x^l)$ and $h(x) = \sum_{m=0}^\infty h_m(x^m)$, where $g_l \in L_S^l(\mathbb{C}^n, \mathbb{C}^n)$ for $l \geq 0$ and $h_m \in L_S^m(\mathbb{C}^n, \mathbb{C}^n)$ for $m \geq 0$, respectively. By Taylor's Theorem

$$f(x + y) = \sum_{k=0}^\infty \frac{D^k f(x)}{k!} (y^k), \qquad |x| \leq \alpha, \quad |y| < r - \alpha.$$

Thus

$$E(f, g + h)(x) - E(f, g)(x) = \sum_{k=1}^\infty \frac{D^k f(g(x))}{k!} ((h(x))^k).$$

From Lemma 2.5,

$$E(f, g + h)(x) - E(f, g)(x)$$

$$= \sum_{k=1}^\infty \sum_{i=k}^\infty \binom{i}{k} f_i\big((g(x))^{i-k}, (h(x))^k\big)$$

$$= \sum_{i=1}^\infty \sum_{k=i}^\infty \binom{k}{i} f_k\big((g(x))^{k-i}, (h(x))^i\big)$$

$$= \sum_{k=1}^\infty \sum_{i=1}^k \binom{k}{i} f_k\big((g(x))^{k-i}, (h(x))^i\big)$$

$$= \sum_{k=1}^\infty \sum_{i=1}^\infty \binom{k}{i} \sum_{j=0}^\infty \sum_{|l|+|m|=j} f_k\big(g_{l_1}(x^{l_1}), \ldots, g_{l_{k-i}}(x^{l_{k-i}}),$$

$$h_{m_1}(x^{m_1}), \ldots, h_{m_i}(x^{m_i})\big),$$

where $|l| = l_1 + \cdots + l_{k-i}$, $|m| = m_1 + \cdots + m_i$. Hence

$$|E(f, g+h) - E(f, g)|_{0, \delta}$$

$$\leq \sum_{j=0}^{\infty} \left( \sum_{k=1}^{\infty} |f_k|_k \sum_{i=1}^{k} \binom{k}{i} \right.$$

$$\left. \times \sum_{|l|+|m|=j} |g_{l_1}|_{l_1} \cdots |g_{l_{k-i}}|_{l_{k-i}} |h_{m_1}|_{m_1} \cdots |h_{m_i}|_{m_i} \right) \delta^j$$

$$= \sum_{k=1}^{\infty} |f_k|_k \sum_{i=1}^{k} \binom{k}{i} \left( \sum_{l=0}^{\infty} |g_l|_l \delta^l \right)^{k-i} \left( \sum_{m=0}^{\infty} |h_m|_m \delta^m \right)^{i}$$

$$\leq \sum_{k=1}^{\infty} k|f_k|_k (|g|_{0, \delta} + |h|_{0, \delta})^{k-1} |h|_{0, \delta}$$

$$\leq \left( \sum_{k=1}^{\infty} k|f_k|_k (r - \beta)^{k-1} \right) |h|_{0, \delta}$$

$$\leq \frac{1}{\beta} |f|_{0, r} |h|_{1, \delta}.$$

Thus $E$ is continuous in the second argument. $\qquad\square$

**Lemma 2.8.** *Let $E$ be defined as in Lemma 2.7. Then $E$ is $C^\infty$.*

*Proof.* Let $f \in A_{0, r}(\mathbb{C}^n, X)$ and $g \in B_{1, \delta}(r)$ with $|g|_{1, \delta} = \alpha < r$ and let $\beta = (r - \alpha)/3$. Let $h \in B_{1, \delta}(\beta)$. Since by Taylor's Theorem, for $|x| \leq \alpha$ and $|y| \leq r - \alpha$,

$$f(x + y) = \sum_{k=0}^{\infty} \frac{D^k f(x)}{k!} (y^k),$$

for any positive integer $N$,

$$E(f, g + h)(x) = \sum_{k=0}^{N} \frac{D^k f(g(x))(h(x))^k}{k!} + R_{N+1}(x),$$

where

$$R_{N+1}(x) = \sum_{k=N+1}^{\infty} \frac{D^k f(g(x))(h(x))^k}{k!}.$$

From Lemma 2.5,

$$R_{N+1}(x) = \sum_{k=N+1}^{\infty} \sum_{j=k}^{\infty} \binom{j}{k} f_j\big((g(x))^{j-k}, (h(x))^k\big).$$

By a similar argument as in Lemma 2.7, we have

$$|R_{N+1}|_{0,\delta} \leq \frac{|f|_{0,r}}{\beta^{N+1}} (|h|_{1,\delta})^{N+1}.$$

From the converse of Taylor's Theorem, $D_2^k E$ exists for $0 \leq k \leq N$ and $D_2^k E(f, g) = D^k f \circ g$. By Lemma 2.7, $D_2^k E(f, g)$ is continuous and $D_2^k E(f, g)$ is linear in its first argument. It follows that $D_1 D_2^k E(f, g)$ exists and $D_1 D_2^k E(f, g) = D_2^k E(\cdot, g)$. The lemma may now be proved by induction.  $\square$

**Lemma 2.9.** *If $g \in A_{1,r}(\mathbb{C}^n, X)$ and $M \in \mathbb{C}^{n \times n}$, then*

$$f(x) = Dg(x) Mx \in A_{0,r}(\mathbb{C}^n, X)$$

*and*

$$|f|_{0,r} \leq r |M| |Dg|_{0,r}.$$

*Proof.* Let $g(x) = \sum_{k=0}^{\infty} g_k(x^k)$, where $g_k \in L_S^k(\mathbb{C}^n, X)$, $k = 0, 1, 2, \ldots$ . From Lemma 2.5, $Dg(x) = \sum_{k=1}^{\infty} k g_k(x^{k-1}, \cdot)$ and $|Dg|_{0,r} =$

$\sum_{k=1}^{\infty} k|g_k|_k r^{k-1}$. Then $f(x) = \sum_{k=1}^{\infty} k g_k(x^{k-1}, Mx)$. Let $f_k \colon \mathbb{C}^n \times \cdots \times \mathbb{C}^n$ ($k$ times) $\to X$ be defined by

$$f_k(v_1, \ldots, v_k) = \sum_{i=1}^{k} g_k(v_1, \ldots, Mv_i, \ldots, v_k),$$

$$v_i \in \mathbb{C}^n, \quad i = 1, \ldots, k.$$

Then $f_k \in L_S^k(\mathbb{C}^n, X)$, $|f_k|_k \le k|M| \|g_k\|_k$ and $f(x) = \sum_{k=1}^{\infty} f_k(x^k)$. Thus

$$|f|_{0,r} = \sum_{k=1}^{\infty} |f_k|_k r^k \le |M| \sum_{k=1}^{\infty} k|g_k|_k r^k = r|M| \|Dg\|_{0,r}. \qquad \square$$

*Proof of Poincaré's Theorem.* Let $r > 0$. We define $V_{0,r} = \{g \mid g \in A_{0,r}(\mathbb{C}^n, \mathbb{C}^n), \ g(0) = 0, \ Dg(0) = 0\}$ and $V_{1,r} = \{g \mid g \in A_{1,r}(\mathbb{C}^n, \mathbb{C}^n), \ g(0) = 0, \ Dg(0) = 0\}$. Then $V_{0,r}$ and $V_{1,r}$ are closed linear subspaces of Banach spaces $A_{0,r}$ and $A_{1,r}$ respectively. Let $\bar{B}_{1,r}(\delta) = \{g \in V_{1,r} \mid \|g\|_{1,r} \le \delta\}$. We define $F \colon V_{0,r} \times \bar{B}_{1,r/2}(r/2) \to V_{0,r/2}$ by

$$F(f, \xi)(y) = D\xi(y)Ay - A\xi(y) - f(y + \xi(y)). \qquad (2.7)$$

Then $F(0,0) = 0$ and from Lemma 2.8 $F$ is $C^1$. Equation (2.3) can be expressed as

$$F(f, \xi)(y) \equiv 0, \qquad y \in \Omega. \qquad (2.8)$$

Let $K = F_\xi(0,0)$. Then $K \colon V_{1,r/2} \to V_{0,r/2}$ is defined by

$$(Kv)(y) = Dv(y)Ay - Av(y).$$

For any $g(x) = \sum_{j=1}^{n} (\sum_{k=2}^{\infty} \sum_{|\alpha|=k} c_\alpha^j x^\alpha) e_j \in V_{0,r/2}$ we define

$$(\tilde{K}g)(x) = \sum_{j=1}^{n} \left( \sum_{k=2}^{\infty} \sum_{|\alpha|=k} \frac{c_\alpha^j}{\alpha \cdot \lambda - \lambda_j} x^\alpha \right) e_j. \qquad (2.9)$$

Since $\sigma(A)$ is in the Poincaré domain and there are no resonant monomials, by Corollary 2.3 there exists a constant $C_0 > 0$ such that $|\alpha \cdot \lambda - \lambda_j| \ge C_0 |\alpha|$ for any $\alpha$ and $j$ with $|\alpha| \ge 2$, $1 \le j \le n$. Let $C_\alpha = (c_\alpha^1, \ldots, c_\alpha^n)^T$ and $\tilde{C}_\alpha = (c_\alpha^1/(\alpha \cdot \lambda - \lambda_1), \ldots, c_\alpha^n/(\alpha \cdot \lambda - \lambda_n))^T$. Denote $v(x) = (\tilde{K}g)(x) = \sum_{k=2}^{\infty} v_k(x^k)$ for a given $g \in V_{0,r/2}$. Then

$v_k(x^k) = \sum_{|\alpha|=k} \tilde{C}_\alpha x^\alpha$ and $|v_k|_k = \sum_{|\alpha|=k} |\tilde{C}_\alpha| \leq \frac{1}{C_0 k} \sum_{|\alpha|=k} |C_\alpha| = \frac{1}{C_0 k} |g_k|_k$
for each $k \geq 2$, where $g_k(x^k) = \sum_{|\alpha|=k} C_\alpha x^\alpha$. Hence

$$|v|_{0,r/2} = \sum_{k=2}^{\infty} |v_k|_k (r/2)^k \leq \frac{1}{C_0} \sum_{k=2}^{\infty} \frac{1}{k} |g_k|_k (r/2)^k \leq \frac{1}{C_0} |g|_{0,r/2} < \infty,$$

and from Lemma 2.5, we have

$$|Dv|_{0,r/2} = \sum_{k=2}^{\infty} k|v_k|_k (r/2)^{k-1} \leq \frac{2}{C_0 r} \sum_{k=2}^{\infty} |g_k|_k (r/2)^k$$

$$= \frac{2}{C_0 r} |g|_{0,r/2} < \infty.$$

These imply that $\tilde{K}g \in V_{1,r/2}$ and $\tilde{K}$ is a bounded linear operator from $V_{0,r/2}$ to $V_{1,r/2}$. A calculation shows

$$(Kv)(x) = \sum_{j=1}^{n} \left( \sum_{k=2}^{\infty} \sum_{|\alpha|=k} c_\alpha^j x^\alpha \right) e_j = g(x).$$

Hence $\tilde{K} = K^{-1}$, that is, $F_\xi(0,0)$ has a bounded inverse.

By the Implicit Function Theorem, there exists an $\epsilon > 0$ such that for any $f \in V_{0,r}$, if $|f|_{0,r} \leq \epsilon$, there exists $\xi = \xi(f) \in \bar{B}_{1,r/2}(r/2)$ such that $F(f,\xi) = 0$ and $\xi(0) = 0$.

For a given $f \in V_{0,r}$, let $\tilde{f}(x) = \gamma^{-1} f(\gamma x)$, where $0 < \gamma < 1$. It is obvious that $\tilde{f} \in V_{0,r}$. Since $f(x) = O(|x|^2)$ as $x \to 0$, we can choose $\gamma > 0$ such that $|\tilde{f}|_{0,r} < \epsilon$. From the above discussion, there exists a $\bar{\xi} \in \bar{B}_{1,r/2}(r/2)$ such that $F(\tilde{f}, \bar{\xi}) = 0$. Let $\xi(x) = \gamma \bar{\xi}(\gamma^{-1} x)$, $|x| \leq \gamma r/2$. Then $\xi(x) \in V_{1,\gamma r/2}$ and

$$F(f,\xi)(x) = D\xi(x) Ax - A\xi(x) - f(x + \xi(x))$$

$$= \gamma F(\tilde{f}, \bar{\xi})(\gamma^{-1} x) = 0, \qquad |x| \leq \gamma r/2.$$

Thus the theorem is proved.                                                   $\square$

**Corollary 2.10.** *Poincaré's Theorem is valid even if $A$ is not diagonalizable.*

*Proof.* It is sufficient to show that the operator $K$ defined in the proof of the theorem has also a bounded inverse even though $A$ is not diagonalizable. Without loss of generality, we assume that $A$ is in upper triangular Jordan normal form and $A = S + \epsilon N$, where $S = \text{diag}(\lambda_1, \ldots, \lambda_n)$, $\epsilon > 0$ is arbitrarily small and $N = (a_{ij})$ satisfies $a_{ij} = 0$ if $j \neq i + 1$, $a_{i, i+1} = 1$ or $0$, $i = 1, \ldots, n - 1$. We see that $K = K_S + K_{\epsilon N}$, where $K_S$ and $K_{\epsilon N}$ are defined by

$$(K_S v)(y) = Dv(y)Sy - Sv(y)$$

and

$$(K_{\epsilon N} v)(y) = Dv(y)(\epsilon N)y - (\epsilon N)v(y)$$

for $v \in V_{1, r}$. From the proof of the theorem we see that $K_S$ has a bounded inverse on $V_{0, r}$. Since $\|K_{\epsilon N}\| = \epsilon \|K_N\|$, we can always choose $\epsilon$ to be so small that

$$\|K_{\epsilon N}\| < \frac{1}{\|K_S^{-1}\|}.$$

Thus $K = K_S + K_{\epsilon N}$ has a bounded inverse.                              □


**Corollary 2.11.** *If $\sigma(A)$ is in the Poincaré domain, then there exists an analytic change of variables $x = y + \xi(y)$, where $\xi(y) = O(|y|^2)$ as $y \to 0$, and $y$ is in a neighborhood of the origin in $\mathbb{C}^n$, such that it transforms* (2.1) *into*

$$\dot{x} = Ax + h(x), \tag{2.10}$$

*where $h(x)$ consists of at most finitely many monomials and $h$ commutes with $e^S$ with $S$ being the semisimple part of $A$.*


*Proof.* Without loss of generality, we assume that $A = S + \epsilon N$ is in the Jordan normal form, where $S = \text{diag}(\lambda_1, \ldots, \lambda_n)$, $N$ is nilpotent, and $\epsilon$ is arbitrary small. Since $\sigma(A)$ is in the Poincaré domain, there are at most a finite number of resonant monomials. Suppose that there are no resonant monomials of order bigger than $m$. We note that $m$ is independent of $\epsilon$. By the normal form theory (see Section 2.1), there is an analytic change of variables which transforms (2.1) to

$$\dot{x} = Ax + h(x) + f(x), \tag{2.11}$$

where $h(x)$ is a linear combination of resonant monomials and $f(x) = O(|x|^{m+1})$ as $x \to 0$.

We define $V_{0,r}^m = \{g \mid g \in A_{0,r}(\mathbb{C}^n, \mathbb{C}^n), \; g(x) = O(|x|^{m+1})$ as $x \to 0$, $|g|_{0,r} < \infty\}$ and $V_{1,r}^m = \{g \mid g \in A_{1,r}(\mathbb{C}^n, \mathbb{C}^n), \; g(x) = O(|x|^{m+1})$ as $x \to 0$, $|g|_{1,r} < \infty\}$. Then $V_{0,r}^m$ and $V_{1,r}^m$ are closed linear subspaces of $A_{0,r}$ and $A_{1,r}$ respectively. Let $\bar{B}_{1,\delta}^m(r) = \{g \in V_{1,\delta}^m \mid |g|_{1,\delta} \le r\}$. We define $F$: $V_{0,r}^m \times V_{0,r}^m \times \bar{B}_{1,r/2}^m(r/2) \to V_{0,r/2}^m$ by

$$F(f, h, \xi)(y) = D\xi(y) Ay - A\xi(y) - f(y + \xi(y))$$

$$+ h(y) + D\xi(y)h(y) - h(y + \xi(y)).$$

Then $F(0, 0, 0) = 0$. Let $K = F_\xi(0, 0, 0)$. Then $K: V_{1,r/2}^m \to V_{0,r/2}^m$ is defined by

$$(Kv)(y) = Dv(y) Ay - Av(y).$$

Since there are no resonant monomials of order greater than $m$, in a similar way as in the proof of the Poincaré Theorem and Corollary 2.11, the corollary can be proved.                                    □

**Theorem 2.12.** (*Siegel*) *Let* $\sigma(A) = \{\lambda_1, \ldots, \lambda_n\}$ *be the spectrum of* $A$. *If there exist* $C_0 > 0$ *and* $\mu > 0$ *such that for any* $\alpha = (\alpha_1, \ldots, \alpha_n)$ *with* $|\alpha| \ge 2$

$$|\lambda \cdot \alpha - \lambda_j| \ge \frac{C_0}{|\alpha|^\mu}, \qquad 1 \le j \le n, \tag{2.12}$$

*then the equation* (2.1) *can be transformed to* (2.2) *by an analytic transformation.*

As in the proof of Poincaré's Theorem, we need to find an analytic solution $\xi$ with $\xi(z) = O(|z|^2)$ as $z \to 0$ of equation (2.3). However, since there does not exist a positive lower bound for $\{|\lambda \cdot \alpha - \lambda_j|, |\alpha| \ge 2, 1 \le j \le n\}$ and there does not exist a bounded inverse of $F_\xi(0, 0)$, we are not able to apply the Implicit Function Theorem to solve (2.3) as we did in the case of Poincaré's Theorem.

We shall prove Siegel's Theorem in the case $A = \text{diag}(\lambda_1, \ldots, \lambda_n)$. The idea of the proof is the following: We find first an "approximate bounded inverse of $F_\xi(0,0)$ by which we construct a sequence of approximate solutions $\{\phi_j\}$ of (2.3). The domains of $\{\phi_j\}$ will shrink as $j$ increases, each $\phi_{j+1}$ is a better approximate solution than $\phi_j$, and $\{\phi_j\}$ tends to an analytic solution of (2.3) which is defined in a neighborhood of the origin in $\mathbb{C}^n$.

First we need some notation and lemmas.

$H_{0,r} = \{g \mid g: \mathscr{D}_r \to \mathbb{C}^n$ is analytic and $g(z) = O(|z|^2)$ as $z \to 0, \sup_{z \in \mathscr{D}_r} |g(z)| < +\infty\}$.

$H_{1,r} = \{g \mid g \in H_{0,r}$ and $\|Dg\|_{0,r} = \sup_{z \in \mathscr{D}_r} |Dg(z)| < +\infty\}$.

For $g \in H_{0,r}$, we define $\|g\|_{0,r} = \sup_{z \in \mathscr{D}_r} |g(z)|$. For $g \in H_{1,r}$, we define $\|g\|_{1,r} = \|g\|_{0,r} + \|Dg\|_{0,r}$. Then $\{H_{0,r}, \|\cdot\|_{0,r}\}$ and $\{H_{1,r}, \|\cdot\|_{1,r}\}$ are Banach spaces.

We denote $\bar{B}_{i,r}(\delta) = \{g \in H_{i,r} \mid \|g\|_{i,r} \leq \delta\}$, $\delta > 0$, the closed ball with radius $\delta$ in $H_{i,r}$, $i = 0, 1$.

Define a linear operator $K_r: H_{1,r} \to H_{0,r}$ by

$$K_r v(z) = D_z v(z) \cdot Az - Av(z), \qquad v \in H_{1,r}, \qquad z \in D_r.$$

**Lemma 2.13.** *Let $r \in (0,1)$. If $A = \text{diag}(\lambda_1, \ldots, \lambda_n)$ and the small divisor condition (2.12) holds for $A$, then for any $g \in H_{0,r}$ and any $\delta \in (0,r)$, there exist a unique $v \in H_{1,r-\delta}$ and a positive constant $C$ which does not depend on $g$, $r$ and $\delta$ such that $(K_{r-\delta}v)(z) = g(z)$ for $z \in \mathscr{D}_{r-\delta}$, that is,*

$$K_{r-\delta}v = g \qquad in\ H_{0,r-\delta},$$

*and*

$$\|v\|_{1,r-\delta} \leq C \frac{\|g\|_{0,r}}{\delta^{\mu+2}}.$$

*Proof.* Let the Taylor expansion of $g$ at $z = 0$ be

$$g(z) = \sum_{k=2}^{\infty} \sum_{|\alpha|=k} \sum_{j=1}^{n} c_\alpha^j z^\alpha e_j,$$

where the $c_\alpha^j$ are complex constants. For any $w \in H_{1,r-\delta}$ we assume that the Taylor expansion of $w$ at $z = 0$ is

$$w(z) = \sum_{k=2}^{\infty} \sum_{|\alpha|=k} \sum_{j=1}^{n} d_\alpha^j z^\alpha e_j,$$

where the $d_\alpha^j$ are complex constants. If $K_{r-\delta} w = g$ in $H_{0,r-\delta}$ then by a direct calculation we must have

$$d_\alpha^j = \frac{c_\alpha^j}{\lambda \cdot \alpha - \lambda_j}, \qquad \text{for } |\alpha| \geq 2 \text{ and } 1 \leq j \leq n.$$

Let

$$v(z) = \sum_{k=2}^{\infty} \sum_{|\alpha|=k} \sum_{j=1}^{n} \frac{c_\alpha^j}{\lambda \cdot \alpha - \lambda_j} z^\alpha e_j.$$

Let $C_\alpha = (c_\alpha^1, \ldots, c_\alpha^n)^T$ for $|\alpha| \geq 2$. Then by the small divisor condition (2.12) and the Cauchy's inequality

$$\|v\|_{0,r-\delta} \leq \sum_{k=2}^{\infty} \sum_{|\alpha|=k} \frac{|C_\alpha|}{C_0} k^\mu (r-\delta)^k$$

$$\leq \frac{\|g\|_{0,r}}{C_0} \sum_{k=2}^{\infty} k^\mu \frac{(r-\delta)^k}{r^k} \leq \frac{\|g\|_{0,r}}{C_0} \sum_{k=2}^{\infty} k^\mu e^{-k\delta/r}$$

$$\leq \frac{\|g\|_{0,r}}{C_0} \delta^{-(\mu+1)} \left( \int_0^\infty y^\mu e^{-y} dy + (\mu+1)^\mu \right)$$

$$\leq C_1 \|g\|_{0,r} \delta^{-(\mu+1)},$$

where $C_1 = C_0^{-1}(\Gamma(\mu+1) + (\mu+1)^\mu)$, which is a positive constant and does not depend on $g$, $r$ and $\delta$. Thus $v \in H_{0,r-\delta}$. Furthermore, from the Cauchy's inequality

$$\|D_z v\|_{0,r-\delta} \leq \frac{n\|v\|_{0,r-\delta/2}}{\left( \dfrac{\delta}{2} \right)} \leq nC_1 \|g\|_{0,r} \left( \frac{\delta}{2} \right)^{-(\mu+2)} \leq C_2 \|g\|_{0,r} \delta^{-(\mu+2)},$$

where $C_2$ is a positive constant which does not depend on $g$, $r$, and $\delta$

either. Therefore $v \in H_{1,r-\delta}$ and

$$\|v\|_{1,r-\delta} \le C\|g\|_{0,r}\delta^{-(\mu+2)},$$

where $C$ is a positive constant which does not depend on $g$, $r$ and $\delta$. This proves the lemma. $\qquad\square$

The above lemma says that even though $K_r$ has no bounded inverse, it has an "approximate bounded inverse $\check{K}_{r-\delta}^{-1}$ from $H_{0,r}$ to $H_{1,r-\delta}$ which is defined as follows: For any $g \in H_{0,r}$ and $\delta \in (0,r)$, where $0 < r < 1$, let $v \in H_{1,r-\delta}$ be the unique solution of equation $K_{r-\delta}v = g$ in $H_{0,r-\delta}$. We define $\check{K}_{r-\delta}^{-1}g = v$. Then by Lemma 2.13, $\check{K}_{r-\delta}^{-1}$ is a bounded linear operator from $H_{0,r}$ to $H_{1,r-\delta}$.

Let $r \in (0,1)$ and $\beta \in (0,r/2)$. Now we consider a mapping $\mathscr{F}(\cdot\,;r,\beta)\colon \overline{B}_{1,\beta}(r/2) \to H_{0,\beta}$ defined by

$$\mathscr{F}(\xi;r,\beta) = D_z\xi \cdot A - A\xi - f\circ(I + \xi).$$

Since $\|I + \xi\|_{0,\beta} \le \beta + r/2 < r$ and $D^k f$ is uniformly continuous on $\overline{\mathscr{D}}_{\beta+r/2}$ for any $k \ge 0$, $\mathscr{F}(\cdot\,;r,\beta)$ is $C^2$ from $\overline{B}_{1,\beta}(r/2)$ to $H_{0,\beta}$ for any fixed $r \in (0,1)$ and $\beta \in (0,r/2)$. Thus equation (2.3) can be written as

$$\mathscr{F}(\xi;r,\beta) = 0, \qquad \xi \in \overline{B}_{1,\beta}(r/2), \quad 0 < r < 1 \text{ and } \beta \in (0,r/2).$$

**Lemma 2.14.** *Let* $r \in (0,1)$, $\beta \in (0,r/2)$, $\delta \in (0,\beta)$, *and* $\phi \in \overline{B}_{1,\beta}(r/2)$. *Then for any* $u \in H_{1,\beta-\delta}$,

$$D_\xi\,\mathscr{F}(\phi;r,\beta-\delta)(I + D_z\phi)u - D_z(\mathscr{F}(\phi;r,\beta-\delta))u$$

$$= (I + D_z\phi)K_{\beta-\delta}u. \tag{2.13}$$

*Proof.* In fact (2.13) holds if and only if

$$(D_z\phi)K_{\beta-\delta}u + D_z(K_{\beta-\delta}\phi)u = K_{\beta-\delta}((D_z\phi)u)$$

or

$$D_z((D_z\phi)A)u + (D_z\phi)(D_zu)A = D_z((D_z\phi)u)A + (D_z\phi)Au,$$

$$\text{for } u \in H_{1,\beta-\delta}.$$

The last equality follows from a direct calculation. $\qquad\square$

*Proof of Siegel's Theorem.* Since the small divisor condition (2.12) implies that there exist no resonant monomials, by the normal-form theory introduced in Section 2.1, we can make near identity transformations

$$x = z + \xi^k(z), \qquad k = 2, 3, \ldots, 4[n + \mu + 3],$$

such that (2.1) becomes

$$\dot{z} = Az + \tilde{f}(z), \qquad z \in \mathcal{D}_r,$$

where $\tilde{f}(z) = O(|z|^{4(n+\mu+3)+1})$ as $z \to 0$, and $r > 0$ is sufficiently small. Therefore we may assume for (2.1) that $f \in H_{0,r}$ with $r \in (0, 1)$ sufficiently small and $f(z) = O(|z|^{4(n+\mu+3)+1})$ as $z \to 0$. We can also assume that $\|D_z^2 f\|_{0,r} \leq \frac{1}{2}$ if $r$ is sufficiently small.

We consider the following sequences of real numbers $\{r_j\}$ and $\{\delta_j\}$:

$$r_j = \frac{1}{4}r\left(1 + \frac{1}{2^j}\right), \qquad \delta_j = \frac{(r_j - r_{j+1})}{2}, \qquad j = 0, 1, \ldots.$$

Define sequences of functions $\{\phi_j\}$ and $\{u_j\}$, $j = 0, 1, 2, \ldots$, inductively as follows:

$$\phi_0 = 0;$$

$$u_j = -(I + D_z\phi_j)\tilde{K}_{r_j-\delta_j}^{-1}(I + D_z\phi_j)^{-1} \mathcal{F}(\phi_j; r, r_j);$$

$$\phi_{j+1} = \phi_j + u_j.$$

We will show that if $r$ is sufficiently small, then the sequences $\{\phi_j\}$ and $\{u_j\}$ are well defined.

From Lemma 2.13, for any $\delta \in (0, r)$, there is a constant $C$ such that $\|\tilde{K}_{r-\delta}^{-1}\| \leq C\delta^{-(\mu+n+2)}$. We may assume that $C \geq 1$. Let $\bar{C}_0 = 8nC(16/r)^{n+\mu+3}$ and $\epsilon_0 = 1/(2\bar{C}_0^2)$. We note that $r_0 = r/2$ and $\epsilon_0 = O(r^{2(n+\mu+3)})$ as $r \to 0$. We may assume that $r$ is so small that $\epsilon_0 \leq \frac{r}{2}$. Let the sequence $\{\epsilon_j\}$, $j = 0, 1, 2, \ldots$, be defined recursively as follows: $\epsilon_{j+1} = \bar{C}_0^{j+1}\epsilon_j^2$. It is clear that (1) $\epsilon_j = \bar{C}_0^{-(j+2)}(\frac{1}{2})^{2^j}$, (2) $\epsilon_j \to 0$ as $j \to \infty$, and (3) $\epsilon_{j+1} \leq \frac{1}{2}\epsilon_j \leq \epsilon_j - \epsilon_{j+1}$. We claim that if $r > 0$ is small enough then $\{\phi_j\}$ and $\{u_j\}$ have the following properties:

$(A_j)$: $\phi_j \in H_{1,r_j}$ and $\|\phi\|_{1,r_j} \leq \epsilon_0 - \epsilon_j$,

$(B_j)$: $\mathcal{F}(\phi_j; r, r_j) \in H_{0,r_j}$ and $\|\mathcal{F}(\phi_j; r, r_j)\|_{0,r_j} \leq \epsilon_j^2$,

$(C_j)$: $u_j \in H_{1,r_{j+1}}$ and $\|u_j\|_{1,r_{j+1}} \leq \epsilon_{j+1}$,

for $j = 0, 1, 2, \ldots$.

We prove this claim by induction on $j$. It is sufficient to show the following statements:

(1) $(A_0)$ and $(B_0)$ are true;

(2) $(A_j)$ and $(B_j)$ imply $(C_j)$;

(3) $(A_j)$ and $(C_j)$ imply $(A_{j+1})$;

(4) $(A_j)$, $(B_j)$, $(C_j)$, and $(A_{j+1})$ imply $(B_{j+1})$.

*Proof of (1).* $(A_0)$ is trivial since $\phi_0 = 0$. We note that $\mathcal{F}(0; r, r_0)(z) = -f(z)$ for $z \in \mathcal{D}_{r_0}$ and $r_0 = \frac{1}{2}$. Thus $\mathcal{F}(0; r, r_0) \in H_{0,r_0}$. Since $\|f\|_{0,r} = O(r^{4(\mu+n+3)+1})$ and $\epsilon_0^2 = O(r^{4(\mu+n+3)})$, we may choose $r > 0$ so small that $\|f\|_{0,r_0} \leq \epsilon_0^2$. Hence $(B_0)$ holds.

*Proof of (2).* We note that $\delta_j = (r_j - r_{j+1})/2 = (1/2^{j+4})r$. By using $(A_j)$, we have

$$\|D_z \phi_j\|_{0,r_j} \leq \frac{1}{2}.$$

Hence $I + D_z \phi_j$ has a bounded inverse for $z \in \mathcal{D}_{r_j}$ and

$$\|(I + D_z \phi_j)^{-1}\|_{0,r_j} \leq 2.$$

Then by the definition of $u_j$,

$$\|u_j\|_{0,r_j-\delta_j} = \|(I + D_z \phi_j) \tilde{K}_{r_j-\delta_j}^{-1} (I + D_z \phi_j)^{-1} \mathcal{F}(\phi_j; r, r_j)\|_{0,r_j-\delta_j}$$

$$\leq 4C\|\mathcal{F}(\phi_j; r, r_j)\|_{0,r_j} \delta_j^{-(n+\mu+2)} \leq 4C\delta_j^{-(n+\mu+2)}\epsilon_j^2.$$

$$\|D_z u_j\|_{0,r_{j+1}} \leq n|u_j|_{0,r_j-\delta_j}\delta_j^{-1} \leq 4nC\delta_j^{-(n+\mu+3)}\epsilon_j^2.$$

Thus

$$\|u_j\|_{1,r_{j+1}} \le 8nC\delta_j^{-(n+\mu+3)}\epsilon_j^2 = 8nC\left(\frac{2^{j+4}}{r}\right)^{n+\mu+3}\epsilon_j^2 \le \overline{C}_0^{j+1}\epsilon_j^2 = \epsilon_{j+1}.$$

*Proof of (3).* By using $(A_j)$ and $(C_j)$, $\phi_{j+1}$ is obviously analytic in $\mathscr{D}_{r_{j+1}}$ and

$$\|\phi_{j+1}\|_{1,r_{j+1}} \le \|\phi_j\|_{1,r_j} + \|u_j\|_{1,r_{j+1}} \le \epsilon_0 - \epsilon_j + \epsilon_{j+1} \le \epsilon_0 - \epsilon_{j+1}.$$

*Proof of (4).* By Taylor expansion, for any $\beta \in (0, r/2)$, $\phi \in \overline{B}_{1,\beta}(r/2)$ and $u \in H_{1,\beta}$ such that $\phi + u \in \overline{B}_{1,\beta}(r/2)$, we have the following equality,

$$\mathscr{F}(\phi + u; r, \beta) = \mathscr{F}(\phi; r, \beta) + D_\xi\mathscr{F}(\phi; r, \beta)u + R(\phi, u), \quad (2.14)$$

where

$$R(\phi, u) = \int_0^1 (1 - t)D_\xi^2\mathscr{F}(\phi + tu; r, \beta)(u^2)\,dt$$

$$= \int_0^1 (1 - t)D_z^2 f(I + \phi + tu)(u^2)\,dt,$$

$D_\xi^2\mathscr{F}$ is the second derivative of $\mathscr{F}$ with respect to $\xi$, and $D_z^2 f$ is the second derivative of $f$ with respect to $z$.

By Lemma 2.14,

$$D_\xi\mathscr{F}(\phi_j; r, r_{j+1})u_j - D_z\big(\mathscr{F}(\phi_j; r, r_{j+1})\big)(I + D_z\phi_j)^{-1}u_j$$

$$= (I + D_z\phi_j)K_{r_{j+1}}(I + D_z\phi_j)^{-1}u_j \quad \text{in } H_0, r_{j+1}.$$

Thus by the definition of $\{u_j\}$ we have

$$D_\xi\mathscr{F}(\phi_j; r, r_{j+1})u_j - D_z\big(\mathscr{F}(\phi_j; r, r_{j+1})\big)(I + D_z\phi_j)^{-1}u_j$$

$$= -\mathscr{F}(\phi_j; r, r_j) \quad \text{in } H_{0,r_{j+1}}. \quad (2.15)$$

Hence from (2.14),

$$\mathcal{F}(\phi_{j+1}; r, r_{j+1}) = D_z\big(\mathcal{F}(\phi_j; r, r_{j+1})\big)\big(I + D_z\phi_j\big)^{-1} u_j + R(\phi_j, u_j).$$

Then we have the following estimates:

$$\|D_z\big(\mathcal{F}(\phi_j; r, r_{j+1})\big)\big(I + D_z\phi_j\big)^{-1} u_j\|_{0, r_{j+1}}$$

$$\leq 2n\|\mathcal{F}(\phi_j; r, r_j)\|_{0, r_j}(r_j - r_{j+1})^{-1}\epsilon_{j+1}$$

$$\leq n\left(\frac{2^{j+4}}{r}\right)\epsilon_j^2 \epsilon_{j+1} \leq \frac{1}{2}\,\overline{C}_0^{j+1}\epsilon_j^2\epsilon_{j+1} = \frac{1}{2}\,\epsilon_{j+1}^2.$$

Since $\|I + \phi_j + tu_j\|_{0, r_{j+1}} < r$ for $t \in [0, 1]$, $R(\phi_j, u_j)$ is well defined. We note that $\|D_z^2 f\| \leq \frac{1}{2}$. Therefore

$$\|R(\phi_j, u_j)\|_{0, r_{j+1}} \leq \frac{1}{2}\,\epsilon_{j+1}^2.$$

Hence

$$\|\mathcal{F}(\phi_{j+1}; r, r_{j+1})\|_{0, r_{j+1}} \leq \epsilon_{j+1}^2.$$

Thus the claim is proved.

We note that $r_j > r/4$ for every $j \geq 0$. Then by Claim $(A_j)$, every $\phi_j \in \overline{B}_{1, r/4}(r/2)$. By Claim $(C_j)$, $\{\phi_j\}$ is a Cauchy sequence in $\overline{B}_{1, r/4}(r/2)$. Thus there exists $\overline{\xi} \in \overline{B}_{1, r/4}(r/2)$ such that $\overline{\xi} = \lim_{j \to \infty} \phi_j$. From $(B_j)$ we conclude that $\lim_{j \to \infty} \mathcal{F}(\phi_j; r, r/4) = 0$. Therefore $\mathcal{F}(\overline{\xi}; r, r/4) = 0$ since $\mathcal{F}$ is continuous in $\xi \in \overline{B}_{1, r/4}(r/2)$. This completes the proof of Siegel's Theorem. $\qquad\square$

## 2.3 Normal Forms of Equations with Periodic Coefficients

Consider the $T$-periodic differential equation

$$\dot{x} = f(t, x), \qquad x \in \mathbb{C}^n, \quad t \in \mathbb{R}, \tag{3.1}$$

where $f$ is continuous, $f(t, \cdot) \in C^{r+1}(\mathbb{C}^n, \mathbb{C}^n)$, $r \geq 2$, $f(t, 0) = 0$ for all

$t \in \mathbb{R}$, and there is $T > 0$ such that $f(t + T, x) = f(t, x)$ for all $t \in \mathbb{R}$, $x \in \mathbb{C}^n$. We may make a change of variables in (3.1) such that the resulting equation is simpler than (3.1).

Let $H_n^k$ be as in Section 2.1, and

$$H_{n,T}^{k,s} = \{f \in C^s(\mathbb{R} \times \mathbb{C}^n, \mathbb{C}^n) \mid f(t, \cdot) \in H_n^k \quad \text{for each} \quad t \in \mathbb{R};$$

$$f(t + T, x) = f(t, x) \quad \text{for all} \quad t \in \mathbb{R} \quad \text{and} \quad x \in \mathbb{C}^n\},$$

where $s$ is a nonnegative integer. When $s = 0$, we use $H_{n,T}^k$ instead of $H_{n,T}^{k,0}$. Each $H_{n,T}^{k,s}$ is a linear space. Suppose (3.1) is in the following form

$$\dot{x} = B(t)x + f^2(t, x) + \cdots + f^r(t, x) + O(|x|^{r+1}),$$

$$x \in \mathbb{C}^n, \quad (3.2)$$

where $B(t)$ is an $n \times n$ matrix with continuous $T$-periodic entries and $f^k \in H_{n,T}^k$, $k = 2, \ldots, r$, $r \geq 2$.

The linear part of equation (3.2) is

$$\dot{x} = B(t)x, \qquad x \in \mathbb{C}^n. \qquad (3.3)$$

Let $X(t)$ be the fundamental matrix of (3.3) with $X(0) = I$. Then $J = X(T)$ is a monodromy matrix of the $T$-periodic linear equation (3.3). Let $A$ be a constant $n \times n$ matrix such that $e^{AT} = J$. It is well known from Floquet theory that the nonsingular $T$-periodic transformation

$$x = P(t)y, \qquad (3.4)$$

where $P(t) = X(t)e^{-At}$, converts (3.3) to a linear system with constant coefficients, $\dot{y} = Ay$.

By transformation (3.4), (3.2) changes into the following:

$$\dot{x} = Ax + f^2(t, x) + \cdots + f^r(t, x) + O(|x|^{r+1}), \qquad x \in \mathbb{C}^n, \quad (3.5)$$

where $f^k \in H_{n,T}^k$, $k = 2, \ldots, r$. We note that the $f^k(t, x)$, $k = 2, \ldots, r$, in (3.4) may be different from those in (3.2). In the following, we discuss (3.5) instead of (3.2) since they are equivalent.

Now we change variables in (3.5) by a $T$-periodic transformation

$$x = y + h^k(t, y), \qquad y \in \Omega, \quad t \in \mathbb{R}, \qquad (3.6)$$

where $h^k \in H_{n,T}^{k,1}$, $2 \le k \le r$, and $\Omega$ is a neighborhood of the origin in $\mathbb{C}^n$ on which $I + h^k(t, \cdot)$ is invertible for each $t \in \mathbb{R}$. Substituting (3.6) into (3.5) we obtain:

$$\dot{y} = Ay + f^2(t, y) + \cdots + f^{k-1}(t, y)$$

$$+ \left( f^k(t, y) + Ah^k(t, y) - h_y^k(t, y) Ay - \frac{\partial}{\partial t} h^k(t, y) \right)$$

$$+ O(|y|^{k+1}), \qquad y \in \Omega. \tag{3.7}$$

Notice that transformation (3.6) does not affect the terms in (3.5) of order less than $k$ in $x$.

We define for each $k \ge 2$ an operator $\mathscr{L}_A^k \colon H_{n,T}^{k,1} \to H_{n,T}^k$ by

$$\mathscr{L}_A^k h(t, y) = \frac{\partial}{\partial t} h(t, y) + h_y(t, y) Ay - Ah(t, y), \qquad h \in H_{n,T}^{k,1}. \tag{3.8}$$

It is clear that $\mathscr{L}_A^k$ is linear. We recall that the operator $L_A^k \colon H_n^k \to H_n^k$ is defined by

$$L_A^k h(y) = h_y(y) Ay - Ah(y), \qquad h \in H_n^k.$$

Thus

$$\mathscr{L}_A^k h(t, \cdot) = \frac{\partial}{\partial t} h(t, \cdot) + L_A^k h(t, \cdot),$$

and (3.7) can be rewritten as

$$\dot{y} = Ay + f^2(t, y) + \cdots + f^{k-1}(t, y)$$

$$+ \left( f^k(t, y) - \mathscr{L}_A^k h^k(t, y) \right) + O(|y|^{k+1}), \qquad y \in \Omega.$$

Let $\mathscr{R}_T^k$ be the range of $\mathscr{L}_A^k$ in $H_{n,T}^k$, and $\mathscr{C}_T^k$ be a complementary subspace to $\mathscr{R}_T^k$ in $H_{n,T}^k$, that is,

$$H_{n,T}^k = \mathscr{R}_T^k \oplus \mathscr{C}_T^k. \tag{3.9}$$

We have then the following theorem.

**Theorem 3.1.** *Let the decompositions (3.9) be given for* $k = 2, \ldots, r$. *There exist a neighborhood* $\Omega$ *of the origin and a sequence of near identity* $T$-*periodic transformations* $x = y + h^k(t, y)$, $y \in \Omega$, $k = 2, \ldots, r$, *such that the resulting equation of (3.5) is of the form*:

$$\dot{y} = Ay + g^2(t, y) + \cdots + g^r(t, y) + O(|y|^{r+1}), \qquad y \in \Omega, \quad (3.10)$$

*where* $g^k \in \mathscr{C}_T^k$, $k = 2, \ldots, r$.

**Definition 3.2.** The truncated equation of (3.10)

$$\dot{y} = Ay + g^2(t, y) + \cdots + g^r(t, y)$$

is called an $A$-normal form up to order $r$ of (3.5).

As in Section 2.1, we may find $A$-normal forms by solving a system of partial differential equations or by using the matrix representation method. Let

$$C_T(\mathbb{R}, \mathbb{C}^d) = \{ f \in C^0(\mathbb{R}, \mathbb{C}^d) \mid f(t + T) = f(t) \text{ for all } t \in \mathbb{R} \}$$

and $C_T^r(\mathbb{R}, \mathbb{C}^d)$ be the linear space of all $C^r$ functions in $C_T(\mathbb{R}, \mathbb{C}^d)$. Recall also that

$$\{u_i(x)\}_{i=1}^{d_k} = \left\{ \frac{1}{\sqrt{\alpha!}} x^\alpha e_j \,\middle|\, |\alpha| = k, j = 1, \ldots, n \right\},$$

is an orthonormal basis for $H_n^k$, where $d_k = \dim H_n^k$ (see (1.18)) and $i \sim (j, \alpha)$ is in the reverse lexicographic ordering. Any element of $H_{n,T}^k$ is of the form

$$\sum_{i=1}^{d_k} p_i(t) u_i(x),$$

where $p_i \in C_T(\mathbb{R}, \mathbb{C})$. We then identify $H_{n,T}^k$ with $C_T(\mathbb{R}, \mathbb{C}^{d_k})$ in the following way:

$$\sum_{i=1}^{d_k} p_i(t) u_i(x) \rightarrow \sum_{i=1}^{d_k} p_i(t) \bar{e}_i, \qquad (3.11)$$

where $\{\bar{e}_i\}_{i=1}^{d_k}$ is the standard basis of $\mathbb{C}^{d_k}$. Then, the linear operator

$$\mathscr{L}_A^k : H_{n,T}^{k,1} \to H_{n,T}^k$$

gives, by the identification (3.11), the linear operator

$$\tilde{\mathscr{L}}_A^k : C_T^1(\mathbb{R}, \mathbb{C}^{d_k}) \to C_T(\mathbb{R}, \mathbb{C}^{d_k})$$

defined by

$$(\tilde{\mathscr{L}}_A^k f)(t) = \frac{d}{dt} f(t) + \tilde{L}_A^k f(t), \qquad f \in C_T^1(\mathbb{R}, \mathbb{C}^{d_k}), \quad (3.12)$$

where $\tilde{L}_A^k : \mathbb{C}^{d_k} \to \mathbb{C}^{d_k}$ is the matrix representation of $L_A^k$ with respect to the basis $\{(1/\sqrt{\alpha!}) x^\alpha e_j\}$. Let $\tilde{\mathscr{R}}_T^k$ be the range of $\tilde{\mathscr{L}}_A^k$ and $\tilde{\mathscr{C}}_T^k$ be a complementary subspace to $\tilde{\mathscr{R}}_T^k$ in $C_T(\mathbb{R}, \mathbb{C}^{d_k})$. Then the range of $\mathscr{L}_A^k$ is

$$\mathscr{R}_T^k = \left\{ f(t,x) = \sum_{i=1}^{d_k} p_i(t) u_i(x) \,\middle|\, (p_1(t), \ldots, p_{d_k}(t))^T \in \tilde{\mathscr{R}}_T^k \right\}, \quad (3.13)$$

and

$$\mathscr{C}_T^k = \left\{ f(t,x) = \sum_{i=1}^{d_k} p_i(t) u_i(x) \,\middle|\, (p_1(t), \ldots, p_{d_k}(t))^T \in \tilde{\mathscr{C}}_T^k \right\} \quad (3.14)$$

is a complementary subspace to $\mathscr{R}_T^k$ in $H_{n,T}^k$.

Let $(\cdot, \cdot)_T$ be an inner product on $C_T(\mathbb{R}, \mathbb{C}^{d_k})$ that is defined as follows: For any $f, g \in C_T(\mathbb{R}, \mathbb{C}^{d_k})$,

$$(f, g)_T = \frac{1}{T} \int_0^T (f(t), g(t)) \, dt,$$

where $(\cdot, \cdot)$ is the usual inner product in $\mathbb{C}^{d_k}$.

**Theorem 3.3.** *The space of T-periodic solutions of the equation*

$$\dot{g}(t) = (\tilde{L}_A^k)^* g(t), \qquad g \in C_T^1(\mathbb{R}, \mathbb{C}^{d_k}), \quad (3.15)$$

*is an orthogonal complementary subspace $\tilde{\mathscr{C}}_T^k$ to $\tilde{\mathscr{R}}_T^k$ with respect to the inner product $(\cdot, \cdot)_T$ in $C_T(\mathbb{R}, \mathbb{C}^{d_k})$, $k = 2, \ldots, r$.*

Recall that $L_{A*}^k$ is the adjoint operator of $L_A^k$ with respect to the inner product $\langle \cdot, \cdot \rangle$ in $H_n^k$. We define the inner product $\langle \cdot, \cdot \rangle_T$ on $H_{n,T}^k$ by

$$\langle f(t,x), g(t,x) \rangle_T = \frac{1}{T} \int_0^T \langle f(t,x), g(t,x) \rangle dt, \qquad f, g \in H_{n,T}^k.$$

We have the following:

**Theorem 3.4.** *The linear operator* $(\mathscr{L}_A^k)^*\colon H_{n,T}^{k,1} \to H_{n,T}^k$ *defined by*

$$((\mathscr{L}_A^k)^* h)(t,x) = -\frac{\partial}{\partial t} h(t,x) + L_{A*}^k h(t,x), \qquad h \in H_{n,T}^{k,1}, \quad (3.16)$$

*is the adjoint operator of* $\mathscr{L}_A^k$ *with respect to the inner product* $\langle \cdot, \cdot \rangle_T$ *in* $H_{n,T}^k$.

**Lemma 3.5.** *Let* $A = \mathrm{diag}(\lambda_1, \ldots, \lambda_n)$ *and* $f \in C_T^1(\mathbb{R}, \mathbb{C})$. *Then* $f(t)x^\alpha e_j \in \mathrm{Ker}(\mathscr{L}_A^k)^*$ *if and only if there exists an integer m such that*

$$\lambda \cdot \alpha - \lambda_j = \frac{2m\pi}{T} i, \qquad i = \sqrt{-1},$$

*and*

$$f(t) = ce^{\frac{2m\pi}{T} it},$$

*where c is a constant.*

**Definition 3.6.** *If* $\sigma(A) = \{\lambda_1, \ldots, \lambda_n\}$ *is the spectrum of* $A$, *then the following relations are called resonant conditions:*

$$\lambda \cdot \alpha - \lambda_j = \frac{2m\pi}{T} i, \qquad i = \sqrt{-1}, \quad m \in \mathbb{Z}, \quad |\alpha| \geq 2, \quad (3.17)$$

*where* $\mathbb{Z}$ *denotes the set of all integers.* Let $(x_1, x_2, \ldots, x_n)$ be coordinates with respect to the standard basis $\{e_1, \ldots, e_n\}$ of $\mathbb{C}^n$ in which the matrix $A$ has a Jordan normal form with diagonal elements $\{\lambda_1, \ldots, \lambda_n\}$.

Then a monomial $\exp(\frac{2m\pi it}{T})x^\alpha e_j(|\alpha| = k \geq 2, 1 \leq j \leq n)$ is called a resonant monomial of order $k$ if and only if there exists an integer $m$ such that (3.17) is satisfied for $\alpha$ and $j$.

Let $A = S + N$ be the $S - N$ decomposition of $A$. Then it is easy to see that $\mathscr{L}_A^k = \mathscr{L}_S^k + L_N^k$ is the $S - N$ decomposition of $\mathscr{L}_A^k$. So we have the following:

**Theorem 3.7.** *If $A = \mathrm{diag}(\lambda_1, \ldots, \lambda_n)$, then an A-normal form up to order $r \geq 2$ can be chosen so that its nonlinear part consists of all resonant monomials up to order $r$. If $A$ is upper (or lower) triangular with diagonal elements $\{\lambda_1, \ldots, \lambda_n\}$, then an A-normal form up to order $r \geq 2$ can be chosen so that its nonlinear part is spanned by resonant monomials up to order $r$.*

**Remark 3.8.** If we consider a $T$-periodic system over the reals, then the above discussion is valid except for the following. It is well known that we cannot always find a real matrix $A$ such that $e^{AT} = J$, but we can always find a real matrix $A$ such that $e^{2AT} = J^2$. In this case there is a real $2T$-periodic transformation $x = P(t)y$ such that the equation (3.2) is changed to $\dot{y} = Ay$ and (3.1) is changed to a $2T$-periodic system over reals. We note that such a $2T$-periodic system has some kind of symmetry. We will discuss normal forms of equations with symmetry in Section 2.5.

If $A$ is diagonalizable over the complex numbers, then we cannot apply Theorem 3.7 directly. But we can use the method for getting $A$-normal forms of real equations described in Section 2.1. We illustrate the method in the following example.

**Example 3.9.** Assume that the $T$-periodic system (3.1) is real and two-dimensional and the monodromy matrix of (3.3) is

$$J = \begin{bmatrix} \cos \omega T & -\sin \omega T \\ \sin \omega T & \cos \omega T \end{bmatrix}, \qquad 0 \leq \frac{\omega T}{2\pi} = \frac{p}{q} < 1,$$

where $p$ and $q$ are positive integers with $(p, q) = 1$. Then after a Floquet change of coordinates, equation (3.1) is transformed into (3.5)

where

$$A = \begin{bmatrix} 0 & -\omega \\ \omega & 0 \end{bmatrix}.$$

We make a transformation to complex coordinates by $z_1 = x_1 + ix_2$, $z_2 = x_1 - ix_2$. Then the resulting equation of the form of (3.5) is

$$\dot{z} = \Lambda z + g^2(t, z) + \cdots + g^r(t, z) + O(|z|^{r+1}), \qquad (3.18)$$

where $z = (z_1, z_2)^T$, $\Lambda = \begin{bmatrix} \omega i & 0 \\ 0 & -\omega i \end{bmatrix}$, and $g^k(t, z) \in H_{2,T}^k$ for $2 \le k \le r$. We note that the second equation of (3.18) is conjugate to the first one. The resonance conditions in our case are

$$\begin{cases} \alpha_1 \omega i - \alpha_2 \omega i - \omega i = m \dfrac{2\pi i}{T} & \text{for } j = 1, \quad |\alpha| = k \ge 2, \quad m \in \mathbb{Z}, \\[2mm] \alpha_1 \omega i - \alpha_2 \omega i + \omega i = m \dfrac{2\pi i}{T} & \text{for } j = 2, \quad |\alpha| = k \ge 2, \quad m \in \mathbb{Z}, \end{cases}$$

which are equivalent to

$$(\alpha_1 - \alpha_2 - 1)\frac{p}{q} = m \qquad \text{for } j = 1, \quad |\alpha| = k \ge 2, \quad m \in \mathbb{Z},$$

$$(\alpha_1 - \alpha_2 + 1)\frac{p}{q} = m \qquad \text{for } j = 2, \quad |\alpha| = k \ge 2, \quad m \in \mathbb{Z}.$$

The only possibilities to get resonant monomials when $2 \le k \le q$ are: for $j = 1$, when $\alpha_1 - \alpha_2 - 1 = 0$ and then $m = 0$, or $\alpha_1 - \alpha_2 - 1 = -q$ and then $k = q - 1$, $m = -p$; for $j = 2$, when $\alpha_1 - \alpha_2 + 1 = 0$ and then $m = 0$, or $\alpha_1 - \alpha_2 + 1 = q$ and then $k = q - 1$, $m = p$. Therefore the resonant monomials up to order $q$ are: $\{z_1^{\alpha_1} z_2^{\alpha_2} e_1 \mid \alpha_1 - \alpha_2 = 1, 2 \le \alpha_1 + \alpha_2 \le q\} \cup \{z_1^{\alpha_1} z_2^{\alpha_2} e_2 \mid \alpha_2 - \alpha_1 = 1, 2 \le \alpha_1 + \alpha_2 \le q\} \cup \{z_2^{q-1} e_1, z_1^{q-1} e_2\}$. The coefficient of any resonant monomial will be $c \exp(-m\frac{q}{p}\omega i t)$, where $c$ is a complex constant. Hence a $\Lambda$-normal form of (3.18) up to order $q$ is:

$$\begin{cases} \dot{z}_1 = \omega i z_1 + c_1 z_1^2 z_2 + \cdots + c_k z_1^{k+1} z_2^k + d e^{q\omega i t} z_2^{q-1}, \\ \dot{z}_2 = -\omega i z_2 + \bar{c}_1 z_1 z_2^2 + \cdots + \bar{c}_k z_1^k z_2^{k+1} + \bar{d} e^{-q\omega i t} z_1^{q-1}, \end{cases} \quad (3.19)$$

where $\bar{z}_2 = z_1, c_1, \ldots, c_k,$ $d$ are complex constants, $q - 1 \le 2k + 1 \le q$.

If we let $w = z_1$, $\bar{w} = z_2$, then from (3.19) we get

$$\dot{w} = \omega i w + c_1 |w|^2 w + \cdots + c_k |w|^{2k} w + d e^{q\omega i t} \bar{w}^{q-1}. \quad (3.20)$$

The equation for $\bar{w}$ is omitted since it is conjugate to the equation for $w$.

We can obtain a real $A$-normal form from (3.20) by applying the transformation $w = x_1 + i x_2$. We can also apply the transformation $w = r e^{i\theta}$ to (3.20) to get real normal forms in polar coordinates. For example, the normal form in polar coordinates is

$$\begin{cases} \dot{r} = a_1 r^3 + \cdots + a_k r^{2k+1} \\ \qquad + r^{q-1} \left( d_1 \cos\left( \frac{2p\pi}{T} t - q\theta \right) - d_2 \sin\left( \frac{2p\pi}{T} t - q\theta \right) \right), \\ \dot{\theta} = \omega + b_1 r^2 + \cdots + b_k r^{2k} \\ \qquad + r^{q-2} \left( d_2 \cos\left( \frac{2p\pi}{T} t - q\theta \right) + d_1 \sin\left( \frac{2p\pi}{T} t - q\theta \right) \right), \end{cases}$$

where $a_i = \mathrm{Re}(c_i)$, $b_i = \mathrm{Im}(c_i)$, $d_1 = \mathrm{Re}(d)$, and $d_2 = \mathrm{Im}(d)$.

Now if we let $w = v e^{\omega i t}$, (3.20) becomes

$$\dot{v} = c_1 |v|^2 v + \cdots + c_k |v|^{2k} v + d \bar{v}^{q-1}. \quad (3.21)$$

It is simpler than (3.20), but we note that the original equation (3.1) is a $qT$-periodic perturbation of (3.21). If we change (3.21) to polar coordinates, then (3.21) becomes

$$\begin{cases} \dot{r} = a_1 r^3 + \cdots + a_k r^{2k+1} + r^{q-1}(d_1 \cos(q\theta) + d_2 \sin(q\theta)), \\ \dot{\theta} = b_1 r^2 + \cdots + b_k r^{2k} + r^{q-2}(d_2 \cos(q\theta) - d_1 \sin(q\theta)), \end{cases}$$

where $a_i = \mathrm{Re}(c_i)$, $b_i = \mathrm{Im}(c_i)$, $d_1 = \mathrm{Re}(d)$, and $d_2 = \mathrm{Im}(d)$.

## 2.4 Normal Forms of Maps near a Fixed Point

Consider a $C^{r+1}$ map $F: \Omega \subseteq \mathbb{C}^n \to \mathbb{C}^n$, where $r \geq 2$ and $\Omega$ is a neighborhood of the origin in $\mathbb{C}^n$. We assume that the origin is a fixed point of $F(x)$, that is, $F(0) = 0$. Then $F(x)$ can be written as

$$F(x) = Ax + f^2(x) + f^3(x) + \cdots + f^r(x) + O(|x|^{r+1}),$$

$$x \in \Omega, \quad \text{as } x \to 0, \quad (4.1)$$

where $A$ is an $n \times n$ constant matrix and $f^k \in H_n^k$ for $2 \leq k \leq r$.

Now we change variables in (4.1) by

$$x = H(y) \equiv y + h^k(y), \qquad y \in \Omega_k, \quad (4.2)$$

where $h^k \in H_n^k$, $2 \leq k \leq r$, and $\Omega_k \subseteq \Omega$ is a neighborhood of the origin in $\mathbb{C}^n$ on which $I + h^k(\cdot)$ is invertible. The inverse transformation to (4.2),

$$y = x - h^k(x) + O(|x|^{k+1}), \quad \text{as } x \to 0,$$

is a smooth diffeomorphism in $\Omega_k$. The transformed map of (4.1), $G = H^{-1} \circ F \circ H$, will take the form:

$$G(y) = Ay + f^2(y) + \cdots + f^{k-1}(y)$$

$$(4.3)$$

$$+ \left[ f^k(y) + Ah^k(y) - h^k(Ay) \right] + O(|y|^{k+1}), \quad \text{as } y \to 0,$$

where $G(y)$ is defined in the neighborhood $\Omega_k$. We note that transformation (4.2) does not affect the terms in (4.1) with order $\leq k - 1$.

We define the operator $L_A^k: H_n^k \to H_n^k$ by

$$L_A^k h(x) = h(Ax) - Ah(x), \qquad h \in H_n^k, \quad (4.4)$$

and let $\mathcal{R}^k$ be the range of $L_A^k$ in $H_n^k$, and $\mathcal{C}^k$ be any complementary subspace to $\mathcal{R}^k$ in $H_n^k$, that is,

$$H_n^k = \mathcal{R}^k \oplus \mathcal{C}^k. \quad (4.5)$$

Notice that the operator $L_A^k$ is different from the one in Section 2.1. We have the following theorem.

**Theorem 4.1.** *Suppose that the decompositions (4.5) are given for $k = 2, \ldots, r$. There exists a sequence of near identity transformations,*

$$x = y + h^k(y), \qquad y \in \tilde{\Omega}, \quad 2 \le k \le r,$$

*where $h^k \in H_n^k$ and $\tilde{\Omega} \subseteq \Omega$ is a neighborhood of the origin in $\mathbb{C}^n$, such that the map (4.1) takes the form*

$$G(y) = Ay + g^2(y) + \cdots + g^r(y) + O(|y|^{r+1}), \qquad y \in \tilde{\Omega}, \quad (4.6)$$

*where $g^k(y) \in \mathscr{C}^k, 2 \le k \le r$.*

**Definition 4.2.** The truncated form of the map (4.6),

$$G(y) = Ay + g^2(y) + \cdots + g^r(y),$$

is called an $A$-normal form of (4.1) up to order $r$.

**Lemma 4.3.** *If $p, q \in H_n^k$ and $A$ is an $n \times n$ matrix, then*

*(1)* $$\langle p(Ax), q(x) \rangle = \langle p(x), q(A^*x) \rangle,$$

*(2)* $$\langle Ap(x), q(x) \rangle = \langle p(x), A^*q(x) \rangle,$$

*where $A^*$ is the adjoint operator of $A$ with respect to the inner product $(\cdot, \cdot)$ in $\mathbb{C}^n$.*

*Proof.* To prove (1) it is sufficient to show that

$$\langle (Ax)^\alpha e_i, x^\beta e_i \rangle = \langle x^\alpha e_i, (A^*x)^\beta e_i \rangle$$

or equivalently that

$$\langle (Ax)^\alpha e_i, x^\beta e_i \rangle = \langle (A^Tx)^\beta e_i, x^\alpha e_i \rangle$$

for $|\alpha| = k, |\beta| = k$, and $1 \le i \le n$.

We have

$$\langle (Ax)^\alpha e_i, x^\beta e_i \rangle$$

$$= \left\langle \prod_{j=1}^n (a_{j1}x_1 + a_{j2}x_2 + \cdots + a_{jn}x_n)^{\alpha_j} e_i, x^\beta e_i \right\rangle = \beta! c_\alpha,$$

where $c_\alpha$ is the coefficient of $x^\beta$ in the expansion of $(Ax)^\alpha$. Similarly,

$$\langle (A^T x)^\beta e_i, x^\alpha e_i \rangle$$

$$= \left\langle \prod_{j=1}^n (a_{1j}x_1 + a_{2j}x_2 + \cdots + a_{nj}x_n)^{\beta_j} e_i, x^\alpha e_i \right\rangle = \alpha! c_\beta,$$

where $c_\beta$ is the coefficient of $x^\alpha$ in the expansion of $(A^T x)^\beta$. It can be shown that $\beta! c_\alpha = \alpha! c_\beta$ by the Binomial Theorem and elementary calculations. The proof of (1) is then complete. The proof of (2) is trivial. $\qquad\square$

**Theorem 4.4.** $L^k_{A*}$ *is the adjoint operator of* $L^k_A$ *with respect to the inner product* $\langle \cdot, \cdot \rangle$ *in* $H^k_n$ *for each* $k \geq 2$*, where* $A^*$ *is the adjoint operator of* $A$ *with respect to the inner product* $(\cdot, \cdot)$ *in* $\mathbb{C}^n$*.*

*Proof.* By Lemma 4.3, for any $p, q \in H^k_n$ we have

$$\langle L^k_A p(x), q(x) \rangle = \langle p(Ax) - Ap(x), q(x) \rangle$$

$$= \langle p(Ax), q(x) \rangle - \langle Ap(x), q(x) \rangle$$

$$= \langle p(x), q(A^*x) \rangle - \langle p(x), A^*q(x) \rangle$$

$$= \langle p(x), L^k_{A*} q(x) \rangle. \qquad\square$$

**Corollary 4.5.** Ker$(L^k_{A*})$ *is the orthogonal complementary subspace to* $\mathscr{R}^k$ *with respect to the inner product* $\langle \cdot, \cdot \rangle$ *in* $H^k_n$ *for* $k \geq 2$*.*

**Definition 4.6.** Let $\sigma(A) = \{\lambda_1, \ldots, \lambda_n\} \subset \mathbb{C}$ be the spectrum of $A$. Then the following relations are called resonant conditions:

$$\lambda^\alpha = \lambda_j, \tag{4.7}$$

where $\lambda^\alpha = \lambda_1^{\alpha_1} \cdots \lambda_n^{\alpha_n}$, $|\alpha| \geq 2$. Let $(x_1, \ldots, x_j)$ be coordinates with respect to the standard basis $\{e_1, \ldots, e_n\}$ of $\mathbb{C}^n$ in which the matrix $A$ has a Jordan normal form with diagonal elements $\{\lambda_1, \ldots, \lambda_n\}$. Then a monomial $x^\alpha e_j$ ($|\alpha| = k \geq 2$ and $1 \leq j \leq n$) is called a resonant monomial of order $k$ if and only if (4.7) holds for $\alpha$ and $j$.

**Theorem 4.7.** *If* $A = \mathrm{diag}(\lambda_1, \ldots, \lambda_n)$, *then an $A$-normal form up to order $r \geq 2$ can be chosen so that its nonlinear part consists of all resonant monomials up to order $r$.*

As in Section 2.1, we can apply also the matrix representation method to compute $A$-normal forms of maps. Let $\tilde{L}_A^k$ be the matrix representation of $L_A^k$ with respect to the basis $U_k$ (see Section 2.1) of $H_n^k$. Then we have the following:

**Theorem 4.8.** *If* $A = \mathrm{diag}(\lambda_1, \ldots, \lambda_n)$ *then $\tilde{L}_A^k$ is diagonal; if $A$ is upper (or lower) triangular with the diagonal elements $\{\lambda_1, \ldots, \lambda_n\}$, then $\tilde{L}_A^k$ is lower (or upper) triangular and if $u_i(x) = x^\alpha e_j$ is the ith element of basis $U_k$, then the ith element of the diagonal of $\tilde{L}_A^k$ is $\lambda^\alpha - \lambda_j$.*

Let $A = S + N$ be the $S - N$ decomposition of $A$. We define the operator $\mathcal{N}^k \colon H_n^k \to H_n^k$ by

$$\mathcal{N}^k h(x) = h(Ax) - h(Sx) - Nh(x), \qquad h \in H_n^k.$$

**Theorem 4.9.** *If* $A = S + N$ *is the $S - N$ decomposition of $A$, then $L_A^k = L_S^k + \mathcal{N}^k$ is the $S - N$ decomposition of $L_A^k$.*

**Corollary 4.10.** *If* $A = S + N$ *is the $S - N$ decomposition of $A$ and $S = \mathrm{diag}(\lambda_1, \ldots, \lambda_n)$, then an $A$ normal form up to order $r$ can be chosen so that its nonlinear part is spanned by resonant monomials up to order $r$.*

Resonant monomials have the following symmetry property.

**Lemma 4.11.** *Let* $A = S + N$ *be the* $S - N$ *decomposition of* $A$ *and* $S = \mathrm{diag}(\lambda_1, \ldots, \lambda_n)$. *Then every resonant monomial* $\gamma(x) = x^\alpha e_j$ *commutes with* $S$, *that is,*

$$S\gamma(x) = \gamma(Sx).$$

**Corollary 4.12.** *If* $A = S + N$ *is the* $S - N$ *decomposition of* $A$, *then an* $A$-*normal form up to order* $r$ *can be chosen so that it commutes with* $S$.

*Proof.* We consider first the case that $S$ is diagonal. By Corollary 4.10, an $A$-normal form can be chosen to contain only resonant monomials in its nonlinear part. Then the desired conclusion follows from Lemma 4.11.

Suppose now that $S$ is not diagonal. Let $P$ be a nonsingular transformation such that $A_0 = P^{-1}AP$ is in upper triangular Jordan form and $A_0 = S_0 + N_0$ is the $S - N$ decomposition of $A_0$, where $S_0$ is diagonal and $N_0$ is strictly upper triangular. Then $A = PS_0P^{-1} + PN_0P^{-1}$ is the $S - N$ decomposition of $A$. By the uniqueness of such a decomposition, $S = PS_0P^{-1}$ and $N = PN_0P^{-1}$. From Corollary 4.10 and Lemma 4.11 there is an $A_0$-normal form

$$F(x) = A_0 x + f^2(x) + \cdots + f^r(x), \tag{4.8}$$

so

$$S_0 f^k(x) = f^k(S_0 x), \qquad k = 2, \ldots, r.$$

Then we change variables in (4.8) by $x = P^{-1}y$. We get an $A$-normal form

$$G(y) = Ay + g^2(y) + \cdots + g^k(y),$$

where $G(y) = PF(P^{-1}y)$ and $g^k(y) = Pf^k(P^{-1}y)$, $k = 2, \ldots, r$. The nonlinear terms of $G$ satisfy

$$g^k(Sy) = Pf^k(P^{-1}Sy) = Pf^k(S_0P^{-1}y)$$

$$= PS_0f^k(P^{-1}y) = SPf^k(P^{-1}y) = Sg^k(y),$$

for $k = 2, \ldots, r$. And the matrix $A$ commutes with $S$ obviously. Thus the theorem is proved.                                    $\square$

The above results are valid for normal forms of maps on $\mathbb{R}^n$.

**Example 4.13.** Consider a mapping $F(x) = -x + O(|x|^2)$ as $x \to 0$ from $\mathbb{R}$ to $\mathbb{R}$. $\lambda = -1$ is the only eigenvalue. Then the resonant conditions are

$$\lambda^k - \lambda = 0, \qquad k \ge 2,$$

that is,

$$(-1)^{k-1} = 1, \qquad k \ge 2.$$

Hence, the resonant monomials are $x^3, x^5, \ldots, x^{2k+1}, \ldots, k \ge 1$. Thus the normal form up to order 4 is

$$G(x) = -x + ax^3,$$

where $a$ is a real constant.

The next example illustrates the use of the matrix representation method.

**Example 4.14.** Suppose that the matrix $A$ of the linear part of a nonlinear map from $\mathbb{R}^2$ to $\mathbb{R}^2$ is

$$A = \begin{bmatrix} -1 & 1 \\ 0 & -1 \end{bmatrix}.$$

The resonant conditions are $(-1)^{\alpha_1+\alpha_2} = -1$, $j = 1, 2$, for $k = \alpha_1 + \alpha_2 \ge 2$. Hence there are no resonant monomials of even order and every monomial of odd order is resonant. Therefore we need only to find a basis of a complementary subspace $\mathscr{C}^k$ for $k$ odd. The matrix representation of $L_A^k$ is taken with respect to the following basis of $H_2^k$:

$$\{u_{\beta+1}(x, y) = x^{k-\beta}y^\beta e_2, \quad \beta = 0, \dots, k;$$

$$u_{\beta+k+2}(x, y) = x^{k-\beta}y^\beta e_1, \qquad \beta = 0, \dots, k\},$$

where $\{e_1, e_2\}$ is the standard basis of $\mathbb{R}^2$. We have to compute $L_A^k u_i$ for $i = 1, \dots, d_2 = 2(k + 1)$.

$$L_A^k\begin{bmatrix} x^{k-\beta}y^\beta \\ 0 \end{bmatrix}$$

$$= \begin{bmatrix} (-x + y)^{k-\beta}(-y)^\beta \\ 0 \end{bmatrix} - \begin{bmatrix} -1 & 1 \\ 0 & -1 \end{bmatrix}\begin{bmatrix} x^{k-\beta}y^\beta \\ 0 \end{bmatrix}$$

$$= (-1)^k\begin{bmatrix} ((-1)^k + 1)x^{k-\beta}y^\beta + \displaystyle\sum_{j=1}^{k-\beta}(-1)^j\binom{k-\beta}{j}x^{k-\beta-j}y^{\beta+j} \\ 0 \end{bmatrix}.$$

$$L_A^k\begin{bmatrix} 0 \\ x^{k-\beta}y^\beta \end{bmatrix}$$

$$= \begin{bmatrix} 0 \\ (-x + y)^{k-\beta}(-y)^\beta \end{bmatrix} - \begin{bmatrix} -1 & 1 \\ 0 & -1 \end{bmatrix}\begin{bmatrix} 0 \\ x^{k-\beta}y^\beta \end{bmatrix}$$

$$= (-1)^k\begin{bmatrix} (-1)^{k+1}x^{k-\beta}y^\beta \\ ((-1)^k + 1)x^{k-\beta}y^\beta + \displaystyle\sum_{j=1}^{k-\beta}(-1)^j\binom{k-\beta}{j}x^{k-\beta-j}y^{\beta+j} \end{bmatrix}.$$

In terms of the basis of $H_2^k$ we have, for odd $k \geq 3$:

$$L_A^k u_{\beta+1} = -u_{\beta+k+2} + \sum_{j=1}^{k-\beta}(-1)^{j+1}\binom{k-\beta}{j}u_{\beta+j+1}, \quad \beta = 0, \dots, k,$$

$$L_A^k u_{\beta+k+1} = \sum_{j=1}^{k-\beta}(-1)^{j+1}\binom{k-\beta}{j}u_{\beta+j+k+2}, \qquad \beta = 0, \dots, k.$$

Therefore we get the following matrix representations of $L_A^k$, for $k$ odd:

$\tilde{L}_A^k$ is the following $2(k+1) \times 2(k+1)$ matrix

$$
\begin{bmatrix}
0 & 0 & 0 & \cdots & 0 & 0 \\
\binom{k}{1} & 0 & 0 & \cdots & 0 & 0 \\
-\binom{k}{2} & \binom{k-1}{1} & 0 & \cdots & 0 & 0 \\
\binom{k}{3} & -\binom{k-1}{2} & \binom{k-2}{1} & \cdots & 0 & 0 & & & & & O \\
\vdots & \vdots & \vdots & \cdots & \vdots & \vdots \\
1 & -1 & 1 & \cdots & 1 & 0 \\
-1 & 0 & 0 & \cdots & 0 & 0 & 0 & 0 & 0 & \cdots & 0 & 0 \\
0 & -1 & 0 & \cdots & 0 & 0 & \binom{k}{1} & 0 & 0 & \cdots & 0 & 0 \\
0 & 0 & -1 & \cdots & 0 & 0 & -\binom{k}{2} & \binom{k-1}{1} & 0 & \cdots & 0 & 0 \\
0 & 0 & 0 & \cdots & 0 & 0 & \binom{k}{3} & -\binom{k-1}{2} & \binom{k-2}{1} & \cdots & 0 & 0 \\
\vdots & \vdots & \vdots & \cdots & \vdots & \vdots & \vdots & \vdots & \vdots & \cdots & \vdots & \vdots \\
0 & 0 & 0 & \cdots & 0 & -1 & 1 & -1 & 1 & \cdots & 1 & 0
\end{bmatrix}.
$$

It follows that for $k$ odd a basis of $\text{Ker}(\tilde{L}_A^k)^*$ can be chosen as $\{\bar{e}_1, \bar{e}_2 + k\bar{e}_{k+2}\}$, where $\{\bar{e}_1, \ldots, \bar{e}_{2(k+1)}\}$ is the standard basis of $\mathbb{R}^{2(k+1)}$. We may also take $\{\bar{e}_1, \bar{e}_2\}$ as a basis for a complementary subspace $\tilde{\mathscr{C}}^k$, $k \geq 3$, odd. Hence a normal form up to order $r \geq 3$ is

$$
G(x, y) = \begin{bmatrix} -x + y \\ -y + \sum_{k=1}^{m} \left( a_k x^{2k+1} + b_k x^{2k} y \right) \end{bmatrix},
$$

where $a_k, b_k \in \mathbb{R}$ are real constants, and $r - 1 \leq 2m + 1 \leq r$.

.

## 2.5 Normal Forms of Equations with Symmetry

In this section, we consider equations with symmetry and their normal forms.

**Definition 5.1.** Let $S$ be an invertible $n \times n$ matrix. We say that the equation

$$
\dot{x} = f(x), \qquad x \in \Omega \subseteq \mathbb{C}^n, \tag{5.1}
$$

where $\Omega$ is a neighborhood of the origin of $\mathbb{C}^n$, and $f \in C^r(\Omega, \mathbb{C}^n)$ $(r \geq 1)$, has $S$-symmetry if and only if

$$f(Sx) = Sf(x), \qquad \text{for all } x \in \Omega. \qquad (5.2)$$

That (5.1) has $S$-symmetry is equivalent to the fact that (5.1) is invariant under the transformation $x \to Sx$.

From the definition, for a fixed $f \in C^r(\Omega, \mathbb{C}^n)$, if matrix $S$ satisfies (5.2), then so does $S^{-1}$; if $S_1, S_2$ both satisfy (5.2), then so does $S_1 \cdot S_2$. Thus the set $\Gamma$ of all $n \times n$ matrices which satisfy the relation (5.2) forms a group under matrix multiplication.

**Definition 5.2.** Let $\Gamma$ be a group of $n \times n$ matrices. If the right-hand side of equation (5.1) satisfies

$$f(Sx) = Sf(x), \qquad \text{for all } x \in \Omega \subseteq \mathbb{C}^n \text{ and any } S \in \Gamma,$$

then we say that (5.1) has the group $\Gamma$-symmetry.

**Example 5.3.** The following are some examples of symmetry groups:
(1) $O(n)$, the $n$-dimensional orthogonal group, which consists of all $n \times n$ orthogonal matrices;
(2) $SO(n)$, the $n$-dimensional special orthogonal group, which consists of all $n \times n$ orthogonal matrices whose determinants are equal to 1, $SO(n)$ is also called the $n$-dimensional rotation group;
(3) $Z_q$, the group generated by $S = K_{2\pi/q} = \begin{bmatrix} \cos\frac{2\pi}{q} & -\sin\frac{2\pi}{q} \\ \sin\frac{2\pi}{q} & \cos\frac{2\pi}{q} \end{bmatrix}$, where $q$ is a positive integer;
(4) the flip group generated by $K_n = \begin{bmatrix} I_{n-1} & 0 \\ 0 & -1 \end{bmatrix}$, where $I_{n-1}$ is the $(n-1) \times (n-1)$ identity matrix;
(5) $D_q$, the dihedral group, generated by $\{K_{2\pi/q}, K_2\}$, where $K_{2\pi/q}$ is defined in (3) and $K_2$ is defined in (4).

In what follows, we will discuss $A$-normal forms of equations with $S$-symmetry, where $S$ is an invertible $n \times n$ matrix.

**Lemma 5.4.** *If equation (5.1) has $S$-symmetry and is of the following form*

$$\dot{x} = Ax + f^2(x) + \cdots + f^k(x) + O(|x|^{k+1}), \qquad x \in \Omega, \quad (5.3)$$

*where $f^k \in H_n^k$, $k = 2, \ldots, r$, then*

(i)  $SA = AS$;
(ii) $f^k(Sx) = Sf^k(x)$, *for any $x \in \Omega$, $k = 2, 3, \ldots, r$.*

*Proof.* It follows from Definition 5.1.                                    □

**Lemma 5.5.** *The S-symmetry of an equation is invariant under S-symmetrical transformations of variables.*

*Proof.* Suppose that (5.1) has $S$-symmetry. We change variables in (5.1) by

$$x = h(y), \qquad y \in \Omega, \tag{5.4}$$

where $h(y)$ is a diffeomorphism on $\Omega$ with the property $h(Sy) = Sh(y)$ for any $y \in \Omega$. The resulting equation after the change of variables (5.4) is

$$\dot{y} = \left(h_y(y)\right)^{-1} f(h(y)), \qquad y \in \Omega.$$

Let

$$g(y) = \left(h_y(y)\right)^{-1} f(h(y)).$$

Then

$$g(Sy) = \left(h_y(Sy)\right)^{-1} f(h(Sy)) = \left(Sh_y(y)S^{-1}\right)^{-1} f(Sh(y))$$

$$= S\left(h_y(y)\right)^{-1} S^{-1} Sf(h(y)) = Sg(y), \quad \text{for any } y \in \Omega.$$

Thus Lemma 5.5 is proved.                                    □

For any $k \geq 2$, the set

$$\overline{H}_n^{k,S} = \left\{ f \in H_n^k \mid f(Sx) = Sf(x), \text{ for any } x \in \mathbb{C}^n \right\}$$

is a linear subspace of $H_n^k$. We shall use the notation $\overline{H}_n^k$ instead of $\overline{H}_n^{k,S}$ whenever it is clear from the context what the matrix $S$ is.

**Lemma 5.6.** *Suppose $AS = SA$. Then $\overline{H}_n^k$ is $L_A^k$-invariant.*

*Proof.* For any $h \in \overline{H}_n^k$, let $g(x) = L_A^k h(x)$. Then

$$g(Sx) = h_x(Sx) \cdot ASx - Ah(Sx)$$

$$= S \cdot h_x(x) \cdot S^{-1} \cdot SAx - ASh(x) = Sg(x), \quad \text{for any } x \in \mathbb{C}^n.$$

Thus Lemma 5.6 is proved. $\qquad\qquad\qquad\qquad\qquad\qquad\qquad\qquad\qquad\qquad\square$

Suppose $AS = SA$. Let $\overline{L}_A^k$ be the restriction of $L_A^k$ to the subspace $\overline{H}_n^k$. Let $\overline{\mathscr{R}}^k$ be the range of $\overline{L}_A^k$ in $\overline{H}_n^k$, and $\overline{\mathscr{C}}^k$ be any complementary subspace to $\overline{\mathscr{R}}^k$ in $\overline{H}_n^k$. Then we have the following theorem.

**Theorem 5.7.** *If equation (5.3) has S-symmetry, then there exists a series of near identity transformations with S-symmetry which bring equation (5.3) into the form*

$$\dot{y} = Ay + g^2(y) + \cdots + g^r(y) + O(|y|^{r+1}), \qquad (5.5)$$

*where $g^k \in \overline{\mathscr{C}}^k$, $k = 2, \ldots, r$.*

**Definition 5.8.** Suppose that equation (5.3) has $S$-symmetry. Then the truncated equation of (5.5)

$$\dot{y} = Ay + g^2(y) + \cdots + g^r(y),$$

where $g^k \in \overline{\mathscr{C}}^k$, $k = 2, \ldots, r$, is called an $A$-normal form with $S$-symmetry up to order $r$ of equation (5.3).

**Lemma 5.9.** *Assume that $AS = SA$ and $A^*S = SA^*$. Then a complementary subspace $\overline{\mathscr{C}}^k$ to the range of $\overline{L}_A^k$ in $\overline{H}_n^k$ is given by $\mathrm{Ker}(L_{A^*}^k) \cap \overline{H}_n^k$.*

*Proof.* It follows from Lemma 5.6 that the subspace $\overline{H}_n^k$ is invariant with respect to both $L_A^k$ and $L_{A^*}^k$. From Theorem 1.7, $(\overline{L}_A^k)^* = L_{A^*}^k|_{\overline{H}_n^k}$. Then the lemma is proved. $\qquad\qquad\qquad\qquad\qquad\qquad\qquad\qquad\qquad\square$

**Remark 5.10.** In the case when $\overline{H}_n^k$ is not $L_{A*}^k$ invariant, we can apply the matrix representation method as discussed in Section 2.1 to find $\overline{\mathscr{C}}^k$. We notice that the matrix representation $\overline{\tilde{L}}_A^k$ of $\overline{L}_A^k$ is an $s \times s$ matrix where $s = \dim(\overline{H}_n^k)$.

To find $A$-normal forms of equations with $S$-symmetry, we have to find the subspaces $\overline{H}_n^k$, $k \geq 2$. To do this we introduce a linear operator $L^{k,S}: H_n^k \to H_n^k$ by

$$L^{k,S}h(x) = h(Sx) - Sh(x), \qquad h \in H_n^k. \tag{5.6}$$

It is clear that $\overline{H}_n^k = \text{Ker}(L^{k,S})$. We may apply the matrix representation method to find $\text{Ker}(L^{k,S})$ in general. In the case when $S$ is diagonal, we can easily find a basis of $\text{Ker}(L^{k,S})$.

**Lemma 5.11.** If $S = \text{diag}(s_1, \ldots, s_n)$, then the set of all $S$-symmetrical monomials of order $k$

$$\{x^\alpha e_j | s^\alpha = s_j, |\alpha| = k, 1 \leq j \leq n\}$$

forms a basis of $\overline{H}_n^k$.

*Proof.* It is easy to see that any $S$-symmetrical monomial of order $k$ belongs to $\text{Ker}(L^{k,S})$ and thus to $\overline{H}_n^k$. If $h(x) = \sum_{j=1}^n \sum_{|\alpha|=k} c_{j\alpha} x^\alpha e_j \in \overline{H}_n^k$, then we have

$$L^{k,S}h(x) = \sum_{j=1}^n \sum_{|\alpha|=k} c_{j\alpha}(s^\alpha - s_j)x^\alpha e_j = 0.$$

Hence $c_{j\alpha} = 0$ for all monomials which are not $S$-symmetrical. It follows that $h(x)$ must be a linear combination of $S$-symmetrical monomials of order $k$. $\qquad \square$

**Example 5.12.** Let

$$S = \begin{bmatrix} -1 & 0 \\ 0 & -1 \end{bmatrix}. \tag{5.7}$$

Then $x^\alpha e_j \in \bar{H}_n^k$ if and only if

$$(-1)^{\alpha_1}(-1)^{\alpha_2} = -1, \qquad \alpha_1 + \alpha_2 = k,$$

that is,

$$(-1)^{\alpha_1 + \alpha_2} = -1, \qquad \alpha_1 + \alpha_2 = k.$$

Hence, the $S$-symmetrical monomials are those for which $|\alpha| = \alpha_1 + \alpha_2 \geq 2$ is odd. Therefore

$$\bar{H}_n^{2k} = \{0\}, \qquad \bar{H}_n^{2k+1} = H_n^{2k+1}, \qquad k = 1, 2, \ldots.$$

If equation (5.1) is two-dimensional with linear part

$$A = \begin{bmatrix} 0 & 1 \\ 0 & 0 \end{bmatrix},$$

and has $S$-symmetry, where $S$ is defined by (5.7), then by Lemma 5.9, Example 1.15, and the above discussion, an $A$-normal form up to order $r \geq 2$ is

$$\begin{cases} \dot{x} = y, \\ \dot{y} = \sum_{k=1}^{m} \left( a_k x^{2k+1} + b_k x^{2k} y \right), \end{cases}$$

where $a_k, b_k$ are all complex constants, $k = 1, \ldots, m$, $r - 1 \leq 2m + 1 \leq r$.

**Theorem 5.13.** *Suppose that*

$$\dot{x} = Ax + f(x), \qquad f(x) = O(|x|^2) \quad \text{as } x \to 0,$$

*is an $A$-normal form with $S$-symmetry. Then the resulting equation of the linear change of coordinates*

$$x = Py$$

*is an $\tilde{A}$-normal form with $\tilde{S}$-symmetry, where $\tilde{A} = P^{-1}AP$, $\tilde{S} = P^{-1}SP$.*

*Proof.* The assertion that the resulting equation is an $\tilde{A}$-normal form is trivial. The $\tilde{S}$-symmetry follows from the following calculation:

$$\tilde{A}\tilde{S} = P^{-1}APP^{-1}SP = P^{-1}ASP = P^{-1}SAP = P^{-1}SPP^{-1}AP = \tilde{S}\tilde{A},$$

and

$$g(\tilde{S}y) = P^{-1}f(PP^{-1}SPy) = P^{-1}Sf(Py) = P^{-1}SPP^{-1}f(Py) = \tilde{S}g(y),$$

where $g(y) = P^{-1}f(Py)$.                                                     $\square$

**Remark 5.14.** For a real equation with $S$-symmetry, if the matrix $A$ is diagonalizable over the complex numbers, then we can also apply the method introduced in Section 2.1 to this case. We illustrate this idea with the following example.

**Example 5.15.** Suppose

$$\dot{x} = f(x), \qquad x = (x_1, x_2)^T \in \mathbb{R}^2,$$

$$f(x) = O(|x|^2) \quad \text{as } x \to 0,$$

$$(5.8)$$

has $Z_q$-symmetry, $q \geq 3$. With the complex change of variables $z = P^{-1}x$, where $z = (z_1, z_2)^T \in \mathbb{C}^2$ and

$$P^{-1} = \begin{bmatrix} 1 & i \\ 1 & -i \end{bmatrix},$$

(5.8) becomes

$$\dot{z} = g(z), \qquad z = (z_1, z_2)^T \in \mathbb{C}^2, \qquad z_2 = \bar{z}_1, \qquad (5.9)$$

where $g(z) = P^{-1}f(Pz)$ and the second component of (5.9) is conjugate to the first one. Let $S$ denote the generator matrix of the $Z_q$-symmetry group ($q \geq 3$):

$$
S = \begin{bmatrix} \cos \dfrac{2\pi}{q} & -\sin \dfrac{2\pi}{q} \\ \sin \dfrac{2\pi}{q} & \cos \dfrac{2\pi}{q} \end{bmatrix}.
$$

Let

$$
\tilde{S} = P^{-1}SP = \begin{bmatrix} e^{i\frac{2\pi}{q}} & 0 \\ 0 & e^{-i\frac{2\pi}{q}} \end{bmatrix}.
$$

Since (5.8) has an $S$-symmetry, (5.9) has an $\tilde{S}$-symmetry by Theorem 5.13. We can find a normal form with $\tilde{S}$-symmetry. The linear part of equation (5.9) has the zero matrix since (5.8) does. Therefore every nonlinear monomial is resonant. We note that $\tilde{S}$ is a diagonal matrix and the $\tilde{S}$-symmetry conditions are

$$
\left( e^{i\frac{2\pi}{q}} \right)^{\alpha_1} \left( e^{-i\frac{2\pi}{q}} \right)^{\alpha_2} = e^{i\frac{2\pi}{q}}, \quad \alpha_1 + \alpha_2 = k \geq 2, \quad \text{for the first equation,}
$$

$$
\left( e^{i\frac{2\pi}{q}} \right)^{\alpha_1} \left( e^{-i\frac{2\pi}{q}} \right)^{\alpha_2} = e^{-i\frac{2\pi}{q}}, \quad \alpha_1 + \alpha_2 = k \geq 2,
$$

$$
\text{for the second equation,}
$$

which are equivalent to

$$
\begin{cases} \alpha_1 - \alpha_2 - 1 = lq, & l \in \mathbb{Z}, \quad \text{for the first equation,} \\ \alpha_1 - \alpha_2 + 1 = lq, & l \in \mathbb{Z}, \quad \text{for the second equation,} \quad (5.10) \\ \alpha_1 + \alpha_2 = k \geq 2. \end{cases}
$$

Therefore

$$
\overline{H}_2^{k,\tilde{S}} = \{0\} \quad \text{for} \quad k \leq q, k \quad \text{even, and} \quad k \neq q - 1;
$$

$$
\overline{H}_2^{k,\tilde{S}} = \text{span}\{ z_1^{m+1} z_2^{m} e_1, z_1^{m} z_2^{m+1} e_2 \},
$$

$$
2m + 1 = k, \quad \text{for } k \leq q, k \text{ odd, and } k \neq q - 1;
$$

$$
\overline{H}_2^{q-1,\tilde{S}} = \text{span}\{ z_2^{q-1} e_1, z_1^{q-1} e_2, z_1^{m+1} z_2^{m} e_1, z_1^{m} z_2^{m+1} e_2 \}
$$

$$
\text{for } q = 2(m+1), m \geq 1;
$$

$$
\overline{H}_2^{q-1,\tilde{S}} = \text{span}\{ z_2^{q-1} e_1, z_1^{q-1} e_2 \} \quad \text{for odd } q \geq 3.
$$

Thus the normal form of (5.9) with $\tilde{S}$-symmetry up to order $q$ is

$$\begin{cases} \dot{z}_1 = c_1 z_1^2 z_2 + \cdots + c_m z_1^{m+1} z_2^m + c_{m+1} z_2^{q-1}, \\ \dot{z}_2 = \bar{c}_1 z_1 z_2^2 + \cdots + \bar{c}_m z_1^m z_2^{m+1} + \bar{c}_{m+1} z_1^{q-1}, \end{cases} \tag{5.11}$$

where $c_k$ are all complex constants, $q - 1 \le 2m + 1 \le q$. Since the second equation of (5.11) is conjugate to the first one, we let $z = z_1$ and omit the second equation of (5.11). Then we say that

$$\dot{z} = c_1 |z|^2 z + \cdots + c_m |z|^{2m} z + c_{m+1} \bar{z}^{q-1} \tag{5.12}$$

is an $A$-normal form with $Z_q$-symmetry up to order $q$.

We can apply to (5.12) the change of coordinates $z = x_1 + ix_2$ to obtain a real normal form with $S$-symmetry of (5.8) up to order $q$.

Another real normal form can be obtained from (5.12) by using polar coordinates $z = re^{i\theta}$:

$$\begin{cases} \dot{r} = a_1 r^3 + \cdots + a_m r^{2m+1} + (a_{m+1}\cos q\theta + b_{m+1}\sin q\theta) r^{q-1}, \\ \dot{\theta} = b_1 r^2 + \cdots + b_m r^{2m} - (a_{m+1}\sin q\theta - b_{m+1}\cos q\theta) r^{q-2}, \end{cases}$$

where $a_k = \mathrm{Re}(c_k)$, $b_k = \mathrm{Im}(c_k)$ for $k = 1, \ldots, m + 1$.

## 2.6 Normal Forms of Linear Hamiltonian Systems

In this and the next sections, we discuss normal forms of Hamiltonian systems over the reals

$$\dot{x} = J\nabla H(x), \qquad x \in \mathbb{R}^{2n}, \tag{6.1}$$

where

$$J = \begin{bmatrix} 0 & I_n \\ -I_n & 0 \end{bmatrix},$$

$I_n$ is the $n \times n$ identity matrix, $H \in C^r(\mathbb{R}^{2n}, \mathbb{R})$, $r \ge 1$, and $\nabla H(x)$ is the gradient of $H(x)$.

We note that $J^T = J^{-1} = -J$, where $J^T$ is the transpose of $J$. Hence $J$ is an orthogonal skew-symmetric matrix.

If $H(x)$ is a quadratic form, then $H(x) = \frac{1}{2}(x, Bx)$, where $(\cdot, \cdot)$ is the usual scalar product in $\mathbb{R}^{2n}$, and $B$ is a $2n \times 2n$ symmetric matrix. We note that $\nabla H(x) = Bx$. For such an $H(x)$, (6.1) can be rewritten as

$$\dot{x} = JBx, \qquad x \in \mathbb{R}^{2n}, \tag{6.2}$$

which is a linear Hamiltonian system.

**Definition 6.1.** A linear operator $A: \mathbb{R}^{2n} \to \mathbb{R}^{2n}$ is called infinitesimally symplectic if and only if

$$A^* = JAJ,$$

where $A^*$ is the adjoint operator of $A$.

The set of all infinitesimally symplectic operators is a vector space, denoted by $sp(2n, \mathbb{R})$.

**Lemma 6.2.** *A linear system of equations*

$$\dot{x} = Ax, \qquad x \in \mathbb{R}^{2n},$$

*is Hamiltonian if and only if $A$ is an infinitesimally symplectic operator.*

*Proof.* Suppose the system is Hamiltonian. Then $A = JB$ for some symmetric matrix $B$. Therefore

$$A^T = B^T J^T = -BJ = -J^{-1}AJ = JAJ.$$

Conversely, suppose $A^T = JAJ$. We define $B = J^{-1}A$. Then

$$B^T = -A^T J^T = JAJ \cdot J = -JA = J^{-1}A = B,$$

that is, $B$ is symmetric. We define $H(x) = \frac{1}{2}(x, Bx)$. Then the system can be rewritten as

$$\dot{x} = J\nabla H(x), \qquad x \in \mathbb{R}^{2n}. \qquad \square$$

**Corollary 6.3.** *If*

$$A = \begin{bmatrix} A_1 & A_2 \\ A_3 & A_4 \end{bmatrix},$$

*where each $A_i$ $(1 \le i \le 4)$ is an $n \times n$ matrix, then $A \in sp(2n, \mathbb{R})$ if and only if $A_2^T = A_2$, $A_3^T = A_3$, and $A_1^T = -A_4$.*

**Theorem 6.4.** *Let $A \in \mathrm{sp}(2n, \mathbb{R})$. If $\lambda$ is an eigenvalue of $A$ with algebraic multiplicity $m$, then $-\lambda$, $\bar{\lambda}$, and $-\bar{\lambda}$ are also eigenvalues of $A$ with the same multiplicity.*

*Proof.* Let $p(\lambda)$ be the characteristic polynomial of $A$. Then by Lemma 6.2,

$$p(\lambda) = \det(\lambda I - A) = \det(\lambda I - J^{-1}A^T J^{-1})$$

$$= \det(J(\lambda I + A^T)J^{-1}) = \det(-J)\det(J^{-1})\det(-\lambda I - A^T)$$

$$= p(-\lambda).$$

Since $A$ is a real matrix we also have $p(\lambda) = \overline{p(\bar{\lambda})}$. This implies the result. □

**Corollary 6.5.** *The characteristic polynomial of an infinitesimally symplectic operator must be a product of factors of the form $\lambda^2$, $(\lambda + \alpha)(\lambda - \alpha)$, $\lambda^2 + \alpha^2$, and $((\lambda - \alpha)^2 + \beta^2)((\lambda + \alpha)^2 + \beta^2)$, where $\alpha, \beta$ are real positive numbers.*

**Definition 6.6.** A linear operator $S: \mathbb{R}^{2n} \to \mathbb{R}^{2n}$ is called a symplectic operator if and only if

$$S^*JS = J,$$

where $S^*$ is the adjoint operator of $S$.

The set of all linear symplectic operators forms a Lie group under the matrix composition, and is denoted by $Sp(2n, \mathbb{R})$.

**Lemma 6.7.** *A linear symplectic transformation brings a linear Hamiltonian system into a linear Hamiltonian system.*

*Proof.* Suppose that $S$ is a linear symplectic operator. We apply a change of variables $x = Sy$ to the linear Hamiltonian system

$$\dot{x} = JBx, \tag{6.3}$$

where $B$ is a symmetric matrix. Then the resulting equation of (6.3) is

$$\dot{y} = S^{-1}JBSy. \qquad (6.4V)$$

Since $S$ is symplectic, $S^{-1}J = JS^T$. Therefore, (6.4) can be expressed as

$$\dot{y} = J\tilde{B}y,$$

where $\tilde{B} = S^TBS$. Since $B$ is symmetric, the matrix $\tilde{B}$ is also symmetric, thus the transformed equation (6.4) is a linear Hamiltonian system.   □

**Definition 6.8.** Two $2n \times 2n$ infinitesimally symplectic matrices $A_1$ and $A_2$ are symplectically similar if there exists a symplectic matrix $S$ such that $A_2 = S^{-1}A_1S$. Two linear Hamiltonian systems

$$\dot{x} = A_1x \quad \text{and} \quad \dot{x} = A_2x, \quad x \in \mathbb{R}^{2n},$$

are symplectically conjugate if matrices $A_1$ and $A_2$ are symplectically similar.

It is easy to see that symplectic conjugacy (or symplectic similarity) is an equivalent relation. In every equivalence class of symplectically conjugate linear Hamiltonian systems we will find one as a representation of this class. We will call this system a normal form. In order to describe these normal forms, we introduce below some basic concepts of symplectic vector spaces.

**Definition 6.9.** Let $V$ be an even-dimensional vector space over the reals. A bilinear form $\tau(\cdot, \cdot)$ on $V$ is called skew-symmetric if

$$\tau(x, y) = -\tau(y, x), \quad \text{for all } x, y \in V;$$

$\tau(\cdot, \cdot)$ is called nondegenerate if $\tau(x, y) = 0$ for all $y \in V$ implies $x = 0$. A nondegenerate, skew-symmetric, bilinear form $\tau(\cdot, \cdot)$ defined on $V$ is called a symplectic form and $(V, \tau)$ is called a symplectic vector space. A basis $\{v_1, \ldots, v_n, w_1, \ldots, w_n\}$ of a symplectic vector space $(V, \tau)$ is called a symplectic basis if

$$\tau(v_i, v_j) = 0, \quad \tau(w_i, w_j) = 0, \quad \text{and} \quad \tau(v_i, w_j) = \delta_{ij}$$

$$\text{for } i, j = 1, 2, \ldots, n,$$

where $\delta_{ij}$ is the Kronecker symbol.

**Example 6.10.** The bilinear form $\omega(\cdot, \cdot): \mathbb{R}^{2n} \times \mathbb{R}^{2n} \to \mathbb{R}$ defined by

$$\omega(x, y) = (x, Jy), \quad \text{for all } x, y \in \mathbb{R}^{2n},$$

where $(\cdot, \cdot)$ is the usual scalar product in $\mathbb{R}^{2n}$, is a symplectic form, $(\mathbb{R}^{2n}, \omega)$ is a symplectic vector space, and the standard basis of $\mathbb{R}^{2n}$ is also a symplectic basis of $(\mathbb{R}^{2n}, \omega)$.

**Definition 6.11.** Let $W$ be a subspace of a symplectic vector space $(V, \tau)$. $W$ is called a symplectic subspace if $\tau|_W$ is nondegenerate. Let $W_1, W_2$ be two symplectic subspaces of a symplectic vector space $(V, \tau)$. $W_1$ and $W_2$ are called $\tau$-orthogonal if $\tau(x, y) = 0$ for all $x \in W_1$ and $y \in W_2$.

**Definition 6.12.** An infinitesimally symplectic mapping $A$ on $(V, \tau)$ is called decomposable if $V = V_1 \oplus V_2$, where $V_1$ and $V_2$ are proper, $A$-invariant, and $\tau$-orthogonal symplectic subspaces of $V$. $A$ is called indecomposable if $A$ is not decomposable.

**Theorem 6.13.** *Let $A$ be an infinitesimally symplectic mapping defined on a symplectic vector space $(V, \tau)$. Suppose $V = V_1 \oplus \cdots \oplus V_s$, where $V_i$, $i = 1, \ldots, s$, are proper, $A$-invariant, mutually $\tau$-orthogonal, symplectic subspaces. If*

$$\Gamma_i = \left\{ v_1^i, \ldots, v_{n_i}^i, w_1^i, \ldots, w_{n_i}^i \right\}$$

*is a symplectic basis of $V_i$ and the matrix representation of $A|_{V_i}$ with respect to $\Gamma_i$ is*

$$\begin{bmatrix} A_i & B_i \\ C_i & -A_i^T \end{bmatrix},$$

*where $B_i^T = B_i$, $C_i^T = C_i$ for $i = 1,\ldots,s$, and $n_1 + \cdots + n_s = n$, then $\{v_1^1,\ldots,v_{n_1}^1,\ldots,v_1^s,\ldots,v_{n_s}^s,w_1^1,\ldots,w_{n_1}^1,\ldots,w_1^s,\ldots,w_{n_s}^s\}$ is a symplectic basis of $(V,\tau)$ and the matrix representation of $A$ under this basis is*

$$\begin{bmatrix}
A_1 & & & & \vdots & B_1 & & & \\
& A_2 & & & \vdots & & B_2 & & \\
& & \ddots & & \vdots & & & \ddots & \\
& & & A_s & \vdots & & & & B_s \\
\cdots & \cdots & \cdots & \cdots & \vdots & \cdots & \cdots & \cdots & \cdots \\
C_1 & & & & \vdots & -A_1^T & & & \\
& C_2 & & & \vdots & & -A_2^T & & \\
& & \ddots & & \vdots & & & \ddots & \\
& & & C_s & \vdots & & & & -A_s^T
\end{bmatrix}.$$

By using Theorem 6.13 we may consider only the cases where $A$ is an indecomposable infinitesimally symplectic operator defined on $(\mathbb{R}^{2n}, \omega)$. We have two cases: (i) $A$ is semisimple; (ii) $A = S + N$ is nonsemi-simple, where $S$ is the semisimple part of $A$, $N$ is the nilpotent part of $A$, and $N \neq 0$. We give below the list of normal forms of infinitesimally symplectic operators, but omit their proofs. The normal forms will be denoted by $A$ and the corresponding Hamiltonian functions by $H(x)$.

*List I. Normal forms of indecomposable semisimple infinitesimally symplectic mappings.*
(1) $\sigma(A) = \{0\}$,

$$A = \begin{bmatrix} 0 & 0 \\ 0 & 0 \end{bmatrix}, \quad H(x,y) = 0, \quad x, y \in \mathbb{R}.$$

(2) $\sigma(A) = \{\pm\alpha, \alpha > 0\}$,

$$A = \begin{bmatrix} \alpha & 0 \\ 0 & -\alpha \end{bmatrix}, \qquad H(x, y) = \alpha xy, \quad x, y \in \mathbb{R}.$$

(3) $\sigma(A) = \{\pm\beta i, \beta > 0\}$,

$$A = \begin{bmatrix} 0 & \pm\beta \\ \mp\beta & 0 \end{bmatrix}, \qquad H(x, y) = \pm\frac{1}{2}\beta(x^2 + y^2), \quad x, y \in \mathbb{R}.$$

(4) $\sigma(A) = \{\pm\alpha \pm \beta i, \alpha > 0, \beta > 0\}$,

$$A = \begin{bmatrix} \alpha & \beta & 0 & 0 \\ -\beta & \alpha & 0 & 0 \\ 0 & 0 & -\alpha & \beta \\ 0 & 0 & -\beta & -\alpha \end{bmatrix},$$

$$H(x) = \alpha(x_1 y_1 + x_2 y_2) + \beta(x_2 y_1 - x_1 y_2), \quad x = (x_1, x_2, y_1, y_2)^T \in \mathbb{R}^4.$$

*List II.* Normal forms of indecomposable nonsemisimple infinitesimally symplectic mappings.

(1) $\sigma(A) = \{0\}$,

$$H(x) = \sum_{i=1}^{n-1} x_i y_{i+1} \mp \frac{1}{2}x_n^2, \quad x = (x_1, \ldots, x_n, y_1, \ldots, y_n)^T \in \mathbb{R}^{2n}.$$

(2) $\sigma(A) = \{0\}$,

$$A = \left[\begin{array}{cccc:cccc} 0 & & & & \vdots & & & \\ 1 & \ddots & & & \vdots & & & \\ & \ddots & \ddots & & \vdots & & & \\ & & 1 & 0 & \vdots & & & \\ \cdots & \cdots & \cdots & \cdots & \cdots & \cdots & \cdots & \cdots \\ & & & & \vdots & 0 & -1 & \\ & & & & \vdots & & \ddots & \ddots \\ & & & & \vdots & & \ddots & -1 \\ & & & & \vdots & & & 0 \end{array}\right] \begin{array}{l} \Big\} n \\[2em] \Big\} n \end{array} , \quad n \text{ odd},$$

$$H(x) = \sum_{i=1}^{n-1} x_i y_{i+1}, \quad x = (x_1, \ldots, x_n, y_1, \ldots, y_n)^T \in \mathbb{R}^{2n}.$$

(3) $\sigma(A) = \{\pm\alpha, \alpha > 0\}$,

$$A = \left[\begin{array}{cccc:cccc} \alpha & & & & \vdots & & & \\ 1 & \alpha & & & \vdots & & & \\ & 1 & \alpha & & \vdots & & & \\ & & \ddots & \ddots & \vdots & & & \\ & & 1 & \alpha & \vdots & & & \\ \cdots & \cdots & \cdots & \cdots & \cdots & \cdots & \cdots & \cdots \\ & & & & \vdots & -\alpha & -1 & \\ & & & & \vdots & & -\alpha & -1 \\ & & & & \vdots & & & -\alpha & \ddots \\ & & & & \vdots & & & & \ddots & -1 \\ & & & & \vdots & & & & & -\alpha \end{array}\right] \begin{array}{l} \Big\} n \\[2em] \Big\} n \end{array} ,$$

$$H(x) = \alpha \sum_{i=1}^{n} x_i y_i + \sum_{i=1}^{n-1} x_i y_{i+1}, \quad x = (x_1, \ldots, x_n, y_1, \ldots, y_n) \in \mathbb{R}^{2n}.$$

(4) $\sigma(A) = \{\pm\beta i, \beta > 0\}$,

$$A = \left[\begin{array}{cccc:cccc} A_2 & & & & \vdots & & & \\ I_2 & \ddots & & & \vdots & & & \\ & \ddots & \ddots & & \vdots & & & 0 \\ & & I_2 & A_2 & \vdots & & & \\ \cdots & \cdots & \cdots & \cdots & \cdots & \cdots & \cdots & \cdots \\ & & & & \vdots & A_2 & -I_2 & \\ & & & & \vdots & & \ddots & \ddots \\ & & & & \vdots & & & \ddots & -I_2 \\ & & & \pm I_2 & \vdots & & & & A_2 \end{array}\right] \begin{array}{l} \Big\} n \\[2em] \Big\} n \end{array} , \quad n \text{ even}$$

where $I_2 = \begin{bmatrix} 0 & 1 \\ 1 & 0 \end{bmatrix}$, $A_2 = \begin{bmatrix} 0 & -\beta \\ \beta & 0 \end{bmatrix}$,

$$H(x) = \beta \sum_{i=1}^{n/2} (x_{2i-1}y_{2i} - x_{2i}y_{2i-1}) + \sum_{i=1}^{n-2} x_i y_{i+2} \mp \frac{1}{2}(x_{n-1}^2 + x_n^2),$$

where $x = (x_1, \ldots, x_n, y_1, \ldots, y_n)^T \in \mathbb{R}^{2n}$.

(5) $\sigma(A) = \{\pm\beta i, \beta > 0\}$,

$$\begin{bmatrix}
0 & & & & & \vdots & & & & -\epsilon\beta \\
1 & 0 & & & & \vdots & & & \epsilon\beta & \\
 & \ddots & & \ddots & & \vdots & & \ddots & & \\
 & & \ddots & & \ddots & \vdots & \epsilon\beta & & & \\
 & & & 1 & 0 & \vdots & -\epsilon\beta & & & \\
\cdots & \cdots & \cdots & \cdots & \cdots & \cdots & \cdots & \cdots & \cdots & \cdots \\
 & & & \epsilon\beta & & \vdots & 0 & -1 & & \\
 & & -\epsilon\beta & & & \vdots & 0 & & \ddots & \\
 & \vdots & & & & \vdots & & & \ddots & -1 \\
 & -\epsilon\beta & & & & \vdots & & & & \\
\epsilon\beta & & & & & \vdots & & & & 0
\end{bmatrix}, \epsilon = \pm 1, n \text{ odd},$$

$$H(x) = -\epsilon\beta \sum_{i=1}^{\frac{n-1}{2}} (-1)^{i+1}(x_i x_{n+1-i} + y_i y_{n+1-i})$$

$$+ \sum_{i=1}^{n-1} x_i y_{i+1} + \frac{1}{2}(-1)^{\left[\frac{n}{2}\right]+1} \epsilon\beta\left(x_{\frac{n+1}{2}}^2 + y_{\frac{n+1}{2}}^2\right),$$

where $x \in (x_1, \ldots, x_n, y_1, \ldots, y_n)^T \in \mathbb{R}^{2n}$.

(6) $\sigma(A) = \{\pm\alpha \pm \beta i, a > 0, \beta > 0\}$,

$$A = \begin{bmatrix}
B_2 & & & & \vdots & & & \\
I_2 & \ddots & & & \vdots & & & \\
 & \ddots & \ddots & & \vdots & & & \\
 & & I_2 & B_2 & \vdots & & & \\
\cdots & \cdots & \cdots & \cdots & \cdots & \cdots & \cdots & \cdots \\
 & & & & \vdots & -B_2^T & -I_2 & \\
 & & & & \vdots & & \ddots & \ddots \\
 & & & & \vdots & & \ddots & -I_2 \\
 & & & & \vdots & & & -B_2^T
\end{bmatrix}, \quad n \text{ even},$$

where $B_2 = \begin{bmatrix} \alpha & -\beta \\ \beta & \alpha \end{bmatrix}$, $I_2 = \begin{bmatrix} 1 & 0 \\ 0 & 1 \end{bmatrix}$,

$$H(x) = \alpha \sum_{i=1}^{n} x_i y_i + \beta \sum_{i=1}^{n/2} (x_{2i-1} y_{2i} - x_{2i} y_{2i-1}) + \sum_{i=1}^{n-2} x_i y_{i+2},$$

where $x = (x_1, \ldots, x_n, y_1, \ldots, y_n) \in \mathbb{R}^{2n}$.

### 2.7 Normal Forms of Nonlinear Hamiltonian Systems

Consider a Hamiltonian system of equations

$$\dot{x} = J \nabla H(x), \qquad x \in \Omega \subseteq \mathbb{R}^{2n}, \tag{7.1}$$

where $\Omega$ is a neighborhood of the origin in $\mathbb{R}^{2n}$, the Hamiltonian function $H(x) = H_2(x) + H_3(x) + \cdots + H_r(x) + O(|x|^{r+1})$, $H_k \in P_{2n}^k$, the linear space of all $k$th order scalar homogeneous polynomials in $2n$ variables, $k = 2, \ldots, r$.

**Definition 7.1.** A diffeomorphism $S: \Omega \subseteq \mathbb{R}^{2n} \to \mathbb{R}^{2n}$ is called a symplectic diffeomorphism if

$$(DS(x))^* J(DS(x)) = J \quad \text{for all } x \in \Omega \subseteq \mathbb{R}^{2n},$$

that is, the linear mapping $DS(x): \mathbb{R}^{2n} \to \mathbb{R}^{2n}$ is symplectic for all $x \in \Omega \subseteq \mathbb{R}^{2n}$, where $\Omega$ is a neighborhood of the origin in $\mathbb{R}^{2n}$. If $S(x)$ is a symplectic diffeomorphism on $\Omega$, then $x = S(y)$, $y \in \Omega$, is called a symplectic transformation.

**Theorem 7.2.** *A symplectic transformation $x = S(y)$, $y \in \Omega$, where $\Omega$ is a neighborhood of the origin in $\mathbb{R}^{2n}$, transforms a Hamiltonian system on $\Omega$ with Hamiltonian function $H(x)$ to a Hamiltonian system on $\bar{\Omega} = S(\Omega)$ with Hamiltonian function $H(S(y))$.*

*Proof.* The resulting system of (7.1) under the transformation $x = S(y)$ is

$$\dot{y} = (DS(y))^{-1} J((DS(y))^*)^{-1} \nabla_y H(S(y)), \qquad y \in \tilde{\Omega}. \quad (7.2)$$

Since $S$ is symplectic,

$$(DS(y))^{-1} J((DS(y))^*)^{-1} = J, \qquad \text{for all } y \in \tilde{\Omega} \subseteq \mathbb{R}^{2n}.$$

Therefore equation (7.2) can be expressed as:

$$\dot{y} = J \nabla_y H(S(y)), \qquad y \in \tilde{\Omega},$$

which is a Hamiltonian system with Hamiltonian function $H(S(y))$. $\quad\square$

Theorem 7.2 says that simplifying a Hamiltonian system by a symplectic transformation is equivalent to simplifying its Hamiltonian function by composing it with this transformation.

**Lemma 7.3.** *If*

$$F(x) = x + F^k(x) + O(|x|^{k+1}), \qquad x \in \Omega \subseteq \mathbb{R}^{2n},$$

*is a symplectic diffeomorphism, where $F^k \in P^k_{2n}$, $k \geq 2$, and $\Omega$ is a neighborhood of the origin in $\mathbb{R}^{2n}$, then*

$$(DF^k(x))^* J + J(DF^k(x)) = 0, \qquad x \in \Omega \subseteq \mathbb{R}^{2n}, \quad (7.3)$$

*that is, $DF^k(x)$ is an infinitesimally symplectic linear map for any $x \in \Omega \subseteq \mathbb{R}^{2n}$.*

*Proof.* By Definition 7.1, we have

$$(I + DF^k(x) + O(|x|^k))^* J(I + DF^k(x) + O(|x|^k)) = J,$$

$$\text{for any } x \in \Omega \subseteq \mathbb{R}^{2n}.$$

Hence (7.3) follows. $\quad\square$

**Corollary 7.4.** *If*

$$F(x) = x + F^k(x) + O(|x|^{k+1}), \qquad x \in \Omega \subseteq \mathbb{R}^{2n},$$

*is a symplectic diffeomorphism, where $F^k \in H_{2n}^k$, $k \geq 2$, and $\Omega$ is a neighborhood of the origin in $\mathbb{R}^{2n}$, then $J(DF^k(x))$ is symmetric for any $x \in \Omega$.*

*Let*

$$\overline{H}_{2n}^k = \{f \in H_{2n}^k \mid Df(x) \text{ is symmetric for any } x \text{ in } \mathbb{R}^{2n}\}.$$

*Then $\overline{H}_{2n}^k$ is a linear subspace of $H_{2n}^k$.*

**Lemma 7.5.** *For any $f \in \overline{H}_{2n}^k$, the equation*

$$\nabla H_{k+1}(x) = f(x), \qquad x \in \mathbb{R}^{2n}, \tag{7.4}$$

*is uniquely solvable for $H_{k+1} \in P_{2n}^{k+1}$.*

*Proof.* Let $\Gamma_i^k = \{\alpha \in \mathbb{R}^{2n}, \alpha = (0, \ldots, 0, \alpha_i, \ldots, \alpha_{2n})$, $\alpha_j$ are nonnegative integers, $i \leq j \leq 2n$, and $|\alpha| = k\}$, $i = 1, \ldots, 2n$, $k \geq 2$. For each $\alpha \in \Gamma_i^k$ we define

$$f_{i,\alpha}(x) = x^\alpha e_i + \sum_{j=i+1}^{2n} \frac{\alpha_j}{\alpha_i + 1} \frac{x^\alpha x_i}{x_j} e_j \quad (1 \leq i \leq 2n - 1),$$

and

$$f_{2n,\alpha}(x) = x_{2n}^k e_{2n}.$$

We will show that the set $\{f_{i,\alpha}\}$ defined above forms a basis of $\overline{H}_{2n}^k$.

Component-wise for each $f_{i,\alpha}$:

$$(f_{i,\alpha})_j = \begin{cases} 0, & j < i, \\ x_i^{\alpha_i} x_{i+1}^{\alpha_{i+1}} \cdots x_{2n}^{\alpha_{2n}}, & j = i, \\ \dfrac{\alpha_j}{\alpha_i + 1} x_i^{\alpha_i+1} x_{i+1}^{\alpha_{i+1}} \cdots x_j^{\alpha_j-1} \cdots x_{2n}^{\alpha_{2n}}, & i < j \le 2n. \end{cases}$$

Each $f_{i,\alpha} \in \bar{H}_{2n}^k$ since as the following calculation shows $Df_{i,\alpha}$ is symmetric for any $x \in \mathbb{R}^{2n}$.

$$(Df_{i,\alpha}(x))_{jl} = \begin{cases} 0, & j < i \quad \text{or} \quad l < i, \\ \alpha_l \dfrac{x^\alpha}{x_l}, & j = i \quad \text{and} \quad l > i, \\ \alpha_j \dfrac{x^\alpha}{x_j}, & j > i \quad \text{and} \quad l = i, \\ \dfrac{\alpha_j \alpha_l}{\alpha_i + 1} \dfrac{x^\alpha x_i}{x_l x_j}, & j > i, l > i, \quad \text{and} \quad l \ne j. \end{cases}$$

Any $f \in \bar{H}_{2n}^k$ is a linear combination of the $\{f_{i,\alpha}\}$. To see this we suppose $f \in \bar{H}_{2n}^k$ and $f(x) = \sum_{i=1}^n \sum_{|\alpha|=k} c_{i\alpha} x^\alpha e_i$, where the $c_{i\alpha}$ are real constants. Then we define

$$\tilde{f}(x) \equiv f(x) - \sum_{\alpha \in \Gamma_1^k} c_{1\alpha} f_{1,\alpha}(x) = \sum_{i=2}^{2n} \sum_{|\alpha|=k} c_{i\alpha}^{(1)} x^\alpha e_i.$$

It is clear that $\tilde{f} \in \bar{H}_{2n}^k$. Since $D\tilde{f}(x)$ is a symmetric matrix and its elements in the first row are all zero,

$$0 \equiv (D\tilde{f}(x))_{1j} = (D\tilde{f}(x))_{j1} = \sum_{|\alpha|=k} c_{j\alpha}^{(1)} \alpha_1 \frac{x^\alpha}{x_1}, \qquad 2 \le j \le n,$$

it follows that each monomial $x^\alpha e_i$ in $\tilde{f}(x)$ with nonzero coefficient is such that $\alpha_1 = 0$. Hence

$$\tilde{f}(x) = \sum_{i=2}^{2n} \sum_{\alpha \in \Gamma_2^k} c_{i\alpha}^{(1)} x^\alpha e_i.$$

By induction, after $2n - 1$ steps we obtain:

$$f(x) - \sum_{\alpha \in \Gamma_1^k} c_{1\alpha} f_{1,\alpha}(x) - \cdots - \sum_{\alpha \in \Gamma_{2n-1}^k} c_{2n-1}^{(2n-2)} f_{2n-1,\alpha}(x)$$

$$- c_{2n}^{(2n-1)} x_{2n}^k = 0.$$

To show that the $\{f_{i,\alpha}\}$ are linearly independent, we assume that $\sum_{i=1}^{2n} \sum_{\alpha \in \Gamma_i^k} c_{i\alpha} f_{i,\alpha}(x) = 0$. The first component of this equation is

$$\sum_{|\alpha|=k} c_{1\alpha} x^\alpha = 0.$$

Therefore all $c_{1\alpha} = 0$. The second component of the equation is

$$\sum_{\alpha \in \Gamma_2^k} c_{2\alpha} x^\alpha + \sum_{\alpha \in \Gamma_1^k} \frac{c_{1\alpha} \alpha_2}{\alpha_1 + 1} \frac{x^\alpha x_1}{x_2} = \sum_{\alpha \in \Gamma_2^k} c_{2\alpha} x^\alpha = 0.$$

Therefore all $c_{2\alpha} = 0$. An induction argument shows that all $c_{i\alpha} = 0$ for $\alpha \in \Gamma_i^k$ and $1 \le i \le 2n$.

Consider now the equation:

$$\nabla H_{k+1}(x) = f_{i,\alpha}(x), \qquad \alpha \in \Gamma_i^k, \quad 1 \le i \le 2n. \qquad (7.5)$$

It is obvious that $H_{k+1}(x) = \frac{1}{\alpha_i + 1} x^\alpha x_i$, $\alpha \in \Gamma_i^k$, is the unique solution of (7.5) in $P_{2n}^{k+1}$. Since any $f \in \bar{H}_{2n}^k$ can be uniquely represented as a linear combination of the $\{f_{i,\alpha}\}$ and $\nabla$ is a linear operator, it follows that (7.4) is uniquely solvable for $H_{k+1} \in P_{2n}^{k+1}$.                    □

Let $j^k$ be the truncation operator that keeps only the terms up to order $k$ in a Taylor expansion.

**Lemma 7.6.** *For any symplectic diffeomorphism of the form*

$$F(x) = x + F^k(x) + O(|x|^{k+1}), \qquad x \in \Omega \subseteq \mathbb{R}^{2n},$$

*where $F^k(x) \in H_{2n}^k$, $k \geq 2$, and $\Omega$ is a neighborhood of the origin in $\mathbb{R}^{2n}$, there exists a Hamiltonian system on $\Omega$ with Hamiltonian function $H(x)$ whose time-one mapping $\Phi_H(x)$ satisfies*

$$x + F^k(x) = j^k \Phi_H(x). \tag{7.6}$$

*In particular, we can choose $H \in P_{2n}^{k+1}$.*

*Proof.* We note that $JF^k(x) \in \overline{H}_{2n}^k$ by Corollary 7.4. Then we define $H(x) = H_{k+1}(x)$ which is the unique solution of equation

$$J\nabla H_{k+1}(x) = F^k(x)$$

in $P_{2n}^{k+1}$ (by Lemma 7.5). Let $\Phi(t, x)$ be the flow of the system

$$\dot{x} = J\nabla H(x).$$

Then $\Phi(t, x)$ can be expanded in a Taylor series

$$\Phi(t, x) = \Phi_1(t)x + \Phi_2(t, x) + \cdots + \Phi_k(t, x) + O(|x|^{k+1}),$$

where $\Phi_1 \in C^1(\mathbb{R}, \mathbb{R}^{2n \times 2n})$, $\Phi_j \in C^1(\mathbb{R} \times \mathbb{R}^{2n}, \mathbb{R}^{2n})$ and $\Phi_j(t, \cdot) \in H_{2n}^j$ for each $t \in \mathbb{R}$, $j = 2, \ldots, k$. By definition, $\Phi(t, x)$ is the solution of

$$\begin{cases} \dot{\Phi}(t, x) = J\nabla H(\Phi(t, x)), \\ \Phi(0, x) = x. \end{cases} \tag{7.7}$$

From (7.7), expanding $\Phi(t, x)$ in its Taylor expansion and equating coefficients, we get

$$\begin{cases} \dot{\Phi}_1(t) = 0, \\ \Phi_1(0) = I_{2n}, \end{cases} \tag{7.8}_1$$

$$\begin{cases} \dot{\Phi}_j(t, x) = 0, \\ \Phi_j(0, x) = 0, \quad j = 2, \ldots, k - 1, \end{cases} \tag{7.8}_j$$

and

$$\begin{cases} \dot{\Phi}_k(t, x) = J\nabla H(\Phi_1(t)x), \\ \Phi_k(0, x) = 0. \end{cases} \tag{7.9}$$

From $(7.8)_1$, we get $\Phi_1(t) \equiv I_{2n}$. From $(7.8)_j$, $2 \leq j \leq k - 1$, it is easy to see that $\Phi_j(t, x) \equiv 0$, $j = 2, \ldots, k - 1$. Accordingly, (7.9) becomes

$$\begin{cases} \dot{\Phi}_k(t, x) = J\nabla H(x) = F^k(x), \\ \Phi_k(0, x) = 0. \end{cases}$$

Hence $\Phi_k(t, x) = F^k(x)t$, for any $t \in \mathbb{R}$, and $x \in \mathbb{R}^{2n}$. It is clear that $\Phi_k(1, x) = F^k(x)$. Thus

$$j^k\Phi_H(x) = j^k\Phi(1, x) = x + F^k(x). \qquad \square$$

**Lemma 7.7.** *The time-one mapping of a Hamiltonian system is a symplectic diffeomorphism.*

*Proof.* Let $\Phi(t, x)$ be the flow of the Hamiltonian system

$$\dot{x} = J\nabla H(x), \qquad x \in \Omega \subseteq \mathbb{R}^{2n},$$

where $\Omega$ is a neighborhood of the origin in $\mathbb{R}^{2n}$. Then we have

$$\frac{d}{dt} \Phi(t, x) = J\nabla H(\Phi(t, x)), \qquad t \in I_x, \quad x \in \Omega,$$

where $I_x$ is the maximal interval containing 0 such that $\Phi(t, x) \in \Omega$ if $t \in I_x$ for any $x \in \Omega$. And

$$\frac{d}{dt} D_x\Phi(t, x)$$

$$= J(D\nabla H)(\Phi(t, x)) \cdot (D_x\Phi(t, x)), \qquad t \in I_x, \quad x \in \Omega. \quad (7.10)$$

For any $x \in \Omega \subseteq \mathbb{R}^{2n}$, let us consider the function

$$\psi_x(t) = (D_x\Phi(t, x))^* J(D_x\Phi(t, x)).$$

Since

$$\frac{d}{dt}\psi_x(t) = \left(\frac{d}{dt}(D_x\Phi(t, x))^*\right) J(D_x\Phi(t, x))$$

$$+ (D_x\Phi(t, x))^* J\left(\frac{d}{dt}(D_x\Phi(t, x))\right),$$

from (7.10),

$$\frac{d}{dt}\psi_x(t) = (D_x\Phi(t, x))^* ((D\nabla H)(\Phi(t, x)))^* (-J) \cdot J(D_x\Phi(t, x))$$

$$+ (D_x\Phi(t, x))^* J \cdot J((D\nabla H)(\Phi(t, x)))(D_x\Phi(t, x))$$

$$= (D_x\Phi(t, x))^* ((D\nabla H)(\Phi(t, x)))^* (D_x\Phi(t, x))$$

$$- (D_x\Phi(t, x))^* ((D\nabla H)(\Phi(t, x)))(D_x\Phi(t, x)),$$

$$t \in I_x, \quad x \in \Omega.$$

We note that $D\nabla H(x)$ is symmetric, being equal to the Hessian matrix of $H(x)$. Therefore $(d/dt)\psi_x(t) \equiv 0$, for $t \in I_x$ and $x \in \Omega$. Hence $(D_x\Phi(t, x))^* J(D_x\Phi(t, x)) = (D_x\Phi(0, x))^* J(D_x\Phi(0, x)) = J$, $t \in I_x$, $x \in \Omega$. By Definition 7.1, $\Phi(t, x)$ is a symplectic diffeomorphism on $\Omega$ for any $t \in I_x$.                                  $\square$

Let $H(x) = H_2(x) + H_3(x) + \cdots + H_r(x) + O(|x|^{r+1})$ be the Hamiltonian function of a Hamiltonian system on a neighborhood of the origin in $\mathbb{R}^{2n}$, where $H_j \in P_{2n}^j$, $j = 2, \ldots, r$. From the results of Section 2.6, we may assume $H_2(x)$ is already in normal form. Thus to simplify $H(x)$ we may apply only near identity symplectic transformations. From Lemmas 7.3–7.7, any near identity symplectic transformation is of the form

$$x = y + J\nabla F_{k+1}(y) + O(|y|^{k+1}), \qquad x \in \Omega,$$

where $F_{k+1}(y) \in P_{2n}^{k+1}$, $k \geq 2$, and $\Omega$ is a neighborhood of the origin in $\mathbb{R}^{2n}$.

**Lemma 7.8.** *If $H_2(x)$ is a quadratic form in $x_1, \ldots, x_{2n}$, then for any $k \geq 3$ and $F_k \in P_{2n}^k$, the kth-order homogeneous polynomial in the expansion of $H_2(y + J\nabla F_k(y) + O(|y|^k))$ is*

$$\sum_{i=1}^{n} \left( \frac{\partial F_k}{\partial y_{n+i}} \frac{\partial H_2}{\partial y_i} - \frac{\partial H_2}{\partial y_{n+i}} \frac{\partial F_k}{\partial y_i} \right).$$

*Proof.* Let $H_2(x) = \frac{1}{2}(x, Bx)$, where $B$ is a $2n \times 2n$ symmetric matrix, and $(\cdot, \cdot)$ is the usual scalar product in $\mathbb{R}^{2n}$. We note that $\nabla H_2(x) = Bx$. Then

$$H_2\big(y + J\nabla F_k(y) + O(|y|^k)\big)$$

$$= \tfrac{1}{2}\big(y + J\nabla F_k(y) + O(|y|^k), By + BJ\nabla F_k(y) + O(|y|^k)\big)$$

$$= \tfrac{1}{2}(y, By) + \tfrac{1}{2}(J\nabla F_k(y), By)$$

$$+ \tfrac{1}{2}(y, BJ\nabla F_k(y)) + O(|y|^{k+1}).$$

The $k$th-order terms in the above expansion are

$$\tfrac{1}{2}(J\nabla F_k(y), By) + \tfrac{1}{2}(y, BJ\nabla F_k(y))$$

$$= \tfrac{1}{2}(J\nabla F_k(y), By) + \tfrac{1}{2}(By, J\nabla F_k(y))$$

$$= (J\nabla F_k(y), By) = (J\nabla F_k(y), \nabla H_2(y))$$

$$= \sum_{i=1}^{n} \left( \frac{\partial P}{\partial y_{n+i}} \frac{\partial H_2}{\partial y_i} - \frac{\partial H_2}{\partial y_{n+i}} \frac{\partial F_k}{\partial y_i} \right). \qquad \square$$

**Definition 7.9.** Let $P, Q \in C^1(\mathbb{R}^{2n}, \mathbb{R})$.

$$[P, Q] = (J\nabla P, \nabla Q) = \sum_{i=1}^{n} \left( \frac{\partial P}{\partial x_{n+i}} \frac{\partial Q}{\partial x_i} - \frac{\partial Q}{\partial x_{n+i}} \frac{\partial P}{\partial x_i} \right)$$

is called the Poisson bracket of $P$ and $Q$.

If $H_2$ and $F_k$ are as in Lemma 7.8, then we have

$$H_2\big(x + J\nabla F_k(x) + O(|x|^k)\big)$$

$$= H_2(x) + [F_k(x), H_2(x)] + O(|x|^{k+1}). \qquad (7.11)$$

For a given quadratic form $H_2(x)$ in $2n$ variables $x_1, \ldots, x_{2n}$, we define a linear operator $\mathrm{ad}_{H_2}^k \colon P_{2n}^k \to P_{2n}^k$ by

$$\mathrm{ad}_{H_2}^k F(x) = [H_2(x), F(x)], \qquad F \in P_{2n}^k, \qquad (7.12)$$

where $k = 3, 4, \ldots$ . Thus (7.11) can be rewritten as

$$j^k H_2\big(x + J\nabla F_k(x) + O(|x|^k)\big) = H_2(x) - \mathrm{ad}_{H_2}^k F_k(x).$$

Let $R^k$ be the range of $\mathrm{ad}_{H_2}^k$ and $C^k$ be any complement of $R^k$ in $P_{2n}^k$.

**Theorem 7.10.** *There exist a series of near identity symplectic transformations*

$$x = y + J\nabla F_k(y) + O(|y|^k), \qquad y \in \Omega_r,$$

*where $F_k \in P_{2n}^k$, $k = 3, \ldots, r$, and $\Omega_r$ is a neighborhood of the origin in $\mathbb{R}^{2n}$, such that the Hamiltonian function $K(y)$ of the resulting system of (7.1) is of the form*

$$K(y) = H_2(y) + K_3(y) + \cdots + K_r(y) + O(|y|^{r+1}), \qquad y \in \Omega_r,$$

*where $K_j(y) \in C^j$, $j = 3, \ldots, r$.*

*Proof.* The theorem is proved by induction.                     □

**Definition 7.11.** A Hamiltonian function $H(x)$ is called an $H_2$-normal form up to order $r \geq 3$ if $H(x) = H_2(x) + K_3(x) + \cdots K_r(x)$, where $H_2(x)$ is a quadratic form and $K_j(x) \in C^j$, $j = 3, \ldots, r$.

To find normal forms of a Hamiltonian system, it is sufficient to determine the structure of a complementary subspace $C^k$ to the range of $\text{ad}_{H_2}^k$ for $3 \leq k \leq r$. We note that if $J \nabla H_2(x) = Ax$, where $A$ is an infinitesimally symplectic matrix, then $\text{ad}_{H_2}^k F^k(x) = (Ax, \nabla F^k(x))$, where $(\cdot, \cdot)$ is the usual inner product in $\mathbb{R}^{2n}$. We define $\text{ad}_A^k: P_{2n}^k \to P_{2n}^k$ by $\text{ad}_A^k F(x) = (Ax, \nabla F(x))$ for any $F \in P_{2n}^k$, $k \geq 2$. Then $\text{ad}_{H_2}^k = \text{ad}_A^k$. Hence we may use $\text{ad}_A^k$ to study the normal forms instead of $\text{ad}_{H_2}^k$.

**Definition 7.12.** Let $\sigma(A) = \{\lambda_1, \ldots, \lambda_n, -\lambda_1, \ldots, -\lambda_n\}$ be the spectrum of $A$. Then the following relations are called resonant conditions:

$$\sum_{i=1}^{n} \lambda_i(\alpha_i - \alpha_{i+n}) = 0, \quad |\alpha| \geq 3. \tag{7.13}$$

Let $(x_1, x_2, \ldots, x_{2n})$ be symplectic coordinates with respect to the standard basis of $\mathbb{R}^{2n}$, in which the semisimple part of matrix $A$ is $\text{diag}(\lambda_1, \ldots, \lambda_n, -\lambda_1, \ldots, -\lambda_n)$. Then a monomial $x^\alpha$ with $|\alpha| = k \geq 3$ is called a resonant monomial of order $k$ if and only if the multi-index $\alpha = (\alpha_1, \ldots, \alpha_{2n})$ satisfies (7.13).

**Theorem 7.13.** *If $A = \text{diag}(\lambda_1, \ldots, \lambda_n, -\lambda_1, \ldots, -\lambda_n)$ is the matrix of linear Hamiltonian system $\dot{x} = J \nabla H_2(x)$, then an $H_2$-normal form up to order $r \geq 3$ can be chosen so that its $k$th-order homogeneous terms are linear combinations of all resonant monomials of order $k$, $k = 3, \ldots, r$.*

*Proof.* For any monomial $x^\alpha \in P_{2n}^k$, a calculation shows that

$$\text{ad}_A^k x^\alpha = \sum_{i=1}^{n} \lambda_i(\alpha_i - \alpha_{i+n}) x^\alpha = 0.$$

This implies that $\text{Ker}(\text{ad}_A^k)$ is a complementary subspace to $\text{Im}(\text{ad}_A^k)$. Thus the result follows from the definitions of the $H_2$-normal form and resonant monomials.    □

**Corollary 7.14.** Suppose that $A = \text{diag}(\lambda_1, \ldots, \lambda_n, -\lambda_n, \ldots, -\lambda_n)$ is the matrix associated to the Hamiltonian function $H_2(x, y)$, $(x, y) \in \mathbb{R}^n \times \mathbb{R}^n$. If $\lambda_1, \ldots, \lambda_n$ are rationally independent, then the $H_2$-normal form up to order $r \geq 3$ is

$$H(x, y) = \sum_{i=1}^{n} \lambda_i x_i y_i + \sum_{2 \leq |\alpha| \leq [r/2]} a_\alpha x^\alpha y^\alpha,$$

where $\alpha = (\alpha_1, \ldots, \alpha_n)$ is a multi-index and all $a_\alpha$ are real constants.

**Example 7.15.** Suppose $H_2(x_1, x_2, y_1, y_2) = x_1 y_1 - x_2 y_2$. Then

$$A = \begin{bmatrix} 1 & & & \\ & -1 & & \\ & & -1 & \\ & & & 1 \end{bmatrix}, \quad \lambda_1 = 1, \quad \lambda_2 = -1.$$

This is the case called 1: $-1$ resonance. The resonant conditions are

$$(\alpha_1 - \alpha_3) - (\alpha_2 - \alpha_4) = 0, \quad |\alpha| \geq 3,$$

that is,

$$\alpha_1 + \alpha_4 = \alpha_2 + \alpha_3, \quad |\alpha| \geq 3.$$

This implies that any resonant monomial is of even order. Hence an $H_2$-normal form up to order $r$ is

$$H(x) = x_1 y_1 - x_2 y_2 + \sum_{k=2}^{[r/2]} \sum_{i=0}^{k} \sum_{j=0}^{k} c_{ij} x_1^i x_2^j y_1^{k-j} y_2^{k-i},$$

where $c_{ij}$ are real constants, $0 \leq i, j \leq k$, $k = 2, \ldots, [r/2]$.

**Example 7.16.** Suppose $H_2(x, y) = xy$. Then

$$A = \begin{bmatrix} 1 & 0 \\ 0 & -1 \end{bmatrix}.$$

$\lambda = 1$ is clearly rationally independent. Thus, by Corollary 7.14, an $H_2$-normal form of $H(x, y)$ up to order $r \geq 3$ is

$$H(x, y) = xy + \sum_{k=2}^{[r/2]} a_k x^k y^k,$$

where $a_k$ are real constants, $k = 2, \ldots, [r/2]$.

In general we can apply the following adjoint operator method to get the $H_2$-normal forms. In linear space $P_{2n}^k$ we define an inner product $\langle \cdot, \cdot \rangle_1$ as the following. If $p(x) = \sum_{|\alpha|=k} a_\alpha x^\alpha$, $q(x) = \sum_{|\alpha|=k} b_\alpha x^\alpha$, then $\langle p(x), q(x) \rangle_1 = \sum_{|\alpha|=k} a_\alpha b_\alpha \alpha!$. In fact, $\langle \cdot, \cdot \rangle_1$ is the inner product $\langle \cdot, \cdot \rangle$ defined in Section 2.1 for the case dimension $= 1$.

**Theorem 7.17.** *Under the scalar product $\langle \cdot, \cdot \rangle_1$ in the space $P_{2n}^k$, the linear operator $\mathrm{ad}_{A^*}^k \colon P_{2n}^k \to P_{2n}^k$ is the adjoint operator of $\mathrm{ad}_A^k$.*

*Proof.* Let

$$F(x) = \sum_{|\alpha|=k} f_\alpha x^\alpha \in P_{2n}^k, \qquad G(x) = \sum_{|\beta|=k} g_\beta x^\beta \in P_{2n}^k$$

where the $f_\alpha$ and $g_\beta$ are real constants. Then

$$\langle \mathrm{ad}_A^k F(x), G(x) \rangle_1 = \sum_{|\alpha|=k} \sum_{|\beta|=k} f_\alpha g_\beta \langle (Ax, \nabla x^\alpha), x^\beta \rangle_1.$$

$$\langle F(x), \mathrm{ad}_{A^*}^k G(x) \rangle_1 = \sum_{|\alpha|=k} \sum_{|\beta|=k} f_\alpha g_\beta \langle x^\alpha, (A^*x, \nabla x^\beta) \rangle_1.$$

Therefore it will be enough to show that the following is true for any two monomials $x^\alpha$ and $x^\beta$ in $P_{2n}^k$:

$$\langle (Ax, \nabla x^\alpha), x^\beta \rangle_1 = \langle x^\alpha, (A^*x, \nabla x^\beta) \rangle_1.$$

We have

$$\langle (Ax, \nabla x^\alpha), x^\beta \rangle_1$$

$$= \left\langle \left( \left( \sum_{j=1}^{n} a_{ij} x_j \right)_{i=1}^{n}, \left( \alpha_i \frac{x^\alpha}{x_i} \right)_{i=1}^{n} \right), x^\beta \right\rangle_1$$

$$= \left\langle \sum_{i=1}^{n} \alpha_i \sum_{j=1}^{n} a_{ij} \frac{x^\alpha x_j}{x_i}, x^\beta \right\rangle_1$$

$$= \begin{cases} \left( \sum_{i=1}^{n} \alpha_i a_{ii} \right) \alpha!, & \text{if } \beta = \alpha, \\ a_{ij} \alpha_i \beta!, & \text{if } \beta_i = \alpha_i - 1, \beta_j = \alpha_j + 1 \\ & \quad \text{for some } i \neq j, \\ & \quad \text{and } \beta_k = \alpha_k \text{ for } k \neq i \text{ and } j, \\ 0, & \text{otherwise.} \end{cases}$$

$$\langle x^\alpha, (A^*x, \nabla x^\beta) \rangle_1$$

$$= \left\langle x^\alpha, \left( \left( \sum_{j=1}^{n} a_{ji} x_j \right)_{i=1}^{n}, \left( \beta_i \frac{x^\beta}{x_i} \right)_{i=1}^{n} \right) \right\rangle_1$$

$$= \left\langle x^\alpha, \sum_{i=1}^{n} \beta_i \sum_{j=1}^{n} a_{ji} \frac{x^\beta x_j}{x_i} \right\rangle_1 = \left\langle x^\alpha, \sum_{j=1}^{n} \beta_j \sum_{i=1}^{n} a_{ij} \frac{x^\beta x_i}{x_j} \right\rangle_1$$

$$= \begin{cases} \left( \sum_{j=1}^{n} \beta_j a_{jj} \right) \alpha!, & \text{if } \alpha = \beta, \\ a_{ij} \beta_j \alpha!, & \text{if } \alpha_i = \beta_i + 1, \alpha_j = \beta_j - 1 \\ & \quad \text{for some } i \neq j, \\ & \quad \text{and } \alpha_k = \beta_k \text{ for } k \neq i \text{ and } j, \\ 0, & \text{otherwise.} \end{cases}$$

Both expressions are equal. Thus the lemma is proved.    □

**Corollary 7.18.** $\mathrm{Ker}(\mathrm{ad}_{A*}^k)$ is a complementary subspace to the range of $\mathrm{ad}_A^k$ in $P_{2n}^k$ for $k \geq 3$.

**Example 7.19.** Suppose $H_2(x, y) = \frac{1}{2}y^2$. Then

$$A = \begin{bmatrix} 0 & 1 \\ 0 & 0 \end{bmatrix}.$$

Consider the system of linear partial differential equations

$$\left( A^*\begin{pmatrix} x \\ y \end{pmatrix}, \nabla F(x, y) \right) = 0,$$

that is,

$$x \frac{\partial F}{\partial y} = 0. \tag{7.14}$$

The homogeneous polynomial solutions of (7.14) are

$$F(x, y) = c_k x^k, \qquad k \geq 1,$$

where $c_k$ are real constants. Hence an $H_2$-normal form up to order $r \geq 3$ is

$$H(x, y) = \frac{1}{2}y^2 + \sum_{k=3}^{r} c_k x^k,$$

where $c_k$ are real constants.

**Example 7.20.** Suppose $H_2(x_1, x_2, y_1, y_2) = \frac{1}{2}(y_1^2 + y_2^2)$. Then

$$A = \begin{bmatrix} 0 & 0 & 1 & 0 \\ 0 & 0 & 0 & 1 \\ 0 & 0 & 0 & 0 \\ 0 & 0 & 0 & 0 \end{bmatrix}.$$

If we solve the partial differential equation

$$(A^*x, \nabla F(x)) = x_1 \frac{\partial F}{\partial y_1} + x_2 \frac{\partial F}{\partial y_2} = 0,$$

where $x = (x_1, x_2, y_1, y_2)^T$, then any polynomial solution is of the following form:

$$F(x) = \Phi(x_1, x_2, x_1 y_2 - x_2 y_1),$$

where $\Phi$ is an arbitrary differentiable function. In order to show $\Phi$ is a polynomial in its arguments, without loss of generality, we assume $F$ is a homogeneous polynomial of order $m$. Let

$$F(x_1, x_2, y_1, y_2) = \sum_{i+j+k+l=m} c_{ijkl} x_1^i x_2^j y_1^k y_2^l.$$

Denote $z = x_1 y_2 - x_2 y_1$. If $x_1 \neq 0$, then we have

$$\Phi(x_1, x_2, z) = \sum_{i+j+k+l=m} c_{ijkl} x_1^{i-l} x_2^j y_1^k (z + x_2 y_1)^l.$$

Note that the left-hand side of the above equality is independent of $y_1$. Taking $y_1 = 0$, we have

$$F(x_1, x_2, y_1, y_2) = \Phi(x_1, x_2, z) = \sum_{i+j+l=m} c_{ij0l} x_1^{i-l} x_2^j z^l$$

$$= \sum_{i+j+l=m} c_{ij0l} x_1^{i-l} x_2^j (x_1 y_2 - x_2 y_1)^l.$$

Since $F(x_1, x_2, y_1, y_2)$ is a polynomial in $x_1, x_2, y_1, y_2$, $c_{ij0l} = 0$ if $i < l$. Hence

$$\Phi(x_1, x_2, z) = \sum_{\substack{i+j+l=m \\ i \geq l}} c_{ij0l} x_1^{i-l} x_2^j z^l.$$

By reindexing, we get

$$\Phi(x_1, x_2, z) = \sum_{k+j+2l=m} c_{jkl} x_1^k x_2^j z^l,$$

where $c_{jkl}$ are real constants. So $\Phi$ is a polynomial in its arguments and has the form shown above. Therefore, an $H_2$-normal form up to order $r$ is

$$H(x_1, x_2, y_1, y_2) = \tfrac{1}{2}\left(y_1^2 + y_2^2\right) + \sum_{i+j+2k=3}^{r} c_{ijk} x_1^i x_2^j (x_1 y_2 - x_2 y_1)^k,$$

where $c_{ijk}$ are real constants. For $r = 3$,

$$\text{Ker}\left(\widetilde{\text{ad}}^3_{A*}\right)$$

$$= \text{span}\left\{x_1^3, x_1^2 x_2, x_1 x_2^2, x_2^3, x_1(x_1 y_2 - x_2 y_1), x_2(x_1 y_2 - x_2 y_1)\right\}.$$

By an elementary argument we can choose also the following as a complementary subspace to the range of $\text{ad}_A^3$ in $P_4^3$:

$$C^3 = \text{span}\left\{x_1^3, x_1^2 x_2, x_1 x_2^2, x_2^3, x_1^2 y_2, x_2^2 y_1\right\}.$$

Hence an $H_2$-normal form up to order 3 is

$$H(x) = \tfrac{1}{2}\left(y_1^2 + y_2^2\right) + a_1 x_1^3 + a_2 x_1^2 x_2$$

$$+ a_3 x_1 x_2^2 + a_4 x_2^3 + a_5 x_1^2 y_2 + a_6 x_2^2 y_2,$$

where $a_1, \ldots, a_6$ are all real constants.

We suppose that the standard basis $\{x^\alpha \mid |\alpha| = k\}$ of $P_{2n}^k$ is in the reverse lexicographic ordering. We denote by $\widetilde{\text{ad}}_A^k$ the matrix representation of $\text{ad}_A^k$ with respect to the basis $\{x^\alpha \mid |\alpha| = k\}$.

**Lemma 7.21.** *If* $A = \text{diag}(\lambda_1, \ldots, \lambda_n, -\lambda_1, \ldots, -\lambda_n)$, *then* $\widetilde{\text{ad}}_A^k$ *is also diagonal. If $A$ is upper (or lower) triangular with diagonal elements* $\{\lambda_1, \ldots, \lambda_n, -\lambda_1, \ldots, -\lambda_n\}$, *then* $\widetilde{\text{ad}}_A^k$ *is lower (or upper) triangular. Furthermore, if $x^\alpha$ is the ith basis element of $P_{2n}^k$, then for both cases the ith element of the diagonal of* $\widetilde{\text{ad}}_A^k$ *is* $\sum_{i=1}^n \lambda_i(\alpha_i - \alpha_{i+n})$.

The proof of Lemma 7.21 is similar to that of Lemma 1.13.

**Theorem 7.22.** *Suppose $A = S + N$ is the $S - N$ decomposition of $A$. Then* $\text{ad}_A^k = \text{ad}_S^k + \text{ad}_N^k$ *is the $S - N$ decomposition of* $\text{ad}_A^k$ *and* $\text{Ker}(\text{ad}_{S*}^k) \cap \text{Ker}(\text{ad}_{N*}^k)$ *is a complementary subspace to the range of* $\text{ad}_A^k$ *in $P_{2n}^k$ for each $k \geq 3$.*

The proof of Theorem 7.22 is similar to that of Theorem 1.18.

We can also choose $\text{Ker}(\text{ad}_S^k) \cap \text{Ker}(\text{ad}_{N*}^k)$ as a complementary subspace to the range of $\text{ad}_A^k$ in $P_{2n}^k$ for $k \geq 3$.

**Corollary 7.23.** *Let $A$ be the matrix of linear Hamiltonian system $\dot{x} = J\nabla H_2(x)$, $x \in \mathbb{R}^{2n}$. If $A = S + N$ is the $S - N$ decomposition of $A$ and $S = \mathrm{diag}(\lambda_1, \ldots, \lambda_n, -\lambda_1, \ldots, -\lambda_n)$, then an $H_2$-normal form up to order $r \geq 3$ can be chosen so that its nonlinear part of order greater than 2 is spanned by resonant monomials up to order $r$.*

**Corollary 7.24.** *Suppose $A$ is the matrix of the linear Hamiltonian system $\dot{x} = J\nabla H_2(x)$, $x \in \mathbb{R}^{2n}$. Let $A = S + N$ be the $S - N$ decomposition of $A$. An $H_2$-normal form can be chosen so that it satisfies the relation $H(e^S x) = H(x)$ for any $x$ in a neighborhood of the origin in $\mathbb{R}^{2n}$. Consequently an $H_2$-normal form of the Hamiltonian system can be chosen so that it has group $\Gamma$-symmetry, where $\Gamma$ is the group generated by $e^S$.*

For the case when the semisimple part of the matrix $A$ is diagonalizable over the complex numbers, we can apply a method similar to that introduced in Section 2.1 for real normal forms to get a real basis of $\mathrm{Ker}(\mathrm{ad}_{A^*}^k)$. We illustrate this idea by Examples 7.25 and 7.28.

**Example 7.25.** Let $H_2(x, y) = \frac{1}{2}(x^2 + y^2)$. Then

$$A = \begin{bmatrix} 0 & 1 \\ -1 & 0 \end{bmatrix}.$$

We change variables to $z_1 = x + iy$, $z_2 = x - iy$. Then the matrix of the linear transformation is

$$P = \frac{1}{2}\begin{bmatrix} 1 & 1 \\ -i & i \end{bmatrix},$$

and the matrix of the linear part of the transformed equation is $\tilde{A} = P^{-1}AP = \mathrm{diag}(i, -i)$ with respect to the new basis. The resonant monomials are $\{z_1^k z_2^k, k \geq 2\}$. Since $\bar{z}_2 = z_1$, $z_1^k z_2^k$ is real. We change coordinates by $z_1 = x + iy$ and $z_2 = x - iy$. Then the corresponding real basis of $\mathrm{Ker}(\mathrm{ad}_A^{2k})$ is $\{(x^2 + y^2)^k\}$ for each $k \geq 2$ and $\mathrm{Ker}(\mathrm{ad}_A^{2k+1}) = \{0\}$ for any $k \geq 1$. Hence an $H_2$-normal form up to order $r$ is

$$H(x, y) = \frac{1}{2}(x^2 + y^2) + a_2(x^2 + y^2)^2 + \cdots + a_k(x^2 + y^2)^k,$$

where $a_2, \ldots, a_k$ are real constants, $r - 1 \leq 2k \leq r$.

**Definition 7.26.** If Hamiltonian $H$ is a polynomial of degree $r$ in the symplectic variables $x_1, \ldots, x_n, y_1, \ldots, y_n$, that is actually a polynomial of degree $[r/2]$ in the variables $\rho_i = (x_i^2 + y_i^2)/2$, $i = 1, \ldots, n$, then $H$ is called a Birkhoff normal form of degree $r$.

Then by a similar argument to that in example 7.25, we have the following theorem.

**Theorem 7.27.** *Let* $H_2(x_1, \ldots, x_n, y_1, \ldots, y_n) = \frac{1}{2}\lambda_1(x_1^2 + y_1^2) + \cdots + \frac{1}{2}\lambda_n(x_n^2 + y_n^2)$, *where the* $\lambda_i$ *are real constants. If* $\lambda_1, \ldots, \lambda_n$ *are rationallly independent, then an* $H_2$-*normal form up to order* $r \geq 3$ *is a Birkhoff normal form of degree* $r$.

**Example 7.28.** Suppose

$$H_2(x_1, x_2, y_1, y_2) = \alpha(x_1 y_2 - x_2 y_1) - \tfrac{1}{2}\rho(x_1^2 + x_2^2),$$

where $\alpha > 0$, $\rho = \pm 1$. Then

$$A = \begin{bmatrix} 0 & -\alpha & 0 & 0 \\ \alpha & 0 & 0 & 0 \\ \rho & 0 & 0 & -\alpha \\ 0 & \rho & \alpha & 0 \end{bmatrix}.$$

Let $z_1 = x_1 + ix_2$, $z_2 = y_1 + iy_2$, $z_3 = x_1 - ix_2$, $z_4 = y_1 - iy_2$. Then we have

$$\tilde{A} = \begin{bmatrix} \alpha i & & & \\ \rho & \alpha i & & \\ & & -\alpha i & \\ & & \rho & -\alpha i \end{bmatrix}.$$

Let $\tilde{A} = \tilde{S} + \tilde{N}$ be the $S - N$ decomposition of $\tilde{A}$, where

$$\tilde{S} = \begin{bmatrix} \alpha i & & & \\ & \alpha i & & \\ & & -\alpha i & \\ & & & -\alpha i \end{bmatrix} \quad \text{and} \quad \tilde{N} = \begin{bmatrix} 0 & & & \\ \rho & 0 & & \\ & & 0 & \\ & & \rho & 0 \end{bmatrix}.$$

The set of all resonant monomials of order $k$ forms a basis of $\text{Ker}(\text{ad}_{\tilde{S}}^k)$. The resonant conditions are $\alpha_1 + \alpha_2 = \alpha_3 + \alpha_4$ with $|\alpha| = k \geq 3$. Obviously, all resonant monomials must be of even order. All the fourth-order resonant monomials are $z_1^2 z_3^2$, $z_1 z_2 z_3^2$, $z_2^2 z_3^2$, $z_1^2 z_3 z_4$, $z_1 z_2 z_3 z_4$, $z_2^2 z_3 z_4$, $z_1^2 z_4^2$, $z_1 z_2 z_4^2$, and $z_2^2 z_4^2$.

To find a basis of

$$\tilde{C}^4 = \text{Ker}\big(\text{ad}_{\tilde{S}}^4\big) \cap \text{Ker}\big(\text{ad}_{\tilde{N}*}^4\big),$$

we apply the undetermined coefficient method. Since the system of linear partial differential equations for $\text{Ker}(\text{ad}_{\tilde{N}*}^4)$ is

$$z_2 \frac{\partial F}{\partial z_1} + z_4 \frac{\partial F}{\partial z_3} = 0,$$

we get for a basis of $\tilde{C}^4$: $z_2^2 z_4^2$, $i(z_1 z_2 z_4^2 - z_2^2 z_3 z_4)$, $(z_2 z_3 - z_1 z_4)^2$. We note that $\bar{z}_3 = z_1$ and $\bar{z}_4 = z_2$, whence this basis is real. After changing variables by $z_1 = x_1 + ix_2$, $z_2 = y_1 + iy_2$, $z_3 = x_1 - ix_2$, $z_4 = y_1 - iy_2$, we get a real basis of $C^4$: $(y_1^2 + y_2^2)^2$, $(y_1^2 + y_2^2)(x_1 y_2 - x_2 y_1)$, $(x_1 y_2 - x_2 y_1)^2$. Thus the $H_2$-normal form up to order 4 is

$$H(x_1, x_2, y_1, y_2) = \alpha(x_1 y_2 - x_2 y_1) - \tfrac{1}{2}\rho(x_1^2 + x_2^2) + a(y_1^2 + y_2^2)^2$$

$$+ b(y_1^2 + y_2^2)(x_1 y_2 - x_2 y_1) + c(x_1 y_2 - x_2 y_1)^2,$$

where $a, b, c$ are real constants.

Since $\text{ad}_A^k$ is a linear operator on $P_{2n}^k$, we can apply also a matrix representation method to find complement $C^k$ for each $k \geq 3$. We illustrate this method with the following example.

**Example 7.29.** Suppose $H_2(x_1, x_2, y_1, y_2) = x_1 y_2 \mp \tfrac{1}{2} x_2^2$. Then

$$A = \begin{bmatrix} 0 & 0 & 0 & 0 \\ 1 & 0 & 0 & 0 \\ 0 & 0 & 0 & -1 \\ 0 & \pm 1 & 0 & 0 \end{bmatrix}.$$

For $k = 3$, under the basis $\{x^\alpha, |\alpha| = 3\}$ of $P_4^3$ which is in the reverse lexicographic ordering, the matrix representation $\widetilde{\mathrm{ad}_A^3}$ of operator $\mathrm{ad}_A^3$ is shown in the table below.

$$
\begin{bmatrix}
0 & 1 & 0 & 0 & 0 & 0 & 0 & 0 & 0 & 0 & 0 & 0 & 0 & 0 & 0 & 0 & 0 & 0 & 0 & 0 \\
0 & 0 & 0 & \pm1 & 2 & 0 & 0 & 0 & 0 & 0 & 0 & 0 & 0 & 0 & 0 & 0 & 0 & 0 & 0 & 0 \\
0 & 0 & 0 & 0 & 0 & 1 & 0 & 0 & 0 & 0 & 0 & 0 & 0 & 0 & 0 & 0 & 0 & 0 & 0 & 0 \\
0 & 0 & -1 & 0 & 0 & 0 & 1 & 0 & -1 & 0 & 0 & 0 & 0 & 0 & 0 & 0 & 0 & 0 & 0 & 0 \\
0 & 0 & 0 & 0 & 0 & 0 & \pm1 & 0 & 0 & 0 & 3 & 0 & 0 & 0 & 0 & 0 & 0 & 0 & 0 & 0 \\
0 & 0 & 0 & 0 & 0 & 0 & 0 & 0 & \pm1 & 0 & 0 & 2 & 0 & 0 & 0 & 0 & 0 & 0 & 0 & 0 \\
0 & 0 & 0 & 0 & 0 & -1 & 0 & 0 & 0 & \pm2 & 0 & 0 & 2 & 0 & 0 & 0 & 0 & 0 & 0 & 0 \\
0 & 0 & 0 & 0 & 0 & 0 & 0 & 0 & 0 & 0 & 0 & 0 & 0 & 1 & 0 & 0 & 0 & 0 & 0 & 0 \\
0 & 0 & 0 & 0 & 0 & 0 & 0 & -2 & 0 & 0 & 0 & 0 & 0 & 0 & 1 & 0 & 0 & 0 & 0 & 0 \\
0 & 0 & 0 & 0 & 0 & 0 & 0 & 0 & 0 & 0 & 0 & 0 & 0 & 0 & 0 & 1 & 0 & 0 & 0 & 0 \\
0 & 0 & 0 & 0 & 0 & 0 & 0 & 0 & 0 & 0 & 0 & 0 & \pm1 & 0 & 0 & 0 & 0 & 0 & 0 & 0 \\
0 & 0 & 0 & 0 & 0 & 0 & 0 & 0 & 0 & 0 & 0 & 0 & 0 & 0 & \pm1 & 0 & 0 & 0 & 0 & 0 \\
0 & 0 & 0 & 0 & 0 & 0 & 0 & 0 & 0 & 0 & 0 & -1 & 0 & 0 & 0 & \pm1 & 0 & 0 & 0 & 0 \\
0 & 0 & 0 & 0 & 0 & 0 & 0 & 0 & 0 & 0 & 0 & 0 & 0 & 0 & 0 & 0 & 0 & \pm1 & 0 & 0 \\
0 & 0 & 0 & 0 & 0 & 0 & 0 & 0 & 0 & 0 & 0 & 0 & 0 & -2 & 0 & 0 & 0 & 0 & \pm2 & 0 \\
0 & 0 & 0 & 0 & 0 & 0 & 0 & 0 & 0 & 0 & 0 & 0 & 0 & 0 & -1 & 0 & 0 & 0 & 0 & \pm3 \\
0 & 0 & 0 & 0 & 0 & 0 & 0 & 0 & 0 & 0 & 0 & 0 & 0 & 0 & 0 & 0 & 0 & 0 & 0 & 0 \\
0 & 0 & 0 & 0 & 0 & 0 & 0 & 0 & 0 & 0 & 0 & 0 & 0 & 0 & 0 & 0 & -3 & 0 & 0 & 0 \\
0 & 0 & 0 & 0 & 0 & 0 & 0 & 0 & 0 & 0 & 0 & 0 & 0 & 0 & 0 & 0 & 0 & -2 & 0 & 0 \\
0 & 0 & 0 & 0 & 0 & 0 & 0 & 0 & 0 & 0 & 0 & 0 & 0 & 0 & 0 & 0 & 0 & 0 & -1 & 0
\end{bmatrix}
$$

Then $\mathrm{Ker}((\widetilde{\mathrm{ad}_A^3})^*) = \mathrm{span}\{e_{14} \pm \frac{1}{2}e_{19}, 2e_8 + e_{15} \pm 2e_{20}, e_{17}\}$. We can choose $\tilde{C}^3 = \mathrm{span}\{e_{14}, e_{15}, e_{17}\}$ as a complementary space to the range of $\widetilde{\mathrm{ad}_A^3}$ in $\mathbb{R}^{20}$. Then an $H_2$-normal form up to order 3 is

$$
H(x_1, x_2, y_1, y_2) = x_1 y_2 \mp \tfrac{1}{2}x_2^2 + a_1 x_2 y_1^2 + a_2 x_2 y_1 y_2 + a_3 y_1^3,
$$

where $a_1, a_2, a_3$ are all real constants.

## 2.8 Takens's Theorem

Takens's Theorem gives relationships between diffeomorphisms and vector fields in neighborhoods of the origin in $\mathbb{R}^n$. In the simple case where $F(x) = Ax$ is a linear diffeomorphism on $\mathbb{R}^n$, if there exists a real $n \times n$ matrix $B$ such that $\exp(B) = A$, then $F(x)$ is the time-one map of the flow generated by the vector field $X(x) = Bx$. Takens's Theorem generalizes this simple case to nonlinear diffeomorphisms.

We assume that $F: \mathbb{R}^n \to \mathbb{R}^n$ is a $C^r$-diffeomorphism ($r \geq 2$), with $F(0) = 0$. The Taylor expansion of $F(x)$ at the origin is

$$
F(x) = Ax + F^2(x) + F^3(x) + \cdots + F^r(x) + o(|x|^r),
$$

$$\text{as } x \to 0, \quad (8.1)$$

where $F^k \in H_n^k$ for $k = 2, \ldots, r$. We assume that

$$A = S(I + N), \qquad (8.2)$$

where $S$ is semisimple, $N$ is nilpotent, $SN = NS$ and $I$ is the identity. There is no loss of generality in this assumption since any linear transformation on $\mathbb{R}^n$ has the $S - N$ decomposition and if the transformation is invertible so is its semisimple part. Then Takens's Theorem can be stated as follows:

**Theorem 8.1.** (*Takens*) *Given the diffeomorphism* $F(x)$ *defined by* (8.1) *and* (8.2) *and any integer* $1 \leq l \leq r$, *there exist a diffeomorphism* $\psi_l$: $\Omega \subseteq \mathbb{R}^n \to \mathbb{R}^n$, *where* $\Omega$ *is a neighborhood of the origin in* $\mathbb{R}^n$, *and a vector field* $X(x)$ *on* $\mathbb{R}^n$ *such that*
(i)   $j^l(\psi_l \circ F \circ \psi_l^{-1})$ *is an A-normal form of diffeomorphism* $F(x)$ *up to order* $l$,
(ii)  $X(Sx) = SX(x)$ *for any* $x \in \mathbb{R}^n$,
(iii) $j^l(\psi_l \circ F \circ \psi_l^{-1})(x) = j^l(\Phi_X(1, Sx))$,
    where $j^l$ is the truncation operator up to order $l$ and $\Phi_X(t, x)$ is the flow of $X(x)$. Furthermore for such a vector field $X(x), j^l X(x)$ is uniquely determined by $j^l F(x)$.
    Theorem 8.1 will be proved using several lemmas. First we consider the simple case when $A = I + N$, where $N$ is nilpotent and upper triangular, that is, the diffeomorphism is given by

$$F_1(x) = (I + N)x + F^2(x) + F^3(x) + \cdots + F^r(x) + o(|x|^r).$$

$$(8.3)$$

**Lemma 8.2.** *There exists a vector field* $X(x)$ *on* $\mathbb{R}^n$ *such that* $j^r X(x)$ *is uniquely determined by* $j^r F_1(x)$, *and its flow* $\Phi_X(t, x)$ *satisfies*:

$$j^r \Phi_X(1, x) = j^r F_1(x). \qquad (8.4)$$

*Proof.* Assume that $X(x)$ is of the form:

$$X(x) = Bx + X^2(x) + \cdots + X^r(x) + O(|x|^{r+1}), \qquad (8.5)$$

where $B \in \mathbb{R}^{n \times n}$ and $X^k \in H_n^k$, $k = 2, \ldots, r$ and that its flow is of the form:

$$\Phi_X(t, x) = \Phi^1(t)x + \Phi^2(t, x) + \cdots + \Phi^r(t, x) + O(|x|^{r+1}),$$

$$(8.6)$$

where $\Phi^k(t, \cdot) \in H_n^k$ for any $t \in \mathbb{R}$; and $\Phi^1(t)$ and $\Phi^k(t, x)$ for $k = 2, \ldots, r$ are $C^r$ in $t$. Then we have to determine the vector field $X(x)$ such that its flow $\Phi_X(t, x)$ satisfies (8.4). Since $\Phi_X$ is the flow of $X$:

$$\begin{cases} \dot{\Phi}_X(t, x) = X(\Phi_X(t, x)) \\ \qquad = B\Phi_X(t, x) + X^2(\Phi_X(t, x)) \\ \qquad + \cdots + X^r(\Phi_X(t, x)) + O(|x|^{r+1}), \\ \Phi_X(0, x) = x. \end{cases}$$

$$(8.7)$$

We denote by $p^{ij}(t, x)$ ($j > 1$, $i = 2, \ldots, r - 1$) the homogeneous polynomial of order $j$ in the expansion of $X^i(\Phi_X(t, x))$. From (8.6) and (8.7), comparing equal-order homogeneous polynomials, we have

$$\begin{cases} \dot{\Phi}^1(t) = B\Phi^1(t), \\ \Phi^1(0) = I, \end{cases}$$

$$(8.8)$$

$$\begin{cases} \dot{\Phi}^k(t, x) = B\Phi^k(t, x) + \sum_{i=2}^{k-1} p^{ik}(t, x) + X^k(\Phi^1(t, x)), \\ \Phi^k(t, 0) = 0. \end{cases}$$

$$2 \le k \le r, \quad (8.9)$$

The solution of (8.8) is $\Phi^1(t) = e^{Bt}$ and therefore $j^1 \Phi_X(1, x) = j^1 F_1(x)$ implies $e^B = I + N$. Then $B = \log(I + N) \equiv \tilde{N}$ is also upper triangular. So, the solution of (8.8) is now determined as:

$$\Phi^1(t) = e^{\tilde{N}t} \quad \text{with} \quad \tilde{N} = \log(I + N). \qquad (8.10)$$

Assume now by induction that $X^2, \ldots, X^k$, $\Phi^2, \ldots, \Phi^k$ have already been determined. Then

$$p^{k+1}(t, x) \equiv \sum_{i=2}^{k} p^{i, k+1}(t, x)$$

is known and (8.9) (with $k + 1$ instead of $k$) becomes

$$\begin{cases} \dot{\Phi}^{k+1}(t, x) = \tilde{N}\Phi^{k+1}(t, x) + p^{k+1}(t, x) + X^{k+1}(e^{\tilde{N}t}x) \\ \Phi^{k+1}(0, x) = 0. \end{cases}$$

By the variation of constants formula, we get

$$\Phi^{k+1}(t, x) = e^{\tilde{N}t} \int_0^t e^{-\tilde{N}s} \left[ X^{k+1}(e^{\tilde{N}s}x) + p^{k+1}(s, x) \right] ds. \quad (8.11)$$

From (8.4), $\Phi^{k+1}(1, x) = F^{k+1}(x)$, which gives:

$$\int_0^1 e^{-\tilde{N}t} X^{k+1}(e^{\tilde{N}t}x) dt = e^{-\tilde{N}} F^{k+1}(x) - \int_0^1 e^{-\tilde{N}t} p^{k+1}(t, x) dt. \quad (8.12)$$

We note that both sides of (8.12) belong to $H_n^{k+1}$ and $X^{k+1}$ is unknown. We define the operator $T^k \colon H_n^k \to H_n^k$ by

$$(T^k h)(x) = \int_0^1 e^{-\tilde{N}t} h(e^{\tilde{N}t}x) dt, \qquad h \in H_n^k. \quad (8.13)$$

**Claim 8.3.** $T^k$ is a linear invertible operator for $k \geq 2$.

From Claim 8.3 it follows that (8.12) is uniquely solvable for $X^{k+1}$ and consequently $\Phi^{k+1}$ is determined by (8.11). $\qquad \Box$

*Proof of Claim 8.3.* It is obvious that $T^k$ is linear. Since $H_n^k$ is a finite dimensional linear space, to show that $T^k$ is invertible is equivalent to showing that the null space of $T^k$ is $\{0\}$. Since $N$ is upper triangular and nilpotent,

$$\tilde{N} = \log(I + N) = \sum_{j=1}^m \frac{(-N)^j}{j}$$

for some integer $m$ is also upper triangular and nilpotent. Thus, $e^{\tilde{N}t}$ and $e^{-\tilde{N}t}$ are of the form:

$$\begin{bmatrix} 1 & & & * \\ & 1 & & \\ & & \ddots & \\ 0 & & & 1 \end{bmatrix},$$

where (*) denotes entries depending on $t$. Let $x^\alpha e_j$ be any monomial in $H_n^k$. Then $(e^{\tilde{N}t}x)^\alpha e_j = x^\alpha e_j + r(t, x)e_j$, where $r(t, x)e_j$ consists of terms "bigger" than $x^\alpha e_j$ in the reverse lexicographic ordering for any $t \in \mathbb{R}$. Similarly, $e^{-\tilde{N}t}(e^{\tilde{N}t}x)^\alpha e_j = x^\alpha e_j + \tilde{r}(t, x)$, where $\tilde{r}(t, x)$ consists only of terms "bigger" than $x^\alpha e_j$ for any $t \in \mathbb{R}$. Let $0 \ne h \in H_n^k$ and $cx^\alpha e_j$ with $c \ne 0$, $|\alpha| = k$, $1 \le j \le n$, be the "smallest" monomial of $h$ with respect to the reverse lexicographic ordering. Thus from above discussion, $(T^k h)(x) = cx^\alpha e_j + \hat{r}(x)$, where $\hat{r}(x)$ consists only of terms "bigger" than $x^\alpha e_j$ in the reverse lexicographic ordering. This implies that $(T^k h)(x) \ne 0$ if $h \ne 0$.                                         □

**Corollary 8.4.** *Lemma 8.2 is still valid in the case where N is not upper triangular.*

*Proof.* Let $P$ be a nonsingular $n \times n$ matrix. It is easy to see that (1) $\log(I + P^{-1}NP) = P^{-1}\log(I + N)P$ and (2) $h(x)$ is a solution of $T^k(h) = f$ if and only if $\tilde{h}(x) = P^{-1}h(Px)$ is the solution of

$$\int_0^1 e^{-P^{-1}\tilde{N}Pt}\tilde{h}(e^{P^{-1}\tilde{N}Pt}x)dt = P^{-1}f(Px).$$

Hence $T^k$ is also invertible in the case where $\tilde{N}$ is not upper triangular.                                         □

**Lemma 8.5.** *Let $X(x)$ be a $C^r$ vector field on $\mathbb{R}^n$ and $\Phi_X(t, x)$ be its flow. Suppose that $\sigma$ is an invertible $n \times n$ matrix. Then $\sigma^{-1}\Phi_X(t, \sigma x)$ is the flow of the vector field $\sigma^{-1}X(\sigma x)$, that is,*

$$\Phi_{\sigma^{-1}X\sigma}(t, x) = \sigma^{-1}\Phi_X(t, \sigma x).$$

*Proof.* Let $\Phi_X(t, x)$ be the flow of $\dot{y} = X(y)$. Then $\sigma^{-1}\Phi_X(t, \sigma x)$ satisfies the equation $\dot{z} = \sigma^{-1}X(\sigma z)$ and the initial condition at $t = 0$ is $\sigma^{-1}\Phi_X(0, \sigma x) = \sigma^{-1}\sigma x = x$. Therefore $\sigma^{-1}\Phi_X(t, \sigma x)$ is the flow of the vector field $\sigma^{-1}X(\sigma x)$.                                         □

**Lemma 8.6.** *Suppose that $F(x)$ is a $C^r$ diffeomorphism with linear part $(I + N)x$, where $N$ is nilpotent, and that $F(x)$ commutes with a nonsingular matrix $\sigma \in \mathbb{R}^{n \times n}$. Then the vector field $X(x)$ defined by Lemma 8.2 has $\sigma$ symmetry up to order $r$, that is,*

$$j^r \sigma^{-1} X(\sigma x) = j^r X(x) \quad \text{for any } x \in \mathbb{R}^n.$$

*Proof.* By Lemma 8.2, $j^r \Phi_X(1, x) = j^r F(x)$. Therefore, $\sigma^{-1} j^r \Phi_X(1, \sigma x) = \sigma^{-1} j^r F(\sigma x)$ and since $F$ has $\sigma$-symmetry, $j^r \sigma^{-1} \Phi_X(1, \sigma x) = j^r F(x)$. By Lemma 8.5, $j^r \Phi_{\sigma^{-1} X \sigma}(1, x) = j^r F(x)$ and by Lemma 8.2, $j^r X(x)$ is uniquely determined by $j^r F(x)$. Thus

$$j^r \sigma^{-1} X(\sigma x) = j^r X(x), \quad \text{for any } x \in \mathbb{R}^n. \qquad \square$$

*Proof of Theorem 8.1.* From Corollary 4.12 an $A$-normal form of diffeomorphism (8.1) can be chosen so that it commutes with $S$, the semisimple part of $A$. Let $\psi_l$ be a diffeomorphism in neighborhood $\Omega$ of the origin in $\mathbb{R}^n$ for $2 \le l \le r$ such that

$$j^l \big( \psi_l \circ F \circ \psi_l^{-1} \big)(x) = S(I + N)x + \bar{F}^2(x) + \cdots + \bar{F}^l(x)$$

is the $A$-normal form of (8.1) which commutes with $S$. We factor out $S$. Then we get

$$j^l \big( \psi_l \circ F \circ \psi_l^{-1} \big)(x) = S\big[ (I + N)x + \hat{F}^2(x) + \cdots + \hat{F}^l(x) \big], \quad (8.14)$$

where $\hat{F}^k(x) = S^{-1} \bar{F}^k(x)$, $2 \le k \le l$, and $\hat{F}^k$ commutes with $S$ since $\bar{F}^k$ commutes with $S$. By Lemma 8.2, Corollary 8.4, and Lemma 8.6 there exists a vector field $X(x)$ which commutes with $S$ such that

$$j^l \big( \psi_l \circ F \circ \psi_l^{-1} \big) = S j^l \big( \Phi_X(1, x) \big),$$

where $\Phi_X(t, x)$ is the flow of $X(x)$. Since $X(x)$ commutes with $S$ it follows from Lemma 8.5 that $\Phi_X(1, x)$ also commutes with $S$. Therefore $j^l(\psi_l \circ F \circ \psi_l^{-1})(x) = j^l(\Phi_X(1, Sx))$. $\qquad \square$

We show next that if the diffeomorphism satisfies some group symmetry, then the vector field given by Theorem 8.1 also satisfies the same group symmetry.

**Theorem 8.7.** *Let* $\Gamma$ *be a group of invertible* $n \times n$ *matrices. If the diffeomorphism* (8.1) *has* $\Gamma$-*symmetry, that is,* $F(\gamma x) = \gamma F(x)$ *for any* $\gamma \in \Gamma$, *then the diffeomorphism* $\psi_r$ *and the vector field* $X$ *given by Theorem* 8.1 *also have* $\Gamma$-*symmetry up to order* $r$, *that is,*

$$j^r \psi_r(\gamma x) = \gamma j^r \psi_r(x) \quad \text{and} \quad j^r X(\gamma x) = \gamma j^r X(x),$$

$$\text{for any } x \in \mathbb{R}^n, \quad \gamma \in \Gamma.$$

*Proof.* The linear part of diffeomorphism (8.1) is $Ax = S(I + N)x = (S + N_1)x$, where $N_1 = SN$ is nilpotent and commutes with $S$. From the symmetry assumption it follows that

$$S + N_1 = \gamma^{-1}(S + N_1)\gamma = \gamma^{-1}S\gamma + \gamma^{-1}N_1\gamma, \qquad \gamma \in \Gamma.$$

But $\gamma^{-1}S\gamma$ is semisimple, $\gamma^{-1}N_1\gamma$ is nilpotent, and the two commute. By the uniqueness of the $S - N$ decomposition it follows that $S = \gamma^{-1}S\gamma$ and $N_1 = \gamma^{-1}N_1\gamma$. Thus, $S\gamma = \gamma S$ and $N_1\gamma = \gamma N_1$ for any $\gamma \in \Gamma$. Therefore also $N\gamma = \gamma N$ for any $\gamma \in \Gamma$. From the results in Section 2.5, we can find an $A$-normal form of (8.1) with $\Gamma$-symmetry by a transformation with $\Gamma$-symmetry. It follows that the nonlinear terms in (8.14) commute with every $\gamma \in \Gamma$ and therefore by Lemma 8.6 the vector field $X(x)$ given by Theorem 8.1 also commutes with each $\gamma \in \Gamma$ up to order $r$. $\qquad\qquad\square$

## 2.9 Versal Deformations of Matrices

Let $A: \Lambda \subseteq \mathbb{R}^k \to \mathbb{R}^{n \times n}$ be a $C^1$ mapping, where $\Lambda$ is a neighborhood of the origin in $\mathbb{R}^k$. To find a canonical form of $A(\lambda)$, $\lambda \in \Lambda$, we may try to find the Jordan canonical form for each $A(\lambda)$ by a linear change of coordinates depending on $\lambda$. However, the linear transformations which change $A(\lambda)$ to its Jordan form may not depend smoothly on $\lambda$. As an example, let

$$A(\lambda) = \begin{bmatrix} 0 & \lambda \\ 0 & 0 \end{bmatrix}, \qquad \lambda \in \mathbb{R}.$$

The Jordan form of $A(\lambda)$ for each $\lambda \in \mathbb{R}$ is

$$\begin{bmatrix} 0 & 0 \\ 0 & 0 \end{bmatrix}, \text{ if } \lambda = 0, \quad \text{and} \quad \begin{bmatrix} 0 & 1 \\ 0 & 0 \end{bmatrix}, \text{ if } \lambda \neq 0.$$

It is easy to see that any linear transformation which changes $A(\lambda)$ to its Jordan form must be discontinuous at $\lambda = 0$.

In this section we will derive a canonical form (or normal form) for a family of matrices that depend smoothly on parameters by a linear transformation also depending smoothly on parameters.

**Definition 9.1.** Let $A_0 \in \mathbb{R}^{n \times n}$ be fixed. A family of matrices $A(\lambda)$, $\lambda \in \Lambda \subseteq \mathbb{R}^k$, is called a deformation of $A_0$ if $A: \Lambda \to \mathbb{R}^{n \times n}$ is $C^1$ and $A(0) = A_0$.

**Definition 9.2.** Let both $A(\lambda)$ and $B(\mu)$ be deformations of $A_0$, where $\lambda \in \Lambda \subseteq \mathbb{R}^k$ and $\mu \in \Delta \subseteq \mathbb{R}^l$. If there exist a deformation $C(\mu)$ of the identity matrix $I$ with $\mu \in \tilde{\Delta} \subseteq \Delta$ and a $C^1$ mapping $\phi: \tilde{\Delta} \to \Lambda$ with $\phi(0) = 0$ such that

$$B(\mu) = C(\mu) A(\phi(\mu)) C(\mu)^{-1}, \qquad \mu \in \tilde{\Delta}, \qquad (9.1)$$

then we say that $B(\mu)$ is induced from the deformation $A(\lambda)$ by $C(\mu)$ and $\phi(\mu)$.

**Definition 9.3.** A deformation $A(\lambda)$ of $A_0$ is called a versal deformation of $A_0$ if any deformation $B(\mu)$ of $A_0$ can be induced from $A(\lambda)$. A versal deformation of $A_0$ is called a miniversal deformation if the dimension of its parameter space is the smallest among all versal deformations of $A_0$.

**Example 9.4.** Let $A_0 = \begin{bmatrix} 1 & 0 \\ 0 & 0 \end{bmatrix}$. Consider the following three deformations of $A_0$:

$$A_1(\lambda) = \begin{bmatrix} 1 + \lambda_1 & \lambda_2 \\ \lambda_3 & \lambda_4 \end{bmatrix}, \qquad A_2(\lambda) = \begin{bmatrix} 1 & 0 \\ 0 & \lambda_1 \end{bmatrix},$$

$$A_3(\lambda) = \begin{bmatrix} 1 + \lambda_1 & 0 \\ 0 & \lambda_2 \end{bmatrix},$$

where $\lambda_1, \ldots, \lambda_4$ are real parameters. We will show that $A_1(\lambda)$ is a versal deformation but it is not a miniversal deformation, $A_2(\lambda)$ is not a versal deformation, and $A_3(\lambda)$ is a miniversal deformation.

Let $B(\mu)$ be an arbitrary deformation $A_0$ with $\mu \in \Delta \subseteq \mathbb{R}^l$. Then $B(\mu)$ must have the following form:

$$B(\mu) = \begin{bmatrix} 1 + b_1(\mu) & b_2(\mu) \\ b_3(\mu) & b_4(\mu) \end{bmatrix}, \qquad \mu \in \Delta,$$

where the $b_i : \Delta \to \mathbb{R}$ are $C^1$ functions with $b_i(0) = 0$, $1 \leq i \leq 4$. Thus, $B(\mu)$ is induced from $A_1(\lambda)$ by $C(\mu) = I$ and $\phi(\mu) = (b_1(\mu), b_2(\mu), b_3(\mu), b_4(\mu))^T$. Hence, $A_1(\lambda)$ is a versal deformation.

To show that $A_2(\lambda)$ is not a versal deformation, we consider the following deformation of $A_0$:

$$B(\mu) = \begin{bmatrix} 1 + \mu & 0 \\ 0 & 0 \end{bmatrix}, \qquad \mu \in \mathbb{R}.$$

Suppose $B(\mu)$ is induced from $A_2(\lambda)$ by a deformation $C(\mu)$ of $I$ and a $C^1$ mapping $\phi : \Delta \to \mathbb{R}$, that is,

$$B(\mu) = C(\mu) A_2(\phi(\mu)) C(\mu)^{-1}, \qquad \mu \in \Delta, \qquad (9.2)$$

where $\Delta$ is a neighborhood of the origin in $\mathbb{R}$. Let

$$C(\mu) = \begin{bmatrix} c_1(\mu) & c_2(\mu) \\ c_3(\mu) & c_4(\mu) \end{bmatrix}, \qquad \mu \in \Delta,$$

where the $c_i(\mu)$ are $C^1$ functions. Since $C(\lambda)$ is a deformation of $I$, $c_1(0) = c_4(0) = 1$, $c_2(0) = c_3(0) = 0$. Then from (9.2) one must have

$$\begin{bmatrix} 1 + \mu & 0 \\ 0 & 0 \end{bmatrix} \begin{bmatrix} c_1(\mu) & c_2(\mu) \\ c_3(\mu) & c_4(\mu) \end{bmatrix} = \begin{bmatrix} c_1(\mu) & c_2(\mu) \\ c_3(\mu) & c_4(\mu) \end{bmatrix} \begin{bmatrix} 1 & 0 \\ 0 & \phi(\mu) \end{bmatrix},$$

$$\mu \in \Delta.$$

This implies $(1 + \mu) c_1(\mu) = c_1(\mu)$ and therefore $c_1(\mu) = 0$, $\mu \in \Delta$, which contradicts $c_1(0) = 1$.

It will be shown later by Theorem 9.17 that $A_3(\lambda)$ is a miniversal deformation of $A_0$ and $A_1(\lambda)$ is not a miniversal deformation of $A_0$.

Now we consider a characterization of versal deformations. We first introduce some notation. Let $GL(n, \mathbb{R})$ be the set of all invertible real $n \times n$ matrices. It is well known that $GL(n, \mathbb{R})$ is a Lie group and $gl(n, \mathbb{R})$ is the Lie algebra associated with $GL(n, \mathbb{R})$ with respect to the Lie bracket:

$$[u, v] = uv - vu, \qquad u, v \in gl(n, \mathbb{R}).$$

We define a group action $\pi: GL(n, \mathbb{R}) \times gl(n, \mathbb{R}) \to gl(n, \mathbb{R})$ by

$$\pi(g, u) = gug^{-1}, \qquad g \in GL(n, \mathbb{R}), \quad u \in gl(n, \mathbb{R}).$$

**Definition 9.5.** Let $A_0 \in gl(n, \mathbb{R})$. The set

$$\gamma(A_0) = \{\pi(g, A_0) \mid g \in GL(n, \mathbb{R})\}$$

is called the orbit through $A_0$.

In other words, the orbit through $A_0$ under the action $\pi$ is the set of all real $n \times n$ matrices similar to $A_0$. It is well known that $\gamma(A_0)$ is a submanifold of $gl(n, \mathbb{R})$. For a fixed $u \in gl(n, \mathbb{R})$, we define a linear operator $L_u: gl(n, \mathbb{R}) \to gl(n, \mathbb{R})$ by

$$L_u v = [v, u], \qquad v \in gl(n, \mathbb{R}).$$

**Definition 9.6.** Let $A_0 \in gl(n, \mathbb{R})$. The set

$$Z_{A_0} = \text{Ker}(L_{A_0})$$

is called the centralizer of $A_0$.

We note that $Z_{A_0}$ is the set of all real $n \times n$ matrices that commute with $A_0$, and it is a subspace of $gl(n, \mathbb{R})$.

Let $T_{A_0}(\gamma(A_0))$ be the tangent space to $\gamma(A_0)$ at $A_0$ and $\text{Im}(L_{A_0})$ be the image of $L_{A_0}$ in $gl(n, \mathbb{R})$. Then we have the following lemma.

**Lemma 9.7.** $T_{A_0}(\gamma(A_0)) = \text{Im}(L_{A_0})$.

*Proof.* Let $u \in gl(n, \mathbb{R})$ and $|u|$ be sufficiently small. Hence $I + u \in GL(n, \mathbb{R})$. Then as $|u| \to 0$,

$$\pi(I + u, A_0) = (I + u)A_0(I + u)^{-1} = (I + u)A_0(I - u + o(|u|))$$

$$= A_0 + uA_0 - A_0 u + o(|u|) = \pi(I, A_0) + L_{A_0} u + o(|u|).$$

Therefore $D_g \pi(I, A_0) = L_{A_0}$ and hence $T_{A_0}(\gamma(A_0)) = \text{Im}(L_{A_0})$.   □

**Corollary 9.8.** $\text{codim}(\gamma(A_0)) = \dim(Z_{A_0})$.

*Proof*

$$\dim(gl(n, \mathbb{R})) = \dim(\text{Im}(L_{A_0})) + \dim(\text{Ker}(L_{A_0}))$$

$$= \dim(T_{A_0}(\gamma(A_0))) + \dim(Z_{A_0})$$

$$= \dim(\gamma(A_0)) + \dim(Z_{A_0}).$$   □

**Corollary 9.9.** *Let $V$ be a submanifold of $GL(n, \mathbb{R})$. If $V$ is transversal to $Z_{A_0}$ at $I$, then*

$$T_{A_0}(\gamma(A_0)) = L_{A_0}(T_I V).$$

*Proof.* By the definition of the transversality, for any $u \in gl(n, \mathbb{R})$ there exist $u_1 \in T_I V$ and $u_2 \in T_I(Z_{A_0})$ such that $u = u_1 + u_2$. Hence

$$L_{A_0} u = [u, A_0] = [u_1, A_0] + [u_2, A_0] = [u_1, A_0] = L_{A_0} u_1.$$   □

**Lemma 9.10.** *Let $A: \Lambda \to gl(n, \mathbb{R})$ be $C^1$, where $\Lambda$ is a neighborhood of the origin in $\mathbb{R}^k$, and $V$ be a submanifold of $GL(n, \mathbb{R})$. Assume that $A(\lambda)$ is transversal to $\gamma(A_0)$ at $\lambda = 0$, $I \in V$ and $V$ is transversal to $Z_{A_0}$ at $I$, $k = \text{codim}(\gamma(A_0))$, and $\dim(V) = \dim(\gamma(A_0))$. Then the mapping $\Phi: V \times \Lambda \to gl(n, \mathbb{R})$ defined by*

$$\Phi(v, \lambda) = vA(\lambda)v^{-1}, \qquad v \in V, \quad \lambda \in \Lambda,$$

*is a local diffeomorphism in a neighborhood of $(I, 0)$ in $V \times \Lambda$.*

*Proof.* From Lemma 9.7, $D_v\Phi(I, 0) = D_g\pi(I, A_0) = L_{A_0}$, $D_\lambda\Phi(I, 0) = DA(0)$. Therefore

$$D\Phi(I, 0)(u, v) = L_{A_0}u + DA(0)v, \qquad u \in T_IV, \quad v \in T_0\Lambda.$$

By the transversality hypothesis on $A(\lambda)$,

$$T_{A_0}(\gamma(A_0)) + DA(0)(T_0\Lambda) = gl(n, \mathbb{R}). \tag{9.3}$$

By Corollary 9.9, $L_{A_0}(T_IV) = T_{A_0}(\gamma(A_0))$. Therefore $D\Phi(I, 0)$ is surjective. We note that $\dim(V \times \Lambda) = \dim(gl(n, \mathbb{R}))$. Hence the conclusion follows from the Inverse Function Theorem. $\square$

**Theorem 9.11.** *A deformation $A(\lambda)$ of $A_0$ is a versal deformation if and only if $A(\lambda)$ is transversal to the submanifold $\gamma(A_0)$ at $\lambda = 0$ in $gl(n, \mathbb{R})$.*

*Proof.* Let $\lambda$ be in a neigborhood $\Lambda$ of the origin in $\mathbb{R}^k$. Assume that $A(\lambda)$ is a versal deformation of $A_0$. Let $B(\mu)$ be an arbitrary deformation of $A_0$ with $\mu \in \Delta \subseteq \mathbb{R}^l$. Then there exists a deformation $C(\mu)$ of $I$ with $\mu \in \tilde{\Delta} \subseteq \Delta$ and a $C^1$ mapping $\phi: \tilde{\Delta} \to \Lambda$ with $\phi(0) = 0$ such that

$$B(\mu) = C(\mu)A(\phi(\mu))C(\mu)^{-1}, \qquad \mu \in \tilde{\Delta}.$$

By taking the derivative with respect to $\mu$ at $\mu = 0$ we get

$$DB(0)v = DA(0) \cdot D\phi(0)v + [DC(0)v, A_0], \qquad v \in \mathbb{R}^l.$$

Since $B(\mu)$ is an arbitrary deformation of $A_0$, $DB(0)v$ can be any element in $gl(n, \mathbb{R})$. On the other hand, by Lemma 9.7, $[DC(0)v, A_0] \in T_{A_0}(\gamma(A_0))$. Therefore

$$T_{A(0)}(gl(n, \mathbb{R})) = DA(0)(T_0\Lambda) + T_{A(0)}(\gamma(A_0)).$$

This says that $A(\lambda)$ is transversal to the submanifold $\gamma(A_0)$ at $\lambda = 0$ in $gl(n, \mathbb{R})$.

Conversely, assume that $A(\lambda)$ is transversal to $\gamma(A_0)$ at $\lambda = 0$. By the definition of transversality, $k \geq \text{codim}(\gamma(A_0))$. First we consider the case $k = \text{codim}(\gamma(A_0))$. Let $V$ and $\Phi$ be as in Lemma 9.10. Let $B(\mu)$ be an arbitrary deformation of $A_0$ with $\mu \in \Delta \subseteq \mathbb{R}^l$. Then by Lemma 9.10 the equation

$$B(\mu) = \Phi(v, \lambda), \qquad v \in V, \quad \lambda \in \Lambda$$

has a unique solution $v = C(\mu)$, $\lambda = \phi(\mu)$ for $\mu$ in a sufficiently small neighborhood $\tilde{\Delta} \subseteq \Delta$ of the origin in $\mathbb{R}^k$, where $C(\mu)$ and $\phi(\mu)$ are $C^1$ with $C(0) = I$ and $\phi(0) = 0$. Thus

$$B(\mu) = \Phi(C(\mu), \phi(\mu)) = C(\mu)A(\phi(\mu))C(\mu)^{-1}, \qquad \mu \in \tilde{\Delta}.$$

This says that $A(\lambda)$ is a versal deformation of $A_0$. For the case where $k > \text{codim}(\gamma(A_0)) = d$, there exists a $C^1$ mapping $\lambda(\mu)$ defined in a neighborhood $\tilde{\Lambda}$ of the origin in $\mathbb{R}^d$ such that $A(\lambda(\mu))$ is a deformation of $A_0$ induced from $A(\lambda)$ and it is transversal to $\gamma(A_0)$ at $\mu = 0$. From the above discussion, $A(\lambda(\mu))$ is a versal deformation of $A_0$. Hence $A(\lambda)$ is a versal deformation of $A_0$. □

From Theorem 9.11, the dimension of a parameter space of a miniversal deformation of $A_0$ is equal to $\text{codim}(\gamma(A_0)) = \dim(Z_{A_0})$. On the other hand, a miniversal deformation of $A_0$ is not unique since the mapping $A(\lambda)$ which is transversal to $\gamma(A_0)$ at $\lambda = 0$ is not unique.

Theorem 9.11 shows also that the problem of finding a miniversal deformation of $A_0$ can be reduced to an algebraic problem of finding a complementary subspace to $T_{A_0}(\gamma(A_0)) = \text{Im}(L_{A_0})$ in $gl(n, \mathbb{R})$. We show next how to construct a complementary subspace to $\text{Im}(L_{A_0})$ in $gl(n, \mathbb{R})$. We introduce first a Hermitian scalar product in $gl(n, \mathbb{R})$:

$$\langle\langle u, v \rangle\rangle = tr(uv^*), \qquad u, v \in gl(n, \mathbb{R}).$$

where $tr(uv^*)$ denotes the trace of $uv^*$ and $v^*$ is the transpose of the matrix $v$. If $u = (u_{ij})$ and $v = (v_{ij})$, then by definition $\langle\langle u, v \rangle\rangle = \Sigma_{i,j} u_{ij} v_{ij}$.

**Theorem 9.12.** $L_{A_0^*}$ *is the adjoint operator of* $L_{A_0}$ *with respect to the inner product* $\langle\langle \cdot , \cdot \rangle\rangle$ *in* $gl(n, \mathbb{R})$.

*Proof.* For any $u, v \in gl(n, \mathbb{R})$,

$$\langle\langle L_{A_0} u, v \rangle\rangle = \langle\langle uA_0, v \rangle\rangle - \langle\langle A_0 u, v \rangle\rangle$$

$$= tr(uA_0 v^*) - tr(A_0 uv^*) = \langle\langle u, L_{A_0^*} v \rangle\rangle. \quad \square$$

**Corollary 9.13.** $\mathrm{Ker}(L_{A_0^*})$ *is the orthogonal complementary subspace to* $\mathrm{Im}(L_{A_0})$ *with respect to the inner product* $\langle\langle \cdot, \cdot \rangle\rangle$ *in* $gl(n, \mathbb{R})$.

**Corollary 9.14.** *Let* $\{v_1, \ldots, v_k\}$ *be a basis of* $\mathrm{Ker}(L_{A_0^*})$. *Then*

$$A_0 + \sum_{i=1}^{k} \lambda_i v_i,$$

*where* $\lambda_1, \ldots, \lambda_k$ *are real parameters, is a miniversal deformation of* $A_0$.

By Corollaries 9.13 and 9.14, the problem of finding a miniversal deformation of $A_0$ can be reduced to finding a basis of the centralizer $Z_{A_0^*}$ of $A_0^*$. In the case where $A_0$ is an upper triangular Jordan matrix, we have the following lemma.

**Lemma 9.15.** *Suppose* $A_0$ *is an upper triangular Jordan matrix with only one eigenvalue and a sequence of Jordan blocks of sizes* $n_1 \geq n_2 \geq \cdots \geq n_s$. *Then* $Z_{A_0^*}$ *consists precisely of the matrices of the form shown in Figure* 9.1, *where each oblique segment stands for a sequence of identical entries and the blank part consists of zero entries and* $\dim(Z_{A_0^*})$ $= n_1 + 3n_2 + 5n_3 + \cdots + (2s - 1)n_s =$ *number of oblique segments.*

*Proof.* For simplicity, we consider the case where $s = 2$. Then the matrix $A_0$ has the following form

$$A_0 = \lambda I + \begin{bmatrix} N_1 & 0 \\ 0 & N_2 \end{bmatrix},$$

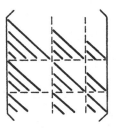

Figure 9.1. Case $s = 3$.

where

$$N_1 = \begin{bmatrix} 0 & 1 & & \\ & 0 & \ddots & \\ & & \ddots & 1 \\ & & & 0 \end{bmatrix}_{n_1 \times n_1}, \qquad N_2 = \begin{bmatrix} 0 & 1 & & \\ & 0 & \ddots & \\ & & \ddots & 1 \\ & & & 0 \end{bmatrix}_{n_2 \times n_2},$$

and $n_1 + n_2 = n$.

Let $u \in \operatorname{Ker}(L_{A_0^*})$ and

$$u = \begin{bmatrix} u_1 & u_2 \\ u_3 & u_4 \end{bmatrix},$$

where the sizes of $u_1$, $u_2$, $u_3$, and $u_4$ are $n_1 \times n_1$, $n_1 \times n_2$, $n_2 \times n_1$, and $n_2 \times n_2$, respectively. Since

$$\begin{bmatrix} u_1 & u_2 \\ u_3 & u_4 \end{bmatrix} \begin{bmatrix} N_1^* & 0 \\ 0 & N_2^* \end{bmatrix} - \begin{bmatrix} N_1^* & 0 \\ 0 & N_2^* \end{bmatrix} \begin{bmatrix} u_1 & u_2 \\ u_3 & u_4 \end{bmatrix} = 0,$$

one must have

$$u_1 N_1^* - N_1^* u_1 = 0, \qquad u_2 N_1^* - N_2^* u_2 = 0,$$

$$u_3 N_1^* - N_2^* u_3 = 0, \qquad u_4 N_2^* - N_2^* u_4 = 0. \tag{9.4}$$

By solving (9.4) for $u_1$, $u_2$, $u_3$, and $u_4$, one obtains the desired result. $\square$

**Remark 9.16.** A basis $\{v_1, \ldots, v_k\}$ of $\operatorname{Ker}(L_{A_0^*})$ can be chosen such that each $v_i$ is a matrix with all entries on one of the oblique segments

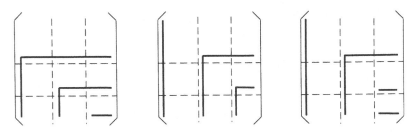

Figure 9.2. Case $s = 3$.

described in Lemma 9.15 equal to 1 and all other entries being zero. By an elementary algebraic discussion, we can show that a basis $\{w_1, \ldots, w_k\}$ of a complementary subspace to $\mathrm{Im}(L_{A_0})$ can be chosen such that each $w_i$ is a matrix with only one entry on one of the oblique segments described in Lemma 9.15 equal to 1 and all other entries being zero. Thus

$$A_0 + \sum_{i=1}^{k} \lambda_i w_i,$$

where each $\lambda_i$ is a real parameter, is also a miniversal deformation of $A_0$.

Some of the matrices generated by such matrices $\{w_1, \ldots, w_k\}$ are shown in Figure 9.2, in which all entries that are not on the black segments are equal to zero.

**Theorem 9.17.** *If $A_0$ is an upper triangular Jordan matrix with $m$ distinct eigenvalues $\omega_1, \ldots, \omega_m$, and the sequence of blocks corresponding to the eigenvalue $\omega_i$ are of sizes $n_1(\omega_i) \geq n_2(\omega_i) \geq \cdots \geq n_{s_i}(\omega_i)$, $1 \leq i \leq m$, then the dimension of the parameter space of a miniversal deformation of $A_0$ is*

$$d = \sum_{i=1}^{m} \left( n_1(\omega_i) + 3n_2(\omega_i) + 5n_3(\omega_i) + \cdots + (2s_i - 1)n_{s_i}(\omega_i) \right),$$

*and a miniversal deformation of $A_0$ is block diagonal, each of the blocks being a miniversal deformation described in Remark 9.16 for the block of $A_0$ corresponding to each eigenvalue.*

*Proof.* The proof is similar to that of Lemma 9.15.

**Example 9.18.** (1) Let

$$
A_0 = \begin{bmatrix} \omega_1 & & & \\ & \omega_2 & & \\ & & \ddots & \\ & & & \omega_n \end{bmatrix}, \qquad \omega_i \neq \omega_j \quad \text{if} \quad i \neq j, i, j = 1, \ldots, n.
$$

Then a miniversal deformation of $A_0$ is

$$
\begin{bmatrix} \omega_1 & & & \\ & \omega_2 & & \\ & & \ddots & \\ & & & \omega_n \end{bmatrix} + \begin{bmatrix} \lambda_1 & & & \\ & \lambda_2 & & \\ & & \ddots & \\ & & & \lambda_n \end{bmatrix},
$$

where $\lambda_i$, $i = 1, \ldots, n$, are parameters.
  (2) Let

$$
A_0 = 0.
$$

Then a miniversal deformation of $A_0$ is

$$
\begin{bmatrix} \lambda_{11} & \cdots & \lambda_{1n} \\ \lambda_{21} & \cdots & \lambda_{2n} \\ \vdots & & \\ \lambda_{n1} & \cdots & \lambda_{nn} \end{bmatrix},
$$

where the $\lambda_{ij}$, $i, j = 1, \ldots, n$, are parameters.
  (3) Let

$$
A_0 = \begin{bmatrix} 0 & 1 & & \\ & 0 & \ddots & \\ & & \ddots & 1 \\ & & & 0 \end{bmatrix}.
$$

Then a miniversal deformation of $A_0$ can be chosen as one of the following:

$$
\begin{bmatrix}
\lambda_1 & 1 & & & \\
\lambda_2 & \lambda_1 & \ddots & & \\
\vdots & \ddots & \ddots & 1 \\
\lambda_n & \cdots & \lambda_2 & \lambda_1
\end{bmatrix}
\quad \text{or} \quad
\begin{bmatrix}
0 & 1 & & & \\
0 & 0 & \ddots & & \\
0 & \ddots & & 0 & 1 \\
\lambda_1 & \lambda_2 & \cdots & & \lambda_n
\end{bmatrix},
$$

where the $\lambda_i$, $i = 1, \ldots, n$, are parameters.

(4) Let

$$
A_0 = \begin{bmatrix}
0 & 1 & 0 \\
0 & 0 & 1 \\
0 & 0 & 0
\end{bmatrix}.
$$

A miniversal deformation of $A_0$ can be chosen as one of the following:

$$
\begin{bmatrix}
0 & 1 & 0 \\
0 & 0 & 1 \\
\lambda_1 & \lambda_2 & \lambda_3
\end{bmatrix}
\quad \text{or} \quad
\begin{bmatrix}
\lambda_1 & 1 & 0 \\
\lambda_2 & 0 & 1 \\
\lambda_3 & 0 & 0
\end{bmatrix}
$$

$$
\text{or} \quad
\begin{bmatrix}
0 & 1 & 0 \\
\lambda_1 & \lambda_2 & 1 \\
\lambda_3 & 0 & 0
\end{bmatrix}
\quad \text{or} \quad
\begin{bmatrix}
\lambda_1 & 1 & 0 \\
\lambda_2 & \lambda_1 & 1 \\
\lambda_3 & \lambda_2 & \lambda_1
\end{bmatrix},
$$

where $\lambda_1$, $\lambda_2$, and $\lambda_3$ are parameters.

(5) Let

$$
A_0 = \begin{bmatrix}
0 & 1 & 0 \\
0 & 0 & 0 \\
0 & 0 & 0
\end{bmatrix}.
$$

Here we have two blocks corresponding to the same eigenvalue. Therefore, a miniversal deformation of $A_0$ can be chosen as one of the following:

$$\begin{bmatrix} 0 & 1 & 0 \\ \lambda_1 & \lambda_2 & \lambda_3 \\ \lambda_4 & 0 & \lambda_5 \end{bmatrix} \quad \text{or} \quad \begin{bmatrix} \lambda_1 & 1 & 0 \\ \lambda_2 & 0 & \lambda_4 \\ \lambda_3 & 0 & \lambda_5 \end{bmatrix},$$

where $\lambda_i, i = 1, \ldots, 5$, are parameters.

(6) Let

$$A_0 = \begin{bmatrix} \alpha & 1 & \vdots & 0 \\ 0 & \alpha & \vdots & 0 \\ \cdots & \cdots & \vdots & \cdots \\ 0 & 0 & \vdots & \beta \end{bmatrix}, \quad \alpha \neq \beta.$$

Here we have two blocks corresponding to distinct eigenvalues. Therefore, a miniversal deformation of $A_0$ can be chosen as one of the following:

$$A_0 + \begin{bmatrix} \lambda_1 & 0 & 0 \\ \lambda_2 & \lambda_1 & 0 \\ 0 & 0 & \lambda_3 \end{bmatrix} \quad \text{or} \quad A_0 + \begin{bmatrix} 0 & 0 & 0 \\ \lambda_1 & \lambda_2 & 0 \\ 0 & 0 & \lambda_3 \end{bmatrix},$$

where $\lambda_1$, $\lambda_2$, and $\lambda_3$ are parameters.

(7) Let

$$A_0 = \begin{bmatrix} 0 & 1 \\ -1 & 0 \end{bmatrix}.$$

By solving $L_{A_0^*} u = 0$ directly, a miniversal deformation of $A_0$ can be found as one of the following:

$$A_0 + \begin{bmatrix} \lambda_1 & \lambda_2 \\ -\lambda_2 & \lambda_1 \end{bmatrix} \quad \text{or} \quad A_0 + \begin{bmatrix} 0 & 0 \\ \lambda_1 & \lambda_2 \end{bmatrix},$$

where $\lambda_1$ and $\lambda_2$ are parameters.

**Remark 9.19.** For the case where the matrix $A_0$ has nonreal eigenvalues, one can find a nonsingular matrix $p \in \mathbb{C}^{n \times n}$ such that $p^{-1} A_0 p$ is in an upper triangular Jordan form. Then one can apply Lemma 9.15 or Theorem 9.17 to find a basis of the linear space of all complex matrices commuting with $(p^{-1} A_0 p)^*$ and then to find miniversal deformations of $A_0$. We illustrate this idea by the following example.

**Example 9.20.** Let

$$
A_0 = \begin{bmatrix} \alpha & \beta & 1 & 0 \\ -\beta & \alpha & 0 & 1 \\ 0 & 0 & \alpha & \beta \\ 0 & 0 & -\beta & \alpha \end{bmatrix}, \quad \beta \neq 0.
$$

We take

$$
p = \begin{bmatrix} 1 & 0 & 1 & 0 \\ -i & 0 & i & 0 \\ 0 & 1 & 0 & 1 \\ 0 & -i & 0 & i \end{bmatrix}.
$$

Then

$$
p^{-1}A_0 p = \begin{bmatrix} \alpha + \beta i & 1 & 0 & 0 \\ 0 & \alpha + \beta i & 0 & 0 \\ 0 & 0 & \alpha - \beta i & 1 \\ 0 & 0 & 0 & \alpha - \beta i \end{bmatrix}.
$$

Any complex $4 \times 4$ matrix commuting with $(p^{-1}A_0 p)^*$ is of the form

$$
B(\epsilon) = \begin{bmatrix} \epsilon_1 & 0 & 0 & 0 \\ \epsilon_2 & \epsilon_1 & 0 & 0 \\ 0 & 0 & \epsilon_3 & 0 \\ 0 & 0 & \epsilon_4 & \epsilon_3 \end{bmatrix}, \quad \epsilon_i \in \mathbb{C}, \quad i = 1, \dots, 4.
$$

Since

$$
(p^{-1})^* B(\epsilon) p^*
$$

$$
= \begin{bmatrix} \epsilon_1 + \epsilon_3 & (\epsilon_1 - \epsilon_3)i & 0 & 0 \\ -(\epsilon_1 - \epsilon_3)i & \epsilon_1 + \epsilon_3 & 0 & 0 \\ \epsilon_2 + \epsilon_4 & (\epsilon_2 - \epsilon_4)i & \epsilon_1 + \epsilon_3 & (\epsilon_1 - \epsilon_3)i \\ -(\epsilon_2 - \epsilon_4)i & \epsilon_2 + \epsilon_4 & -(\epsilon_1 - \epsilon_3)i & \epsilon_1 + \epsilon_3 \end{bmatrix},
$$

$$
\mathrm{Ker}(L_{A_0^*}) = \left\{ \begin{bmatrix} \lambda_1 & \lambda_2 & 0 & 0 \\ -\lambda_2 & \lambda_1 & 0 & 0 \\ \lambda_3 & \lambda_4 & \lambda_1 & \lambda_2 \\ -\lambda_4 & \lambda_3 & -\lambda_2 & \lambda_1 \end{bmatrix} \lambda_1, \dots, \lambda_4 \in \mathbb{R} \right\}.
$$

Hence a miniversal deformation of $A_0$ can be chosen as one of the following:

$$A_0 + \begin{bmatrix} \lambda_1 & \lambda_2 & 0 & 0 \\ -\lambda_2 & \lambda_1 & 0 & 0 \\ \lambda_3 & \lambda_4 & \lambda_1 & \lambda_2 \\ -\lambda_4 & \lambda_3 & -\lambda_2 & \lambda_1 \end{bmatrix},$$

or

$$A_0 + \begin{bmatrix} \lambda_1 & 0 & 0 & 0 \\ \lambda_2 & 0 & 0 & 0 \\ \lambda_3 & 0 & 0 & 0 \\ \lambda_4 & 0 & 0 & 0 \end{bmatrix},$$

or

$$A_0 + \begin{bmatrix} 0 & 0 & 0 & 0 \\ 0 & 0 & 0 & 0 \\ 0 & 0 & 0 & 0 \\ \lambda_1 & \lambda_2 & \lambda_3 & \lambda_4 \end{bmatrix},$$

where $\lambda_i$'s are real parameters.

## 2.10 Versal Deformations of Infinitesimally Symplectic Matrices

In this section we discuss versal deformations of infinitesimally symplectic matrices. We recall that

$$Sp(2n, \mathbb{R}) = \{u \in GL(2n, \mathbb{R}) \mid u^*Ju = J\},$$

$$sp(2n, \mathbb{R}) = \{u \in gl(2n, \mathbb{R}) \mid u^*J + Ju = 0\},$$

where

$$J = \begin{bmatrix} 0 & I_n \\ -I_n & 0 \end{bmatrix},$$

$I_n$ is the identity matrix in $\mathbb{R}^{n \times n}$.

**Definition 10.1.** Let $A_0 \in sp(2n, \mathbb{R})$ be given. A deformation $A(\lambda)$ of $A_0$ with $\lambda \in \Lambda \subseteq \mathbb{R}^k$ is called an infinitesimally symplectic deformation of $A_0$ if every $A(\lambda)$ is infinitesimally symplectic.

**Definition 10.2.** An infinitesimally symplectic deformation $A(\lambda)$ of $A_0$ is called an infinitesimally symplectic versal deformation of $A_0$ if and only if any infinitesimally symplectic deformation $B(\mu)$ of $A_0$ can be induced from $A(\lambda)$ by an infinitesimally symplectic deformation $C(\mu)$ of the identity matrix $I$ in $\mathbb{R}^{2n \times 2n}$ and a $C^1$ mapping $\phi(\mu)$. An infinitesimally symplectic versal deformation of $A_0$ is called an infinitesimally symplectic miniversal deformation of $A_0$ if the dimension of its parameter space is the smallest among all infinitesimally symplectic versal deformations of $A_0$.

**Lemma 10.3.** $Sp(2n, \mathbb{R})$ is a $C^\infty$-submanifold of $GL(2n, \mathbb{R})$.

*Proof.* We define a mapping $F: GL(2n, \mathbb{R}) \to GL(2n, \mathbb{R})$ by

$$F(p) = p^*Jp, \qquad p \in GL(2n, \mathbb{R}).$$

Hence $Sp(2n, \mathbb{R}) = \{p \in GL(2n, \mathbb{R}) \mid F(p) = J\}$. We note that, for any fixed $p \in GL(2n, \mathbb{R})$, $DF(p)u = u^*Jp + p^*Ju = p^*((up^{-1})^*J + J(up^{-1}))p = p^*DF(I)(up^{-1})p$. Hence $\operatorname{rank}(DF(p)) = \operatorname{rank}(DF(I))$. Since $F$ is $C^\infty$, $Sp(2n, \mathbb{R})$ is a $C^\infty$-submanifold of $GL(2n, \mathbb{R})$. □

It is easy to see that $sp(2n, \mathbb{R})$ is a linear subspace of $gl(2n, \mathbb{R})$.

**Lemma 10.4.** $T_I(Sp(2n, \mathbb{R})) = sp(2n, \mathbb{R})$.

*Proof.* Let $c(t)$ be any curve in $Sp(2n, \mathbb{R})$ with $c(0) = I$, $t \in [-1, 1]$. Since $c(t)^*Jc(t) = J$, for any $t \in [-1, 1]$,

$$c'(0)^*J + Jc'(0) = 0.$$

This says that the tangent vector $c'(0) \in sp(2n, \mathbb{R})$.

On the other hand, if $A_0 \in sp(2n, \mathbb{R})$, then we define $c(t) = \exp(A_0 t)$ and $B(t) = c(t)^* J c(t)$, $t \in [-1, 1]$. Then

$$B'(t) = \exp((A_0 t)^*) A_0^* J \exp(A_0 t) + \exp((A_0 t)^*) J A_0 \exp(A_0 t)$$

$$= \exp((A_0 t)^*)(A_0^* J + J A_0) \exp(A_0 t) = 0, \qquad t \in [-1, 1].$$

Hence $B(t) \equiv J$. This says that $c(t)$ is a curve in $sp(2n, \mathbb{R})$. Since $c'(0) = A_0$, $A_0 \in T_I(Sp(2n, \mathbb{R}))$.                                   $\square$

**Lemma 10.5.** $\dim(Sp(2n, \mathbb{R})) = \dim(sp(2n, \mathbb{R})) = 2n^2 + n$.

*Proof.* For any $p \in sp(2n, \mathbb{R})$, we may assume that

$$p = \begin{bmatrix} p_1 & p_2 \\ p_3 & p_4 \end{bmatrix},$$

where $p_i \in gl(n, \mathbb{R})$, $i = 1, 2, 3, 4$. Then from Corollary 6.3, $p \in sp(2n, \mathbb{R})$ if and only if $p_4 = -p_1^*$, $p_2^* = p_2$, $p_3^* = p_3$. Therefore

$$\dim(sp(2n, \mathbb{R})) = 2n^2 + n.$$

From Lemma 10.4, $\dim(Sp(2n, \mathbb{R})) = \dim(sp(2n, \mathbb{R}))$.                    $\square$

We define a group action $\pi^s$: $Sp(2n, \mathbb{R}) \times sp(2n, \mathbb{R}) \to sp(2n, \mathbb{R})$ by

$$\pi^s(g, u) = g u g^{-1}, \qquad g \in Sp(2n, \mathbb{R}), \qquad u \in sp(2n, \mathbb{R}).$$

Let $A_0 \in sp(2n, \mathbb{R})$ be fixed. Then the orbit through $A_0$ in $sp(2n, \mathbb{R})$ is

$$\gamma^s(A_0) = \{\pi^s(g, A_0) \mid g \in Sp(2n, \mathbb{R})\}.$$

Then $\gamma^s(A_0)$ is a submanifold of $Sp(2n, \mathbb{R})$.

For a given $u \in sp(2n, \mathbb{R})$, we define a linear operator $L_u^s$: $sp(2n, \mathbb{R}) \to sp(2n, \mathbb{R})$ by

$$L_u^s v = [v, u], \qquad v \in sp(2n, \mathbb{R}).$$

Then for given $A_0 \in sp(2n, \mathbb{R})$, the centralizer of $A_0$ in $sp(2n, \mathbb{R})$ is

$$Z^s_{A_0} = \text{Ker}(L^s_{A_0}).$$

If we replace $gl(n, \mathbb{R})$ and $GL(n, \mathbb{R})$ by $sp(2n, \mathbb{R})$ and $Sp(2n, \mathbb{R})$ respectively, derivatives for corresponding maps by tangent maps, $\gamma(A_0)$, $L_{A_0}$, and $Z_{A_0}$ by $\gamma^s(A_0)$, $L^s_{A_0}$, and $Z^s_{A_0}$, respectively, and the condition that $V$ is transversal to $Z_{A_0}$ in $gl(n, \mathbb{R})$ by the assumption that $sp(2n, \mathbb{R}) = T_I V \oplus Z^s_{A_0}$, then it is not hard to see that Lemmas 9.7 and 9.10 and Corollaries 9.8 and 9.9 hold for infinitesimally symplectic matrices. Thus we have the following.

**Theorem 10.6.** *An infinitesimally symplectic deformation $A(\lambda)$ of $A_0$ is versal if and only if $A(\lambda)$ is transversal to $\gamma^s(A_0)$ in $sp(2n, \mathbb{R})$ at $\lambda = 0$.*

Let $\langle\langle \cdot, \cdot \rangle\rangle$ be the inner product in $sp(2n, \mathbb{R})$ which is induced from the Hermitian inner product in $gl(2n, \mathbb{R})$ defined in Section 2.9. It is easy to show the following.

**Theorem 10.7.** $L^s_{A_0^*}$ *is the adjoint operator of $L^s_{A_0}$ with respect to the inner product $\langle\langle \cdot, \cdot \rangle\rangle$ in $sp(2n, \mathbb{R})$.*

**Corollary 10.8.** $\text{Ker}(L^s_{A_0^*})$ *is the orthogonal complementary subspace to $\text{Im}(L^s_{A_0})$ with respect to the inner product $\langle\langle \cdot, \cdot \rangle\rangle$ in $sp(2n, \mathbb{R})$.*

We note that $\text{Ker}(L^s_{A_0^*}) = \text{Ker}(L_{A_0^*}) \cap sp(2n, \mathbb{R})$, where $L_{A_0^*}$ is defined in Section 2.9. We can first apply the results in Section 2.9 to find a basis of $\text{Ker}(L_{A_0^*})$ and then by restricting $\text{Ker}(L_{A_0^*})$ to $sp(2n, \mathbb{R})$ to obtain $\text{Ker}(L^s_{A_0^*})$.

We can also derive other complementary subspaces to $Im L^s_{A_0}$ in $sp(2n, \mathbb{R})$ from $\text{Ker}(L^s_{A_0^*})$ by some elementary algebraic methods.

**Example 10.9.**

(i) Let

$$A_0 = \begin{bmatrix} 0 & 0 \\ 0 & 0 \end{bmatrix}.$$

Then

$$\mathrm{Ker}(L_{A_0^*}) = \left\{ \begin{bmatrix} \lambda_1 & \lambda_2 \\ \lambda_3 & \lambda_4 \end{bmatrix} \middle| \lambda_1, \dots, \lambda_4 \in \mathbb{R} \right\}.$$

Hence

$$\mathrm{Ker}(L_{A_0^*}^s) = \left\{ \begin{bmatrix} \lambda_1 & \lambda_2 \\ \lambda_3 & -\lambda_1 \end{bmatrix} \middle| \lambda_1, \lambda_2, \lambda_3 \in \mathbb{R} \right\}.$$

An infinitesimally symplectic miniversal deformation of $A_0$ is

$$A(\lambda) = \begin{bmatrix} \lambda_1 & \lambda_2 \\ \lambda_3 & -\lambda_1 \end{bmatrix}, \qquad \lambda_1, \lambda_2, \lambda_3 \in \mathbb{R}.$$

(ii) Let

$$A_0 = \begin{bmatrix} \alpha & 0 \\ 0 & -\alpha \end{bmatrix}, \qquad \alpha > 0.$$

Then

$$\mathrm{Ker}(L_{A_0^*}) = \left\{ \begin{bmatrix} \lambda_1 & 0 \\ 0 & \lambda_2 \end{bmatrix} \middle| \lambda_1, \lambda_2 \in \mathbb{R} \right\}.$$

Hence

$$\mathrm{Ker}(L_{A_0^*}^s) = \left\{ \begin{bmatrix} \lambda & 0 \\ 0 & -\lambda \end{bmatrix} \middle| \lambda \in \mathbb{R} \right\}.$$

An infinitesimally symplectic miniversal deformation of $A_0$ is

$$A(\lambda) = \begin{bmatrix} \alpha + \lambda & 0 \\ 0 & -\alpha - \lambda \end{bmatrix}, \qquad \lambda \in \mathbb{R}.$$

(iii) Let

$$A_0 = \begin{bmatrix} 0 & \beta \\ -\beta & 0 \end{bmatrix}, \qquad \beta > 0.$$

A direct calculation shows that

$$\mathrm{Ker}(L_{A_0^*}) = \left\{ \begin{bmatrix} \lambda_1 & \lambda_2 \\ -\lambda_2 & \lambda_1 \end{bmatrix} \middle| \lambda_1, \lambda_2 \in \mathbb{R} \right\}.$$

Hence

$$\mathrm{Ker}(L_{A_0^*}^s) = \left\{ \begin{bmatrix} 0 & \lambda \\ -\lambda & 0 \end{bmatrix} \middle| \lambda \in \mathbb{R} \right\}.$$

An infinitesimally symplectic miniversal deformation of $A_0$ is

$$A(\lambda) = \begin{bmatrix} 0 & \beta + \lambda \\ -\beta - \lambda & 0 \end{bmatrix}, \qquad \lambda \in \mathbb{R}.$$

(iv) Let

$$A_0 = \begin{bmatrix} \alpha & \beta & 0 & 0 \\ -\beta & \alpha & 0 & 0 \\ 0 & 0 & -\alpha & \beta \\ 0 & 0 & -\beta & -\alpha \end{bmatrix}, \qquad \alpha > 0, \quad \beta > 0.$$

We take

$$p = \frac{1}{\sqrt{2}} \begin{bmatrix} 1 & 1 & 0 & 0 \\ -i & i & 0 & 0 \\ 0 & 0 & 1 & 1 \\ 0 & 0 & -i & i \end{bmatrix}.$$

Then $p^{-1} = p^*$

$$p^{-1}A_0 p = \begin{bmatrix} \alpha - \beta i & & & \\ & \alpha + \beta i & & \\ & & -\alpha - \beta i & \\ & & & -\alpha + \beta i \end{bmatrix}.$$

Any $4 \times 4$ complex matrix commuting with $(p^{-1}A_0 p)^*$ is of the form

$$
B(\epsilon) = \begin{bmatrix} \epsilon_1 & & & \\ & \epsilon_2 & & \\ & & \epsilon_3 & \\ & & & \epsilon_4 \end{bmatrix}, \qquad \epsilon_i \in \mathbb{C}, \quad i = 1,\dots,4,
$$

and

$$
pB(\epsilon)p^{-1}
$$

$$
= \tfrac{1}{2} \begin{bmatrix} \epsilon_1 + \epsilon_2 & (\epsilon_1 - \epsilon_2)i & 0 & 0 \\ -(\epsilon_1 - \epsilon_2)i & \epsilon_1 + \epsilon_2 & 0 & 0 \\ 0 & 0 & \epsilon_3 + \epsilon_4 & (\epsilon_3 - \epsilon_4)i \\ 0 & 0 & -(\epsilon_3 - \epsilon_4)i & \epsilon_3 + \epsilon_4 \end{bmatrix}.
$$

Since $pB(\epsilon)p^{-1} \in sp(4,\mathbb{R})$, $\epsilon_1 + \epsilon_2 = -(\epsilon_3 + \epsilon_4) \in \mathbb{R}$ and $(\epsilon_1 - \epsilon_2)i = (\epsilon_3 - \epsilon_4)i \in \mathbb{R}$. Thus an infinitesimally symplectic miniversal deformation of $A_0$ is

$$
A(\lambda) = A_0 + \begin{bmatrix} \lambda_1 & \lambda_2 & 0 & 0 \\ -\lambda_2 & \lambda_1 & 0 & 0 \\ 0 & 0 & -\lambda_1 & \lambda_2 \\ 0 & 0 & -\lambda_2 & -\lambda_1 \end{bmatrix}, \qquad \lambda_1, \lambda_2 \in \mathbb{R}.
$$

We can also take the following as infinitesimally symplectic miniversal deformations of $A_0$:

$$
A(\lambda) = A_0 + \begin{bmatrix} \lambda_1 & \lambda_2 & 0 & 0 \\ 0 & 0 & 0 & 0 \\ 0 & 0 & -\lambda_1 & 0 \\ 0 & 0 & -\lambda_2 & 0 \end{bmatrix},
$$

or

$$
A(\lambda) = A_0 + \begin{bmatrix} 0 & 0 & 0 & 0 \\ \lambda_1 & \lambda_2 & 0 & 0 \\ 0 & 0 & 0 & -\lambda_1 \\ 0 & 0 & 0 & -\lambda_2 \end{bmatrix},
$$

where $\lambda_1, \lambda_2 \in \mathbb{R}$.

(v) Let

$$A_0 = \begin{bmatrix} 0 & & & & & \vdots & & & \\ 1 & \ddots & & & & \vdots & & & \\ & \ddots & \ddots & & & \vdots & & & \\ & & 1 & & 0 & \vdots & & & n \\ \cdots & \cdots & \cdots & \cdots & \cdots & \vdots & 0 & -1 & \cdots \cdots \cdots \\ & & & & & \vdots & & \ddots & \\ & & & & & \vdots & & \ddots & -1 \\ & & & & & \vdots & & & 0 \end{bmatrix} \begin{matrix} n \\ \\ n \end{matrix}, \qquad m = 0 \text{ or } 1.$$

$$(-1)^m$$

We take

$$p = \begin{bmatrix} 1 & & & \vdots & & & & \\ & \ddots & & \vdots & & & & \\ & & 1 & \vdots & & & & n \\ \cdots & \cdots & \cdots & \vdots & \cdots & \cdots & \cdots & \cdots \\ & & & \vdots & & & (-1)^{m+n-1} & \\ & & & \vdots & & (-1)^{m+1} & & \\ & & & \vdots & (-1)^m & & & n \end{bmatrix}.$$

Then

$$p^{-1}A_0 p = \begin{bmatrix} 0 & & & & \vdots & & & & \\ 1 & \ddots & & & \vdots & & & & n \\ & \ddots & \ddots & & \vdots & & & & \\ & & 1 & 0 & \vdots & & & & \\ \cdots & \cdots & \cdots & \cdots & \vdots & \cdots & \cdots & \cdots & \cdots \\ & & & & \vdots & 1 & 0 & & \\ & & & & \vdots & & 1 & \ddots & \\ & & & & \vdots & & & \ddots & \ddots \\ & & & & \vdots & & & 1 & 0 \end{bmatrix} \begin{matrix} n \\ \\ \\ n \end{matrix}.$$

Any complex $2n \times 2n$ matrix commuting with $(p^{-1}A_0 p)^*$ is of the form

$$\begin{bmatrix} \epsilon_1 & \epsilon_2 & \cdots & \epsilon_{2n} \\ & \ddots & & \vdots \\ & & \ddots & \epsilon_2 \\ & & & \epsilon_1 \end{bmatrix}, \qquad \epsilon_i \in \mathbb{C}, \quad i = 1, \ldots, 2n.$$

By a similar argument as in (iv), we find an infinitesimally symplectic miniversal deformation of $A_0$ is

$$A(\lambda) = A_0 + \left[\begin{array}{c:c} A & B \\ \hdashline C & D \end{array}\right],$$

where

$$A = \begin{bmatrix} 0 & \lambda_1 & 0 & \lambda_2 & 0 & & \cdots & \beta_2 & \beta_1 \\ & & & & & & & & \beta_2 \\ & & & & & & & & \vdots \\ & & & & & & & & 0 \\ & & & & & & & & \lambda_2 \\ & & & & & & & & 0 \\ & & & & & & & & \lambda_1 \\ & & & & & & & & 0 \end{bmatrix},$$

$$B = \begin{bmatrix} \lambda_n & 0 & \lambda_{n-1} & 0 & \cdots & & \gamma \\ 0 & & -\lambda_{n-1} & & & & -\gamma & \vdots \\ \lambda_{n-1} & & & & & \gamma & & \vdots \\ 0 & & & & & & & 0 \\ \vdots & & & & & & & \alpha\lambda_2 \\ & & -\gamma & & & & -\alpha\lambda_2 & 0 \\ \gamma & \cdots & & 0 & \alpha\lambda_2 & 0 & \alpha\lambda_1 \end{bmatrix},$$

$C$ is a zero matrix,

$$D = \begin{bmatrix} 0 \\ -\lambda_1 \\ 0 \\ -\lambda_2 \\ 0 \\ \vdots \\ \vdots \\ -\beta_2 \\ -\beta_1 & -\beta_2 & \cdots & & \cdots & 0 & -\lambda_2 & 0 & -\lambda_1 & 0 \end{bmatrix},$$

where

$$\beta_1 = \begin{cases} \lambda_{\frac{n}{2}}, & n \text{ even,} \\ 0, & n \text{ odd,} \end{cases} \qquad \beta_2 = \begin{cases} 0, & n \text{ even,} \\ \lambda_{\frac{n-1}{2}}, & n \text{ odd,} \end{cases}$$

$$\gamma = \begin{cases} 0, & n \text{ even,} \\ \lambda_{\frac{n+1}{2}}, & n \text{ odd,} \end{cases} \qquad \alpha = (-1)^m,$$

and the $\{\lambda_i\}$ are real parameters; or

$$A(\lambda) = A_0 + \begin{bmatrix} & \vdots & \lambda_1 & & \\ 0 & \vdots & & \lambda_2 & \\ & \vdots & & & \ddots \\ & \vdots & & & & \lambda_n \\ \cdots & \cdots & \cdots & \cdots & \cdots \\ & 0 & \vdots & & 0 \end{bmatrix}$$

where $\lambda_i \in \mathbb{R}$, $i = 1, \ldots, n$.
(vi) Let

$$A_0 = \begin{bmatrix} 0 & & & \vdots & & & \\ 1 & \ddots & & \vdots & & & \\ & \ddots & \ddots & \vdots & & & \\ & & 1 & 0 & \vdots & & \\ \cdots & \cdots & \cdots & \cdots & 0 & -1 & \\ & & & \vdots & & \ddots & \ddots \\ & & & \vdots & & & \ddots & -1 \\ & & & \vdots & & & & 0 \end{bmatrix} \begin{matrix} n \\ \\ \\ \\ \\ \\ n \end{matrix} \quad , \qquad n \text{ odd.}$$

We take

$$p = \begin{bmatrix} 1 & & & \vdots & & & \\ & \ddots & & \vdots & & & \\ & & 1 & \vdots & & & \\ \cdots & \cdots & \cdots & \cdots & & & 1 \\ & & & \vdots & & -1 & \\ & & & \vdots & \ddots & & \\ & & & \vdots & -1 & & \\ & & & \vdots & 1 & & \end{bmatrix}.$$

Then

$$
p^{-1}A_0 p =
\begin{bmatrix}
0 & & & & & \vdots & & & \\
1 & \ddots & & & & \vdots & & & \\
& \ddots & \ddots & & & \vdots & & & \\
& & \ddots & 1 & 0 & \vdots & & & \\
\cdots & \cdots & \cdots & \cdots & \cdots & \cdots & \cdots & \cdots & \cdots \\
& & & & & 0 & & & \\
& & & & & 1 & \ddots & & \\
& & & & & \vdots & \ddots & \ddots & \\
& & & & & \vdots & & 1 & 0
\end{bmatrix}.
$$

The complex matrices commuting with $(p^{-1}A_0 p)^*$ are of the form

$$
\begin{bmatrix}
\epsilon_1 & \cdots & \epsilon_n & \vdots & \epsilon_{n+1} & \cdots & \epsilon_{2n} \\
 & \ddots & & \vdots & & \ddots & \\
 & & \epsilon_1 & \vdots & & & \epsilon_{n+1} \\
\cdots & \cdots & \cdots & \cdots & \cdots & \cdots & \cdots \\
\epsilon_{2n+1} & \cdots & \epsilon_{3n} & \vdots & \epsilon_{3n+1} & \cdots & \epsilon_{4n} \\
 & \ddots & & \vdots & & \ddots & \\
 & & \epsilon_{2n+1} & \vdots & & & \epsilon_{3n+1}
\end{bmatrix},
\qquad \epsilon_i \in \mathbb{C}, \quad i = 1, \ldots, 4n.
$$

Hence, an infinitesimally symplectic miniversal deformation of $A_0$ is $A(\lambda) = A_0 +$ the matrix:

$$
\begin{bmatrix}
\lambda_1 & \lambda_2 & \cdots & \cdots & \cdots & \cdots & \lambda_n & \vdots & \lambda_{n+1} & 0 & \lambda_c & 0 & & \lambda_a \\
 & \ddots & \ddots & & & & & \vdots & 0 & -\lambda_c & 0 & & \ddots & -\lambda_a \\
 & & \ddots & \ddots & & & & \vdots & \lambda_c & 0 & & \ddots & & \\
 & & & \ddots & \ddots & & & \vdots & 0 & & \ddots & & & \\
 & \bigcirc & & & \ddots & & \lambda_2 & \vdots & \vdots & & & -\lambda_a & \bigcirc & \\
 & & & & & & \lambda_1 & \vdots & \lambda_a & & & & & \\
\cdots & \cdots & \cdots & \cdots & \cdots & \cdots & \cdots & \cdots & \lambda_d & \vdots & -\lambda_1 & & & \\
 & & & -\lambda_d & & & 0 & \vdots & -\lambda_2 & \ddots & & & \\
 & \bigcirc & & & \ddots & & \vdots & \vdots & 0 & & \ddots & & \bigcirc & \\
 & & & & \ddots & & 0 & \vdots & \vdots & & & \ddots & & \\
 & & \ddots & & & \ddots & \lambda_{a+2} & \vdots & \vdots & & & & \ddots & \\
 & -\lambda_d & \ddots & & & -\lambda_{a+2} & 0 & \vdots & \vdots & & & & & \ddots \\
\lambda_d & 0 & \cdots & 0 & \lambda_{a+2} & 0 & \lambda_{a+1} & \vdots & -\lambda_n & \cdots & \cdots & \cdots & -\lambda_2 & -\lambda_1
\end{bmatrix}
$$

where $a = (3n + 1)/2$, $c = n + 2$, $d = 2n + 1$, $\{\lambda_i\}$ are real parameters; or $A(\lambda) = A_0 +$ the matrix

$$
\begin{bmatrix}
\lambda_1 & \lambda_2 & \cdots & \cdots & \cdots & \lambda_n & \vdots & \lambda_{n+1} & & 0 & \cdots & \cdots & \cdots & 0 \\
 & & & & & & \vdots & 0 & \lambda_{n+2} & & & & & \\
 & & \bigcirc & & & & \vdots & & & \ddots & & \lambda_a & & 0 \\
 & & & & & & \vdots & & & 0 & & 0 & & \ddots \\
 & & & & & & \vdots & 0 & & & \ddots & & & 0 \\
\cdots\cdots\cdots\cdots\cdots\cdots\cdots\cdots\cdots\cdots\cdots\cdots\cdots\cdots\cdots\cdots\cdots\cdots \\
0 & & & & & 0 & \vdots & -\lambda_1 & & & & & & \\
 & \ddots & & & & & \vdots & -\lambda_2 & & & & & & \\
 & & 0 & & 0 & & \vdots & & & & & \bigcirc & & \\
 & & & \lambda_{2n+1} & & & \vdots & & & & & & & \\
 & & 0 & & \lambda_{2n} & & \vdots & & & & & & & \\
 & \ddots & & & \ddots & 0 & \vdots & & & & & & & \\
0 & & & & 0 & \lambda_{a+1} & \vdots & -\lambda_n & & & & & &
\end{bmatrix}
$$

where $a = (3n + 1)/2$ and $\{\lambda_i\}$ are real parameters.
(vii) Let

$$
A_0 = \begin{bmatrix}
0 & & & & \vdots & & & & \beta \\
1 & \ddots & & & \vdots & & & \ddots & \\
 & \ddots & \ddots & & \vdots & & -\beta & & \\
 & & 1 & 0 & \vdots & \beta & & & \\
\cdots\cdots\cdots\cdots\cdots\cdots\cdots\cdots\cdots\cdots\cdots\cdots\cdots\cdots \\
 & & & -\beta & \vdots & 0 & -1 & & \\
 & & \beta & & \vdots & & \ddots & \ddots & \\
 & \ddots & & & \vdots & & & \ddots & -1 \\
-\beta & & & & \vdots & & & & 0
\end{bmatrix}
\begin{matrix} \\ \\ n \\ \\ \\ \\ n \\ \\ \end{matrix}
\quad , \quad \beta \neq 0, n \text{ odd.}
$$

We take

$$
p = \frac{1}{\sqrt{2}} \begin{bmatrix}
1 & & & \vdots & 1 & & \\
 & \ddots & & \vdots & & \ddots & \\
 & & 1 & \vdots & & & 1 \\
\cdots\cdots\cdots\cdots\cdots\cdots\cdots\cdots\cdots\cdots\cdots \\
 & & -i & \vdots & & & i \\
 & \ddots & & \vdots & & \ddots & \\
-i & & & \vdots & i & &
\end{bmatrix}
\begin{matrix} \\ \\ n \\ \\ \\ n \\ \\ \end{matrix}
\quad .
$$

Then

$$p^{-1}A_0p = \left[\begin{array}{ccccccc}
-\beta i & & & & \vdots & & \\
1 & -\beta i & & & \vdots & & \\
& & \ddots & & \vdots & & \\
& & 1 & -\beta i & \vdots & & \\
\hdotsfor{7} \\
& & & & \vdots & \beta i & \\
& & & & \vdots & 1 & \beta i \\
& & & & \vdots & & \ddots \\
& & & & \vdots & & 1 & \beta i
\end{array}\right] \begin{array}{c} \\ \\ \left.\rule{0pt}{2.2em}\right\} n \\ \\ \\ \left.\rule{0pt}{2.2em}\right\} n \\ \\ \end{array}.$$

The complex $2n \times 2n$ matrices commuting with $(p^{-1}A_0p)^*$ are of the form

$$\left[\begin{array}{ccccccc}
\epsilon_1 & \cdots & \epsilon_n & \vdots & & & \\
& \ddots & \vdots & \vdots & & & \\
& & \epsilon_1 & \vdots & & & \\
\hdotsfor{7} \\
& & \vdots & \epsilon_{n+1} & \cdots & \epsilon_{2n} \\
& & \vdots & & \ddots & \vdots \\
& & \vdots & & & \epsilon_{n+1}
\end{array}\right], \qquad \epsilon_i \in \mathbb{C}, \quad i = 1,\ldots,2n.$$

Thus, an infinitesimally symplectic miniversal deformation of $A_0$ is $A(\lambda) = A_0 +$ the matrix:

or $A(\lambda) = A_0 +$ the matrix

$$
\left[
\begin{array}{ccccccc:cccccc}
0 & \lambda_1 & 0 & \cdots & 0 & \lambda_g & 0 & \lambda_{g+1} & 0 & \cdots & \cdots & \cdots & \cdots & 0 \\
 & & & & & & & 0 & & & & & & \\
 & & \bigcirc & & & & & \vdots & & & & 0 & & \\
 & & & & & & & \vdots & & & \lambda_n & & & \\
 & & & & & & & 0 & & 0 & & 0 & & \\
 & & & & & & & \vdots & & & & & & 0 \\
 & & & & & & & 0 & \cdots & \cdots & \cdots & \cdots & \cdots & 0 \\
\hdashline
0 & \cdots & \cdots & \cdots & \cdots & \cdots & 0 & 0 & & & & & & \\
\vdots & & & & & & \vdots & -\lambda_1 & & & & & & \\
 & & 0 & & 0 & & \vdots & 0 & & & & & & \\
 & & & -\lambda_n & & & \vdots & \vdots & & & \bigcirc & & & \\
 & & 0 & & \ddots & & \vdots & 0 & & & & & & \\
 & & & & & & \vdots & -\lambda_g & & & & & & \\
0 & \cdots & \cdots & \cdots & 0 & -\lambda_{g+1} & 0 & & & & & & &
\end{array}
\right]
$$

where $g = (n - 1)/2$ and $\{\lambda_i\}$ are real parameters.

(viii) Let

$$
A_0 = \left[
\begin{array}{cccc:cccc}
\alpha & & & & \vdots & & & \\
1 & \ddots & & & \vdots & & & \\
 & \ddots & 1 & \alpha & \vdots & & & \\
\hdashline
 & & & & \vdots & -\alpha & -1 & \\
 & & & & \vdots & & \ddots & \ddots \\
 & & & & \vdots & & & \ddots & -1 \\
 & & & & \vdots & & & & -\alpha
\end{array}
\right] \begin{array}{l} \\ n \\ \\ \\ \\ \\ n \end{array}, \qquad \alpha > 0.
$$

We take

$$
p = \left[
\begin{array}{cccc:cccc}
1 & & & & \vdots & & & \\
 & \ddots & & & \vdots & & & \\
 & & & 1 & \vdots & & & \\
\hdashline
 & & & & \vdots & & & (-1)^{n-1} \\
 & & & & \vdots & & \ddots & \\
 & & & & \vdots & -1 & & \\
 & & & & \vdots & 1 & &
\end{array}
\right] \begin{array}{l} \\ n \\ \\ \\ \\ \\ n \end{array} .
$$

Then

$$
p^{-1}A_0 p = \left[
\begin{array}{cccc:cccc}
\alpha & & & & \vdots & & & \\
1 & \ddots & & & \vdots & & & \\
 & \ddots & \ddots & & \vdots & & & \\
 & & 1 & \alpha & \vdots & & & \\
\hdashline
 & & & & -\alpha & & & \\
 & & & & 1 & \ddots & & \\
 & & & & & \ddots & \ddots & \\
 & & & & & & 1 & -\alpha
\end{array}
\right]
\begin{array}{l} \left. \rule{0pt}{1.8em}\right\} n \\[2.2em] \left. \rule{0pt}{1.8em}\right\} n \end{array}
$$

The complex matrices commuting with $(p^{-1}A_0 p)^*$ are of the form

$$
\left[
\begin{array}{cccc:cccc}
\epsilon_1 & \cdots & \epsilon_n & & \vdots & & & \\
 & \ddots & \vdots & & \vdots & & & \\
 & & \epsilon_1 & & \vdots & & & \\
\hdashline
 & & & & \epsilon_{n+1} & \cdots & \epsilon_{2n} & \\
 & & & & & \ddots & \vdots & \\
 & & & & & & \epsilon_{n+1} &
\end{array}
\right]
\begin{array}{l} \left. \rule{0pt}{1.4em}\right\} n \\[1.6em] \left. \rule{0pt}{1.4em}\right\} n \end{array}
\qquad \epsilon_i \in \mathbb{C}, \quad i = 1,\ldots,2n.
$$

Thus an infinitesimally symplectic miniversal deformation of $A_0$ is

$$
A(\lambda) = A_0 + \left[
\begin{array}{cccc:cccc}
\lambda_1 & \cdots & \lambda_n & & \vdots & & & \\
 & \ddots & \vdots & & \vdots & & & \\
 & & \lambda_1 & & \vdots & & & \\
\hdashline
 & & & & -\lambda_1 & & & \\
 & & & & \vdots & \ddots & & \\
 & & & & -\lambda_n & & -\lambda_1 &
\end{array}
\right],
$$

or

$$
A(\lambda) = A_0 + \left[
\begin{array}{cccc:cccc}
\lambda_1 & \cdots & \lambda_n & & \vdots & & & \\
 & 0 & & & \vdots & & & 0 \\
\hdashline
 & & & & -\lambda_1 & & & \\
 & 0 & & & \vdots & & 0 & \\
 & & & & -\lambda_n & & &
\end{array}
\right],
$$

where $\lambda_i \in \mathbb{R}$, $i = 1, 2, \ldots, n$.

## 2.11 Normal Forms with Codimension One or Two

We consider the following parameter-dependent equations as our models for bifurcation problems

$$\dot{x} = A(\epsilon)x + f(x), \qquad x \in \Omega \subseteq \mathbb{R}^n, \quad \epsilon \in \Lambda \subseteq \mathbb{R}^k, \quad (11.1)$$

where $A(\epsilon)$ is a versal deformation of the matrix $A(0)$ (see Section 2.9) and the equation $\dot{x} = A(0)x + f(x)$ is an $A(0)$-normal form (in some cases $A(\epsilon)$ may be simplified further).

Since we consider equations on center manifolds, we may assume that the linear part of the equation has only eigenvalues with zero real parts when the parameters are zero.

*List 1.* The following are normal forms of codimension 1.

$$\dot{x} = \epsilon - x^2, \qquad x \in \mathbb{R}, \quad \epsilon \in \mathbb{R} \quad \text{(saddle-node)},$$

$$\dot{x} = \epsilon x - x^2, \qquad x \in \mathbb{R}, \quad \epsilon \in \mathbb{R} \quad \text{(transcritical)},$$
(i)

$$\dot{x} = \epsilon x \pm x^3, \qquad x \in \mathbb{R}, \quad \epsilon \in \mathbb{R} \quad \text{(pitchfork)}, \tag{ii}$$

$$\begin{cases} \dot{x} = \epsilon x - y + (\pm x - by)(x^2 + y^2), \\ \dot{y} = x + \epsilon y + (bx \pm y)(x^2 + y^2), \end{cases} \quad \text{(Hopf)}, \tag{iii}$$

where $b$ is a real constant and $\epsilon$ is a real parameter.

*List 2.* The following are normal forms of codimension 2.
(i) Double zero eigenvalues:

$$\begin{cases} \dot{x} = y, \\ \dot{y} = \epsilon_1 + \epsilon_2 y + x^2 \pm xy, \end{cases}$$

where $\epsilon_1$ and $\epsilon_2$ are real parameters.
(ii) Double zero eigenvalues with $Z_2$-symmetry:

$$\begin{cases} \dot{x} = y, \\ \dot{y} = \epsilon_1 x + \epsilon_2 y \pm x^3 - x^2 y; \end{cases}$$

where $\epsilon_1$ and $\epsilon_2$ are real parameters.

(iii) Double zero eigenvalues with $Z_q$-symmetry $(q \geq 3)$:

$$\dot{z} = \epsilon z + A_1 |z|^2 z + A_2 |z|^4 z + \cdots + A_k |z|^{2k} z + \bar{z}^{q-1},$$

$$q \geq 3, \quad q - 1 \leq 2k + 1 \leq q;$$

where $z \in \mathbb{C}$, $\epsilon$ is a complex parameter, the $A_i$ are complex constants, and Re $A_1 \neq 0$.

(iv) Double zero eigenvalues with flip symmetry or one pair of purely imaginary eigenvalues and one zero eigenvalue:

$$\begin{cases} \dot{x} = \epsilon_1 x + a_1 xy + a_2 x^3 + a_3 xy^2, \\ \dot{y} = \epsilon_2 + b_1 x^2 + b_2 y^2 + b_3 x^2 y + b_4 y^3, \end{cases}$$

where $\epsilon_1$ and $\epsilon_2$ are real parameters, and the $a_i$ and $b_i$ are real constants with some restrictions (see Section 4.6).

(v) One pair of imaginary eigenvalues and one zero eigenvalue with flip symmetry or two different pairs of purely imaginary eigenvalues:

$$\begin{cases} \dot{x} = \epsilon_1 x + a_1 x^3 + a_2 xy^2 + a_3 x^5 + a_4 x^3 y^2 + a_5 xy^4, \\ \dot{y} = \epsilon_2 y + b_1 x^2 y + b_2 y^3 + b_3 x^4 y + b_4 x^2 y^3 + b_5 y^5, \end{cases}$$

where $\epsilon_1$ and $\epsilon_2$ are real parameters, and the $a_i$ and $b_i$ are real constants with some restrictions (see Section 4.7).

## Calculations for Normal Forms in List 1

(i) Let $A_0 = 0 \in \mathbb{R}$. Then an $A_0$-normal form up to order 2 is

$$\dot{x} = ax^2, \quad x \in \mathbb{R}, \quad a \in \mathbb{R}. \tag{11.2}$$

We assume that $a \neq 0$. On changing variables $x \to -(1/a) x$, (11.2) becomes

$$\dot{x} = -x^2. \tag{11.3}$$

Since a miniversal deformation of $A_0$ is $A(\lambda) = \lambda$, $\lambda \in \mathbb{R}$, we may take the following as a codimension-one normal form:

$$\dot{x} = \lambda x - x^2. \tag{11.4}$$

If we change variables by $x \to x + \lambda/2$, then (11.4) becomes

$$\dot{x} = -\lambda^2/4 - x^2, \tag{11.5}$$

which is induced from the following equation by $\epsilon = -\lambda^2/4$:

$$\dot{x} = \epsilon - x^2, \quad \epsilon \in \mathbb{R}. \tag{11.6}$$

Therefore we take (11.6) instead of (11.4) as a codimension-one normal form.

(ii) Let $A_0 = 0 \in \mathbb{R}$ and the equation satisfy the reflection symmetry. An $A_0$-normal form up to order 3 is

$$\dot{x} = ax^3, \quad x \in \mathbb{R}, \quad a \in \mathbb{R}. \tag{11.7}$$

We assume that $a \neq 0$. By a similar argument, we get

$$\dot{x} = \epsilon x \pm x^3, \quad \epsilon \in \mathbb{R}.$$

(iii) Let

$$A_0 = \begin{bmatrix} 0 & -\omega \\ \omega & 0 \end{bmatrix}, \quad \omega > 0.$$

An $A_0$-normal form up to order 3 (see Example 1.22) is

$$\begin{bmatrix} \dot{x} \\ \dot{y} \end{bmatrix} = \begin{bmatrix} 0 & -\omega \\ \omega & 0 \end{bmatrix}\begin{bmatrix} x \\ y \end{bmatrix} + (x^2 + y^2)\begin{bmatrix} ax - by \\ bx + ay \end{bmatrix}, \tag{11.8}$$

where $a, b \in \mathbb{R}$. We assume that $a \neq 0$. Since a miniversal deformation of $A_0$ is

$$A(\lambda) = \begin{bmatrix} \lambda_1 & -\omega - \lambda_2 \\ \omega + \lambda_2 & \lambda_1 \end{bmatrix}, \quad \lambda_1, \lambda_2 \in \mathbb{R},$$

we may consider the following equation as a model of bifurcation problems:

$$\begin{bmatrix} \dot{x} \\ \dot{y} \end{bmatrix} = \begin{bmatrix} \lambda_1 & -\omega - \lambda_2 \\ \omega + \lambda_2 & \lambda_1 \end{bmatrix} \begin{bmatrix} x \\ y \end{bmatrix} + (x^2 + y^2) \begin{bmatrix} ax - by \\ bx + ay \end{bmatrix}. \quad (11.9)$$

We rescale time by $t \to t/(\omega + \lambda_2)$ when $\lambda_2$ is close to zero and define a new parameter $\epsilon = \lambda_1/(\omega + \lambda_2)$. Then (11.9) becomes

$$\begin{bmatrix} \dot{x} \\ \dot{y} \end{bmatrix} = \begin{bmatrix} \epsilon & -1 \\ 1 & \epsilon \end{bmatrix} \begin{bmatrix} x \\ y \end{bmatrix} + \frac{(x^2 + y^2)}{\omega + \lambda_2} \begin{bmatrix} ax - by \\ bx + ay \end{bmatrix}. \quad (11.10)$$

If we change variables by $x \to (\sqrt{\omega + \lambda_2} / \sqrt{|a|})x$, $y \to (\sqrt{\omega + \lambda_2} / \sqrt{|a|})y$, then (11.10) becomes

$$\begin{bmatrix} \dot{x} \\ \dot{y} \end{bmatrix} = \begin{bmatrix} \epsilon & -1 \\ 1 & \epsilon \end{bmatrix} \begin{bmatrix} x \\ y \end{bmatrix} + (x^2 + y^2) \begin{bmatrix} \pm x - cy \\ cx \pm y \end{bmatrix}, \quad (11.11)$$

where $c = b/|a|$.

### Calculations for Normal Forms in List 2

(i) Let

$$A_0 = \begin{bmatrix} 0 & 1 \\ 0 & 0 \end{bmatrix}.$$

Then an $A_0$-normal form up to order 2 (see Example 1.15) is

$$\begin{cases} \dot{x} = y, \\ \dot{y} = ax^2 + bxy, \end{cases}$$

where $a, b$ are real constants. We assume that $a \cdot b \neq 0$. On the other hand, a miniversal deformation of $A_0$ (see Example 9.18(3)) is

$$A(\lambda) = \begin{bmatrix} 0 & 1 \\ \lambda_1 & \lambda_2 \end{bmatrix}, \quad \lambda_1, \lambda_2 \in \mathbb{R}.$$

Hence we may take the following as one of the models of bifurcation problems:

$$\begin{cases} \dot{x} = y, \\ \dot{y} = \lambda_1 x + \lambda_2 y + ax^2 + bxy. \end{cases} \quad (11.12)$$

We change variables by $x \to x - \lambda_1/2a$, $y \to y$ and define new parameters by $\mu_1 = -\lambda_1^2/4a$, $\mu_2 = (2a\lambda_2 - b\lambda_1)/2a$. Then (11.12) becomes

$$\begin{cases} \dot{x} = y, \\ \dot{y} = \mu_1 + \mu_2 y + ax^2 + bxy. \end{cases} \quad (11.13)$$

Let

$$x \mapsto \frac{a}{b^2} x, \quad y \mapsto \frac{|a|a}{|b|^3} y, \quad t \mapsto \left|\frac{b}{a}\right| t, \quad \mu_1 \mapsto \frac{a^3}{b^4} \epsilon_1, \quad \mu_2 \mapsto \left|\frac{a}{b}\right| \epsilon_2.$$

Then (11.13) can be changed to

$$\begin{cases} \dot{x} = y, \\ \dot{y} = \epsilon_1 + \epsilon_2 y + x^2 \pm xy, \end{cases} \quad (11.14)$$

where $\epsilon_1, \epsilon_2$ are real parameters.

We note that we can also change variables in (11.12) by $x \to x - \lambda_2/b$, $y \to y$ and define new parameters by $\mu_1 = (a\lambda_2^2 - b\lambda_1\lambda_2)/b^2$, $\mu_2 = (b\lambda_1 - 2a\lambda_2)/b$. Then (11.12) becomes

$$\begin{cases} \dot{x} = y, \\ \dot{y} = \mu_1 + \mu_2 x + ax^2 + bxy. \end{cases} \quad (11.15)$$

It can be changed to

$$\begin{cases} \dot{x} = y, \\ \dot{y} = \epsilon_1 + \epsilon_2 x + x^2 \pm xy. \end{cases} \quad (11.16)$$

(ii) Let

$$A_0 = \begin{bmatrix} 0 & 1 \\ 0 & 0 \end{bmatrix}.$$

An $A_0$-normal form satisfying $Z_2$-symmetry up to order 3 (see Example 5.12) is

$$\begin{cases} \dot{x} = y, \\ \dot{y} = ax^3 + bx^2y, \end{cases}$$

where $a, b$ are real constants. We assume that $a \cdot b \neq 0$. A miniversal deformation of $A_0$ is

$$A(\lambda) = \begin{bmatrix} 0 & 1 \\ \lambda_1 & \lambda_2 \end{bmatrix}.$$

We note that $A(\lambda)$ commutes with

$$S = \begin{bmatrix} -1 & 0 \\ 0 & -1 \end{bmatrix},$$

which is a generator of the $Z_2$-symmetry group. Thus we may take the following as one of the models of bifurcation problems:

$$\begin{cases} \dot{x} = y, \\ \dot{y} = \lambda_1 x + \lambda_2 y + ax^3 + bx^2y, \end{cases} \tag{11.17}$$

where $\lambda_1, \lambda_2$ are real parameters. Let $x \to (\sqrt{|a|}/b)\, x$, $y \to (|a|\sqrt{|a|b}/|b|^3)\, y$, $\epsilon_1 = b^2\lambda_1/a^2$, $\epsilon_2 = |b|\lambda_2/|a|$, and $t \to |b|t/|a|$. Then (11.17) becomes

$$\begin{cases} \dot{x} = y, \\ \dot{y} = \epsilon_1 x + \epsilon_2 y \pm x^3 - x^2y. \end{cases} \tag{11.18}$$

This is the case (ii).

  (iii) Let

$$A_0 = \begin{bmatrix} 0 & 0 \\ 0 & 0 \end{bmatrix}.$$

An $A_0$-normal form with $Z_q$-symmetry ($q \geq 3$) up to order $q$ (see Example 5.15) is

$$\dot{z} = A_1|z|^2z + \cdots + A_k|z|^{2k}z + B\bar{z}^{q-1}, \tag{11.19}$$

where $z \in \mathbb{C}$, the $A_i$ and $B$ are complex constants, $q - 1 \leq 2k + 1 \leq q$. We assume that $B \neq 0$. Suppose that $B = |B|e^{i\phi}$. We change variable by $z \to e^{i\phi/q}z$, rescale time by $t \to t/|B|$, and change coefficients by $A_i \to A_i|B|$. Then (11.19) becomes

$$\dot{z} = A_1|z|^2 z + \cdots + A_k|z|^{2k}z + \bar{z}^{q-1}. \qquad (11.20)$$

A miniversal deformation commuting with matrix $K_{2\pi/q}$ of $A_0$ is

$$A(\epsilon) = \begin{bmatrix} \epsilon_1 & -\epsilon_2 \\ \epsilon_2 & \epsilon_1 \end{bmatrix}, \qquad \epsilon_1, \epsilon_2 \text{ are real parameters.}$$

Since the complex form of $A(\epsilon)$ is

$$\begin{bmatrix} \epsilon & 0 \\ 0 & \epsilon \end{bmatrix}, \qquad \epsilon \in \mathbb{C}, \qquad (11.21)$$

combining (11.20) and (11.21) we obtain

$$\dot{z} = \epsilon z + A_1|z|^2 z + \cdots + A_k|z|^{2k}z + \bar{z}^{q-1}.$$

This is the case (iii).

(iv) Let

$$A_0 = \begin{bmatrix} 0 & 0 \\ 0 & 0 \end{bmatrix}.$$

An $A_0$-normal form with flip symmetry up to order 3 (see Section 2.5) is

$$\begin{cases} \dot{x} = a_1 x^2 + a_2 y^2 + a_3 x^3 + a_4 xy^2, \\ \dot{y} = b_1 xy + b_2 x^2 y + b_3 y^3, \end{cases}$$

where the $a_i$ and $b_i$ are real constants. Let

$$S = \begin{bmatrix} 1 & 0 \\ 0 & -1 \end{bmatrix}.$$

$S$ is a generator of the flip-symmetry group. A miniversal deformation of $A_0$ commuting with $S$ is

$$A(\lambda) = \begin{bmatrix} \lambda_1 & 0 \\ 0 & \lambda_2 \end{bmatrix}, \qquad \lambda_1, \lambda_2 \in \mathbb{R}.$$

Thus we obtain the following system:

$$\begin{cases} \dot{x} = \lambda_1 x + a_1 x^2 + a_2 y^2 + a_3 x^3 + a_4 xy^2, \\ \dot{y} = \lambda_2 y + b_1 xy + b_2 x^2 y + b_3 y^3. \end{cases} \qquad (11.22)$$

We change variables in (11.22) by $x \to x + \alpha$, $y \to y$, where $\alpha$ will be determined later (note that this transformation satisfies flip symmetry with respect to $y$). Then (11.22) becomes

$$\begin{cases} \dot{x} = \left( \lambda_1 \alpha + a_1 \alpha^2 + a_3 \alpha^3 \right) + \left( \lambda_1 + 2a_1 \alpha + 3a_3 \alpha^2 \right) x \\ \qquad + (a_1 + 3a_3 \alpha) x^2 + (a_2 + a_4 \alpha) y^2 + a_3 x^3 + a_4 xy^2, \\ \dot{y} = \left( \lambda_2 + b_1 \alpha + b_2 \alpha^2 \right) y + (b_1 + 2b_2 \alpha) xy + b_2 x^2 y + b_3 y^3. \end{cases}$$

$$(11.23)$$

We can choose an $\alpha = \alpha(\lambda_1)$ such that for each small $\lambda_1$

$$\lambda_1 + 2a_1 \alpha + 3a_3 \alpha^2 = 0$$

provided $a_1 \neq 0$, and $\alpha(\lambda_1)$ is continuous and $\alpha(0) = 0$. We define new parameters

$$\epsilon_1 = \lambda_1 \alpha + a_1 \alpha^2 + a_3 \alpha^3, \qquad \epsilon_2 = \lambda_2 + b_1 \alpha + b_2 \alpha^2$$

and omit the terms with factor $\alpha$ in the coefficients of the higher-order terms in (11.40). We obtain the following:

$$\begin{cases} \dot{x} = \epsilon_1 + a_1 x^2 + a_2 y^2 + a_3 x^3 + a_4 xy^2, \\ \dot{y} = \epsilon_2 y + b_1 xy + b_2 x^2 y + b_3 y^3. \end{cases} \qquad (11.24)$$

After exchanging $x$ and $y$, (11.24) becomes the normal form of the case (iv).

Let the matrix of the linear part of a system be

$$A_0 = \begin{bmatrix} 0 & -1 & 0 \\ 1 & 0 & 0 \\ 0 & 0 & 0 \end{bmatrix}.$$

We change variables by $z = x_1 + ix_2$, $y = x_3$. Then the matrix of the linear part of the resulting system is

$$\tilde{A}_0 = \begin{bmatrix} i & 0 & 0 \\ 0 & -i & 0 \\ 0 & 0 & 0 \end{bmatrix}.$$

The resonant monomials with respect to $\tilde{A}_0$ are $z_1 z_3 e_1$, $z_2 z_3 e_2$, $z_1 z_2 e_3$, $z_3^2 e_3$, $z_1^2 z_2 e_1$, $z_1 z_3^2 e_1$, $z_1 z_2^2 e_1$, $z_2^3 e_3$, and $z_1 z_2 z_3 e_3$. We note that $z_1 = z$, $z_2 = \bar{z}$, and $z_3 = y$ in this case. A miniversal deformation of $A_0$ is

$$\tilde{A}(\lambda) = \begin{bmatrix} \lambda_1 + i & 0 & 0 \\ 0 & \lambda_2 - i & 0 \\ 0 & 0 & \lambda_3 \end{bmatrix}, \qquad \lambda_1, \lambda_2, \lambda_3 \in \mathbb{C}.$$

But $\lambda_2 = \bar{\lambda}_1$, and $\lambda_3 \in \mathbb{R}$ in this case. Thus we obtain the following

$$\begin{cases} \dot{z} = (\lambda_1 + i)z + A_1 zy + A_2 |z|^2 z + A_3 zy^2, \\ \dot{y} = \lambda_3 y + a_2 |z|^2 + b_2 y^2 + c_2 |z|^2 y + d_2 y^3, \end{cases} \qquad (11.25)$$

where $A_i$ are complex constants, $a_2, b_2, c_2, d_2$ are real constants, and the equation for $\dot{\bar{z}}$ is omitted. Let $z = re^{i\theta}$. Then (11.25) becomes

$$\begin{cases} \dot{r} = \epsilon_1 r + a_1 ry + b_1 r^3 + c_1 ry^2, \\ \dot{y} = \epsilon_2 y + a_2 r^2 + b_2 y^2 + c_2 r^2 y + d_2 y^3, \end{cases} \qquad (11.26)$$

where $\epsilon_1 = \text{Re } \lambda_1$, $\epsilon_2 = \lambda_3$, $a_1 = \text{Re } A_1$, $b_1 = \text{Re } A_2$, $c_1 = \text{Re } A_3$, and the equation for $\dot{\theta}$ is omitted since $\dot{\theta} = 1 + O(|(\text{Im } \lambda_1, r, y)|)$. If we replace $y$ by $x$ and replace $r$ by $y$ then (11.26) is the same as (11.22). This is the case (iv) with $x \geq 0$.

(v) Let

$$A_0 = \begin{bmatrix} 0 & -1 & 0 \\ 1 & 0 & 0 \\ 0 & 0 & 0 \end{bmatrix}.$$

We assume that the equation satisfies flip symmetry. More precisely, the equation satisfies the group $\Gamma$-symmetry, where $\Gamma$ is generated by

$$S = \begin{bmatrix} 1 & 0 & 0 \\ 0 & 1 & 0 \\ 0 & 0 & -1 \end{bmatrix}.$$

By a similar argument as that for the second subcase of case (iv), and noticing that the equation in complex form satisfies $S$-symmetry, we obtain the following:

$$\begin{cases} \dot{z} = (\lambda_1 + i)z + A_1|z|^2 z + A_2 zy^2 + A_3|z|^4 z + A_4|z|^2 zy^2 + A_5 y^4, \\ \dot{y} = \lambda_2 y + b_1 y^3 + b_2|z|^2 y + b_3|z|^4 y + b_4|z|^2 y^3 + b_5 y^5, \end{cases}$$

$$(11.27)$$

where $z \in \mathbb{C}$, $y \in \mathbb{R}$, $\lambda_1$ is a complex parameter, $\lambda_2$ is a real parameter and the $A_i$ and $b_i$ are complex and real constants. Changing variables by $z = re^{i\theta}$, $y = y$ in (11.27) and omitting the equation for $\dot{\theta}$ we get the normal form of case (v) with $x \geq 0$.

Let the matrix of the linear part of a system be

$$A_0 = \begin{bmatrix} 0 & -\omega_1 & 0 & 0 \\ \omega_1 & 0 & 0 & 0 \\ 0 & 0 & 0 & -\omega_2 \\ 0 & 0 & \omega_2 & 0 \end{bmatrix},$$

where $\omega_1, \omega_2 > 0$ and $\omega_1$ and $\omega_2$ are rationally independent. We change variables by $z_1 = x_1 + ix_2$, $z_2 = x_1 - ix_2$, $z_3 = x_3 + ix_4$, $z_4 = x_3 - ix_4$ in the system. Then the matrix of the linear part of the resulting system is

$$A_0 = \begin{bmatrix} \omega_1 i & 0 & 0 & 0 \\ 0 & -\omega_1 i & 0 & 0 \\ 0 & 0 & \omega_2 i & 0 \\ 0 & 0 & 0 & -\omega_2 i \end{bmatrix}.$$

The resonant conditions for $\tilde{A}_0$ are $\omega_j = \alpha_1 \omega_1 - \alpha_2 \omega_1 + \alpha_3 \omega_2 - \alpha_4 \omega_2$, or

$$(\alpha_1 - \alpha_2 - 1)\omega_1 + (\alpha_3 - \alpha_4)\omega_2 = 0, \qquad j = 1,$$

$$(\alpha_1 - \alpha_2)\omega_1 + (\alpha_3 - \alpha_4 - 1)\omega_2 = 0, \qquad j = 3.$$

Since $\omega_1$ and $\omega_2$ are rationally independent, we must have $\alpha_1 - \alpha_2 = 1$, $\alpha_3 = \alpha_4$ for $j = 1$ and $\alpha_1 = \alpha_2$, $\alpha_3 - \alpha_4 = 1$ for $j = 3$. Furthermore a miniversal deformation of $\tilde{A}_0$ is

$$\begin{bmatrix} \lambda_1 + \omega_1 i & 0 & 0 & 0 \\ 0 & \bar{\lambda}_1 - \omega_1 i & 0 & 0 \\ 0 & 0 & \lambda_2 + \omega_2 i & 0 \\ 0 & 0 & 0 & \bar{\lambda}_2 - \omega_2 i \end{bmatrix}, \qquad \lambda_1, \lambda_2 \in \mathbb{C}.$$

Thus we obtain the following:

$$\begin{cases} \dot{z}_1 = (\omega_1 i + \lambda_1)z_1 + A_1|z_1|^2 z_1 + A_2 z_1 |z_3|^2 + A_3 |z_1|^4 z_1 \\ \qquad + A_4 |z_1|^2 |z_3|^2 z_1 + A_5 z_1 |z_3|^4, \\ \dot{z}_3 = (\omega_2 i + \lambda_2)z_3 + B_2 |z_1|^2 z_3 + B_2 |z_3|^2 z_3 + B_3 |z_1|^4 z_3 \\ \qquad + B_4 |z_1|^2 |z_3|^2 z_3 + B_5 |z_3|^4 z_3, \end{cases} \qquad (11.28)$$

where the $A_i$ and $B_i$ are complex constants. Let $z_1 = r_1 e^{i\theta_1}$, $z_3 = r_2 e^{i\theta_2}$. Then (11.28) becomes

$$\begin{cases} \dot{r}_1 = \epsilon_1 r_1 + a_1 r_1^3 + a_2 r_1 r_2^2 + a_3 r_1^5 + a_4 r_1^3 r_2^2 + a_5 r_1 r_2^4, \\ \dot{r}_2 = \epsilon_2 r_2 + b_1 r_1^2 r_2 + b_2 r_2^3 + b_3 r_1^4 r_2 + b_4 r_1^2 r_2^3 + b_5 r_2^5, \end{cases}$$

where $\epsilon_1, \epsilon_2$ are real parameters, the $a_i$ and $b_i$ are real constants, and the equations for $\dot{\theta}_1$ and $\dot{\theta}_2$ are omitted since $\dot{\theta}_1$ and $\dot{\theta}_2$ are close to 1 when $r_1$ and $r_2$ are close to zero. This is the case (v) with $x \geq 0$ and $y \geq 0$.

### 2.12 Bibliographical Notes

Normal form theory has a long history. The basic idea of simplifying ordinary differential equations through changes of variables can be found in the early work of Briot and Bouquet [1]. Poincaré [2] made very important contributions to normal form theory. After Poincaré, this theory was developed by Liapunov [1], Dulac [2], Birkhoff [1], Cherry [1], Siegel [1], Sternberg [1–4], Chen [1], Gustavson [1], Moser [1], and many others. Recently normal form theory has been developed very rapidly since it plays a very important role in bifurcation theory. We mention the work of Arnold [1, 6, 7], Belitskii [1–5], Bruno [1–6], Takens [1, 3, 4], Cushman [1–3], Cushman and Sanders [1–5], Elphick, et al. [1], van der Meer [1–2], and so forth. For an outline of the normal form theory, we also refer our readers to the books by Arnold [4], Arnold and Il'yashenko [1], Bruno [6], Chow and Hale [1], Golubitsky and Schaeffer [1], Guckenheimer and Holmes [1], and Wiggins [1].

Most transformations which lead the equations to their normal forms are formal series. Only in a very few cases are the transformations convergent series (see Siegel [1] and Bruno [6]). From the point of view of the applications, one only needs terms of lower orders in normal forms. That is the main reason we consider here only normal forms up to a certain order.

In Section 2.1, we introduced two methods of computing normal forms. The matrix representation method is based on linear algebra and is well known. The method of adjoints was first given by Belitskii [3–4], and then by Elphick et al. [1]. Our treatment is similar to that of Elphick et al. [1]. Another important method is based on the representation theory of Lie algebra $sl(2, R)$. This method was used in the computation of normal forms of nonlinear Hamiltonian systems (see Cushman, Deprit, and Mosak [1] and Cushman, Sanders, and White [1]). General applications of the method of representation theory of $sl(2, R)$ to normal form theory were studied by Arnold and Il'yashenko [1], Cushman and Sanders [2], Meyer [3], and Wang [2–3].

The applications of $S$–$N$ decomposition technique to normal form theory were first used in Arnold [4] and in van der Meer [1].

In Section 2.2 we introduced linearization theory for analytic systems. Poincaré's and Siegel's Theorem were proved in Arnold [4]. Our treatment for the proof of Poincaré's theorem is essentially the same as that given by Meyer [2] for diffeomorphisms. The proof of Siegel's theorem in this chapter is given by Chow, Lu, and Shen [1] in a similar way as that used in the proof of Siegel's theorem for mappings given by Zehnder [1]. Results of linearization for general cases have been given by Chen [1], Hartman [1], Nagumo and Ise [1], Sell [1], Sternberg [1], and others.

Normal forms of equations with periodic coefficients and normal forms of maps near fixed points were studied in Arnold [4]. Normal forms of equations with symmetry were studied in Elphick et al. [1] and Golubitsky and Schaeffer [1].

Normal forms of linear Hamiltonian systems were studied first by Williamson [1]. Lists I and II in Section 2.6 for normal forms of indecomposable infinitesimally symplectic matrices are from Burgoyne and Cushman [1] in which one can find the complete proofs for the two lists of normal forms.

Normal forms of nonlinear Hamiltonian systems were studied first by Whittaker [1] and then by Birkhoff [1], Cherry [1], and Siegel [2]. See also Arnold [3], Bruno [3], Gustavson [1], and Meyer [1]. For more references we recommend van der Meer [1]. The main results in Section 2.7 can be found in Elphick et al. [1] and Wang [4].

Takens's Theorem was given by Takens [2]. In Section 2.8 we give an elementary proof of this theorem. The idea of our proof is motivated by Rousseau [1].

The main results of versal deformations of matrices were given by Arnold [6]. Following Arnold's idea, Galin [2] and Koçak [1] gave versal deformations of infinitesimally symplectic matrices.

The general method for computing normal forms of systems with parameters $\epsilon \in \mathbb{R}^k$ is to add an equation $\dot{\epsilon} = 0$ to the original system and to get normal forms for the extended system. See, for example, Guckenheimer and Holmes [1]. Different treatments were used in Chow and Wang [1], Elphick et al. [1] and Vanderbauwhede [3]. The lists of codimension-one and codimension-two normal forms can also be found in Guckenheimer [3], and Elphick et al. [1].

The computation of coefficients of normal forms for a given system was studied by many authors; see, for example, Chow, Drachman, and

Wang [1], Cushman and Sanders [1–5], Knobloch [1], Rand and Keith [1], and Rand and Armbruster [1]. They gave computer programs by using symbolic computation softwares such as MACSYMA. Recursive formulas for coefficients of normal forms were given by Bruno [6], Hsu and Favretto [1], and Wang [5].

Other topics in normal form theory for volume preserving vector fields, reversible systems, constrained systems, Hamiltonian systems, and stochastic systems can be found in Arnold [3], Baider and Churchill [1], Broer [1–2], Broer et al. [1], Chow, Lu, and Shen [1], Chua and Oka [1], Cushman, Sanders, and White [1], Khazin and Shnol [1], Moser [1], Starzhinskii [1–2], Sternberg [3], Tirapegui [1], Ushiki [1], and many others.

# 3

# Codimension One Bifurcations

In Chapters 3–5 we will study bifurcation phenomena of vector fields. In order to consider these problems, we should consider not only a single vector field, but also its "nearby" vector fields. This means we need to consider a suitable space of vector fields $\mathscr{X}$, for example, $\mathscr{X}^r(\mathbb{R}^n)$, the Banach space of all $C^r$ $(r \geq 1)$ vector fields on $\mathbb{R}^n$, and investigate the qualitative behavior of all vector fields in a small neighborhood of a fixed vector field in $\mathscr{X}$.

We say two vector fields $X$ and $Y$ in $\mathscr{X}$ are *topologically equivalent*, denoted by $X \sim Y$, if there exists a homeomorphism on $\mathbb{R}^n$ which maps the phase orbits of $X$ onto the phase orbits of $Y$ and preserves the direction of the orbits in time. We say $X \in \mathscr{X}$ is *structurally stable* if there is a neighborhood $V$ of $X$ in $\mathscr{X}$ such that $Y \sim X$ for all $Y \in V$. We say $X \in \mathscr{X}$ is a *bifurcation point* if $X$ is not structurally stable. All bifurcation points form a *bifurcation set* $\mathscr{B}$ in $\mathscr{X}$.

Suppose $X \in \mathscr{B}$. This means that in any neighborhood $V$ of $X$ in $\mathscr{X}$, no matter how small it is, we can find $Y \in V$ such that the orbit structures of $X$ and $Y$ are different ($Y$ is not equivalent to $X$). There is a basic question: Is it possible to find a neighborhood $W$ of $X$ in $\mathscr{X}$ such that we can give a complete description of the phase portraits for all $Y \in W$? Related to this question, we need to know the structure of the bifurcation set in $W$ (i.e., $\mathscr{B} \cap W$) which may be very complicated. In some cases, if we put a suitable "nondegenerate condition" on $X$ we can make the structure of $\mathscr{B}$ (near $X$) simple. For example, we may find a finite-dimensional surface $S$ in $W$ passing through $X$ such that for every $Y \in W$ there is a $Z \in S$, $Z \sim Y$. Thus, to answer the above question, we can restrict our study to the family $S$. If we can find such a family with dimension $k$, but such that any $(k-1)$-dimensional family (passing through $X$) has no such property, then $X$ is called a *codimen-*

*sion-k bifurcation point.* In fact, under certain conditions the surface $S$ is transversal to a submanifold at $X$, with codimension $k$ in $\mathscr{X}$. Intuitively, the codimension indicates the degree of complexity of the bifurcation problem.

### 3.1 Definitions and Jet Transversality Theorem

We first define the space of jets. Suppose that $M$ and $N$ are $C^r$-manifolds $(1 \leq r \leq \infty)$ with finite dimensions, and $f, g \in C^r(M, N)$.

**Definition 1.1.** $f$ and $g$ are said to be $k$-tangent at $x \in M$ $(1 \leq k \leq r)$ if in a local coordinate system $f$ and $g$ have the same Taylor coefficients up to order $k$ at $x$.

The $k$-tangency is independent of the choice of local coordinates, and it gives an equivalence relation in $C^r(M, N)$.

**Definition 1.2.** A $k$-jet of a $C^r$ mapping $f$ at $x$ is given by

$$j_x^k(f) = \{g \in C^r(M, N) \,|\, g \text{ is } k\text{-tangent to } f \text{ at } x\},$$

and we define

$$j^k(f) = \{j_x^k(f) \,|\, \forall\, x \in M\}$$

and

$$J^k(M, N) = \{j^k(f) \,|\, \forall\, f \in C^k(M, N)\}.$$

Since the dimensions of $M$ and $N$ are finite, $J^k(M, N)$ is finite dimensional. Moreover, if $M$ and $N$ are $C^r$-manifolds, then $J^k(M, N)$ is a $C^{r-k}$-manifold (see Hirsch [1]).

**Definition 1.3.** Suppose $L$ is a linear space, and $A$ and $B$ are its linear subspaces. $A$ and $B$ are said to be transversal if

$$L = A + B.$$

Suppose $M$ and $N$ are smooth manifolds, $A$ is a smooth submanifold of $N$, $f \in C^\infty(M, N)$, and $p \in M$. Since the tangent space of a manifold at a point is linear, and the tangent mapping $T_p f$ maps the tangent space $T_p M$ into the tangent space $T_{f(p)} N$, we can define the transversality between $f$ and $A$ at $p$ as follows.

**Definition 1.4.** The mapping $f$ and the submanifold $A$ are said to be transversal at $p \in M$ and denoted by $f \pitchfork_p A$, if $f(p) \notin A$ or

$$(T_p f)(T_p M) + T_{f(p)} A = T_{f(p)} N;$$

$f$ and $A$ are said to be transversal and denoted by $f \pitchfork A$ if they are transversal at every point $p \in M$.

The following result is obvious, but it is useful.

**Theorem 1.5.** *Suppose that $M$ and $N$ are smooth manifolds with dimensions $m$ and $n$, respectively, and $A$ is a smooth submanifold of $N$ with codimension $r$. Let $(x^1, \dots, x^m)$ be the local coordinates of $M$ near $x_0$, and $(y^1, \dots, y^n)$ be the local coordinates of $N$ near $f(x_0)$. Suppose in a neighborhood $U$ of $f(x_0)$ in $N$, the set $A \cap U$ can be expressed by $y^1 = \cdots = y^r = 0$. If $f(x_0) \in A$, then $f \pitchfork_{x_0} A$ if and only if the rank of the matrix $(\partial y^i / \partial x^j)|_{\substack{i=1,2,\dots,r \\ j=1,2,\dots,m}}$ at $x_0$ is $r$.*

The following theorem gives a residual set in $C^r(M, N)$ by using the transversality condition.

**Theorem 1.6.** (*Jet Transversality Theorem*) *Suppose $M$ and $N$ are finite-dimensional smooth manifolds without boundaries, and $A$ is a smooth submanifold of $J^k(M, N)$. If $1 \le k < r \le \infty$, then the set of mappings*

$$\mathcal{F} = \{ f \in C^r(M, N) \mid j^k(f) \pitchfork A \}$$

*is residual in $C^r(M, N)$. If, in addition, $A$ is closed, then $\mathcal{F}$ is open in $C^r(M, N)$.*

**Remark 1.7.** A residual subset of $C^r(M, N)$ is one that is the intersection of countably many dense open sets of $C^r(M, N)$. Hence, it is still a dense set. The power of Theorem 1.6 is that by a consideration of the transversal property in a finite-dimensional space $J^k(M, N)$ (for example, using Theorem 1.5) we can obtain a residual set in the infinite-dimensional space $C^r(M, N)$. By this theorem, we will get some generic families of vector fields in some bifurcation problems.

We state two more theorems which will be needed in future discussions.

**Theorem 1.8.** *Suppose that* $f \in C^r(M, N)$ *and* $A$ *is a submanifold of* $N$. *If* $f \pitchfork A$, *then* $f^{-1}(A)$ *is a submanifold of* $M$. *If, in addition,* $A$ *has a finite codimension in* $N$, *then* $f^{-1}(A)$ *has the same codimension in* $M$, *that is,*

$$\operatorname{codim}(f^{-1}(A)) = \operatorname{codim}(A).$$

**Remark 1.9.** If $f: M \to N$ is a submersion, then the condition $f \pitchfork A$ is satisfied.

**Theorem 1.10.** (*Malgrange Preparation Theorem*) *Suppose that* $U \subset \mathbb{R} \times \mathbb{R}^n$ *is an open set with* $(0, 0) \in U$, *and* $f \in C^\infty(U, \mathbb{R})$ *satisfies*

$$f(x, 0) = x^k g(x)$$

*for some integer* $k \geq 1$, *where* $g$ *is smooth in a neighborhood of* $x = 0$, *and* $g(0) \neq 0$. *Then there exist a smooth function* $q$ *defined in a neighborhood* $V$ *of* $(0, 0)$ *in* $\mathbb{R} \times \mathbb{R}^n$ *and* $C^\infty$ *functions* $a_0(\epsilon), \ldots, a_{k-1}(\epsilon)$ *in a neighborhood of the origin in* $\mathbb{R}^n$ *such that* $q(0, 0) \neq 0$, $a_0(0) = \cdots = a_{k-1}(0) = 0$, *and*

$$q(x, \epsilon) f(x, \epsilon) = x^k + \sum_{i=0}^{k-1} a_i(\epsilon) x^i, \qquad (x, \epsilon) \in V.$$

**Remark 1.11.** The previous result is, in some sense, a generalization of the Implicit Function Theorem, and it is obtained by using the Division

Theorem which has been proved for $C^r$ functions by Barbancon [1] and Lasalle [1]. See also Schecter [2].

In the last part of this section, we define the versal deformation of a "singular vector field" (i.e., it is not structurally stable) and define the codimension of a local bifurcation problem, and then try to give a procedure to determine them in Sections 3.2 and 4.1.

If we consider a bifurcation problem locally, we need the following concept.

**Definition 1.12.** Two mappings defined near a point $x$ have a common germ at $x$ if one can find a neighborhood $U$ of $x$ such that they coincide in $U$.

Obviously, the germ of a mapping at a point is an equivalence class, and we call any mapping in this class a representative of this germ.

Similarly to Definitions 1.1 and 1.2, we can define the $k$-jet of a germ at a point. If we consider the problem locally, we usually use the same notation $J^k(M, N)$ to denote the set of all $k$-jets of germs of mappings from $M$ into $N$.

We denote by $V(x_0)$ the space of germs of $C^\infty$ vector fields at $x_0 \in \mathbb{R}^n$. We give $V(x_0)$ the topology induced from inclusion maps. See Hirsch [1, p. 36]. We take a small open set $U$ of the origin in $\mathbb{R}^n$, and define

$$\mathscr{X} = \{(x, X) | X \in V(x), x \in U\}. \tag{1.1}$$

The natural projection

$$\pi_k(x_0): V(x_0) \to J^k_{x_0} := \{j^k_{x_0}(X) | X \in V(x_0)\} \tag{1.2}$$

induces a projection

$$\pi_k: \mathscr{X} \to J^k := \{(x, j^k_x(X)) | x \in U, X \in V(x)\}. \tag{1.3}$$

Now we consider the family of vector fields

$$\dot{x} = X(x, \epsilon), \tag{1.4}$$

where $x \in \mathbb{R}^n$, $\epsilon \in \mathbb{R}^k$, and $X \in C^\infty(\mathbb{R}^n \times \mathbb{R}^k, \mathbb{R}^n)$. If $\epsilon$ varies in a small neighborhood of $\epsilon_0$ in $\mathbb{R}^k$, (1.4) is said to be a *deformation* of vector field $\dot{x} = X(x, \epsilon_0)$. The family of vector fields (1.4) can be expressed briefly by the mapping $X: \mathbb{R}^n \times \mathbb{R}^k \to \mathbb{R}^n$.

**Definition 1.13.** A local family $(X; x_0, \epsilon_0)$ is the germ of the mapping $X$ at the point $(x_0, \epsilon_0)$ of the direct product of the phase space and the parameter space.

**Definition 1.14.** Two local families $(X; x_0, \epsilon_0)$ and $(Y; y_0, \epsilon_0)$ are said to be equivalent if there is a germ of a continuous mapping $y = h(x, \epsilon)$ at the point $(x_0, \epsilon_0)$ such that for every $\epsilon$, $h(\cdot, \epsilon)$ (the representative of the germ) is a homeomorphism mapping the phase orbits of $(X; x_0, \epsilon_0)$ onto the phase orbits of $(Y; y_0, \epsilon_0)$ with $h(x_0, \epsilon_0) = y_0$ and preserving the direction of orbits in time.

**Definition 1.15.** A local family $(Z; x_0, \mu_0)$ is induced from the local family $(X; x_0, \epsilon_0)$, if there is a germ of a continuous mapping $\phi$ at $\mu_0$, $\epsilon = \phi(\mu)$, such that $\epsilon_0 = \phi(\mu_0)$ and $Z(x, \mu) = X(x, \phi(\mu))$.

**Definition 1.16.** A local family $(X; x_0, \epsilon_0)$ is called a versal deformation of the germ of $X(\cdot, \epsilon_0)$ at the point $x_0$ if every other local family containing the same germ of $X(\cdot, \epsilon_0)$ is equivalent to a local family induced from $(X; x_0, \epsilon_0)$.

**Definition 1.17.** The bifurcation of $X(\cdot, \epsilon_0)$ is said to have codimension $m$ if $X(\cdot, \epsilon_0)$ has a versal deformation with $m$ parameters and any $(m - 1)$-parameter local family is not a versal deformation of $X(\cdot, \epsilon)$. If any local family with a finite number of parameters is not a versal deformation of $X(\cdot, \epsilon_0)$, then the codimension of $X(\cdot, \epsilon_0)$ is infinite.

Now suppose $X = X(\cdot, 0)$. Corresponding to the differential equation

$$\dot{x} = Ax + \cdots$$

is a bifurcation point in $V(0)$. The question is how to determine the codimension of $X$ in $V(0)$ and how to find its versal deformation?

We note that some bifurcation problems have infinite codimension, and there is no efficient method to answer the above question in general.

But for some special cases, it is possible to determine their versal deformations (see §3.2 and §4.1 for details).

### 3.2 Bifurcation of Equilibria

In Section 2.11, we have listed the following codimension 1 normal forms:

(i) $\dot{x} = \epsilon \pm x^2$,

(ii) $\dot{x} = \epsilon x \pm x^2$,

(iii) $\dot{x} = \epsilon x \pm x^3$,

(iv) $\begin{cases} \dot{x} = \epsilon x - y + (ax - by)(x^2 + y^2), \\ \dot{y} = x + \epsilon y + (bx + ay)(x^2 + y^2). \end{cases}$

In this section we discuss the local bifurcations of these normal forms, and give a general Hopf bifurcation theorem. We will only consider the cases "$-$" in (i)–(iii), and leave the cases "$+$" as exercises.

(i) $\dot{x} = \epsilon - x^2$, saddle-node bifurcation.

We first show that

$$\dot{x} = \epsilon - x^2 \qquad (\epsilon \in \mathbb{R}^1) \tag{2.1}$$

is a versal deformation of

$$\dot{x} = -x^2 + \cdots . \tag{2.2}$$

Let $V(x_0)$ be the space of germs of $C^\infty$ vector fields at $x_0 \in \mathbb{R}^1$, and $\mathscr{X} = \{(x, X) | X \in V(x), -\sigma < x < \sigma$ for a small $\sigma > 0\}$. Suppose $X \in V(x)$ has a representative $\dot{x} = f(x)$. The natural projection $\pi_k(x_0)$: $V(x_0) \to J_{x_0}^k$ induces a projection

$$\pi_k : \mathscr{X} \to J^k, \qquad (x, X) \mapsto (x, f, f', \dots, f^{(k)}),$$

where $f, f', \dots, f^{(k)}$ are the Taylor coefficients up to order $k$ at $x$.

Let

$$\Sigma = \{(x_0, X) \in \mathscr{X} | \exists x_0 \text{ such that } f(x_0) = f'(x_0) = 0, f''(x_0) \neq 0\}.$$

If $(x, Y) \in \Sigma$, then $Y$ has the same singular character at $x$ as (2.2) at 0.

**Lemma 2.1.** *If $k \geq 2$, then $\pi_k \Sigma$ is locally a smooth submanifold with codimension 2 in $J^k$ and $\Sigma$ is a codimension 2 submanifold of $\mathscr{X}$.*

*Proof.* Obviously,

$$\pi_1 \Sigma = \{(x, f, f') | f = f' = 0\}$$

is a codimension 2 smooth submanifold in $J^1$. By Theorem 1.8 and Remark 1.9, the natural projection

$$\pi_{21}: J^2 \to J^1, \qquad (x, f, f', f'') \mapsto (x, f, f')$$

yields that $\pi_{21}^{-1}(\pi_1 \Sigma)$ is a codimension 2 submanifold in $J^2$. Since $\pi_2 \Sigma = \pi_{21}^{-1}(\pi_1 \Sigma)|_{f'' \neq 0}$ is locally an open subset of $\pi_{21}^{-1}(\pi_1 \Sigma)$, $\pi_2 \Sigma$ is locally a codimension 2 submanifold in $J^2$.

Using Theorem 1.8, we obtain that $\pi_k \Sigma = \pi_{k2}^{-1}(\pi_2 \Sigma)$ is locally a codimension 2 submanifold in $J^k$ for $k > 2$. Hence, $\Sigma = \pi_k^{-1}(\pi_k \Sigma)$ is locally codimension 2 in $\mathscr{X}$.                                        □

Now we consider a deformation of (2.2)

$$\dot{x} = g(x, \lambda), \tag{2.3}$$

where $g \in C^\infty(\mathbb{R}^1 \times \mathbb{R}^m, \mathbb{R}^1)$ and

$$g(x, 0) = -x^2 + O(|x|^3). \tag{2.4}$$

By Theorem 1.10, there are smooth functions $q(x, \lambda), \alpha(\lambda), \beta(\lambda)$ such that

$$q(0, 0) \neq 0, \qquad \alpha(0) = \beta(0) = 0, \tag{2.5}$$

and

$$g(x, \lambda) = q(x, \lambda)[\alpha(\lambda) + \beta(\lambda)x - x^2]. \tag{2.6}$$

The deformation (2.3) gives a mapping

$$\xi: \mathbb{R} \times \mathbb{R}^m \to J^2, \qquad (x, \lambda) \mapsto \pi_2 g(x, \lambda).$$

**Definition 2.2.** The deformation (2.3) is said to be nondegenerate if $\xi \pitchfork \pi_2 \Sigma$ at $(x, \lambda) = (0, 0)$.

**Theorem 2.3.** *Equation (2.1) is a versal deformation of (2.2) provided we consider only nondegenerate deformations of (2.2).*

*Proof.*

$$\pi_2 \Sigma = \{(x, f, f', f'') | f = f'_x = 0, f''_{xx} \neq 0\}.$$

By Theorem 1.5, if (2.3) is nondegenerate, then

$$\text{rank} \left. \frac{\partial(g(x, \lambda), g_x(x, \lambda))}{\partial(x, \lambda)} \right|_{(0,0)} = 2. \tag{2.7}$$

Under condition (2.4), (2.7) is equivalent to

$$g'_{\lambda_i}(0, 0) \neq 0 \quad \text{for some } i, \qquad 1 \leq i \leq m. \tag{2.8}$$

By (2.5) and (2.6), (2.8) is equivalent to

$$\alpha'_{\lambda_i}(0) \neq 0. \tag{2.9}$$

Without loss of generality we take $i = 1$. Since $q(0,0) \neq 0$, (2.3) is equivalent to

$$\dot{x} = \alpha(\lambda) + \beta(\lambda)x - x^2 = \gamma(\lambda) - \left(x - \frac{\beta(\lambda)}{2}\right)^2, \tag{2.10}$$

where $\gamma(\lambda) = \alpha(\lambda) + \tfrac{1}{4}\beta^2(\lambda)$.

Figure 2.1.

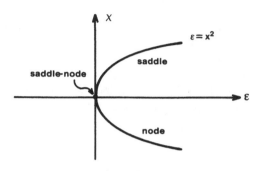

Figure 2.2.

Using (2.9) and (2.5), we have $\gamma'_{\lambda_1}(0) \neq 0$. Hence

$$\begin{cases} \mu_1 = \gamma(\lambda), \\ \mu_2 = \lambda_2, \\ \quad\vdots \\ \mu_m = \lambda_m \end{cases}$$

is a $C^\infty$ transformation and has an inverse $\lambda = \lambda(\mu)$, $\lambda(0) = 0$. Thus, (2.10) becomes

$$\dot{x} = \mu_1 - \left(x - \frac{1}{2}\beta(\lambda(\mu))\right)^2, \qquad (2.11)$$

which is equivalent (by Definition 1.14) to

$$\dot{y} = \mu_1 - y^2. \qquad (2.12)$$

Both (2.11) and (2.12) are $m$-parameter local families. On the other hand, (2.12) is obviously induced from (2.1). And, by Definition 1.16, (2.1) is a versal deformation of (2.2).                    □

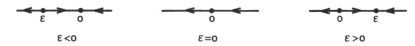

Figure 2.3.

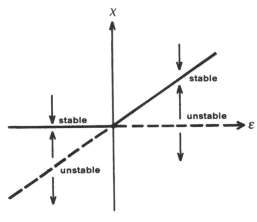

Figure 2.4.

Finally, we study the bifurcations of (2.1). The equilibria of equation (2.1) depend on $\epsilon$. If $\epsilon < 0$, then (2.1) has no equilibria; if $\epsilon = 0$, then $x = 0$ is the only equilibrium; if $\epsilon > 0$, then $x = \pm \sqrt{\epsilon}$ are two equilibria.

The bifurcation diagram consists of a point $\epsilon = 0$ in $\epsilon$-space, and the phase portraits of the vector fields are shown in Figure 2.1.

The relation between equilibria and the parameter $\epsilon$ is given by Figure 2.2.

We remark here that by using the same method we can show that the singular vector field

$$\dot{x} = a_k x^k + \cdots, \qquad a_k \neq 0,$$

has codimension $k - 1$, and

$$\dot{x} = \epsilon_1 + \epsilon_2 x + \cdots + \epsilon_{k-1} x^{k-2} + a_k x^k$$

is a versal deformation restricted to nondegenerate deformations.

(ii) $\dot{x} = \epsilon x - x^2$, transcritical bifurcation.

We note that $x = 0$ is always an equilibrium for all $\epsilon$. When $\epsilon < 0$, it is stable; when $\epsilon > 0$, it is unstable. If $\epsilon = 0$, then $x = 0$ is the unique equilibrium; if $\epsilon \neq 0$, then $x = \epsilon$ is the nonzero equilibrium. The

$\epsilon \leq 0$                                              $\epsilon > 0$

Figure 2.5.

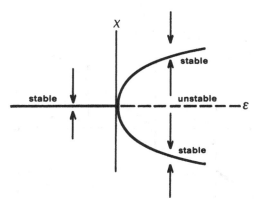

Figure 2.6.

linearized equation at $x = \epsilon$ ($\epsilon \neq 0$) is

$$\dot{x} = -\epsilon x.$$

It is obvious that if $\epsilon < 0$, the critical point $x = \epsilon$ is unstable; if $\epsilon > 0$, it is stable.

The phase portraits are shown in Figure 2.3.

The relation between equilibria and parameter is shown in Figure 2.4.

(iii) $\dot{x} = \epsilon x - x^3$, pitchfork bifurcation.

Note that $x = 0$ is always an equilibrium for all $\epsilon$; it is stable if $\epsilon \leq 0$ and unstable if $\epsilon > 0$. When $\epsilon < 0$, $x = 0$ is the unique equilibrium, which is stable. When $\epsilon > 0$, $x = \pm \sqrt{\epsilon}$ are nonzero equilibria. The linearized equations at $x = \pm \sqrt{\epsilon}$ are the same,

$$\dot{x} = -2\epsilon x.$$

Therefore they are stable.

The phase portraits are shown in Figure 2.5.

The relation between equilibria and parameter is shown in Figure 2.6.

We remark here that the bifurcation of type (i) is generic while (ii) and (iii) are not. However, if we restrict vector fields to the subset of $\mathscr{X}$ in which every system has always at least one equilibrium, then (ii) is generic. There is a similar situation for type (iii).

(iv) Hopf bifurcation.

We consider a $C^\infty$ system defined in a neighborhood of the origin in $\mathbb{R}^2$

$$\begin{cases} \dot{x} = f(x, y; \mu), \\ \dot{y} = g(x, y; \mu), \end{cases} \tag{2.13}$$

where $x, y, \mu \in \mathbb{R}^1$.

Suppose that the origin is an equilibrium of (2.13), and, for $\mu$ near 0, the linear part of (2.13) around the origin has eigenvalues $\alpha(\mu) \pm i\beta(\mu)$. Our first basic hypothesis is

$$\alpha(0) = 0, \qquad \beta(0) = \beta_0 \neq 0. \tag{$H_1$}$$

The second hypothesis is the transversality condition given by the classical Hopf bifurcation theorem

$$\alpha'(0) \neq 0. \tag{$H_2$}$$

In order to describe the third condition, we derive a normal form for equation (2.13) for $\mu = 0$. After we make a suitable linear change of coordinates and $z = x + iy$, equation (2.13) for $\mu = 0$ becomes

$$\dot{z} = i\beta_0 z + F(z, \bar{z}),$$
$$\dot{\bar{z}} = -i\beta_0 \bar{z} + \overline{F(z, \bar{z})}. \tag{2.14}$$

Since the eigenvalues of the linear part of (2.13) for $\mu = 0$ are $\lambda_{1,2} = \pm i\beta_0$, we have the resonances $\lambda_j = k(\lambda_1 + \lambda_2) + \lambda_j$, $j = 1, 2$ and $k = 1, 2, \ldots$. Hence, by a polynomial change of variables

$$z = w + \sum_{2 \leq k+l \leq m} b_{kl} w^k \bar{w}^l,$$

equation (2.14) takes the form (see §2.1, Example 2.1.22)

$$\dot{w} = i\beta_0 w + C_1 w^2 \bar{w} + C_2 w^3 \bar{w}^2 + \cdots + O(|w|^{2k+3}). \tag{2.15}$$

We now make the third hypothesis:

$$\mathrm{Re}(C_1) \ne 0. \tag{$H_3$}$$

**Theorem 2.4.** (*Hopf Bifurcation*) *Suppose that* ($H_1$), ($H_2$), *and* ($H_3$) *hold. Then there are* $\sigma > 0$ *and a neighborhood* $U$ *of* $(x, y) = (0, 0)$ *such that*

(i) *if* $|\mu| < \sigma$ *and* $\mathrm{Re}(C_1)\alpha'(0)\mu < 0$, *the system* (2.13) *has exactly one limit cycle inside* $U$;

(ii) *if* $|\mu| < \sigma$ *and* $\mathrm{Re}(C_1)\alpha'(0)\mu \ge 0$, *the system* (2.13) *has no periodic orbits inside* $U$.

*Moreover, the limit cycle is stable (unstable) if* $\mathrm{Re}(C_1) < 0$ ($\mathrm{Re}(C_1) > 0$), *and it tends to the equilibrium* $(0, 0)$ *as* $\mu \to 0$.

*Proof.* In suitable coordinates, the system (2.13) has the following form:

$$\frac{d}{dt}\begin{bmatrix} x \\ y \end{bmatrix} = \begin{bmatrix} \alpha(\mu) & -\beta(\mu) \\ \beta(\mu) & \alpha(\mu) \end{bmatrix}\begin{bmatrix} x \\ y \end{bmatrix} + \cdots. \tag{2.16}$$

Using the condition ($H_1$), we can make a further change of variables (keeping the linear terms unchanged) to transform (2.16) with $\mu = 0$ to the normal form equation (2.15). By using polar coordinates $(r, \theta)$, we have from (2.15)

$$\begin{cases} \dot{r} = \mathrm{Re}(C_1)r^3 + O(r^5), \\ \dot{\theta} = \beta_0 + O(r^2). \end{cases} \tag{2.17}$$

Since $\beta_0 \ne 0$, in a small neighborhood of $r = 0$ it follows from (2.17) that

$$\frac{dr}{d\theta} = \frac{\mathrm{Re}(C_1)}{\beta_0}r^3 + O(r^5). \tag{2.18}$$

By the same change of variables, we can transform the system (2.16) for $\mu \ne 0$ to the form

$$\begin{cases} \dot{r} = \alpha(\mu)r + a(\mu,\theta)r^2 + b(\mu,\theta)r^3 + O(r^4), \\ \dot{\theta} = \beta(\mu) + O(r^2), \end{cases} \tag{2.19}$$

where $\alpha(\mu)$, $\beta(\mu)$ are the same as in (2.16), and $a(\mu, \theta)$, $b(\mu, \theta) \in C^\infty$, $a(0, \theta) = 0$, $b(0, \theta) = \text{Re}(C_1)$.

In a small neighborhood of $r = 0$ and for small $|\mu|$, we obtain from (2.19)

$$\frac{dr}{d\theta} = \frac{\alpha(\mu)}{\beta(\mu)} r + \bar{a}(\mu, \theta) r^2 + \bar{b}(\mu, \theta) r^3 + O(r^4), \qquad (2.20)$$

where $\bar{a}(0, \theta) = 0$ and $\bar{b}(0, \theta) = \text{Re}(C_1)/\beta_0$.

Suppose that the functions

$$R(r_0, \theta, \mu) = u_1(\theta, \mu) r_0 + u_2(\theta, \mu) r_0^2 + \cdots$$

and

$$h(r_0, \theta) \equiv R(r_0, \theta, 0) = h_1(\theta) r_0 + h_2(\theta) r_0^2 + \cdots$$

are solutions of (2.20) and (2.18), respectively, satisfying the initial conditions

$$R(r_0, 0, \mu) = r_0 \quad \text{and} \quad h(r_0, 0) = r_0.$$

A calculation shows that

$$R(r_0, \theta, \mu) = \exp\left(\frac{\alpha(\mu)}{\beta(\mu)} \theta\right) r_0 + u_2(\theta, \mu) r_0^2 + \cdots, \qquad (2.21)$$

$$h(r_0, \theta) = R(r_0, \theta, 0) = r_0 + \frac{\text{Re}(C_1)}{\beta_0} \theta r_0^3 + \cdots. \qquad (2.22)$$

In fact, by (2.18) we have

$$\frac{\partial}{\partial \theta} \left. \frac{\partial^k h(r_0, \theta)}{\partial r_0^k} \right|_{r_0 = 0} = \lim_{r_0 \to 0} \frac{\partial^k}{\partial r_0^k} \frac{dr}{d\theta} = \begin{cases} 0, & \text{for } k = 1, 2, \\ 6\,\text{Re}(C_1)/\beta_0, & \text{for } k = 3. \end{cases}$$

Then using $h(r_0, 0) = r_0$, we can obtain (2.22).

Now we define the Poincaré map $P(x, \mu)$ along the $x$-axis for the system (2.20), and let

$$V(x, \mu) = P(x, \mu) - x. \qquad (2.23)$$

The number of periodic orbits of (2.20) near $x = 0$ for $|\mu|$ small is determined by the number of zeros of $V(x, \mu)$ for $x > 0$. When $x > 0$, we have

$$V(x, \mu) = R(x, 2\pi, \mu) - x$$

and

$$V(x, 0) = h(x, 2\pi) - x.$$

From (2.21) and (2.22),

$$V(x, \mu) = x\tilde{V}(x, \mu), \tag{2.24}$$

where

$$\tilde{V}(x, \mu) = \left( \left[ \exp\left( 2\pi \frac{\alpha(\mu)}{\beta(\mu)} \right) \right] - 1 \right) + u_2(2\pi, \mu)x + O(x^2), \tag{2.25}$$

and

$$\tilde{V}(x, 0) = 2\pi \frac{\text{Re}(C_1)}{\beta_0} x^2 + O(x^3). \tag{2.26}$$

From (2.25) and condition $(H_2)$ we get

$$\frac{\partial \tilde{V}(0, 0)}{\partial \mu} = \frac{d}{d\mu} \left[ \exp\left( 2\pi \frac{\alpha(\mu)}{\beta(\mu)} \right) \right]_{\mu=0} = \frac{2\pi}{\beta_0} \frac{d\alpha(\mu)}{d\mu} \bigg|_{\mu=0} \neq 0. \tag{2.27}$$

By the Implicit Function Theorem, there exists a unique smooth function $\mu = \mu(x)$, defined for $|x| < \epsilon$, such that $\mu(0) = 0$ and

$$\tilde{V}(x, \mu(x)) \equiv 0. \tag{2.28}$$

Differentiating with respect to $x$, we have

$$\begin{cases} \dfrac{\partial \tilde{V}}{\partial x} + \dfrac{\partial \tilde{V}}{\partial \mu} \mu'(x) = 0, \\[4mm] \dfrac{\partial^2 \tilde{V}}{\partial x^2} + 2 \dfrac{\partial^2 \tilde{V}}{\partial \mu \, \partial x} (\mu'(x)) + \dfrac{\partial^2 \tilde{V}}{\partial \mu^2} (\mu'(x))^2 + \dfrac{\partial \tilde{V}}{\partial \mu} \mu''(x) = 0. \end{cases} \tag{2.29}$$

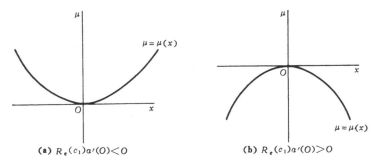

(a) $R_e(c_1)a'(0)<0$                         (b) $R_e(c_1)a'(0)>0$

Figure 2.7.

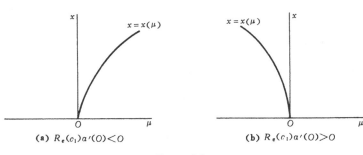

(a) $R_e(c_1)a'(0)<0$                         (b) $R_e(c_1)a'(0)>0$

Figure 2.8.

On the other hand, from (2.26) and the condition ($H_3$) we see that

$$\frac{\partial \tilde{V}(0,0)}{\partial x} = 0, \qquad \frac{\partial^2 \tilde{V}(0,0)}{\partial x^2} = 4\pi \frac{\mathrm{Re}(C_1)}{\beta_0} \neq 0. \qquad (2.30)$$

Substituting (2.30) and (2.27) into (2.29), we obtain

$$\mu'(0) = 0 \quad \text{and} \quad \mu''(0) = -2\,\mathrm{Re}(C_1)(\alpha'(0))^{-1} \neq 0.$$

Therefore, the graph of $\mu = \mu(x)$ is as shown in Figure 2.7 (a) or (b), depending on the sign of $\mathrm{Re}(C_1)\alpha'(0)$. Since we only need to consider the number of zeroes of $V(x, \mu)$ near $(0, 0)$ for $x > 0$, we can determine the inverse function $x = x(\mu) > 0$ of $\mu = \mu(x)$. This is shown in Figure 2.8. It follows that there are $\sigma > 0$ and $\eta > 0$ such that if $|\mu| \in (0, \sigma)$ and $\mu\alpha'(0)\,\mathrm{Re}\,C_1 < 0$, we can find $x = x(\mu) \in (0, \eta)$ such that, among all the orbits of (2.20) passing through the interval $I = \{(x, y)|0 < x < \eta,\ y = 0\}$, the orbit passing through the point $(x, y) = (x(\mu), 0)$, and only this orbit, is periodic. On the other hand, if $|\mu| \in (0, \sigma)$ and $\mu \alpha'(0) \operatorname{Re} C_1 > 0$, the system (2.20) has no periodic orbits passing through the interval $I$. The stability of the periodic orbit is obtained from the first equation of (2.17) (or (2.19)). This completes the proof of Theorem 2.4.                                    □

In Chapter 4 we will apply the Hopf Bifurcation Theorem to the perturbed Hamiltonian system

$$\begin{cases} \dot{x} = -\dfrac{\partial H}{\partial y} + \delta f(x, y, \mu, \delta), \\[2mm] \dot{y} = \dfrac{\partial H}{\partial x} + \delta g(x, y, \mu, \delta), \end{cases} \qquad (2.31)$$

where $H = H(x, y) \in C^{\infty}$ is a Hamiltonian function, $f, g \in C^{\infty}$, the parameters $\mu, \delta \in \mathbb{R}^1$, and $\delta \geq 0$ is small.

Suppose that $x = y = 0$ is an equilibrium of (2.31), and the eigenvalues of the linear part of (2.31) at $x = y = 0$ are $\alpha(\mu, \delta) \pm i\beta(\mu, \delta)$. If there is a function $\mu = \mu(\delta)$, $0 \leq \delta \leq \sigma$, satisfying the condition

$$\alpha(\mu(\delta), \delta) = 0, \qquad \beta(\mu(\delta), \delta) \neq 0, \qquad (\mathrm{H}_1^*)$$

then under some additional conditions Hopf bifurcation may take place at $x = y = 0$ for $\mu = \mu(\delta)$ and $\delta > 0$. This means that for every $\delta > 0$ there is an $\epsilon(\delta) > 0$ such that when $|\mu - \mu(\delta)| < \epsilon(\delta)$, the system has a periodic orbit for $\mu > \mu(\delta)$ (or $\mu < \mu(\delta)$) and has no periodic orbits for $\mu \leq \mu(\delta)$ (or $\mu \geq \mu(\delta)$). When $\delta \to 0$, $\epsilon(\delta)$ may tend to zero (see Figure 2.9(a)). But in many cases, we need to find $\bar{\delta}$ and $\bar{\epsilon}$ such that $\epsilon(\delta) \geq \bar{\epsilon} > 0$ for all $0 < \delta < \bar{\delta}$.

In fact, we can apply Theorem 2.4 to the system (2.31), replacing $\mu$ by $\mu - \mu(\delta)$. Suppose that for $\mu = \mu(\delta)$ the system (2.31) takes the normal form (2.15) at the origin, where $\beta_0 = \beta(\mu(\delta), \delta)$ and $C_i = C_i(\mu(\delta), \delta)$. To obtain a Hopf Bifurcation Theorem for the system (2.31) uniformly with respect to $\delta$ near $\delta = 0$, we need the following conditions instead of conditions $(\mathrm{H}_2)$ and $(\mathrm{H}_3)$, respectively:

$$\alpha^* := \lim_{\delta \to 0} \frac{1}{\delta} \frac{\partial \alpha(\mu(\delta), \delta)}{\partial \mu} \neq 0 \qquad (\mathrm{H}_2^*)$$

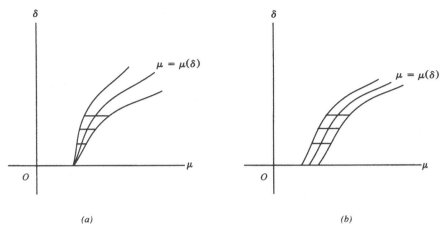

Figure 2.9.

and

$$C_1^* := \lim_{\delta \to 0} \frac{1}{\delta} \operatorname{Re}\big(C_1(\mu(\delta), \delta)\big) \neq 0. \qquad (H_3^*)$$

**Theorem 2.5.** *Suppose that the conditions $(H_1^*)$, $(H_2^*)$, and $(H_3^*)$ are satisfied. Then there are $\bar{\delta} > 0$, $\sigma > 0$, and a neighborhood $U$ of $(x, y) = (0, 0)$ such that*
*(i) the system (2.31) has exactly one limit cycle in $U$ if $0 < \delta < \bar{\delta}$, $|\mu - \mu(\delta)| < \sigma$, and*

$$C_1^* \cdot \alpha^* \cdot (\mu - \mu(\delta)) < 0;$$

*(ii) the system (2.31) has no limit cycles in $U$ if $0 < \delta < \bar{\delta}$, $|\mu - \mu(\delta)| < \sigma$, and*

$$C_1^* \cdot \alpha^* \cdot (\mu - \mu(\delta)) > 0.$$

*Moreover, the limit cycle is asymptotically stable if $C_1^* < 0$, and unstable if $C^* > 0$.*

*Proof.* We replace $\mu$ in Theorem 2.4 by $\mu - \mu(\delta)$, and take $\delta$ as a parameter. Then the proof of Theorem 2.4 is valid for $\delta > 0$, and $V(x, \mu)$ is replaced by $V(x, \mu - \mu(\delta), \delta)$.

System (2.31) is Hamiltonian when $\delta = 0$; this implies

$$V(x, \mu - \mu(\delta), \delta)|_{\delta = 0} \equiv 0$$

for $|x|$ and $|\mu - \mu(\delta)|$ small. Hence

$$V(x, \mu - \mu(\delta), \delta) = \delta x V^*(x, \mu - \mu(\delta), \delta).$$

Conditions $(H_2^*)$ and $(H_3^*)$ imply that

$$\frac{\partial V^*(0,0,0)}{\partial \mu} \neq 0 \quad \text{and} \quad \frac{\partial^2 V^*(0,0,0)}{\partial x^2} \neq 0.$$

Thus we can replace $\tilde{V}$ in (2.24) by $V^*$, and use the Implicit Function Theorem at the initial point $(x, \mu - \mu(\delta), \delta) = (0,0,0)$. This gives the uniform property with respect to $\delta$ near $\delta = 0$.      □

From the proofs of Theorems 2.4 and 2.5 we have the following result which will be used in Chapter 4.

**Theorem 2.6.** *Suppose that the system (2.31) has an equilibrium at $(x_0, y_0)$, the eigenvalues of the linearized equation at $(x_0, y_0)$ are $\alpha(\mu, \delta) \pm \beta(\mu, \delta)$, and there is a function $\mu = \mu(\delta)$ satisfying the conditions $(H_1^*)$, $(H_2^*)$, and $(H_3^*)$. Then there exist $\bar{x} > x_0$, $\bar{\delta} > 0$, and a unique function $\mu = \mu(x, \delta)$ defined in $x_0 \leq x \leq \bar{x}$, $0 \leq \delta \leq \bar{\delta}$ such that*
(i) *when $\mu = \mu(x, \delta)$, $x_0 < x \leq \bar{x}$, and $0 < \delta \leq \bar{\delta}$, the system (2.31) has a periodic orbit passing through the point $(x, 0)$;*
(ii) *$\partial \mu(x, \delta)/\partial x > 0$ if $\alpha^* C_1^* < 0$, and $\partial \mu(x, \delta)/\partial x < 0$ if $\alpha^* C_1^* > 0$.*

**Remark 2.7.** For applications it is convenient to express $\text{Re}(C_1)$ in terms of the coefficients of equation (2.13). In Guckenheimer and Holmes [1], we have the following.
If (2.13) for $\mu = 0$ has the following form

$$\frac{d}{dt}\begin{bmatrix} x \\ y \end{bmatrix} = \begin{bmatrix} 0 & -\beta_0 \\ \beta_0 & 0 \end{bmatrix}\begin{bmatrix} x \\ y \end{bmatrix} + \begin{bmatrix} f(x, y) \\ g(x, y) \end{bmatrix},$$

where $f(0) = g(0) = 0$ and $Df(0) = Dg(0) = 0$, then

$$
\begin{aligned}
\text{Re}(C_1) = \frac{1}{16} \Big\{ & (f_{xxx} + f_{xyy} + g_{xxy} + g_{yyy}) \\
& + \frac{1}{\beta_0} [f_{xy}(f_{xx} + f_{yy}) \\
& - g_{xy}(g_{xx} + g_{yy}) - f_{xx}g_{xx} + f_{yy}g_{yy}] \Big\} \Big|_{x=y=0}. \quad (2.32)
\end{aligned}
$$

More generally, we have the following theorem.

**Theorem 2.8.** *Suppose that equation (2.13) has an equilibrium* $(x(\mu), y(\mu))$ *for* $\mu$ *near* $\mu_0$, *and the linear part at* $(x(\mu), y(\mu))$ *is given by*

$$
\begin{bmatrix} a(\mu) & b(\mu) \\ c(\mu) & d(\mu) \end{bmatrix},
$$

*which satisfies*

$$
\begin{cases} a(\mu_0) + d(\mu_0) = 0, & \beta_0^2 \equiv a(\mu_0)d(\mu_0) - b(\mu_0)c(\mu_0) > 0, \\ a'(\mu_0) + d'(\mu_0) \neq 0. \end{cases}
$$

$$(2.33)$$

*Then the stability of the limit cycle bifurcating from the equilibrium* $(x(\mu_0), y(\mu_0))$ *when* $\mu$ *crosses* $\mu = \mu_0$ *is determined by the following quantity:*

$$
\begin{aligned}
\text{Re}(C_1) = \frac{b}{16\beta_0^4} \Big\{ & \beta_0^2 [b(f_{xxx} + g_{xxy}) + 2d(f_{xxy} + g_{xyy}) \\
& \qquad\qquad - c(f_{xyy} + g_{yyy})] \\
& - bd(f_{xx}^2 - f_{xx}g_{xy} - f_{xy}g_{xx} - g_{xx}g_{yy} - 2g_{xy}^2) \\
& - cd(g_{yy}^2 - g_{yy}f_{xy} - g_{xy}f_{yy} - f_{yy}f_{xx} - 2f_{xy}^2) \\
& + b^2(f_{xx}g_{xx} + g_{xx}g_{xy}) - c^2(f_{yy}g_{yy} + f_{xy}f_{yy}) \\
& - (\beta_0^2 + 3d^2)(f_{xx}f_{xy} - g_{xy}g_{yy}) \Big\} \Big|_{(x(\mu_0), y(\mu_0), \mu_0)}. \quad (2.34)
\end{aligned}
$$

*If* $\mathrm{Re}(C_1) < 0$, *then the limit cycle is stable; if* $\mathrm{Re}(C_1) > 0$, *it is unstable.*

The proof of (2.34) may be found in Wang [1].

**Example 2.9.** Consider a quadratic system

$$\begin{cases} \dot{x} = a(\mu)x + b(\mu)y + a_1x^2 + a_2xy + a_3y^2, \\ \dot{y} = c(\mu)x + d(\mu)y + b_1x^2 + b_2xy + b_3y^2. \end{cases}$$

If $a(\mu)$, $b(\mu)$, $c(\mu)$, and $d(\mu)$ satisfy (2.33), then the stability of the bifurcating limit cycle from $(x, y) = (0, 0)$ when $\mu$ crosses $\mu = \mu_0$ is determined by the quantity

$$v = b\left[ ab\left(2a_1^2 - a_1b_2 - a_2b_1 - 2b_1b_3 - b_2^2\right)\right.$$

$$+ ac\left(2b_3^2 - a_2b_3 - a_3b_2 - 2a_1a_3 - a_2^2\right)$$

$$+ b^2\left(2a_1b_1 + b_1b_2\right) - c^2\left(2a_3b_3 + a_2b_3\right)$$

$$\left. - \left(2a^2 - bc\right)\left(a_1a_2 - b_2b_3\right)\right]\Big|_{\mu = \mu_0}. \tag{2.35}$$

If $v < 0$, then the limit cycle is stable; if $v > 0$, it is unstable.

**Example 2.10.** Consider a cubic system without quadratic terms

$$\begin{cases} \dot{x} = a(\mu)x + b(\mu)y + a_1x^3 + a_2x^2y + a_3xy^2 + a_4y^3, \\ \dot{y} = c(\mu)x + d(\mu)y + b_1x^3 + b_2x^2y + b_3xy^2 + b_4y^3. \end{cases}$$

If (2.33) is satisfied, then the stability of the bifurcating limit cycle from $(0, 0)$ when $\mu$ crosses $\mu = \mu_0$ is determined by the quantity

$$v = b\left[(3a_1 + b_2)b + 2(a_2 + b_3)d - (a_3 + 3b_4)c\right]\Big|_{\mu = \mu_0}. \tag{2.36}$$

If $v < 0$, the limit cycle is stable; if $v > 0$, it is unstable.

**Remark 2.11.** We notice that $\mathrm{Re}(C_1) = 0$ does not imply that the equilibrium is a center. In fact, if $\mathrm{Re}(C_1) = 0$, then we need to consider the coefficients of higher-order terms in (2.15). There may be a Hopf bifurcation of higher order. This will be discussed in Section 5.1 and we will give an alternative proof of Theorem 2.4.

### 3.3  Bifurcation of Homoclinic Orbits

Although we consider bifurcation problems locally, sometimes we need to consider nonlocal bifurcation phenomena such as homoclinic bifurcation.

In this section we consider the system

$$\begin{cases} \dot{x} = f(x, y, \mu), \\ \dot{y} = g(x, y, \mu), \end{cases} \tag{3.1}$$

where $x, y, \mu \in \mathbb{R}$, $f, g \in C^2$, and $f(0,0,0) = g(0,0,0) = 0$. Denote

$$A = \frac{\partial(f, g)}{\partial(x, y)}(0, 0, 0). \tag{3.2}$$

We consider the following conditions:

$$\det A < 0. \tag{$H_1$}$$

This means that the system $(3.1)_{\mu=0}$ has a hyperbolic saddle point at the equilibrium point $O(0, 0)$.

$$(W_0^s \cap W_0^u) \setminus \{0\} \neq \phi, \tag{$H_2$}$$

where $W_0^s$ and $W_0^u$ are, respectively, the stable and unstable manifolds of $(3.1)_{\mu=0}$ at the saddle point $O(0, 0)$.

$$\sigma_0 = \text{trace } A \neq 0. \tag{$H_3$}$$

If $p_0 = (x_0, y_0) \in (W_0^s \cap W_0^u) \setminus \{0\}$, then the orbit $\gamma(p_0)$ through $p_0$ of $(3.1)_{\mu=0}$ approaches the saddle point $O(0, 0)$ as $t \to \pm\infty$. The orbit $\gamma(p_0)$ is called a homoclinic orbit of the saddle point $O(0, 0)$, and the

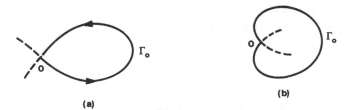

Figure 3.1.

invariant set

$$\Gamma_0 = \gamma(p_0) \cup \{0\}, \qquad p_0 \in (W_0^s \cap W_0^u) \setminus \{0\}$$

is called a homoclinic loop. $\Gamma_0$ can have either of the configurations in Figure 3.1. We only discuss the case of Figure 3.1(a). The case of Figure 3.1(b) can be treated in a similar way.

We will consider (3.1) as a small perturbation of $(3.1)_{\mu = 0}$. The purpose of this section is to determine what happens near $\Gamma_0$ as $\mu$ changes.

We will prove that if $(3.1)_{\mu = 0}$ satisfies the hypotheses $(H_1)$, $(H_2)$, and $(H_3)$, then there are small $\delta > 0$ and $\epsilon > 0$ such that for $|\mu| < \delta$, (3.1) has at most one periodic orbit in the $\epsilon$-neighborhood of the loop $\Gamma_0$.

The case of $\sigma_0 = 0$ is more complicated; we will discuss this case in Section 5.2.

The following questions are crucial in our discussion:
(1) How can we determine the stability of $\Gamma_0$?
(2) How can we determine the relative positions of $W_\mu^s$ and $W_\mu^u$ when $\Gamma_0$ is broken by perturbations (i.e., $\mu \neq 0$)?

In order to define stability of $\Gamma_0$, let us establish first the Poincaré map of $(3.1)_{\mu = 0}$ near $\Gamma_0$. For $p_0 \in W_0^s \cap W_0^u \setminus \{0\}$, let $L_0$ be a transversal to $\Gamma_0$ at $p_0$ and $L_0^+$ be the part of $L_0$ which belongs to the interior region surrounded by $\Gamma_0$ (see Figure 3.2), and $L_0^- = L_0 \setminus L_0^+$. By using the theorem about the continuous dependence of solutions upon the initial conditions and the saddle property, it is easy to see that there is a neighborhood $U$ of $p_0$ such that for any $p \in U \cap L_0^+$, there is $T = T(p) > 0$ with $\phi(T(p), p) \in L_0^+$, $\phi(t, p) \notin L_0^+$ for $0 < t < T(p)$, where $\phi(t, p)$ is the solution of $(3.1)_{\mu = 0}$ through $p$. One can therefore define the Poincaré map $P$ by $P(p) = \phi(T(p), p)$ for $p \in L_0^+ \cap U$. For simplicity of notation, let $L_0^+$ be orthogonal to $\Gamma_0$ at $p_0$, and $n_0$ be a unit

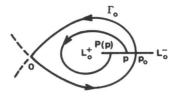

Figure 3.2.

vector along $L_0^+$. Then for $p \in L_0^+ \cap U$,

$$p = \alpha n_0 + p_0, \qquad \alpha > 0 \text{ small}, \qquad (3.3)$$

and

$$P(p) \equiv \phi(T(p), p) = \beta(\alpha)n_0 + p_0, \qquad (3.4)$$

where $\beta(\alpha) \in C^1$ for $\alpha > 0$ small, since $f, g \in C^2$ in (3.1). We consider the function

$$d(\alpha) = \beta(\alpha) - \alpha.$$

**Definition 3.1.** The homoclinic loop $\Gamma_0$ is asymptotically stable (unstable) if $d(\alpha) < 0$ ($> 0$) for all small $\alpha > 0$.

**Remark 3.2.** Note that $\lim_{\alpha \to 0} d(\alpha) = 0$, and the stability of $\Gamma_0$ is determined by the sign of $\lim_{\alpha \to 0} d'(\alpha) = \lim_{\alpha \to 0} \beta'(\alpha) - 1$. $\Gamma_0$ is asymptotically stable (unstable) if $\lim_{\alpha \to 0} d'(\alpha) < 0$ ($> 0$), that is, $\lim_{\alpha \to 0} \beta'(\alpha) < 1$ ($> 1$).

**Theorem 3.3.** *Suppose that $(3.1)_{\mu=0}$ satisfies $(H_1)$, $(H_2)$, and $(H_3)$. Then the homoclinic loop $\Gamma_0$ is asymptotically stable if $\sigma_0 < 0$ and unstable if $\sigma_0 > 0$.*

*Proof.* As in the discussion above, we suppose that $p_0 \in L_0^+ \cap \Gamma_0$, $p \in L_0^+ \cap U$. $p$ and its Poincaré map $P(p)$ are expressed in (3.3) and

(3.4), respectively. Differentiating (3.4) with respect to $\alpha$, we obtain

$$\beta'(\alpha)n_0 = \left[\frac{\partial}{\partial t}\phi(T(p),p)\right]\frac{\partial T(p)}{\partial p}n_0 + \frac{\partial}{\partial p}\phi(T(p),p)n_0$$

$$= F_\beta\left[\left(\frac{\partial T}{\partial x_0}, \frac{\partial T}{\partial y_0}\right)\cdot\left(\begin{matrix}n_{0x}\\n_{0y}\end{matrix}\right)\right] + \frac{\partial}{\partial p}\phi(T(p),p)\cdot n_0, \quad (3.5)$$

where

$$F_\beta = \left(\begin{matrix}f(\beta n_0 + p_0, 0)\\g(\beta n_0 + p_0, 0)\end{matrix}\right), \qquad n_0 = \left(\begin{matrix}n_{0x}\\n_{0y}\end{matrix}\right),$$

and $\beta = \beta(\alpha)$. By taking the inner product with $F_\beta^\perp$, we obtain from (3.5) that

$$\beta'(\alpha) = \frac{\left\langle F_\beta^\perp, \dfrac{\partial\phi(T(p),p)}{\partial p}n_0\right\rangle}{\left\langle F_\beta^\perp, n_0\right\rangle}. \tag{3.6}$$

Note that $\partial\phi(t,p)/\partial p$ is the fundamental matrix of the variational equation

$$\dot u = \frac{\partial(f,g)}{\partial(x,y)}(\phi(t,p),0)\cdot u$$

for which $\dot\phi(t,p)$ is a solution. This implies

$$\frac{\partial\phi}{\partial p}(T(p),p)F_\alpha = F_\beta, \tag{3.7}$$

where $F_\alpha = \left(\begin{matrix}f(\alpha n_0 + p_0, 0)\\g(\alpha n_0 + p_0, 0)\end{matrix}\right)$. Let

$$\frac{\partial\phi}{\partial p}(T(p),p)\cdot n_0 = \xi F_\beta + \eta n_0, \tag{3.8}$$

where $\xi, \eta \in \mathbb{R}$. By the continuity of $f$ and $g$, $F_\beta = (1 + \epsilon_1)F_\alpha + \epsilon_2 n_0$, where $\epsilon_1, \epsilon_2 \to 0$ as $\alpha \to 0$. By (3.7) and (3.8), this implies that

$$\frac{\partial \phi}{\partial p}(T(p), p) \cdot F_\beta = (1 + \epsilon_1 + \epsilon_2 \xi)F_\beta + \epsilon_2 \eta n_0. \qquad (3.9)$$

Equations (3.8) and (3.9) give the following matrix representation of $\partial \phi / \partial p(T(p), p)$ in terms of the basis $\{F_\beta, n_0\}$:

$$\begin{pmatrix} 1 + \epsilon_1 + \epsilon_2 \xi & \xi \\ \epsilon_2 \eta & \eta \end{pmatrix}.$$

Hence,

$$\det \frac{\partial \phi(T(p), p)}{\partial p} = (1 + \epsilon_1)\eta. \qquad (3.10)$$

Therefore, from (3.8), (3.6), and (3.10) we have that

$$\beta'(\alpha) = \eta = \frac{1}{(1 + \epsilon_1)} \left( \det \frac{\partial \phi(T(p), p)}{\partial p} \right), \qquad (3.11)$$

On the other hand, by Lemma 3.4, we have

$$\det \frac{\partial \phi(T(p), p)}{\partial p} = \exp \int_0^{T(p)} \left( \frac{\partial f}{\partial x} + \frac{\partial g}{\partial y} \right)(\phi(t, p), 0) \, dt.$$

Hence, we obtain from (3.11)

$$\beta'(\alpha) = \frac{1}{1 + \epsilon_1} \exp \int_0^{T(p)} \left( \frac{\partial f}{\partial x} + \frac{\partial g}{\partial y} \right)(\phi(t, p), 0) \, dt,$$

where $T(p) \to \infty$ as $\alpha \to 0$. By using the condition $(H_1)$, we can show that $\lim_{\alpha \to 0} \int_0^{T(p)} (\partial f / \partial x + \partial g / \partial y)(\phi(t, p), 0) \, dt$ is equal to either a finite number or infinity.

Suppose now $\sigma_0 = \text{trace } A = (\partial f / \partial x + \partial g / \partial y)(0, 0) < 0 \, (> 0)$. Then by the continuity, there is a neighborhood $V_0$ of the saddle point $O(0, 0)$

such that $(\partial f/\partial x + \partial g/\partial y)(x, y) < \sigma_0/2 < 0 \ (> \sigma_0/2 > 0)$ for $(x, y)$ $\in V_0$. For small $\alpha > 0$, $T(p) = T_1 + T_2$, where $T_1$ and $T_2$ are the times for which the flow $\phi(t, p)$ is inside and outside $V_0$, respectively. Obviously, $T_1 \to \infty$, and $T_2$ is bounded as $\alpha \to 0$. Hence

$$\int_0^{T(p)} \left( \frac{\partial f}{\partial x} + \frac{\partial g}{\partial y} \right) (\phi(t, p), 0) \, dt \to -\infty (+\infty)$$

$$\text{as } \alpha \to 0 \text{ if } \sigma_0 < 0 (\sigma_0 > 0).$$

This implies $\lim_{\alpha \to 0} \beta'(\alpha) = 0 \ (+\infty)$ for $\sigma_0 < 0 \ (\sigma_0 > 0)$ which gives the desired result (see Remark 3.2).                    $\square$

Suppose the solution of $(3.1)_{\mu=0}$ has the form

$$\Phi(t, p) = \begin{bmatrix} \phi(t; t_0, x_0, y_0) \\ \psi(t; t_0, x_0, y_0) \end{bmatrix}$$

satisfying

$$\phi(t_0; t_0, x_0, y_0) = x_0, \qquad \psi(t_0; t_0, x_0, y_0) = y_0. \qquad (3.12)$$

Denote

$$J(t; t_0, x_0, y_0) \equiv J = \det \frac{\partial(\phi, \psi)}{\partial(x_0, y_0)}. \qquad (3.13)$$

**Lemma 3.4.** *We have that*

$$J = \exp \int_{t_0}^t \left[ \frac{\partial f}{\partial x}(\phi, \psi) + \frac{\partial g}{\partial y}(\phi, \psi) \right] dt, \qquad (3.14)$$

*where $J$ is defined in (3.13) and $f = f(x, y, 0)$, $g = g(x, y, 0)$ are the right-hand sides of (3.1) for $\mu = 0$.*

*Proof.* From (3.13) we have

$$
\frac{\partial J}{\partial t} = \begin{vmatrix} \dfrac{\partial}{\partial t}\dfrac{\partial \phi}{\partial x_0} & \dfrac{\partial \phi}{\partial y_0} \\[2ex] \dfrac{\partial}{\partial t}\dfrac{\partial \psi}{\partial x_0} & \dfrac{\partial \psi}{\partial y_0} \end{vmatrix} + \begin{vmatrix} \dfrac{\partial \phi}{\partial x_0} & \dfrac{\partial}{\partial t}\dfrac{\partial \phi}{\partial y_0} \\[2ex] \dfrac{\partial \psi}{\partial x_0} & \dfrac{\partial}{\partial t}\dfrac{\partial \psi}{\partial y_0} \end{vmatrix}
$$

$$
= \begin{vmatrix} \dfrac{\partial f(\phi,\psi)}{\partial x_0} & \dfrac{\partial \phi}{\partial y_0} \\[2ex] \dfrac{\partial g(\phi,\psi)}{\partial x_0} & \dfrac{\partial \psi}{\partial y_0} \end{vmatrix} + \begin{vmatrix} \dfrac{\partial \phi}{\partial x_0} & \dfrac{\partial f(\phi,\psi)}{\partial y_0} \\[2ex] \dfrac{\partial \psi}{\partial x_0} & \dfrac{\partial g(\phi,\psi)}{\partial y_0} \end{vmatrix}
$$

$$
= \begin{vmatrix} \dfrac{\partial f}{\partial x}\dfrac{\partial \phi}{\partial x_0} + \dfrac{\partial f}{\partial y}\dfrac{\partial \psi}{\partial x_0} & \dfrac{\partial \phi}{\partial y_0} \\[2ex] \dfrac{\partial g}{\partial x}\dfrac{\partial \phi}{\partial x_0} + \dfrac{\partial g}{\partial y}\dfrac{\partial \psi}{\partial x_0} & \dfrac{\partial \psi}{\partial y_0} \end{vmatrix} + \begin{vmatrix} \dfrac{\partial \phi}{\partial x_0} & \dfrac{\partial f}{\partial x}\dfrac{\partial \phi}{\partial y_0} + \dfrac{\partial f}{\partial y}\dfrac{\partial \psi}{\partial y_0} \\[2ex] \dfrac{\partial \psi}{\partial x_0} & \dfrac{\partial g}{\partial x}\dfrac{\partial \phi}{\partial y_0} + \dfrac{\partial g}{\partial y}\dfrac{\partial \psi}{\partial y_0} \end{vmatrix} .
$$

Expanding the above determinants, we obtain

$$
\frac{\partial J}{\partial t} = \left[ \frac{\partial f}{\partial x}(\phi,\psi) + \frac{\partial g}{\partial y}(\phi,\psi) \right] \cdot J. \tag{3.15}
$$

From (3.12) we have

$$
J|_{t=t_0} = 1. \tag{3.16}
$$

Equations (3.15) and (3.16) give (3.14). $\qquad\square$

Consider now the system (3.1) for $\mu \neq 0$ small. Condition $(H_1)$ implies that there is a unique equilibrium point in a neighborhood of $(x, y) = (0, 0)$ for $\mu$ in a small neighborhood of zero. By a change of variables, one may suppose without loss of generality that (3.1) satisfies

$$
f(0,0;\mu) = g(0,0;\mu) = 0 \qquad \text{for all } \mu. \tag{3.17}
$$

Let $W_\mu^s$ and $W_\mu^u$ be the stable and unstable manifolds at the saddle point of (3.1).

**Theorem 3.5.** *Suppose that* $(3.1)_{\mu=0}$ *satisfies* $(H_1)$, $(H_2)$, *and* $(H_3)$. *Then there exist* $\delta > 0$ *and a neighborhood* $U$ *of* $\Gamma_0$ *such that for* $|\mu| < \delta$, *if* $(3.1)$ *has a periodic orbit* $\Gamma_\mu$ *in* $U$, *then* $\Gamma_\mu$ *is asymptotically stable for* $\sigma_0 < 0$ *and unstable for* $\sigma_0 > 0$. *Moreover,* $\Gamma_\mu$ *is the unique periodic orbit of* $(3.1)$ *in* $U$.

*Proof.* We only need to consider the periodic orbit $\Gamma_\mu$ of (3.1) with the property that $\Gamma_\mu \to \Gamma_0$ as $\mu \to 0$. For any fixed small $\mu$, let $p_{0\mu} \in W_\mu^u$ and $L_\mu$ be a transversal to $W_\mu^u$ at $p_{0\mu}$. Similarly, we have $L_\mu^+$. Let $n_\mu$ be the unit vector along $L_\mu^+$. Thus

$$L_\mu^+ = \{\alpha n_\mu + p_{0\mu}, \alpha > 0 \text{ small}\}.$$

Suppose that the periodic orbit $\Gamma_\mu$ corresponds to $\alpha = \alpha_\mu^*$. Then by the continuity of solutions with respect to initial conditions, we define the Poincaré map $P_\mu(p) = \beta(\alpha)n_\mu + p_{0\mu}$ for $p = \alpha n_\mu + p_{0\mu}$ with $\alpha$ near $\alpha_\mu^*$. We have that $\beta(\alpha_\mu^*) = \alpha_\mu^*$, and the stability of $\Gamma_\mu$ is determined by the sign of $(\beta'(\alpha_\mu^*) - 1)$. Note that as $\mu \to 0$, $\alpha_\mu^* \to 0$, and the local unstable manifold at $S_\mu$ approaches the local unstable manifold at $S_0$. Hence, by repeating the same arguments as in the proof of Theorem 3.3 we have that $\lim_{\mu \to 0} \beta'(\alpha_\mu^*) = 0$ if $\sigma_0 < 0$ and $\lim_{\mu \to 0} \beta'(\alpha_\mu^*) = +\infty$ if $\sigma_0 > 0$. This gives the stability of $\Gamma_\mu$.

It is impossible for a system to have more than one periodic orbit with the property that all of them are asymptotically stable (or unstable). Therefore the periodic orbit of (3.1) for $|\mu| < \delta$ inside an $\epsilon$-neighborhood of $\Gamma_0$, if it exists, is unique.                                    □

We now show that the existence of a periodic orbit is possible. Suppose $\Gamma_0$ of $(3.1)_{\mu=0}$ is as in Figure 3.1(a) and $\sigma_0 < 0$ (the other case can be considered in a similar manner). By Theorem 3.3, $\Gamma_0$ is asymptotically stable. Let $p_0 \in L_0 \cap \Gamma_0$, $p \in L_0^+$ and near $p_0$, and $P(p)$ be the Poincaré map of $p$. Then $p$ and $P(p)$ have expressions (3.3) and (3.4), respectively. We can find a fixed $p$ which is sufficiently close to $p_0$ such that $\beta(\alpha) < \alpha$. Let $A_\mu$ denote the part of the orbit from $p$ to $P_\mu(p)$ and $B_\mu$ denote the line segment in $L_\mu^+$ joining $P_\mu(p)$ and $p$. Let $\gamma_\mu = A_\mu \cup B_\mu$ (see Figures 3.3 and 3.4). For $(3.1)_{\mu \neq 0}$, the relative positions of the stable and unstable manifolds ($W_\mu^s$ and $W_\mu^u$) have three possibilities which are shown in Figure 3.4(a), (b), and (c). It is easy to see that in case (a), by the Poincaré-Bendixson Theorem, there is a

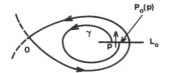

Figure 3.3.

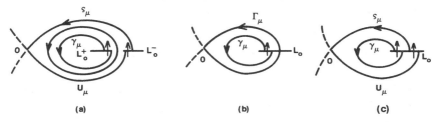

Figure 3.4.

periodic orbit which is a stable limit cycle and is unique (Theorem 3.5). In cases (b) and (c), there are no periodic orbits in the exterior of $\gamma_\mu$. In fact, by Theorem 3.3, the homoclinic loop $\Gamma_\mu$ in case (b) is asymptotically stable. Therefore, in both cases (b) and (c), if there is a periodic orbit in the exterior of $\gamma_\mu$, it must be unstable. This contradicts Theorem 3.5.

Thus, to determine the existence of limit cycles bifurcating from the homoclinic loop $\Gamma_0$, we need to know the relative positions of $W_\mu^u$ and $W_\mu^s$. We will use the Melnikov function to determine the splitting of $W_\mu^u$ and $W_\mu^s$.

Consider the equation

$$\begin{cases} \dot{x} = P_0(x, y) + \epsilon P_1(x, y, \epsilon), \\ \dot{y} = Q_0(x, y) + \epsilon Q_1(x, y, \epsilon), \end{cases} \tag{3.18}$$

where $P_0, Q_0, P_1, Q_1 \in C^r$ and $r \geq 2$. For simplicity, we also consider (3.18) in its vector form:

$$\dot{w} = R(w, \epsilon) = R_0(w) + \epsilon R_1(w, \epsilon), \tag{3.19}$$

where

$$w = \begin{pmatrix} x \\ y \end{pmatrix}, \qquad R_0(w) = \begin{pmatrix} P_0(x, y) \\ Q_0(x, y) \end{pmatrix}, \qquad R_1(w, \epsilon) = \begin{pmatrix} P_1(x, y, \epsilon) \\ Q_1(x, y, \epsilon) \end{pmatrix}.$$

Suppose that $O(x_0, y_0)$ is a hyperbolic saddle point of $(3.18)_{\epsilon=0}$. Assume that $\Gamma_0$ is a homoclinic loop through $O(x_0, y_0)$ and is given by

$$w_0(t) = \begin{pmatrix} \phi(t) \\ \psi(t) \end{pmatrix}.$$

Let $v = w_0(0)$ and $L$ be a transversal line segment through $v$. We assume that $L$ is parallel to the vector $R^{\perp}(v, 0) = R_0^{\perp}(v)$, where

$$a^{\perp} = \begin{pmatrix} -a_2 \\ a_1 \end{pmatrix} \quad \text{for any vector} \quad a = \begin{pmatrix} a_1 \\ a_2 \end{pmatrix} \in \mathbb{R}^2.$$

By the Implicit Function Theorem, equation (3.18) has a hyperbolic saddle point $O(x_{\epsilon}, y_{\epsilon})$ satisfying $O(x_{\epsilon}, y_{\epsilon}) \to O(x_0, y_0)$ as $\epsilon \to 0$. Let $W_{\epsilon}^s$ and $W_{\epsilon}^u$ be the stable and unstable manifolds of $O(x_{\epsilon}, y_{\epsilon})$ of system (3.18).

By the results in Chapter 1, for sufficiently small $\epsilon$ there exists a unique bounded solution $w_{\epsilon}^s(t)$ for $t \geq 0$ such that $w_{\epsilon}^s(t)$ is close to $w_0(t)$ for all $t \geq 0$ and $w_{\epsilon}^s(0) \in L$. Furthermore, $w_{\epsilon}^s(t)$ is $C^r$ with respect to the parameter $\epsilon$ and $w_{\epsilon}^s(t) \in W_{\epsilon}^s$ for all $t \geq 0$. Similarly, we have a unique bounded solution $w_{\epsilon}^u(t)$ for $t \leq 0$ in the unstable manifold $W_{\epsilon}^u$.

Let

$$\sigma(t) = P_{0x}'(\phi(t), \psi(t)) + Q_{0y}'(\phi(t), \psi(t)),$$

$$P_{00}(t) = P_0(\phi(t), \psi(t)), \qquad P_{10}(t) = P_1(\phi(t), \psi(t), 0),$$

$$Q_{00}(t) = Q_0(\phi(t), \psi(t)), \qquad Q_{10}(t) = Q_1(\phi(t), \psi(t), 0),$$

and

$$D(t) = \det \begin{bmatrix} P_{00}(t) & Q_{00}(t) \\ P_{10}(t) & Q_{10}(t) \end{bmatrix}.$$

**Theorem 3.6.** *Assume that the vector field (3.18) is $C^r$, $r \geq 2$, and the above conditions hold. Let*

$$\Delta = \int_{-\infty}^{\infty} D(t) e^{-\int_0^t \sigma(\xi) \, d\xi} \, dt \qquad (3.20)$$

*and*

$$d(\epsilon) = \left\langle w_\epsilon^u(0) - w_\epsilon^s(0), R^\perp(v,0) \right\rangle. \qquad (3.21)$$

*Then for sufficiently small $\epsilon$ we have*

$$d(\epsilon) = \epsilon \Delta + O(|\epsilon|^2).$$

*Proof.* Let

$$\left. \frac{\partial}{\partial \epsilon} w_\epsilon^s(t) \right|_{\epsilon=0} = z^s(t) \qquad \text{for} \quad t \geq 0,$$

and

$$\left. \frac{\partial}{\partial \epsilon} w_\epsilon^u(t) \right|_{\epsilon=0} = z^u(t) \qquad \text{for} \quad t \leq 0.$$

Let

$$\Delta^{s,u}(t) = \left\langle z^{s,u}(t), R_0^\perp(w_0(t)) \right\rangle.$$

By differentiating (3.19) with respect to $\epsilon$, we have that

$$\dot{z}^s(t) = \frac{\partial R_0}{\partial w}(w_0(t))z^s(t) + R_1(w_0(t),0) \qquad \text{for} \quad t \geq 0. \quad (3.22)$$

Next, for any two vectors $a, b \in \mathbb{R}^2$ we define

$$a \wedge b = \langle a, b^\perp \rangle.$$

It is not hard to see that for any $2 \times 2$ matrix $A$, we have

$$(Aa) \wedge b + a \wedge (Ab) = (\operatorname{tr} A)a \wedge b, \qquad a, b \in \mathbb{R}^2. \quad (3.23)$$

For $t \geq 0$, we have the following from (3.22)

$$\frac{d}{dt}\Delta^s(t) = \frac{d}{dt}\left[z^s(t) \wedge R_0(w_0(t))\right]$$

$$= \dot{z}^s(t) \wedge R_0(w_0(t)) + z^s(t) \wedge \frac{d}{dt}R_0(w_0(t))$$

$$= \dot{z}^s(t) \wedge R_0(w_0(t)) + z^s(t) \wedge \frac{\partial R_0}{\partial w}(w_0(t))R_0(w_0(t))$$

$$= \frac{\partial R_0}{\partial w}(w_0(t))z^s(t) \wedge R_0(w_0(t))$$

$$+ R_1(w_0(t),0) \wedge R_0(w_0(t))$$

$$+ z^s(t) \wedge \frac{\partial R_0}{\partial w}(w_0(t))R_0(w_0(t)).$$

By (3.23), for $t \geq 0$,

$$\frac{d}{dt}\Delta^s(t) = \sigma(t)\Delta^s(t) + \left\langle R_1(w_0(t),0), R_0^\perp(w_0(t))\right\rangle.$$

By the variation of constants formula, we have for $t \geq 0$

$$\Delta^s(t) = e^{\int_0^t \sigma(s)\,ds}\left\{\Delta^s(0) + \int_0^t\left[e^{-\int_0^\tau \sigma(s)\,ds}\left\langle R_1(w_0(\tau),0),\right.\right.\right.$$

$$\left.\left.\left. R_0^\perp(w_0(\tau))\right\rangle\right]d\tau\right\}.$$

This implies that

$$\Delta^s(0) + \int_0^t\left[e^{-\int_0^\tau \sigma(s)\,ds}\left\langle R_1(w_0(\tau),0), R_0^\perp(w_0(\tau))\right\rangle\right]d\tau$$

$$= e^{-\int_0^t \sigma(s)\,ds}\Delta^s(t).$$

Since the solution $z_\epsilon^s(t)$ is on the stable manifold $W_\epsilon^s$,

$$\lim_{t \to \infty} e^{-\int_0^t \sigma(s)\,ds} \Delta^s(t) = 0.$$

Hence,

$$\Delta^s(0) = -\int_0^\infty \left[ e^{-\int_0^\tau \sigma(s)\,ds} \langle R_1(w_0(\tau), 0), R_0^\perp(w_0(\tau)) \rangle \right] d\tau.$$

Similarly, we have

$$\Delta^u(0) = -\int_0^{-\infty} \left[ e^{-\int_0^\tau \sigma(s)\,ds} \langle R_1(w_0(\tau), 0), R_0^\perp(w_0(\tau)) \rangle \right] d\tau.$$

Hence $\Delta = \Delta^u(0) - \Delta^s(0)$. This yields the desired result. □

Combining the results of Theorems 3.3, 3.5, and 3.6, we obtain finally:

**Theorem 3.7.** *Suppose that system* $(3.18)_{\epsilon=0}$ *has a hyperbolic saddle at* $(x_0, y_0)$ *with a homoclinic loop* $\Gamma_0$, *and the orientation of* $\Gamma_0$ *is clockwise (counterclockwise). If* $\sigma_0 = P'_{0x}(x_0, y_0) + Q'_{0y}(x_0, y_0) \neq 0$, *then for sufficiently small* $|\epsilon|$ *and in a small neighborhood of* $\Gamma_0$, *we have:*
(i) *if* $\sigma_0 \epsilon \Delta > 0 \, (< 0)$, *system (3.18) has exactly one limit cycle bifurcating from the loop* $\Gamma_0$, *which is asymptotically stable for* $\sigma_0 < 0$ *and unstable for* $\sigma_0 > 0$;
(ii) *if* $\sigma_0 \epsilon \Delta < 0 \, (> 0)$, *(3.18) has no limit cycles.*

**Remark 3.8.** In some cases, we do not need to find the expression for $\Gamma_0$. For example, if $D(t)$ has a fixed sign, then $\Delta$ has the same sign by (3.20).

**Example 3.9.** Suppose that the system

$$\begin{cases} \dot{x} = P(x, y), \\ \dot{y} = Q(x, y) \end{cases}$$

has a hyperbolic saddle point at $(0, 0)$ and a homoclinic loop including $(0, 0)$ with clockwise orientation, and

$$\sigma_0 = P'_{0x}(x_0, y_0) + Q'_{0y}(x_0, y_0) \neq 0.$$

Then for sufficiently small $|\epsilon|$ and in a small neighborhood of $\Gamma_0$, the following perturbed system

$$\begin{cases} \dot{x} = P(x, y) - \epsilon Q(x, y), \\ \dot{y} = Q(x, y) + \epsilon P(x, y), \end{cases}$$

has exactly one limit cycle bifurcating from the loop $\Gamma_0$ when $\epsilon \sigma_0 > 0$; and has no limit cycles when $\epsilon \sigma_0 < 0$.

In fact, it is easy to see that

$$D(t) = \det \begin{bmatrix} P & -Q \\ Q & P \end{bmatrix} > 0,$$

and therefore

$$\Delta = \int_{-\infty}^{\infty} D(t) e^{-\int_0^t \sigma(\xi) \, d\xi} \, dt > 0.$$

By Theorem 3.7, the desired result follows.

### 3.4 Bibliographical Notes

There are many references for codimension one bifurcations; see, for example, the books of Andronov, et al. [1], Arnold [1], Chow and Hale [1], Hassard, Kazarinoff, and Wan [1], Golubitsky and Schaeffer [1], Guckenheimer and Holmes [1], Marsden and McCracken [1], and Wiggens [1, 2].

For a proof of the Jet Transversality Theorem (Theorem 1.6), we refer to Hirsch [1] or Arnold [4]. The proof of Theorem 1.10 can be found in Chow and Hale [1]. Definitions 1.13 through 1.16 are due to Arnold [4].

Theorem 2.6 is different from the classical Hopf Bifurcation Theorem. It gives a uniform property with respect to some parameters. This result will be useful in Chapter 4.

Formula (2.32) is taken from Guckenheimer and Holmes [1], and Marsden and McCracken [1]. Chow and Mallet-Paret [1] presented an alternative by using the method of averaging. Formula (2.34) can be obtained by using (2.32) (see Wang [1]), or by using a formula in Farr et al. [1], which can also be used in more general cases $n > 2$. Formula (2.34) is equivalent to a formula in Andronov et al. [1, p. 253].

The proof of Theorem 3.3 is due to Chow and Hale [1].

Theorem 3.7 was first given by Melnikov [1] and the integral (3.20) is called the Melnikov integral. Our proof of Theorem 3.6 follows basically Guckenheimer and Holmes [1] and Wiggins [2]. For other approaches, see Chow and Hale [1], Chow, Hale, and Mallet-Paret [1], Feng and Qian [1], Ma and Wang [1], and Palmer [2]. For the higher dimension Melnikov method, see Chow and Yamashita [1], Gruendler [1], and Palmer [2]. Results in Section 3.3 can be generalized to the case of heteroclinic orbits. See, for example, Cerkas [1] and Feng [1–2].

The degenerate cases of Hopf bifurcation and homoclinic bifurcation will be discussed in Section 5.1 and Section 5.2.

# 4

# Codimension Two Bifurcations

In this chapter we will introduce some results on codimension two bifurcations of vector fields near nonhyperbolic equilibrium points.

Consider a family of vector fields

$$\dot{x} = f(x, \epsilon), \qquad (X_\epsilon)$$

where $x \in \mathbb{R}^n$, $\epsilon \in \mathbb{R}^m$ ($m \geq 2$), $f \in C^\infty(\mathbb{R}^n \times \mathbb{R}^m, \mathbb{R}^n)$, and $f(0,0,0) = 0$. Suppose that the origin is a nonhyperbolic equilibrium point of $(X_0)$, and the linear part of $(X_0)$ is doubly degenerate. Then, after reduction to a center manifold, the linearized matrix of $(X_0)$ must take one of the following forms:

$$A_1 = \begin{bmatrix} 0 & 1 \\ 0 & 0 \end{bmatrix} \quad \text{or} \quad \begin{bmatrix} 0 & 0 \\ 0 & 0 \end{bmatrix},$$

$$A_2 = \begin{bmatrix} 0 & 1 & 0 \\ -1 & 0 & 0 \\ 0 & 0 & 0 \end{bmatrix},$$

$$A_3 = \begin{bmatrix} 0 & \omega_1 & 0 & 0 \\ -\omega_1 & 0 & 0 & 0 \\ 0 & 0 & 0 & \omega_2 \\ 0 & 0 & -\omega_2 & 0 \end{bmatrix} \quad (\omega_1\omega_2 \neq 0, \ \omega_1 \neq k\omega_2, \ k = 1,\ldots,5).$$

In Sections 4.1–4.5, we discuss the case $A_1$. This means that we will study the bifurcation diagrams in a small neighborhood $U$ of $\epsilon = 0$, and find all possible phase portraits of $(X_\epsilon)$ (corresponding to different $\epsilon \in U$) in a small neighborhood of $x = 0$, under certain nondegenerate conditions on the higher-order terms of $(X_0)$ which will vary for different cases. These conditions make the family essentially a two-parameter family, even though it may contain $m > 2$ parameters. In Sections 4.1–4.5, we suppose that $(X_\epsilon)$ is invariant under a rotation of the phase plane through an angle $2\pi/q$ for $q = 1, 2, 3, 4$ and $q \geq 5$, respectively. These are usually called $1 : q$ resonance problems. For the cases $q = 1, 2$ the matrix $A_1$ is nonzero nilpotent, and in the cases $q \geq 3$, $A_1$ is a zero matrix. The case $q = 1$ is nonsymmetric; it was first studied by Bogdanov [1, 2] and Takens [2], and is usually called the Bogdanov–Takens system. The complete results for the cases $q = 2, 3$ (with codimension two) were first given by Khorozov [1]. All codimension-two results for $1 : q$ resonances are known, except the case $q = 4$ (see Arnold [4, 5]). In these problems an important and difficult part is the study of the existence of periodic orbits, homoclinic or heteroclinic orbits, and the number of periodic orbits, corresponding to different values of the parameters. For problems of this kind, we need the Hopf bifurcation theory and homoclinic (heteroclinic) bifurcation theory, as well as some special techniques, such as the blowing-up transformations, Abelian integrals, and Picard–Fuchs equations. We will give more details in this chapter about all these techniques.

For the types $A_2$ and $A_3$, the study of bifurcations is far from complete. Since the dimension of the system is greater than or equal to 3, some complicated dynamical behavior can occur. The first step to study these bifurcations is to transform the equations into their normal forms (see Section 2.11), and then to study the truncated normal form equations which have some symmetric properties and can be reduced to planar systems because of the nature of $A_2$ and $A_3$.

In Sections 4.6 and 4.7, we will discuss codimension-two bifurcations of the reduced systems corresponding to $A_2$ and $A_3$, respectively. Zoladek [1, 2] gave the complete results for these two cases. We will use a simpler method to prove the uniqueness of periodic orbits for the first case.

## 4.1 Double Zero Eigenvalue

Consider a family of vector fields

$$\begin{cases} \dot{x} = f(x, y, \epsilon), \\ \dot{y} = g(x, y, \epsilon), \end{cases} \tag{1.1}$$

where $x, y \in \mathbb{R}^1$, $\epsilon \in \mathbb{R}^m$, $m \geq 2$, and $f, g \in C^\infty(x, y, \epsilon)$.

We suppose in this section that for $\epsilon = 0$, (1.1) has an equilibrium point at $x = y = 0$ for which the matrix of the linear part is similar to the Jordan block

$$\begin{bmatrix} 0 & 1 \\ 0 & 0 \end{bmatrix}.$$

Thus, the normal form of (1.1) for $\epsilon = 0$ (see Examples 2.1.10 and 2.1.15) is

$$\begin{cases} \dot{x} = y, \\ \dot{y} = ax^2 + bxy + O\big(|(x, y)|^3\big). \end{cases} \quad (1.2)$$

We remark here that the first normal form equation may have the form $\dot{x} = y + \phi(x, y)$, $\phi = O(|(x, y)|^3)$. By a change of coordinates in a small neighborhood of the origin:

$$\bar{x} = x, \qquad \bar{y} = y + \phi(x, y),$$

it can be transformed into the form (1.2).

Another basic hypothesis in this section is $ab \neq 0$ in (1.2). Under this condition, if we make a change of coordinates and time:

$$x \to \frac{a}{b^2}x, \qquad y \to \frac{a^2}{b^3}y, \qquad t \to \frac{b}{a}t,$$

the equation (1.2) is transformed into the form

$$\begin{cases} \dot{x} = y, \\ \dot{y} = x^2 + xy + O\big(|(x, y)|^3\big). \end{cases} \quad (1.3)$$

The reader should be aware that if $ab < 0$, then time is inverted by the scaling. If one wishes to keep the direction of motion, then (1.2) could be transformed into an equation having the same form as (1.2) with $a = 1$, $b = \pm 1$. We only consider the case $b = 1$. The case $b = -1$ can be discussed in a similar way.

**Definition 1.1.** A family of vector fields (1.1) is called a deformation of equation (1.3) if for $\epsilon = 0$ it has the form (1.3).

Under the above hypothesis we can assume that (1.1) is a family of deformations of (1.3) of the following form:

$$\begin{cases} \dot{x} = y + w_1(x, y, \epsilon), \\ \dot{y} = x^2 + xy + O\big(|(x, y)|^3\big) + w_2(x, y, \epsilon), \end{cases} \tag{1.4}$$

where $x, y \in \mathbb{R}^1$, $\epsilon \in \mathbb{R}^m$, $m \geq 2$, $w_1, w_2 \in C^\infty(x, y, \epsilon)$, and $w_i|_{\epsilon=0} = 0$, $i = 1, 2$.

We will show that the following two-parameter family of vector fields

$$\begin{cases} \dot{x} = y, \\ \dot{y} = \mu_1 + \mu_2 y + x^2 + xy \end{cases} \tag{1.5}$$

is a versal deformation of (1.3). This result is not obvious and its proof has been given by Bogdanov [1, 2] and Takens [1].

The discussion will be divided into three parts. We will study the bifurcation diagram and phase portraits of (1.5) in the first part, reduce (1.4) to a canonical form in the second part, and study the versality of (1.5) in the last part.

### (I) The Bifurcation Diagram and Phase Portraits of the System (1.5)

**Theorem 1.2.** (1) *There is a neighborhood $\Delta$ of $\mu_1 = \mu_2 = 0$ in $\mathbb{R}^2$ such that the bifurcation diagram of (1.5) inside $\Delta$ consists of the origin $(\mu_1, \mu_2) = (0, 0)$ and the following curves*:
(a) $SN^+ = \{\mu | \mu_1 = 0, \mu_2 > 0\}$,
(b) $SN^- = \{\mu | \mu_1 = 0, \mu_2 < 0\}$,
(c) $H = \{\mu | \mu_1 = -\mu_2^2, \mu_2 > 0\}$,
(d) $HL = \{\mu | \mu_1 = -\frac{49}{25}\mu_2^2 + O(\mu_2^{5/2}), \mu_2 > 0\}$.

(2) *The bifurcation diagram and phase portraits of (1.5) for $\mu \in \Delta$ are shown in Figure 1.1, where the regions I–IV are formed by the above bifurcation curves.*

A proof of Theorem 1.2 will be given by using the following lemmas.

**Lemma 1.3.** *There is a neighborhood $\Delta_1$ of $\mu_1 = \mu_2 = 0$ such that $SN^+$ and $SN^-$ are saddle-node bifurcation curves while $H$ is a Hopf bifurcation*

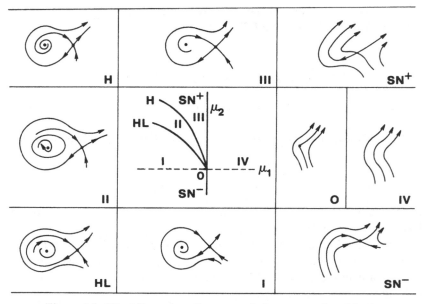

Figure 1.1. The bifurcation diagram and phase portraits of (1.5).

*curve for the system (1.5). Moreover, if $(\mu_1, \mu_2) \in \Delta_1 \cap$ region II and near H, then the system (1.5) has a unique limit cycle in a small neighborhood of the focus $(-\sqrt{-\mu_1}, 0)$. Furthermore, it is unstable, and it tends to the focus as $(\mu_1, \mu_2)$ tends to a point on H. The phase portraits of (1.5) for $(\mu_1, \mu_2) \in \Delta_1 \cap \{\mu | \mu_1 \geq 0\}$ are shown in Figure 1.1.*

*Proof.* If $\mu_1 > 0$, then (1.5) has no equilibria. If $\mu_1 = 0$, then the unique equilibrium $(x, y) = (0, 0)$ is a saddle-node for $\mu_2 \neq 0$, and is a "cusp" type for $\mu_2 = 0$. The phase portraits near $(0, 0)$ for $\mu_1 \geq 0$ are shown in Figure 1.1. (We refer to Zhang et al. [1, p. 130–58] for the details in obtaining the phase portraits.) Finally, if $\mu_1 < 0$, then (1.5) has two equilibria $(x_\pm, 0)$, where $x_\pm = \pm \sqrt{-\mu_1}$. The $2 \times 2$ matrix of the linearized equation at $(x_\pm, 0)$ is

$$A_\pm = \begin{bmatrix} 0 & 1 \\ \pm 2\sqrt{-\mu_1} & \mu_2 \pm \sqrt{-\mu_1} \end{bmatrix}.$$

Since

$$\text{trace}(A_\pm) = \mu_2 \pm \sqrt{-\mu_1} \quad \text{and} \quad \det(A_\pm) = -\left(\pm 2\sqrt{-\mu_1}\right),$$

$(x_+, 0)$ is a saddle point and $(x_-, 0)$ is an unstable focus for $\mu_2 > \sqrt{-\mu_1}$ and a stable focus for $\mu_2 < \sqrt{-\mu_1}$. Therefore, a Hopf bifurcation occurs along the curve $H = \{\mu | \mu_1 = -\mu_2^2, \mu_2 > 0\}$, and a saddle-node bifurcation occurs along the curves $SN^+$ and $SN^-$. By using the formula (3.2.34), it is easy to obtain that

$$16\,\text{Re}(C_1) = \frac{1}{\sqrt{-\mu_1}} > 0.$$

Hence, the focus $(x_-, 0)$ is unstable for $\mu_2 = \sqrt{-\mu_1}$. Moreover, $(x_-, 0)$ will become a stable focus surrounded by an unstable limit cycle for $\mu_2 < \sqrt{-\mu_1}$ and $|\mu_2 - \sqrt{-\mu_1}| \ll 1$, and the cycle tends to the focus as $(\mu_1, \mu_2) \to H$ (Theorem 3.2.4).                    □

In order to discuss the limit cycle and the homoclinic orbit of (1.5), we set

$$\mu_1 = -\delta^4, \qquad \mu_2 = \zeta\delta^2, \qquad x \to \delta^2 x,$$

$$y \to \delta^3 y, \qquad t \to \frac{t}{\delta} \; (\delta > 0), \tag{1.6}$$

where $\zeta$ and $\delta$ are new parameters. Then (1.5) becomes

$$\begin{cases} \dot{x} = y, \\ \dot{y} = -1 + x^2 + \zeta\delta y + \delta xy. \end{cases} \tag{1.7}$$

For $\delta = 0$, (1.7) is a Hamiltonian system:

$$\begin{cases} \dot{x} = y, \\ \dot{y} = -1 + x^2, \end{cases} \tag{1.8}$$

with the first integral

$$H(x, y) = \frac{y^2}{2} + x - \frac{x^3}{3} = h. \tag{1.9}$$

The phase portrait of (1.8) is shown in Figure 1.2.

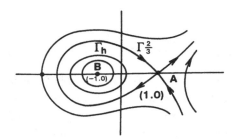

Figure 1.2. The phase portrait of (1.8).

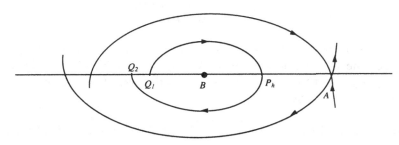

Figure 1.3.

Every closed orbit of (1.8) surrounding $(-1, 0)$ corresponds to a level curve $\Gamma_h = \{(x, y) | H(x, y) = h, \ -\frac{2}{3} < h < \frac{2}{3}\}$. $\Gamma_h$ shrinks to the equilibrium $(-1, 0)$ as $h \to -2/3$, and tends to the homoclinic loop as $h \to 2/3$.

Now we consider (1.7) as a perturbation of (1.8) for $\delta$ small. Note that (1.7) has two equilibria: The point $A(1, 0)$ is a saddle point and the point $B(-1, 0)$ is an equilibrium point with index $+1$ for every $\delta$ and $\zeta$. Hence every closed orbit of (1.7) must cross the line segment $L = \{(x, y) | y = 0, \ -1 \leq x \leq 1\}$ and surround the point $B$.

On the other hand, for every $h \in (-\frac{2}{3}, \frac{2}{3})$, $\Gamma_h$ (the orbit of (1.8)) intersects $L$ at exactly one point $P_h(x(h), 0)$. Therefore, the segment $L$ can be parameterized by $h \in (-\frac{2}{3}, \frac{2}{3})$.

For every $h \in (-\frac{2}{3}, \frac{2}{3})$, we consider the trajectory of (1.7) passing through the point $P_h(x(h), 0) \in L$. Let this trajectory go forward and backward until it intersects the negative $x$-axis at points $Q_2$ and $Q_1$, respectively (Figure 1.3). We denote the piece of trajectory from $Q_1$ to $Q_2$ by $\gamma(h, \delta, \zeta)$. For $h = +2/3$, we take the limiting positions of $\gamma$ by using the local stable and unstable manifolds at $A$.

**Lemma 1.4.** $\gamma(h, \delta, \zeta)$ *is a periodic orbit of (1.7) if and only if*

$$F(h, \delta, \zeta) \equiv \int_{\gamma(h, \delta, \zeta)} (\zeta + x) y \, dx = 0. \qquad (1.10)$$

*Moreover, the system (1.7) has a homoclinic orbit if and only if (1.10) is satisfied for* $h = \frac{2}{3}$.

*Proof.* $\gamma(h, \delta, \zeta)$ is a periodic orbit if and only if $Q_1 = Q_2$. From (1.9) we have

$$\frac{\partial H(x, y)}{\partial x} = 1 - x^2 \neq 0, \text{ if } |x| \neq 1.$$

Hence $Q_1 = Q_2$ if and only if $H(Q_1) = H(Q_2)$.

On the other hand, along the orbits of (1.7) we have that

$$\frac{dH(x, y)}{dt}\bigg|_{(1.7)} dt = \delta(\zeta + x) y^2\big|_{(1.7)} \, dt = \delta(\zeta + x) y \, dx.$$

This implies that

$$H(Q_2) - H(Q_1) = \int_{t(Q_1)}^{t(Q_2)} \frac{dH}{dt}\bigg|_{(1.7)} dt = \delta \int_{\gamma(h, \delta, \zeta)} (\zeta + x) y \, dx. \quad (1.11)$$

This gives the desired results. The homoclinic case can be obtained by taking a limit as $h \to \frac{2}{3} - 0$ (see Lemma 1.5). $\qquad \square$

**Lemma 1.5.** (Bogdanov [1]) *There is* $\delta_0 > 0$ *such that the function* $F(h, \delta, \zeta)$ *given by (1.10) is continuous on the set*

$$U = \{(h, \delta, \zeta) | -\tfrac{2}{3} \leq h \leq \tfrac{2}{3}, 0 \leq \delta \leq \delta_0, \zeta_1 \leq \zeta \leq \zeta_2\},$$

*where* $\zeta_1 < \zeta_2$ *are arbitrary constants. Moreover, F is* $C^\infty$ *in* $\delta$ *and* $\zeta$ *on U, and* $C^\infty$ *in h on the set*

$$V = \{(h, \delta, \zeta) | -\tfrac{2}{3} < h < \tfrac{2}{3}, 0 \leq \delta \leq \delta_0, \zeta_1 \leq \zeta \leq \zeta_2\}.$$

*Proof.* From the theorem about the continuous and differentiable dependence on solutions upon initial conditions and parameters, we know that $F \in C^0$ on $U$ and $\in C^\infty$ on $V$. To prove that it is $C^\infty$ in $\delta$ and $\zeta$ at $h = -2/3$, we can use a theorem of Andronov et al. [1] about the smooth dependence of solutions upon parameters near a focus. To prove $F \in C^\infty$ in $\delta$ and $\zeta$ at $h = 2/3$, we can use a theorem of Shoshitaishvili [1] about the smooth dependence of the separatrix upon the parameters.                                              $\square$

We will consider $F(h, \delta, \zeta)$ as a perturbation of $F(h, 0, \zeta)$. The function $F(h, 0, \zeta)$ is given by

$$F(h, 0, \zeta) = \zeta I_0(h) + I_1(h), \qquad (1.12)$$

where

$$I_i(h) = \int_{\Gamma_h} x^i y \, dx, \qquad i = 0, 1,$$

and $\Gamma_h$ is the level curve of $H(x, y) = h$. The orientation of $\Gamma_h$ is determined by the direction of the vector field (1.8). By Green's formula

$$I_0(h) = \int_{\Gamma_h} y \, dx = \iint_{D(h)} dx \, dy > 0, \qquad h \in \left( -\frac{2}{3}, \frac{2}{3} \right],$$

where $D(h)$ is the region surrounded by $\Gamma_h$. It is easy to show that

$$\lim_{h \to -2/3} I_0(h) = \lim_{h \to -2/3} I_1(h) = 0.$$

By the Mean Value Theorem of integrals, we have

$$\lim_{h \to -2/3} \frac{I_1(h)}{I_0(h)} = \lim_{h \to -2/3} \frac{\iint_{D(h)} x \, dx \, dy}{\iint_{D(h)} dx \, dy} = \lim_{h \to -2/3} \bar{x}(h) = -1,$$

where $(\bar{x}(h), \bar{y}(h)) \in D(h)$ and $D(h)$ shrinks to the point $(-1, 0)$ as $h \to -2/3$.

Now we define

$$
P(h) = \begin{cases} -\dfrac{I_1(h)}{I_0(h)}, & -\tfrac{2}{3} < h \le \tfrac{2}{3}, \\ 1, & h = -\tfrac{2}{3}. \end{cases} \tag{1.13}
$$

It is continuous in $h \in [-2/3, 2/3]$.

We remark here that by Lemma 1.4, to determine the existence and the number of periodic orbits for (1.7), we only need to study the existence and the number of zeros for the function $F(h, \delta, \zeta)$ with respect to $h \in (-\tfrac{2}{3}, \tfrac{2}{3})$. On the other hand, $F(h, \delta, \zeta)$ can be approximated by $F(h, 0, \zeta) = I_0(h)(\zeta - P(h))$. Hence, the behavior of the function $P = P(h)$, as a ratio of two Abelian integrals, is crucial in our discussion.

**Lemma 1.6.** *If* $-2/3 < h < 2/3$, *then* $P(h)$ *satisfies the following Riccati equation*:

$$
(9h^2 - 4)P'(h) = 7P^2 + 3hP - 5. \tag{1.14}
$$

*Proof.* We have that

$$
I_i(h) = \int_{\Gamma_h} x^i y\, dx = 2\int_{\xi(h)}^{\eta(h)} x^i y\, dx, \qquad i = 0, 1, 2, \dots, \tag{1.15}
$$

where $\eta(h)$ and $\xi(h)$ are shown in Figure 1.4 and

$$
y = \left[ 2\left( h - x + \frac{x^3}{3} \right) \right]^{1/2} \tag{1.16}
$$

From (1.9), we obtain that

$$
I_i'(h) = 2\int_{\xi(h)}^{\eta(h)} \frac{x^i}{y}\, dx, \qquad i = 0, 1, 2, \dots. \tag{1.17}
$$

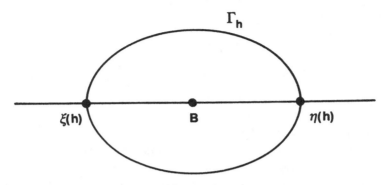

Figure 1.4.

Using (1.15), (1.16), and (1.17), we have

$$I_i(h) = 2\int_{\xi(h)}^{\eta(h)} \frac{x^i y^2}{y}\, dx = 2hI_i'(h) - 2I_{i+1}'(h) + \frac{2}{3}I_{i+3}'(h). \quad (1.18)$$

On the other hand, an integration by parts shows that

$$I_i(h) = 2\left[\frac{x^{i+1}}{i+1}y\Big|_{\xi(h)}^{\eta(h)} - \frac{1}{i+1}\int_{\xi(h)}^{\eta(h)} \frac{x^{i+1}(x^2-1)}{y}\, dx\right].$$

Since $y(\xi(h), h) = y(\eta(h), h) = 0$, we obtain by (1.17) that

$$I_i(h) = \frac{1}{i+1}(I_{i+1}'(h) - I_{i+3}'(h)). \quad (1.19)$$

Removing $I_{i+3}'(h)$ from (1.18) and (1.19), we have

$$(2i+5)I_i(h) = -4I_{i+1}'(h) + 6hI_i'(h).$$

In particular, we have

$$\begin{cases} 5I_0 = -4I_1' + 6hI_0', \\ 7I_1 = -4I_2' + 6hI_1'. \end{cases}$$

We claim that $I_2(h) \equiv I_0(h)$. Indeed, from (1.9)

$$dH = y\, dy + (1 - x^2)\, dx,$$

that is,

$$(1 - x^2) y \, dx = y \, dH - y^2 \, dy. \qquad (1.21)$$

Integrating (1.21) along $\Gamma_h$, we have $I_0(h) = I_2(h)$.
   Thus, (1.20) becomes

$$\begin{cases} 5I_0 = 6hI_0' - 4I_1', \\ 7I_1 = -4I_0' + 6hI_1'. \end{cases} \qquad (1.22)$$

For $-2/3 < h < 2/3$, (1.22) is equivalent to the following Picard–Fuchs
equation

$$\begin{cases} (9h^2 - 4) I_0' = \dfrac{15}{2} hI_0 + 7I_1, \\ (9h^2 - 4) I_1' = 5I_0 + \dfrac{21}{2} hI_1. \end{cases} \qquad (1.23)$$

It is easy to obtain (1.14) from (1.23), (1.13), and the following equation

$$P'(h) = \frac{1}{I_0^2} (I_0' I_1 - I_1' I_0). \qquad \square$$

**Lemma 1.7.** $P(h)$ *has the following properties*:
(1) $\lim_{h \to \frac{2}{3}} P(h) = \frac{5}{7}$;
(2) $P'(h) < 0$ *for* $-2/3 < h < 2/3$, $P'(h) \to -1/8$ *as* $h \to -2/3$,
*and* $P'(h) \to -\infty$ *as* $h \to 2/3$.

*Proof.* $P(h)$ is a solution of (1.14) and $P(h) \to 1$ as $h \to -2/3$. We
rewrite (1.14) into the following form:

$$\begin{cases} \dfrac{dP}{dt} = -7P^2 - 3hP + 5, \\ \dfrac{dh}{dt} = -9h^2 + 4. \end{cases} \qquad (1.24)$$

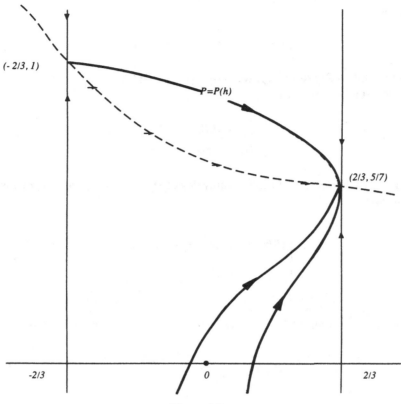

Figure 1.5.

The graph of $P(h)$ is the heteroclinic orbit from the saddle point $(-2/3, 1)$ to the node $(2/3, 5/7)$ in the $hP$-plane (see Figure 1.5). Thus $P(h) \to 5/7$ as $h \to 2/3$.

The graph of the equation

$$7P^2 + 3hP - 5 = 0 \qquad (1.25)$$

has two branches of curves on which the direction of the vector field (1.24) is horizontal. The branch of the hyperbola (1.25) above the $h$-axis is given by

$$C_q: q(h) = \frac{1}{14}\left[-3h + (9h^2 + 140)^{1/2}\right]. \qquad (1.26)$$

Along the curve $C_q$ we have $P'(h) = 0$ and $q'(h) = -3q^2/7q^2 + 5$. Hence the vector field (1.24) is transversal to $C_q$ from the left to the right for $-\frac{2}{3} < h < \frac{2}{3}$. It follows from (1.14) and (1.26) that

$$\lim_{h \to -2/3} P'(h) = -\frac{1}{8} \quad \text{and} \quad q'(-2/3) = -\frac{1}{4}.$$

Therefore, the graph of $P(h)$ is entirely above $C_q$, that is, $P'(h) < 0$ for $-2/3 \le h < 2/3$. The fact that $P'(h) \to -\infty$ as $h \to 2/3$ can be obtained directly from (1.14). □

Now by using the properties of $P(h)$, we continue with the proof of Theorem 1.2.

**Lemma 1.8.** *There is a neighborhood* $\Delta_2$ *of* $\mu_1 = \mu_2 = 0$ *such that for* $(\mu_1, \mu_2) \in \Delta_2$ *there is a curve* $HL = \{\mu | \mu_1 = -\frac{49}{25}\mu_2^2 + O(\mu_2^{5/2}), \mu_2 > 0\}$ *which is a homoclinic loop bifurcation curve of (1.5).*

*Proof.* By Lemma 1.4 and (1.10), the condition for existence of the homoclinic orbit of (1.7) is: $F(2/3, \delta, \zeta) = 0$. From (1.12), (1.13), and Lemma 1.7, $F(2/3, 0, \zeta_0) = 0$, where $\zeta_0 = P(2/3) = 5/7$, and $\partial F/\partial \zeta(\frac{2}{3}, 0, \zeta) = I_0(2/3) > 0$. By the Implicit Function Theorem, there exist a $\delta_0 > 0$ and a function $\zeta = \zeta(\delta)$, defined for $0 \le \delta < \delta_0$, such that $F(2/3, \delta, \zeta(\delta)) = 0$, that is, $\gamma(2/3, \delta, \zeta(\delta))$ is a homoclinic orbit.

Using (1.6), we can change the parameter $(\delta, \zeta)$ back to $(\mu_1, \mu_2)$ to obtain the equation of the bifurcation curve.

In fact, $\mu_2 = \zeta\delta^2$ and $\zeta(\delta) = \zeta_0(1 + O(\delta)) = \frac{5}{7}(1 + O(\delta))$ imply $\delta = O(\mu_2^{1/2})$ as $\mu_2 \to 0^+$. In addition, $\mu_1 = -\delta^4$ and $\mu_2 = \zeta\delta^2$ imply that

$$\mu_1 = -\frac{\mu_2^2}{\zeta^2(\delta)} = -\frac{\mu_2^2}{\zeta_0^2}(1 + O(\delta)) = -\frac{49}{25}\mu_2^2 + O(\mu_2^{5/2}),$$

where $\mu_2 > 0$ and $(\mu_1, \mu_2) \in \Delta_2 = \{(\mu_1, \mu_2) | |\mu_1| + |\mu_2| \le \delta_0^4\}$. This completes the proof of Lemma 1.8. □

**Lemma 1.9.** *For a given* $h_1 \in (-2/3, 2/3)$, *there exist* $\delta_1 > 0$ *and a unique function* $\zeta = \zeta_1(h, \delta)$ *defined in* $h \in [h_1, 2/3]$, $0 \le \delta \le \delta_1$, *such* *that*

$(1^0)$ *the trajectory* $\gamma(h, \delta, \zeta_1(h, \delta))$, $h_1 \leq h < 2/3$ *and* $0 < \delta \leq \delta_1$, *is a periodic orbit of (1.7), which is the unique limit cycle of (1.7); and*

$(2^0)$ $\partial \zeta_1/\partial h < 0$, $h_1 \leq h < 2/3$, $0 < \delta \leq \delta_1$.

*Proof.* We note from (1.12) that

$$F(h, \delta, \zeta)|_{\delta=0} = I_0(h)\zeta + I_1(h) = I_0(h)(\zeta - P(h)), \quad (1.27)$$

where $P(h)$ is defined in (1.13). Hence, for each $h^* \in [h_1, 2/3]$ we have

$$F(h^*, 0, P(h^*)) = 0, \quad \frac{\partial F}{\partial \zeta}\bigg|_{\delta=0} = I_0(h^*) > 0.$$

By the Implicit Function Theorem there exist $\delta^* > 0$, $\sigma^* > 0$, and a function $\zeta = \zeta^*(h, \delta)$ defined in $0 \leq \delta \leq \delta^*$ and $h^* - \sigma^* < h < h^* + \sigma^*$ (if $h^* = 2/3$, then we consider the interval $h^* - \sigma < h \leq h^*$) such that

$$F(h, \delta, \zeta^*(h, \delta)) = 0.$$

This means that $\gamma(h, \delta, \zeta^*(h, \delta))$ is a periodic orbit of (1.7).

Thus, by the compactness of $[h_1, 2/3]$, there exist $\delta_1 > 0$ and a function $\zeta = \zeta_1(h, \delta)$ defined for $0 \leq \delta \leq \delta_1$, $h_1 \leq h \leq 2/3$ such that

$$\zeta_1(h, 0) = P(h), \qquad F(h, \delta, \zeta_1(h, \delta)) = 0, \qquad (1.28)$$

that is, the trajectory $\gamma(h, \delta, \zeta_1(h, \delta))$ is a periodic orbit of (1.7) passing through the point $(x, y) = (x(h), 0)$.

Since $F \in C^\infty$ for all $\delta$, $\zeta$, and $-2/3 < h < 2/3$ (Lemma 1.5), we obtain from (1.28), (1.27), and Lemma 1.7 that

$$\frac{\partial F}{\partial h} + \frac{\partial F}{\partial \zeta_1}\frac{\partial \zeta_1}{\partial h} = 0,$$

$$\frac{\partial F}{\partial \zeta_1}\bigg|_{\delta=0} = I_0(h) > 0,$$

$$\frac{\partial F}{\partial h}\bigg|_{\delta=0, \, \zeta=\zeta_1(h,0)} = I_0'(h)(\zeta_1(h, 0) - P(h)) - I_0(h)P'(h) > 0.$$

This implies that (note that $I_0'(h)$ is finite for $-2/3 \le h \le 2/3$)

$$\left. \frac{\partial \zeta_1(h, \delta)}{\partial h} \right|_{\delta=0} < 0, \qquad h_1 \le h < \frac{2}{3},$$

and

$$\left. \frac{\partial \zeta_1(h, \delta)}{\partial h} \right|_{\delta=0} \to -\infty, \quad \text{as } h \to 2/3.$$

Hence, we can choose $\delta_1$ so small that

$$\frac{\partial \zeta_1(h, \delta)}{\partial h} < 0, \qquad 0 \le \delta \le \delta_1, \qquad h_1 \le h \le \frac{2}{3}.$$

Therefore, for any $\delta_0 \in (0, \delta_1)$ and $\zeta_0 \in (\zeta(\delta_0), P(h_1))$, where $\zeta(\delta) = \frac{5}{7}(1 + O(\delta))$ is the function defined in Lemma 1.8, there exists a unique $h_0 \in (h_1, 2/3)$ such that $\zeta_0 = \zeta_1(h_0, \delta_0)$. Hence

$$F(h_0, \delta_0, \zeta_0) = 0, \tag{1.29}$$

that is, $\gamma(h_0, \delta_0, \zeta_0)$ is a periodic orbit for (1.7).

On the other hand, we consider the trajectory $\gamma(h, \delta_0, \zeta_0)$ for $h$ near $h_0$. From (1.11) and (1.29) we have

$$H(Q_2) - H(Q_1) = \delta F(h, \delta_0, \zeta_0) = \delta \left[ \left. \frac{\partial F}{\partial h} \right|_{(\bar{h}, \delta_0, \zeta_0)} \cdot (h - h_0) \right],$$

$$\tag{1.30}$$

where $Q_1$ and $Q_2$ are the intersection points of $\gamma(h, \delta_0, \zeta_0)$ and the $x$-axis (see Figure 1.3), and $\bar{h}$ is between $h_0$ and $h$. Since $\partial F/\partial h > 0$ for small $\delta$, (1.30) implies that the periodic orbit $\gamma(h_0, \delta_0, \zeta_0)$ is an unstable limit cycle. $\qquad\Box$

**Lemma 1.10.** *There exist $h_2 \in (-2/3, 2/3)$, $\delta_2 > 0$, and a unique function $\zeta = \zeta_2(h, \delta)$ defined for $-2/3 \le h \le h_2$, $0 \le \delta \le \delta_2$ such that*
(i) *if $\zeta = \zeta_2(h, \delta)$, $-2/3 < h \le h_2$ and $0 < \delta \le \delta_2$, then the system (1.7) has an unstable limit cycle $\gamma(h, \delta, \zeta_2(h, \delta))$ passing through the point $(x(h), 0)$;*

(ii) $\partial\zeta_2/\partial h < 0$, $-2/3 < h \le h_2$, $0 \le \delta \le \delta_2$.

*Proof.* The linearized equation of (1.7) at the focus $(-1, 0)$ has the matrix

$$\begin{bmatrix} 0 & 1 \\ -2 & \delta(\zeta - 1) \end{bmatrix}. \tag{1.31}$$

We use the notations as in the Theorem 3.2.6 and replace $\mu$ by $\zeta$, and take $\zeta(\delta) \equiv 1$. The eigenvalues at $(-1, 0)$ are

$$\frac{1}{2}\left\{\delta(\zeta - 1) \pm i\left[8 - \delta^2(\zeta - 1)^2\right]^{1/2}\right\}.$$

Hence the conditions $(H_1^*)$ and $(H_2^*)$ are satisfied, and $\alpha^* > 0$.

Next, we use the formula (3.2.34) and obtain

$$\text{Re}(C_1) = \frac{1}{32}(\delta - 2\delta^2).$$

Hence the condition $(H_3^*)$ is also satisfied (for small $\delta$) and $C_1^* > 0$.

By Theorem 3.2.6 we can find $\bar{x} > -1$, $\delta_2 > 0$, and a function $\zeta = \bar{\zeta}_2(x, \delta)$ defined for $-1 \le x \le \bar{x}$, $0 \le \delta \le \delta_2$ such that $\gamma(h, \delta, \bar{\zeta}_2(x, \delta))$ is a periodic orbit of (1.7) passing through the point $(x, 0)$. Moreover, since $\alpha^* \cdot C_1^* > 0$, we have

$$\frac{\partial\bar{\zeta}_2}{\partial x} < 0, \qquad -1 \le x \le \bar{x}, \qquad 0 \le \delta \le \delta_2. \tag{1.32}$$

From (1.9) we have that $x = x(h)$ satisfies

$$x - \frac{1}{3}x^3 = h.$$

Thus

$$\frac{\partial x(h)}{\partial h} > 0, \quad \text{for } -2/3 \le h < 2/3. \tag{1.33}$$

If we take $h_2$ as the value satisfying $x(h_2) = \bar{x}$, then $-2/3 < h_2 < 2/3$.

Let

$$\zeta_2(h,\delta) \equiv \bar{\zeta}_2(x(h),\delta). \qquad (1.34)$$

Then the conclusion (i) follows.
From (1.34), (1.32), and (1.33), we have

$$\frac{\partial \zeta_2}{\partial h} = \frac{\partial \bar{\zeta}_2}{\partial x} \cdot x'(h) < 0, \qquad -2/3 < h < h_2, \qquad 0 \le \delta \le \delta_2,$$

and thus the conclusion (ii) follows. $\qquad\qquad\qquad\qquad\qquad\qquad$ □

**Lemma 1.11.** *There is a neighborhood $\Delta_3$ of $\mu_1 = \mu_2 = 0$, such that if $(\mu_1, \mu_2) \in \Delta_3$ and is between the curves H and HL (defined in Lemmas 1.3 and 1.8, respectively), then the system (1.5) has a unique periodic orbit and it is an unstable limit cycle. Moreover, as $(\mu_1, \mu_2)$ tends to H, the limit cycle shrinks to the focus; as $(\mu_1, \mu_2)$ tends to HL, the limit cycle tends to the homoclinic loop. The system (1.5) has no limit cycles if $(\mu_1, \mu_2) \in (H \cup HL) \cap \Delta_3$.*

*Proof.* Instead of (1.5) we first consider the system (1.7).
By Lemma 1.10, $\exists h_2 \in (-2/3, 2/3)$, $\delta_2 > 0$, and a function $\zeta = \zeta_2(h, \delta)$ defined in $-2/3 \le h \le h_2, 0 \le \delta \le \delta_2$ and having the properties (i) and (ii).
If we choose $h_1 \in (-2/3, h_2)$, then by Lemma 1.9 $\exists \delta_1 > 0$ and a unique function $\zeta = \zeta_1(h, \delta)$ defined in $h_1 \le h \le 2/3, 0 \le \delta \le \delta_1$ and having the properties ($1^0$) and ($2^0$).
Now let $\delta_3 = \min(\delta_1, \delta_2)$. Then by the uniqueness of $\zeta_1(h, \delta)$ we have

$$\zeta_1(h,\delta) \equiv \zeta_2(h,\delta), \qquad h_1 \le h \le h_2, \qquad 0 \le \delta \le \delta_3.$$

Thus we can define a function in the whole interval $-2/3 \le h \le 2/3$ as follows

$$\zeta = \zeta_3(h,\delta) = \begin{cases} \zeta_2(h,\delta), & \text{if } -2/3 \le h \le h_2, \\ \zeta_1(h,\delta), & \text{if } h_2 \le h \le 2/3, \end{cases} \qquad 0 \le \delta \le \delta_3,$$

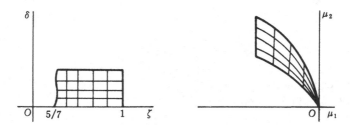

Figure 1.6.

which satisfies

(1) $\gamma(h, \delta, \zeta_3(h, \delta))$ *is a periodic orbit of* (1.7) *passing through the point* $(x(h), 0)$, $-2/3 < h < 2/3$;

(2) $\partial \zeta_3 / \partial h < 0$, $-2/3 \leq h \leq 2/3$, $0 \leq \delta \leq \delta_3$.

The condition (2) implies that for every $\delta_0 \in (0, \delta_3)$ and $\zeta_0 \in [\zeta(\delta_0), 1]$, where $\zeta = \zeta(\delta)$ is the function corresponding to the homoclinic bifurcation and described in Lemma 1.8, $\exists$ a unique $h_0 \in [-2/3, 2/3]$ such that $\zeta_0 = \zeta_3(h_0, \delta_0)$. Hence, if $\zeta_0 \in (\zeta(\delta_0), 1)$, then $\gamma(h_0, \delta_0, \zeta_0)$ is the unique periodic orbit of (1.7). Moreover, it is an unstable limit cycle. If $\zeta_0 \rightarrow \zeta(\delta_0) +$ (or $1 -$ ), then the limit cycle tends to the homoclinic loop (or to the focus).

We finally return from the parameters $\delta$ and $\zeta$ back to $\mu_1$ and $\mu_2$ by using the scaling (1.6).

Since

$$\mu_1 = -\delta^4, \qquad \mu_2 = \zeta \delta^2,$$

the region $0 < \delta \leq \delta_3$, $\zeta(\delta) \leq \zeta \leq 1$ corresponds to a cusp region $0 > \mu_1 \geq -\delta_3^4$ and $(\mu_1, \mu_2)$ is in between the bifurcation curves $H$ and $HL$ (Figure 1.6). Noting that $\zeta_3(-\frac{2}{3}, \delta) = 1$, $\zeta_3(\frac{2}{3}, \delta) = \zeta(\delta)$ (defined in Lemma 1.8), and $\partial \zeta_3 / \partial h < 0$, we conclude that the limit cycle will shrink to the focus or become the homoclinic loop as $(\mu_1, \mu_2)$ goes to $H$ or $HL$, respectively. The existence of $\delta_3$ guarantees the existence of the neighborhood $\Delta_3$. This completes the proof.                     □

**Lemma 1.12.** *There is a neighborhood* $\Delta_4$ *of* $\mu_1 = \mu_2 = 0$ *such that if* $(\mu_1, \mu_2) \in \Delta_4$ *and is above the curve* $H$ *or below the curve* $HL$, *then the system* (1.5) *has no periodic orbits, and has the phase portraits shown in Figure 1.1.*

*Proof.* By Lemma 1.11 we know that if $(\mu_1, \mu_2) \in \Delta_3 \cap (H \cup HL)$, then the system (1.5) has no periodic orbits, and any positive trajectory $\gamma$ starting from the point

$$p \in L = \{(x, y)| -1 < x < 1, y = 0\}$$

is an expanding spiral if $(\mu, \mu_2) \in H$ or a contracting spiral if $(\mu_1, \mu_2) \in HL$ (see Figure 1.1).

We rewrite (1.5) in the following form

$$\begin{cases} \dot{x} = P(x, y) = y, \\ \dot{y} = Q(x, y) = \mu_1 + x^2 + xy + \mu_2 y. \end{cases}$$

We note that

$$\begin{vmatrix} P & Q \\ \dfrac{\partial P}{\partial \mu_2} & \dfrac{\partial Q}{\partial \mu_2} \end{vmatrix} = y^2 > 0, \quad \text{if } y \neq 0. \tag{1.35}$$

This means (1.5) is a family of rotated vector fields with respect to $\mu_2$. For details of rotated vector fields, see Zhang et al. [1].

Now we take $\Delta_4 = \Delta_3$. For any $(\mu_1, \mu_2) \in \Delta_4$ and above $H$ (or below $HL$), we can find $(\mu_1, \bar{\mu}_2) \in H$ (or $\in HL$). Any periodic orbit of $(1.5)_{(\mu_1, \mu_2)}$, if it exists, must cut the segment $L$. The positive trajectory $\bar{\gamma}$ of $(1.5)_{(\mu_1, \bar{\mu}_2)}$ starting from a point $p \in L$ is an expanding (contracting) spiral, and due to (1.35), the positive trajectory $\gamma$ of $(1.5)_{(\mu_1, \mu_2)}$ starting from the same point $p$ must be entirely located outside (inside) $\bar{\gamma}$, and hence $\gamma$ is also an expanding (contracting) spiral. The phase portrait is as shown in Figure 1.1                                                              □

Combining the conclusions of Lemmas 1.3, 1.8, 1.11, and 1.12, we obtain Theorem 1.2, where $\Delta = \Delta_1 \cap \Delta_2 \cap \Delta_3 \cap \Delta_4$.

*(II) A Canonical Form for the System (1.4)*

**Theorem 1.13.** *In a sufficiently small neighborhood of the point $x = y = \epsilon = 0$, there is a $C^\infty$ transformation*

$$\begin{cases} u = u(x, y, \epsilon), \\ v = v(x, y, \epsilon) \end{cases}$$

*such that $u(0,0,0) = v(0,0,0) = 0$. It is nondegenerate at the point $x = y = 0$ and takes the system (1.4) to the form*

$$\begin{cases} \dot{u} = v\theta(u,\epsilon), \\ \dot{v} = \left[\phi(\epsilon) + \psi(\epsilon)u + u^2 + uvQ(u,\epsilon) + v^2\Phi(u,v,\epsilon)\right]\theta(u,\epsilon), \end{cases}$$

$$(1.36)$$

*where $\theta, Q, \Phi, \phi, \psi$ are $C^\infty$ functions, $\phi(0) = \psi(0) = 0$, $Q(0,0) = 1$, and $\theta(0,0) = 1$.*

*Proof.* Let

$$\begin{cases} \xi = x, \\ \eta = y + w_1(x,y,\epsilon), \end{cases}$$

where $(x,y)$ is in a neighborhood of the origin in $\mathbb{R}^2$ and $\epsilon$ is in a neighborhood of the origin in $\mathbb{R}^k$ so that the above transformation is invertible. Then (1.4) is transformed into the following equation defined in a neighborhood of the origin $(0,0,0)$ in $\mathbb{R} \times \mathbb{R} \times \mathbb{R}^k$:

$$\begin{cases} \dot{\xi} = \eta, \\ \dot{\eta} = F(\xi,\epsilon) + \eta G(\xi,\epsilon) + \eta^2 H(\xi,\eta,\epsilon), \end{cases} \qquad (1.37)$$

where $F, G, H$ are $C^\infty$ functions, and

$$F(0,0) = \frac{\partial F}{\partial \xi}(0,0) = 0, \qquad \frac{\partial^2 F}{\partial \xi^2}(0,0) = 2,$$

$$G(0,0) = 0, \qquad \frac{\partial G}{\partial \xi}(0,0) = 1, \qquad H(0,0,0) = 0.$$

Since $G(0,0) = 0$ and $\partial G/\partial \xi(0,0) = 1 \neq 0$, it follows from the Implicit Function Theorem that there exists a $C^\infty$ function $\alpha(\epsilon)$ defined in a neighborhood of $\epsilon = 0$ in $\mathbb{R}^k$ such that $G(\alpha(\epsilon), \epsilon) = 0$ for each $\epsilon$ in this neighborhood. Then the following change of variables

$$\begin{cases} u = \xi - \alpha(\epsilon), \\ v = \eta \end{cases}$$

brings (1.37) to the following equation near the origin $(0,0,0)$ in $\mathbb{R} \times \mathbb{R} \times \mathbb{R}^k$:

$$\begin{cases} \dot{u} = v, \\ \dot{v} = \tilde{F}(u, \epsilon) + uv\tilde{G}(u, \epsilon) + v^2\tilde{H}(u, v, \epsilon), \end{cases} \qquad (1.38)$$

where $\tilde{F}, \tilde{G}, \tilde{H}$ are $C^\infty$ functions and

$$\tilde{F}(0,0) = \frac{\partial \tilde{F}}{\partial u}(0,0) = 0, \qquad \frac{\partial^2 F}{\partial u^2}(0,0) = 2,$$

$$\tilde{G}(0,0) = 1, \qquad \tilde{H}(0,0,0) = 0.$$

By using the Malgrange Preparation Theorem (Theorem 3.1.10), we have

$$\tilde{F}(u, \epsilon) = [\phi_1(\epsilon) + \phi_2(\epsilon)u + u^2]\theta(u, \epsilon),$$

where $\phi_i, \theta \in C^\infty$, $\phi_i(0) = 0$ $(i = 1, 2)$ and $\theta(0, 0) = 1$. Therefore, (1.38) can be rewritten in the following form (in a neighborhood of $(0,0,0)$ in $\mathbb{R} \times \mathbb{R} \times \mathbb{R}^k$)

$$\begin{cases} \dot{u} = v, \\ \dot{v} = \left[\phi_1(\epsilon) + \phi_2(\epsilon)u + u^2 + \dfrac{\tilde{G}(u, \epsilon)}{\theta(u, \epsilon)}uv + \dfrac{\tilde{H}(u, v, \epsilon)}{\theta(u, \epsilon)}v^2\right]\theta(u, \epsilon). \end{cases}$$

$$(1.39)$$

Let

$$\tilde{u} = u, \qquad \tilde{v} = \frac{v}{\sqrt{\theta(u, \epsilon)}},$$

where $(u, v, \epsilon)$ is in a neighborhood of the origin of $\mathbb{R} \times \mathbb{R} \times \mathbb{R}^k$. Then

(1.39) is transformed into the form

$$
\begin{cases}
\dot{\tilde{u}} = \tilde{v}\sqrt{\theta(\tilde{u},\epsilon)}\,, \\
\dot{\tilde{v}} = \left[\phi_1(\epsilon) + \phi_2(\epsilon)\tilde{u} + \tilde{\mu}^2 + \tilde{u}\tilde{v}\tilde{\Phi}(\tilde{u},\epsilon) + \tilde{v}^2\tilde{\Psi}(\tilde{u},\tilde{v},\epsilon)\right]\sqrt{\theta(\tilde{u},\epsilon)}\,,
\end{cases}
$$

$$(1.40)$$

where

$$
\tilde{\Phi}(\tilde{u},\epsilon) = \frac{\tilde{G}(\tilde{u},\epsilon)}{\sqrt{\theta(\tilde{u},\epsilon)}}\,, \qquad \tilde{\Phi}(0,0) = 1,
$$

$$
\tilde{\Psi}(\tilde{u},\tilde{v},\epsilon) = \tilde{H}\!\left(\tilde{u},\tilde{v}\sqrt{\theta(\tilde{u},\epsilon)}\,,\epsilon\right) - \frac{1}{2\theta(\tilde{u},\epsilon)}\frac{\partial\theta}{\partial u}(\tilde{u},\epsilon).
$$

System (1.40) is now in the same form as (1.36), and this completes the proof.  □

**Lemma 1.14.** *In a sufficiently small neighborhood of the point $u = v = \epsilon = 0$ there is a $C^\infty$ transformation*

$$
\begin{cases}
x = x(u,v,\epsilon), \\
y = y(u,v,\epsilon)
\end{cases}
$$

*such that $x(0,0,0) = y(0,0,0) = 0$ and it is nondegenerate at the point $x = y = 0$, and it takes the equation (1.36) to the form*

$$
\begin{cases}
\dot{x} = y\bar{\theta}(x,\epsilon), \\
\dot{y} = \left[\bar{\phi}(\epsilon) + \bar{\psi}(\epsilon)y + x^2 + xy\bar{Q}(x,\epsilon) + y^2\bar{\Phi}(x,y,\epsilon)\right]\bar{\theta}(x,\epsilon),
\end{cases}
$$

$$(1.41)$$

*where $\bar{\theta},\bar{\phi},\bar{\psi},\bar{Q},\bar{\Phi}$ are $C^\infty$ functions, $\bar{\phi}(0) = \bar{\psi}(0) = 0$, $\bar{Q}(0,0) = 1$, and $\bar{\theta}(0,0) = 1$.*

*Proof.* Let

$$\begin{cases} x = u + \dfrac{\psi(\epsilon)}{2}, \\ y = v. \end{cases}$$

Then (1.36) is transformed into the form (1.41), where

$$\bar{\theta}(x, \epsilon) = \theta\left(x - \frac{\psi(\epsilon)}{2}, \epsilon\right),$$

$$\bar{\phi}(\epsilon) = \phi(\epsilon) - \frac{\psi^2(\epsilon)}{4},$$

$$\bar{\psi}(\epsilon) = -\frac{\psi(\epsilon)}{2}(f(\epsilon) + 1),$$

$$\bar{Q}(x, \epsilon) = Q\left(x - \frac{\psi(\epsilon)}{2}, \epsilon\right) - \frac{\psi(\epsilon)}{2}g(\epsilon, x),$$

$$\bar{\Phi}(x, y, \epsilon) = \Phi\left(x - \frac{\psi(\epsilon)}{2}, y, \epsilon\right),$$

and $f(\epsilon)$, $g(\epsilon, x)$ are defined by

$$Q\left(x - \frac{\psi(\epsilon)}{2}, \epsilon\right) = 1 + f(\epsilon) + g(\epsilon, x)x.$$

The conditions $Q(0, 0) = 1$ and $\phi(0) = \psi(0) = 0$ imply $f(0) = 0$, $\bar{\phi}(0) = \bar{\psi}(0) = 0$, and $\bar{Q}(0, 0) = 1$. $\quad\square$

From now on, we focus our attention on equation (1.41). Obviously, the orbits of (1.41) and the orbits of the following equation are the same if we restrict $(x, y, \epsilon)$ to a small neighborhood of $(0, 0, 0)$:

$$\begin{cases} \dot{x} = y, \\ \dot{y} = \bar{\phi}(\epsilon) + \bar{\psi}(\epsilon)y + x^2 + xy\bar{Q}(x, \epsilon) + y^2\bar{\Phi}(x, y, \epsilon), \end{cases} \quad (1.42)$$

where $\bar{\phi}, \bar{\psi}, \bar{Q}, \bar{\Phi}$ are the same as in (1.41). If

$$\text{rank}\left[\frac{\partial(\bar{\phi}(\epsilon), \bar{\psi}(\epsilon))}{\partial \epsilon}\bigg|_{\epsilon=0}\right] = 2\left(\text{say, } \frac{\partial(\bar{\phi}, \bar{\psi})}{\partial(\epsilon_1, \epsilon_2)}\bigg|_{\epsilon=0} \neq 0\right), \quad (1.43)$$

then we can make a change of parameters

$$\begin{cases} \mu_1 = \bar{\phi}(\epsilon), \\ \mu_2 = \bar{\psi}(\epsilon), \\ \mu_3 = \epsilon_3, \\ \vdots \\ \mu_m = \epsilon_m, \end{cases} \quad (1.44)$$

and (1.42) becomes

$$\begin{cases} \dot{x} = y, \\ \dot{y} = \mu_1 + \mu_2 y + x^2 + xyQ(x, \mu) + y^2\Phi(x, y, \mu), \end{cases} \quad (1.45)$$

where $Q(x, \mu) = \bar{Q}(x, \epsilon(\mu))$ and $\Phi(x, y, \mu) = \bar{\Phi}(x, y, \epsilon(\mu))$, and $\epsilon = \epsilon(\mu)$ is the inverse transformation of (1.44) satisfying $\epsilon(0) = 0$. Hence $Q, \Phi \in C^\infty$ and $Q(0, 0) = 1$.

In particular, if we let $Q(x, \mu) \equiv 1$ and $\Phi(x, y, \mu) \equiv 0$, then (1.45) becomes (1.5), which is a two-parameter family of vector fields, and its bifurcation diagrams and phase portraits have been studied in part (I). We will show in the next part that the topological structures of the bifurcation diagrams and the phase portraits of (1.45) are the same for different $Q$ and $\Phi$, as long as $Q, \Phi \in C^\infty$ and $Q(0, 0) = 1$.

If the condition (1.43) is not satisfied, then, for a given $\epsilon$ ($|\epsilon|$ sufficiently small), equation (1.42) is only a special case of the family (1.45). Hence, there is no new kind of phase portrait.

### (III) The Versality of the Deformation (1.5)

As in Section 3.2, we let $V(z_0)$ be the space of germs of $C^\infty$ vector fields at $z_0 \in \mathbb{R}^2$, and

$$\mathcal{X} = \{(z, Z) | Z \in V(z), z \in U\},$$

where $U$ is a small neighborhood of the origin in $\mathbb{R}^2$.

Suppose $Z \in V(z_0)$ has the following representative

$$\dot{z} = H(z), \qquad\qquad (1.46)$$

where

$$z = \begin{pmatrix} x \\ y \end{pmatrix} \in \mathbb{R}^2, \quad H(z) = \begin{pmatrix} f(x, y) \\ g(x, y) \end{pmatrix} \in \mathbb{R}^2, \quad H \in C^\infty.$$

We have the natural projection

$$\pi_k \colon \mathscr{X} \to J^k, (z, Z) \mapsto (z, H, \tilde{D}H, \ldots, \tilde{D}^k H),$$

where each $\tilde{D}^j H$ ($j = 0, 1, \ldots, k$) gives a coordinate expression for the Taylor coefficients of the $k$th-order derivatives of $H$ at $z$. In our case, $\dim(z) = \dim(H) = 2$, $\dim(\tilde{D}H) = 4$, and $\dim(\tilde{D}^2 H) = 6$. We could take the Jacobian and Hessian matrices at $z$ as $\tilde{D}H$ and $\tilde{D}^2 H$, respectively.

We say that (1.46) has the same singular character at $z_0$ as (1.3) at 0, if the following conditions are satisfied:

($H_1$) The matrix of the linear part of $H(z)$ at $z_0$ is similar to $\begin{bmatrix} 0 & 1 \\ 0 & 0 \end{bmatrix}$.

($H_2$) Changing (1.46) to its normal form (1.2) at $z_0$, we have $ab > 0$.

Now consider a subset of $\mathscr{X}$

$$\Sigma = \{(z, Z) \in \mathscr{X} \mid Z \text{ satisfies } (H_1) \text{ and } (H_2) \text{ at } z \in U\},$$

where $U$ is the small neighborhood of $z = 0$.

**Lemma 1.15.** *If $k \geq 2$, then $\pi_k \Sigma$ is locally a smooth codimension-4 submanifold of $J^k$.*

*Proof.* Note that

$$\pi_1 \Sigma = \{(z, H, DH) \mid f = g = \det DH = \operatorname{Tr} DH = 0, \ DH \neq 0\},$$

$$\pi_2 \Sigma = \pi_{21}^{-1}(\pi_1 \Sigma) \big|_{(H_2)},$$

where $\pi_{21}$ is the natural projection from $J^2$ onto $J^1$, and the condition

($H_2$) gives $ab > 0$ which is independent of the conditions $f = g =$ det $DH$ = Tr $DH$ = 0 and $DH \neq 0$ (see Section 2.1, Example 2.1.15). Hence, $\pi_1 \Sigma$ is a smooth submanifold with codimension 4 in $J^1$, and $\pi_2 \Sigma$ is locally an open subset of $\pi_{21}^{-1}(\pi_1(\Sigma))$. By the same arguments as in the proof of Lemma 3.2.1, the desired result follows.          $\square$

We consider a deformation of (1.3)

$$\dot{z} = H(z, \epsilon), \qquad (1.47)$$

where $z \in \mathbb{R}^2$, $\epsilon \in \mathbb{R}^m$, $H \in C^\infty$.

**Definition 1.16.** Equation (1.47) is called a nondegenerate deformation of (1.3) if the mapping

$$(z, \epsilon) \mapsto \pi_2 H(z, \epsilon)$$

is transverse to $\pi_2 \Sigma$ at $(z, \epsilon) = (0, 0)$ in $J^2$.

**Lemma 1.17.** *Any nondegenerate deformation of (1.3) is equivalent to systems (1.45) ($\dim(\mu) = \dim(\epsilon)$).*

*Proof.* By Lemmas 1.13 and 1.14, we only need to show that the nondegenerate condition implies the condition (1.43). Since the nondegeneracy is independent of the choice of coordinates, we can prove the fact by using equation (1.42), that is we consider a nondegenerate deformation (1.47), where

$$\begin{cases} f(x, y, \epsilon) = y, \\ g(x, y, \epsilon) = \bar{\phi}(\epsilon) + \bar{\psi}(\epsilon)y + x^2 + xy\bar{Q}(x, \epsilon) + y^2\bar{\Phi}(x, y, \epsilon). \end{cases}$$

We know that $\pi_2 \Sigma$ can be expressed locally by

$$f = g = \det H = \text{Tr } H = 0,$$

where

$$\det H = \frac{\partial(f, g)}{\partial(x, y)}, \qquad \operatorname{Tr} H = \frac{\partial f}{\partial x} + \frac{\partial g}{\partial y}.$$

By Theorem 3.1.5 and Definition 1.16, the nondegeneracy of (1.47) implies

$$\operatorname{rank} \begin{pmatrix} 0 & 1 & 0 & 0 & \cdots & 0 \\ 0 & 0 & \dfrac{\partial \bar{\phi}}{\partial \epsilon_1} & \dfrac{\partial \bar{\phi}}{\partial \epsilon_2} & \cdots & \dfrac{\partial \bar{\phi}}{\partial \epsilon_m} \\ -2 & -1 & 0 & 0 & \cdots & 0 \\ 1 & 2\Phi & \dfrac{\partial \bar{\psi}}{\partial \epsilon_1} & \dfrac{\partial \bar{\psi}}{\partial \epsilon_2} & \cdots & \dfrac{\partial \bar{\psi}}{\partial \epsilon_m} \end{pmatrix}_{(x, y, \epsilon) = (0,0,0)} = 4,$$

and this implies

$$\operatorname{rank}\left( \frac{\partial(\bar{\phi}, \bar{\psi})}{\partial \epsilon} \right)_{\epsilon = 0} = 2. \qquad \square$$

**Theorem 1.18.** *The family (1.5) is a versal deformation of (1.3) at $(x, y) = (0, 0)$ provided we consider only nondegenerate deformations of (1.3).*

   In order to prove Theorem 1.18, it is sufficient to prove that any two families of (1.45) are equivalent.

**Lemma 1.19.** *For any $Q$, $\Phi \in C^{\infty}(Q(0,0) = 1)$, the conclusions of Theorem 1.2 are true for equation (1.45).*

*Proof.* By the scaling (1.6), (1.45) takes the form

$$\begin{cases} \dot{x} = y, \\ \dot{y} = -1 + x^2 + \delta(\zeta + x)y + O(\delta^2), \end{cases}$$

and the bifurcation function $F(h, \delta, \zeta)$ becomes

$$\bar{F}(h, \delta, \zeta) = \int_{\gamma(h, \delta, \zeta)} ((\zeta + x)y + O(\delta))\, dx.$$

Therefore, all the discussions in part (I) are valid.                    □

**Lemma 1.20.** *For any two families (1.45), there is a $C^\infty$ diffeomorphism in a neighborhood of $\mu = 0$, fixing the point $\mu = 0$ and mapping the bifurcation curves of one to the other.*

In order to prove Lemma 1.20, we need the following lemma. Let $Y_1, Y_2, Y_3$ be three $C^\infty$ curves in a neighborhood of the point $x = y = 0$ in the $xy$-plane, tangent to each other at the point $x = y = 0$. We choose suitable coordinates so that the curves $Y_1, Y_2, Y_3$ are the graphs of the functions $Y_1(x), Y_2(x)$, and $Y_3(x)$, respectively, and

$$Y_i(0) = \frac{dY_i(0)}{dx} = 0.$$

Let

$$I(Y_1, Y_2, Y_3) = [Y_3''(0) - Y_1''(0)]/[Y_2''(0) - Y_1''(0)], \quad (1.48)$$

where $Y_1''(0), Y_2''(0), Y_3''(0)$ are different numbers. We note that $I(Y_1, Y_2, Y_3)$ is a finite number different from zero.

**Lemma 1.21.** *Suppose that $Y_1, Y_2, Y_3$ and $Z_1, Z_2, Z_3$ are two sets of $C^\infty$ curves satisfying the above conditions. Then the condition*

$$I(Y_1, Y_2, Y_3) = I(Z_1, Z_2, Z_3) \quad (1.49)$$

*is necessary and sufficient for the existence of a $C^\infty$ diffeomorphism in a neighborhood of the origin, fixing the origin and mapping $Y_i$ to $Z_i$, $i = 1, 2, 3$.*

*Proof.* To prove the necessity, we suppose that there is a $C^\infty$ transformation

$$z = f(x, y), \qquad u = g(x, y)$$

that transforms $Y_i$: $y = y_i(x)$ to $Z_i$: $z = z_i(u)$, $i = 1, 2, 3$. Therefore

$$f(x, y_i(x)) = z_i\big(g(x, y_i(x))\big), \qquad i = 1, 2, 3.$$

Differentiating the above equality we get

$$f'_x + f'_y y'_i = z'_i(g'_x + g'_y y'_i) \tag{1.50}$$

and

$$f''_{xx} + 2f''_{xy} y'_i + f''_{yy} y'^2_i + f'_y y''_i = z''_i(g'_x + g'_y y'_i)^2$$

$$+ z'_i\big(g''_{xx} + 2g''_{xy} y'_i + g''_{yy} y'^2_i + g'_y y''_i\big). \tag{1.51}$$

Since $f(0,0) = g(0,0) = 0$ and $y_i(0) = z_i(0) = y'_i(0) = z'_i(0) = 0$, from (1.50) we have $f'_x(0,0) = 0$ which implies that $f'_y(0,0) \neq 0$ and $g'_x(0,0) \neq 0$. From (1.51),

$$z''_i(0) = \frac{f''_{xx}(0,0)}{(g'_x(0,0))^2} + \frac{f'_y(0,0)}{(g'_x(0,0))^2} y''_i(0).$$

By using the above equality and (1.48), we have

$$I(Y_1, Y_2, Y_3) = I(Z_1, Z_2, Z_3).$$

For the converse we will prove that both $Y_1, Y_2, Y_3$ and $Z_1, Z_2, Z_3$ can be converted respectively by $C^\infty$ transformations to the set of $C^\infty$ curves $X_1, X_2, X_3$ with $X_1(x) \equiv 0$, $X_2(x) = x^2$, $X_3(x) = cx^2$, where $c = I(Y_1, Y_2, Y_3) = I(Z_1, Z_2, Z_3)$. In what follows, we will give the proof for $Y_1, Y_2, Y_3$ only. The proof for $Z_1, Z_2, Z_3$ is the same. It is easy to find a $C^\infty$ transformation in a neighborhood of the origin to convert $Y_1, Y_2, Y_3$ to $\tilde{Y}_1, \tilde{Y}_2, \tilde{Y}_3$ with $\tilde{y}_1(x) \equiv 0$, $\tilde{y}_2(x) = \alpha x^2 \xi(x)$, and $\tilde{y}_3(x) = \beta x^2 \psi(x)$, where $\alpha$ and $\beta$ are unequal nonzero numbers, and $\xi$ and $\psi$ are $C^\infty$ with $\xi(0) = \psi(0) = 1$. From the necessity part of the lemma, $\beta/\alpha = I(Y_1, Y_2, Y_3)$. We make a $C^\infty$ change of coordinates near the origin again by

$$x \to x, \quad \tilde{y} \to \alpha \xi(x) y.$$

Then $\tilde{Y}_1, \tilde{Y}_2, \tilde{Y}_3$ are mapped to $\bar{Y}_1, \bar{Y}_2, \bar{Y}_3$ with $\bar{y}_1(x) \equiv 0$, $\bar{y}_2(x) = x^2$, and $\bar{y}_3(x) = cx^2\phi(x)$, where $\phi(x)$ is a $C^\infty$ function with $\phi(0) = 1$ and $c = \beta/\alpha = I(Y_1, Y_2, Y_3)$. Now let us find a $C^\infty$ change of coordinates in a neighborhood of the origin to convert $\bar{Y}_1, \bar{Y}_2, \bar{Y}_3$ to $X_1, X_2, X_3$.

Suppose

$$u = y, \qquad v = x + (y - x^2)f(x)$$

transforms the curve $y = cx^2\phi(x)$ into the curve $u = cv^2$. Then we have

$$cx^2\phi(x) = c\left[x + (cx^2\phi(x) - x^2)f(x)\right]^2.$$

Hence

$$f(x) = \frac{\sqrt{\phi(x)} - 1}{x(c\phi(x) - 1)}.$$

Since $\phi(0) = 1$, $\phi \in C^\infty(x)$, and $c \neq 1$, one has that $f \in C^\infty$ in a small neighborhood of $x = 0$. This proves Lemma 1.21. $\qquad\square$

*Proof of Lemma 1.20.* For any family (1.45), the equations of bifurcation curves are $\mu_1 = 0$, $\mu_1 = -\mu_2^2$, and $\mu_1 = -\frac{49}{25}\mu_2^2 + O(\mu_2^{5/2})$ as $\mu_2 \to 0 +$. Hence, by formula (1.48),

$$I(SN_i, H_i, HL) = \frac{49}{25}, \qquad i = 1, 2.$$

Thus Lemma 1.20 follows from Lemma 1.21. $\qquad\square$

**Lemma 1.22.** *Any two families of the form (1.45) are equivalent.*

*Outline of the Proof.* By Lemma 1.20, we may carry out the construction of two families $A$ and $\tilde{A}$ over the same neighborhood $\Delta$ in the parameter space. The neighborhoods of $(x, y) = (0, 0)$ for $A$ and $\tilde{A}$ are denoted by $N(\mu)$ and $\tilde{N}(\mu)$, respectively.

Then we may construct a homeomorphism $\Psi(\mu)$ for a fixed $\mu \in \Delta$, mapping $K(\epsilon)$ (the limit set and singular trajectories of family $A$ in $N(\mu)$) onto $\tilde{K}(\epsilon)$, where $\tilde{K}(\epsilon)$ is a similar set of $\tilde{A}$ in $\tilde{N}(\mu)$.

Finally, we may extend the homeomorphism $\Psi(\mu)$ to obtain a homeomorphism mapping the trajectories of $A$ in $N$ onto the trajectories of $\tilde{A}$ in $\tilde{N}$.

For more details, we refer to Bogdanov [2] and Sotomayor [1].     □

**Remark 1.23.** In Definition 3.1.14, if the mapping $h(\cdot, \epsilon)$ is continuous in $\epsilon$, then $(X; x_0, \epsilon_0)$ and $(Y; y_0, \epsilon_0)$ are said to be *strongly equivalent*. Otherwise, they are *weakly equivalent*. Bogdanov [2] proved the versality of (1.5) in this weak sense. Recently Dumortier and Roussarie [1] gave a proof for the strong versality of (1.5).

### 4.2 Double Zero Eigenvalue with Symmetry of Order 2

In Sections 4.2–4.5, we will study the families of vector fields in the plane that are invariant under a rotation of the plane through an angle $2\pi/q$, $q = 2, 3, \ldots$ (the case $q = 1$ is discussed in Section 4.1). In this section, we consider the case $q = 2$. Khorozov [1] and Carr [1] investigated this case of codimension two. We will introduce their results. However, some proofs may be given in a different way. The normal form of $q = 2$ is (see Section 2.11)

$$\begin{cases} \dot{x} = y, \\ \dot{y} = \epsilon_1 x + \epsilon_2 y \pm x^3 - x^3 y, \end{cases} \qquad (2.1)^{\pm}$$

where $\epsilon_1$ and $\epsilon_2$ are small real parameters. We will give bifurcation diagrams of the vector fields $(2.1)^+$ and $(2.1)^-$, respectively. In Fact, $(2.1)^{\pm}$ is a versal deformation of

$$\begin{cases} \dot{x} = y, \\ \dot{y} = \pm x^3 - x^2 y. \end{cases} \qquad (2.2)^{\pm}$$

It will be shown in Lemma 2.2 that any perturbation of (2.2) with a small parameter $\mu$ can be transformed into the form

$$\begin{cases} \dot{x} = y, \\ \dot{y} = \phi(\mu) x + \psi(\mu) y \pm x^3 + x^2 y \tilde{\Phi}(x, \mu) + y^2 \tilde{\Psi}(x, y, \mu), \end{cases} \qquad (2.3)^{\pm}$$

where $\mu \in \mathbb{R}^k$ ($k \geq 1$), $\bar{\Phi}(x, 0) = -1$, and $\bar{\Psi}(x, y, 0) = 0$. If $k = 2$ and we consider (2.3) as a nondegenerate deformation, then there exists a transformation of parameters

$$\epsilon_1 = \phi(\mu_1, \mu_2), \qquad \epsilon_2 = \psi(\mu_1, \mu_2)$$

such that $(2.3)^{\pm}$ becomes

$$\begin{cases} \dot{x} = y, \\ \dot{y} = \epsilon_1 x + \epsilon_2 y \pm x^3 + x^2 y \Phi(x, \epsilon) + y^2 \Psi(x, y, \epsilon), \end{cases} \quad (2.4)^{\pm}$$

where $\Phi(x, 0) = -1$, $\Psi(x, y, 0) = 0$.

In Lemmas 2.3–2.10, we will discuss $(2.4)^{\pm}$ and will show that the topological structures of the bifurcation diagrams and phase portraits of $(2.4)^{\pm}$ are the same for different $\Phi$ and $\Psi$. If $\Phi = -1$, $\Psi = 0$, then $(2.4)^{\pm}$ becomes $(2.1)^{\pm}$. Moreover, it is not difficult to see from these lemmas that for every small $\mu \in \mathbb{R}^k$ the phase portraits of (2.3) must also be contained in the phase portraits of (2.4).

All the results are local. That means the bifurcation diagrams are in a small neighborhood of parameter space near $(\epsilon_1, \epsilon_2) = (0, 0)$, and the phase portraits are in a small neighborhood of phase space near $(x, y) = (0, 0)$. Therefore, all bifurcation theorems in this chapter should be understood in this sense. Thus, we will obtain the following theorem.

**Theorem 2.1.** *We have*:
(1) *System* $(2.1)^{\pm}$ *is a versal deformation of* $(2.2)^{\pm}$ *among all the nondegenerate deformations of* $(2.2)^{\pm}$ *with symmetry of order* 2.
(2) *The bifurcation diagram of* $(2.1)^{+}$ *consists of the origin and the following curves*:
  (a) $R = \{\epsilon | \epsilon_1 = 0, \epsilon_2 \neq 0\}$,
  (b) $H = \{\epsilon | \epsilon_2 = 0, \epsilon_1 < 0\}$,
  (c) $HL = \{\epsilon | \epsilon_2 = -\frac{1}{5}\epsilon_1 + O(\epsilon_1^{3/2}), \epsilon_1 < 0\}$.
*The bifurcation diagram and phase portraits of* $(2.1)^{+}$ *are shown in Figure 2.1.*
(3) *The bifurcation diagram of* $(2.1)^{-}$ *consists of the origin and the following curves*:
  (a) $R^{+} = \{\epsilon | \epsilon_1 = 0, \epsilon_2 > 0\}$,
  (b) $R^{-} = \{\epsilon | \epsilon_1 = 0, \epsilon_2 < 0\}$,
  (c) $H_1 = \{\epsilon | \epsilon_2 = 0, \epsilon_1 < 0\}$,

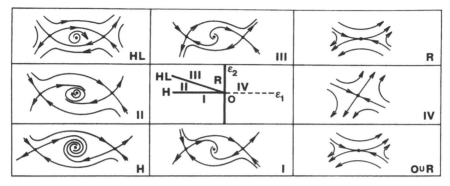

Figure 2.1. The bifurcation diagram and phase portraits of $(2.1)^+$.

(d) $H_2 = \{\epsilon | \epsilon_2 = \epsilon_1 + O(\epsilon_1^2), \epsilon_1 > 0\}$,

(e) $HL = \{\epsilon | \epsilon_2 = 4\epsilon_1/5 + O(\epsilon_1^{3/2}), \epsilon_1 > 0\}$,

(f) $B = \{\epsilon | \epsilon_2 = c\epsilon_1 + O(\epsilon_1^{3/2}), \epsilon_1 > 0, c \approx 0.752\}$.

The bifurcation diagram and phase portraits of $(2.1)^+$ are shown in Figure 2.2.

**Lemma 2.2.** *Consider a family of systems*

$$\begin{cases} \dot{x} = y + w_1(x, y, \mu), \\ \dot{y} = ax^3 + bx^2y + w_2(x, y, \mu), \end{cases} \tag{2.5}$$

*where $ab \neq 0$ and $\mu = (\mu_1, \ldots, \mu_k)$. Suppose $w_i$ $(i = 1, 2)$ satisfies:*

(1) $w_i(x, y, 0) = 0$,

(2) $w_i(x, y, \mu) \in C^\infty$,

(3) $w_i(-x, -y, \mu) = -w_i(x, y, \mu)$.

*Then there exists a smooth mapping $(x, y, \mu) \rightarrow (\bar{x}(x, y, \mu), \bar{y}(x, y, \mu))$ that transforms (2.5) into a system topologically equivalent to $(2.3)^\pm$.*

*If $k = 2$ and (2.5) satisfies the following additional condition:*

(4) *the matrix of the linear part of (2.5) at the origin is a versal deformation of $\begin{bmatrix} 0 & 1 \\ 0 & 0 \end{bmatrix}$, then there exists a smooth mapping*

$$(x, y, \mu_1, \mu_2) \rightarrow (\bar{x}(x, y, \mu), \bar{y}(x, y, \mu), \epsilon_1(\mu), \epsilon_2(\mu))$$

*that transforms (2.5) into a system topologically equivalent to $(2.4)^\pm$.*

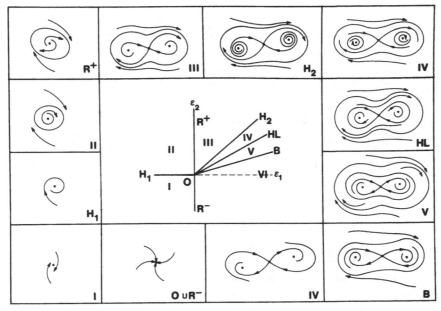

Figure 2.2. The bifurcation diagram and phase portraits of $(2.1)^-$.

*Proof.* Let

$$\begin{cases} \xi = x, \\ \eta = y + w_1(x, y, \mu). \end{cases}$$

Then (2.5) becomes

$$\begin{cases} \dot{\xi} = \eta, \\ \dot{\eta} = a\xi^3(1 + w'_{1y}) + b\xi^2\eta(1 + w'_{1y}) \\ \quad + \left\{ -b\xi^2 w_1(1 + w'_{1y}) + w_2(1 + w'_{1y}) + \eta w'_{1x} \right\}. \end{cases} \tag{2.6}$$

Define functions $h_i$ and $\Psi_i$ $(i = 1, 2, 3)$ in the following way:

$$w'_{1y}(x(\xi, \eta, \mu), y(\xi, \eta, \mu), \mu)$$

$$= h_1(\xi, \mu) + \eta h_2(\xi, \mu) + \eta^2 h_3(\xi, \eta, \mu),$$

$$-b\xi^2 w_1(1 + w'_{1y}) + w_2(1 + w'_{1y}) + \eta w'_{1x}$$

$$= \Psi_1(\xi, \mu) + \eta\Psi_2(\xi, \mu) + \eta^2\Psi_3(\xi, \eta, \mu).$$

Obviously,

$$h_i|_{\mu=0} = \Psi_i|_{\mu=0} = 0, \qquad i = 1,2,3.$$

Thus (2.6) takes the form

$$\begin{cases} \dot{\xi} = \eta, \\ \dot{\eta} = a\xi^3(1 + h_1) + a\xi^3\eta h_2 + b\xi^2\eta(1 + h_1) \\ \qquad + \Psi_1 + \eta\Psi_2 + \eta^2\Phi(\xi,\eta,\mu), \end{cases} \qquad (2.7)$$

where

$$\Phi = a\xi^3 h_3 + b\xi^2(h_2 + \eta h_3) + \Psi_3.$$

Since (2.7) has a symmetry property with respect to $(\xi,\eta)$ under a rotation through $\pi$, we have

$$\Psi_1(\xi,\mu) = \phi_1(\mu)\xi + \beta_1(\xi,\mu)\xi^3,$$

$$ah_2(\xi,\mu)\xi^3 + \Psi_2(\xi,\mu) = \phi_2(\mu) + \beta_2(\xi,\mu)\xi^2.$$

Let

$$a + ah_1 + \beta_1 = F(\xi,\mu),$$

$$b + bh_1 + \beta_2 = G(\xi,\mu).$$

Thus, (2.7) becomes

$$\begin{cases} \dot{\xi} = \eta, \\ \dot{\eta} = \phi_1(\mu)\xi + \phi_2(\mu)\eta + \xi^3 F(\xi,\mu) + \xi^2\eta G(\xi,\mu) + \eta^2\Phi(\xi,\eta,\mu). \end{cases}$$

$$(2.8)$$

Using the Malgrange Preparation Theorem for the symmetric case (see Poénaru [1, p. 64–5)], we have

$$\phi_1(\mu)\xi + F(\xi,\mu)\xi^3 = \left[\tilde{\phi}(\mu)\xi + \text{sgn}\, F(0,0)\xi^3\right]\theta(\xi,\mu),$$

where $F(0,0) = a \neq 0$, $\theta(\xi, 0) = |a| > 0$, $F(-\xi, \mu) = F(\xi, \mu)$, and $\theta(-\xi, \mu) = \theta(\xi, \mu)$. Thus

$$\dot{\eta} = \left[ \tilde{\phi}(\mu)\xi \pm \xi^3 + \frac{\phi_2(\mu)}{\theta(\xi, \mu)} \eta \right.$$

$$\left. + \frac{G(\xi, \mu)}{\theta(\xi, \mu)} \xi^2 \eta + \frac{\Phi(\xi, \eta, \mu)}{\theta(\xi, \mu)} \eta^2 \right] \theta(\xi, \mu).$$

By changing coordinates

$$\begin{cases} u = \xi, \\ v = \eta / \sqrt{\theta(\xi, \mu)}, \end{cases}$$

we obtain

$$\begin{cases} \dot{u} = v\sqrt{\theta}, \\ \dot{v} = \sqrt{\theta} \left[ \tilde{\phi}(\mu)u \pm u^3 + \left( \frac{\phi_2}{\sqrt{\theta}} - \frac{\dot{\theta}}{2\sqrt{\theta} \theta} \right) v \right. \\ \left. + u^2 v \tilde{G}(u, \mu) + v^2 \Phi(u, v, \mu) \right], \end{cases}$$

where $\tilde{G} = G/\sqrt{\theta}$. Using the symmetry property again, we have

$$\frac{\phi_2}{\sqrt{\theta}} - \frac{\dot{\theta}}{2\sqrt{\theta} \cdot \theta} = \tilde{\psi}(\mu) + \sigma(u, \mu)u^2,$$

where $\tilde{\psi}(0) = \sigma(u, 0) = 0$. Denoting $\tilde{G} + \sigma$ by $\overline{G}$, we finally obtain

$$\begin{cases} \dot{u} = \sqrt{\theta} \cdot v, \\ \dot{v} = \sqrt{\theta} \left[ \tilde{\phi}(\mu)u + \tilde{\psi}(\mu)v \pm u^3 + u^2 v \overline{G}(u, \mu) + v^2 \Phi(u, v, \mu) \right], \end{cases}$$

$$(2.9)^{\pm}$$

and $(2.9)^{\pm}$ is topologically equivalent to $(2.3)^{\pm}$. If the condition (4) of

Lemma 2.2 is satisfied, then we can change parameters

$$\begin{cases} \epsilon_1 = \tilde{\phi}(\mu_1, \mu_2), \\ \epsilon_2 = \tilde{\psi}(\mu_1, \mu_2). \end{cases}$$

Thus $(2.9)^{\pm}$ is transformed into a system which is equivalent to $(2.4)^{\pm}$.
In $(2.9)^{\pm}$, $\Phi(u, v, 0) = 0$ and $\overline{G}(u, 0) = b / \sqrt{|a|}$. If $b / \sqrt{|a|} \neq -1$, we can take the scaling in $(2.5)$

$$x \to \frac{\sqrt{|a|}}{|b|} x, \qquad y \to -\frac{\sqrt{|a|}}{\sqrt{|b|}} (\operatorname{sgn} b) y, \qquad t \to -(\operatorname{sgn} b) t,$$

before the first transformation.    □

We will now study $(2.4)^{-}$ in detail. $(2.4)^{+}$ is simpler and can be studied in a similar way.

**Lemma 2.3.** *For $(2.4)^{-}$, $R^{+} \cup R^{-}$ is a bifurcation curve of equilibria while $H_1$ and $H_2$ are Hopf bifurcation curves (see Figure 2.2). When the parameter $\epsilon = (\epsilon_1, \epsilon_2)$ crosses $R^{+} \cup R^{-}$ from the left to the right, the number of equilibria changes from one (a focus or a node) to three (one saddle point and two foci or nodes). When $\epsilon$ crosses $H_1$ from region I to II, the focus changes from stable to unstable (the equilibrium is stable on $H_1$), and a unique limit cycle appears. When $\epsilon$ crosses $H_2$ from region III to IV, the foci change from unstable to stable (the equilibria are unstable on $H_2$), and two unstable limit cycles appear (each of them goes around one focus).*

*Proof.* The coordinates of equilibria satisfy $y = 0$ and $\epsilon_1 x - x^3 = 0$, that is, $(0, 0)$ for $\epsilon_1 \leq 0$, and $(0, 0)$ and $(\pm \sqrt{\epsilon_1}, 0)$ for $\epsilon_1 > 0$. The matrix of the linear part of $(2.4)^{-}$ at $(x, 0)$ is

$$\begin{bmatrix} 0 & 1 \\ \epsilon_1 - 3x^2 & \epsilon_2 + x^2 \Phi(x, \epsilon) \end{bmatrix}.$$

Hence, the first part of Lemma 2.3 is easy to obtain.

In order to prove the second part of the lemma, we use the result of Section 3.2.

On $H_1$, the linear part of $(2.4)^-$ at the equilibrium $(0, 0)$ is

$$\begin{bmatrix} 0 & 1 \\ \epsilon_1 & \epsilon_2 \end{bmatrix},$$

where $\epsilon_1 < 0$, $\epsilon_2 = 0$. By using formula (3.2.34), we have

$$16\beta_0^2 \operatorname{Re}(C_1) = -2 + O(|\epsilon|) < 0.$$

Hence, the equilibrium is a stable (weak) focus, and a unique stable limit cycle appears when the parameters vary across $H_1$ from region I to II (Theorem 3.2.4).

Similarly, on $H_2$ the linear part of $(2.4)^-$ at the equilibrium $(\sqrt{\epsilon_1}, 0)$ is

$$\begin{bmatrix} 0 & 1 \\ -2\epsilon_1 & \epsilon_2 - \epsilon_1 + O(|\epsilon|^2) \end{bmatrix},$$

where $\epsilon_1 > 0$, $\epsilon_2 - \epsilon_1 + O(|\epsilon|^2) = 0$. Again, by using formula (3.1.34), we have

$$16\beta_0^2 \operatorname{Re}(C_1) = (-2 + O(|\epsilon|))$$

$$- \frac{1}{2\epsilon_1}\left[ -\left(-6\sqrt{\epsilon_1}\right)\left(-2\sqrt{\epsilon_1} + O(|\epsilon|)\right)^{3/2} + O(|\epsilon|^2)\right]$$

$$= 4 + O(|\epsilon|) > 0.$$

Hence, $(\sqrt{\epsilon_1}, 0)$ is an unstable focus and a unique unstable limit cycle appears when the parameters cross $H_2$ from III to IV (see Figure 2.2).

By symmetry, we can obtain similar results for the other equilibrium $(-\sqrt{\epsilon_1}, 0)$.                                    □

Now we turn to the discussion of periodic orbits and homoclinic orbits. We first consider the more complicated case: $\epsilon_1 > 0$.

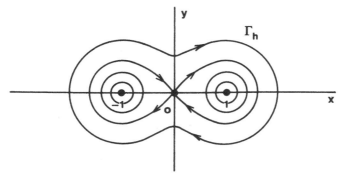

Figure 2.3.

Making the following scaling for $(2.4)^-$:

$$x \to \delta x, \qquad y \to \delta^2 y, \qquad \epsilon_1 = \delta^2,$$

$$\epsilon_2 = \delta^2 \zeta, \qquad t \to t/\delta, \tag{2.10}$$

where $\delta > 0$, we have

$$\begin{cases} \dot{x} = y, \\ \dot{y} = x - x^3 + \delta y(\zeta - x^2) + O(\delta^2). \end{cases} \tag{2.11}$$

For $\delta = 0$, (2.11) is a Hamiltonian system

$$\begin{cases} \dot{x} = y, \\ \dot{y} = x - x^3, \end{cases} \tag{2.12}$$

with the first integral

$$H(x, y) = \frac{y^2}{2} - \frac{x^2}{2} + \frac{x^4}{4}. \tag{2.13}$$

The level curves $\{H(x, y) = h, \ h \geq -\frac{1}{4}\}$ are shown in Figure 2.3. $H = -\frac{1}{4}$ corresponds to the foci $(\pm 1, 0)$; when $-\frac{1}{4} < h < 0$, $H(x, y) = h$ corresponds to two closed curves, each of them surrounding one of the foci; $H = 0$ corresponds to a pair of homoclinic orbits; and when $h > 0$, $H(x, y) = h$ corresponds to a closed curve which surrounds the pair of homoclinic orbits.

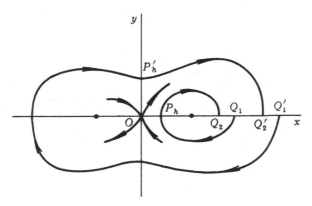

Figure 2.4.

We let $L = L_1 \cup L_2$, where $L_1 = \{(x, y) | y = 0, 0 \leq x < 1\}$ and $L_2 = \{(x, y) | x = 0, y > 0\}$. Then the closed level curve $\Gamma_h$ of (2.12) intersects $L$ at exactly one point $(\alpha(h), 0) = P_h$ if $-1/4 \leq h \leq 0$, or $(0, \beta(h)) = P_h$ if $h > 0$. Hence $L$ can be parameterized by $h$.

For every $h \in (-1/4, \infty)$, we consider the trajectory of (2.11) passing through the point $P_h \in L$. Let this trajectory go forward and backward from $P_h$ until it intersects the positive $x$-axis at points $Q_1$ and $Q_2$, respectively (Figure 2.4).

We denote the piece of trajectory from $Q_1$ to $Q_2$ by $\gamma(h, \delta, \zeta)$.

**Lemma 2.4.** $\gamma = \gamma(h, \delta, \zeta)$ of (2.11) is a closed orbit if and only if

$$\int_\gamma \frac{dH(x, y)}{dt}\bigg|_{(2.11)} dt = 0. \tag{2.14}$$

*Moreover, $\gamma$ is a (or a pair of) homoclinic orbit(s) if and only if (2.14) is satisfied for $h = 0 - (\text{or } 0 +)$.*

*Proof.* It is similar to the proof of Lemma 1.4.                    □

The calculation shows that if $\delta > 0$ then (2.14) is equivalent to

$$F(h, \delta, \zeta) = \int_{\gamma(h, \delta, \zeta)} \left[ (\zeta - x^2) y + O(\delta^2) \right] dx = 0. \quad (2.15)$$

In the same way as for Lemma 1.5, we can prove that the function $F(h, \delta, \zeta)$ is continuous and $C^\infty$ in $\delta$ and $\zeta$ on a set $U = \{(h, \delta, \zeta)| -1/4 \leq h < +\infty, \ 0 \leq \delta \leq \delta_0, \ \zeta_1 \leq \zeta \leq \zeta_2\}$, where $\delta_0$ is some positive number and $\zeta_1 < \zeta_2$ are arbitrary constants. Moreover, $F \in C^\infty$ in $h$ on the set $V = \{(h, \delta, \zeta)| h \in (-1/4, 0) \cup (0, +\infty), \ 0 \leq \delta \leq \delta_0, \ \zeta_1 \leq \zeta \leq \zeta_2\}$.

When $\delta = 0$, (2.15) becomes

$$F(h, 0, \zeta) = \int_{\Gamma_h} (\zeta - x^2) y \, dx = \zeta I_0(h) - I_2(h) = 0, \quad (2.16)$$

where $\Gamma_h$ is the level curve of (2.12), and the Abelian integrals are given by

$$I_i(h) = \int_{\Gamma_h} x^i y \, dx, \qquad i = 0, 2. \quad (2.17)$$

Similar to the discussion in Section 4.1, we have that
(1) $I_0(h) > 0$ for $h > -\frac{1}{4}$, $I_0(-\frac{1}{4}) = I_2(-\frac{1}{4}) = 0$, and
(2) $\lim_{h \to -\frac{1}{4}} I_2(h)/I_0(h) = 1$.
Hence we can define a function

$$P(h) = \begin{cases} \dfrac{I_2(h)}{I_0(h)}, & \text{for } h > -\frac{1}{4}, \\ 1, & \text{for } h = -\frac{1}{4}. \end{cases} \quad (2.18)$$

It is continuous on $-1/4 \leq h < \infty$.

As in Section 4.1, the basic problem is: For given $\zeta$ and $\delta$ small, does there exist $h > -1/4$ such that (2.15) is satisfied? First, we study the properties of $P(h)$.

**Lemma 2.5.** *When $h > -1/4$ and $h \neq 0$, $P(h)$ satisfies the following Riccati equation*

$$4h(4h + 1) P'(h) = 5P^2(h) + 8hP(h) - 4P(h) - 4h. \quad (2.19)$$

*Moreover,* $P'(h) \to -1/2$ *as* $h \to -1/4$, *and* $P'(h) \to -\infty$ *as* $h \to 0$.

*Proof.* Similarly to the proof of Lemma 1.6, we can obtain

$$\begin{cases} 3I_0 = 4hI_0' + I_2', \\ 5I_2 = 4hI_2' + I_4', \end{cases} \tag{2.20}$$

where

$$I_i(h) = \int_{\Gamma_h} x^i y \, dx, \qquad i = 0, 2, 4.$$

From (2.13) we have that along $\Gamma_h(H(x, y) = h)$:

$$0 = xy \, dH = xy^2 \, dy + y(-x^2 + x^4) \, dx.$$

On the other hand, using $y^2 = 2h + x^2 - \frac{1}{2}x^4$, we have that

$$xy^2 \, dy \equiv d\left(\frac{xy^3}{3}\right) - \frac{y^3}{3} \, dx = d\left(\frac{xy^3}{3}\right) - \frac{y}{3}\left(2h + x^2 - \frac{x^4}{2}\right) dx.$$

Hence

$$d\left(\frac{xy^3}{3}\right) - \frac{2}{3}hy \, dx - \frac{4}{3}x^2 y \, dx + \frac{7}{6}x^4 y \, dx = 0.$$

Integrating the above equation along $\Gamma_h$, we get

$$I_4 = \frac{1}{7}(4hI_0 + 8I_2). \tag{2.21}$$

Substituting (2.21) into (2.20) and solving $I_0', I_2'$, we obtain the Picard–Fuchs equation

$$\begin{cases} 3h(4h + 1)I_0' = 3(3h + 1)I_0 - \dfrac{15}{4}I_2, \\ 3h(4h + 1)I_2' = -3hI_0 + 15hI_2. \end{cases} \tag{2.22}$$

Equation (2.22) and $P'(h) = (I_2'I_0 - I_2I_0')/I_0^2$ imply (2.19).

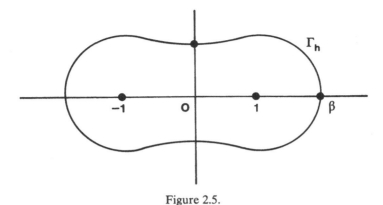

Figure 2.5.

The values of $P'(h)$ as $h \to -1/4$ and as $h \to 0$ can be obtained by direct calculations from (2.19).        □

**Lemma 2.6.** $\lim_{h \to +\infty} P(h) = +\infty$.

*Proof.* Without loss of generality, we assume $h > 0$. From (2.18) and (2.17) we have

$$P(h) = \frac{\int_{\Gamma_h} x^2 y \, dx}{\int_{\Gamma_h} y \, dx} = \frac{\int_0^\beta x^2 y \, dx}{\int_0^\beta y \, dx} \equiv \frac{J_2(\beta)}{J_0(\beta)},$$

where $y = (2h + x^2 - x^4/2)^{1/2}$, and $\beta = \beta(h)$ is the abscissa of the intersection point of $\Gamma_h$ and the $x$-axis (see Figure 2.5). Hence $\beta = \beta(h)$ satisfies

$$\beta^4 - 2\beta^2 = 4h. \tag{2.23}$$

Making the substitution $x = \beta\xi$ in the integral $J_k(\beta)$, we obtain

$$J_k(\beta) = \beta^2 \int_0^1 (\beta\xi)^k g(\xi) \, d\xi, \qquad k = 0, 2,$$

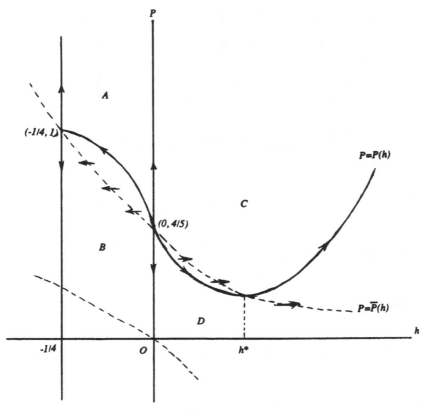

Figure 2.6.

where

$$g(\xi) = \left[ \frac{\beta^2}{2}(1 - \xi^4) + (\xi^2 - 1) \right]^{1/2}.$$

Since $g(\xi) \le g(1/\beta)$ for $0 \le \xi \le 1$, we have $J_0(\beta) \le \alpha_1 \beta^3$ for some positive constant $\alpha_1$. It is easy to obtain that $J_2(\beta) \ge \alpha_2 \beta^5$ for some positive constant $\alpha_2$. Since $h \to \infty \Leftrightarrow \beta \to \infty$ (see (2.23)), we have

$$\lim_{h \to \infty} P(h) = \lim_{\beta \to \infty} \frac{J_2(\beta)}{J_0(\beta)} = +\infty. \qquad \square$$

**Lemma 2.7.** *$P(h)$ has the following properties:*
(1) $\lim_{h \to 0} P(h) = \frac{4}{5}$;
(2) *there exists $h^* > 0$ such that $P'(h) < 0$ for $-1/4 \le h < h^*$ and $P'(h) > 0$ for $h > h^*$;*
(3) *$P(h^*) > 1/2$, $P'(h^*) = 0$, and $P''(h^*) > 0$.*

*Proof.* Rewrite (2.19) in the form

$$
\begin{cases}
\dfrac{dP}{dt} = 5P^2 + 8hP - 4P - 4h, \\[2mm]
\dfrac{dh}{dt} = 4h(4h + 1).
\end{cases}
\tag{2.24}
$$

Since $P(h) \to 1$ as $h \to -1/4$, the graph of $P(h)$ is the heteroclinic orbit from the saddle point $(-\frac{1}{4}, 1)$ to the node $(0, \frac{4}{5})$ in the phase plane (see Figure 2.6). Hence $\lim_{h \to 0} P(h) = \frac{4}{5}$. We denote the two branches of the hyperbola $5P^2 + 8hP - 4P - 4h = 0$ by $\bar{P}(h)$ and $\tilde{P}(h)$.

It is clear that the phase plane is divided into nine parts by the lines $h = 0$, $h = -\frac{1}{4}$ and the curves $P = \bar{P}(h)$, $P = \tilde{P}(h)$. On the two lines the vector field is vertical, and on the two curves the vector field is horizontal. In every one of the nine parts, $dP/dh$ has a fixed sign. From Lemma 2.5 and by calculations we know that

$$
\lim_{h \to -\frac{1}{4}} P'(h) = -\frac{1}{2}, \qquad \lim_{h \to 0} P'(h) = -\infty,
$$

$$
\lim_{h \to -\frac{1}{4}} \bar{P}'(h) = -1, \qquad \lim_{h \to 0} \tilde{P}'(h) = -\frac{3}{5}.
\tag{2.25}
$$

Hence, the graph of $P(h)$ must stay in part $A$ for $-\frac{1}{4} < h < 0$ and must enter part $D$ for $0 < h \ll 1$. In parts $A$ and $D$, $dP/dh$ is negative. But $P(h) \to +\infty$ as $h \to +\infty$ (Lemma 2.6) and $\bar{P}(h) \to \frac{1}{2}$ as $h \to +\infty$, hence there exists $h^* > 0$ such that $P(h^*) = \bar{P}(h^*)$, that is, $P'(h^*) = 0$ and $P'(h) > 0$ for $h > h^*$. Noting $\bar{P}'(h) < 0$ and $\bar{P}(h) \to \frac{1}{2}$ as $h \to +\infty$, we have $P(h^*) = \bar{P}(h^*) > \frac{1}{2}$. Finally, from (2.19) we obtain

$$
4h(4h + 1)P'' = (10P - 24h - 8)P' + 8\left(P - \frac{1}{2}\right).
$$

Hence,

$$4h^*(4h^* + 1)P''(h^*) = 8\left(P(h^*) - \frac{1}{2}\right) > 0,$$

which implies

$$P''(h^*) > 0. \qquad \qquad \square$$

**Lemma 2.8.** *For (2.4)$^-$, HL is a homoclinic loop bifurcation curve and B is a double limit cycle bifurcation curve. The phase portraits in regions III, IV, V, and VI are shown in Figure 2.2.*

*Proof.* The idea is similar to the proof of Theorem 1.2. Consider first system (2.11) instead of (2.4)$^-$. For given $\zeta$ and small $\delta$, the periodic orbits of (2.11) are determined by the zeros of equation (2.15) (Lemma 2.4) which can be approximated well by the zeros of equation (2.16).

More precisely, suppose $h_0$ is one of the solutions of the equation

$$P(h) = \zeta_0, \qquad \qquad (2.26)$$

that is, $F(h_0, 0, P(h_0)) = 0$ (see (2.16) and (2.18)). Since $\partial F/\partial \zeta|_{(h_0, 0, P(h_0))} = I_0(h_0) \neq 0$ if $h_0 > -\frac{1}{4}$, the Implicit Function Theorem implies that there are $\delta_0 > 0$, $\sigma_0 > 0$, and a function $\zeta = \zeta(h, \delta)$ defined in $U_0 = \{(h, \delta): 0 \leq \delta \leq \delta_0, |h - h_0| \leq \sigma_0\}$ such that $\zeta(h_0, 0) = P(h_0)$ and $F(h, \delta, \zeta(h, \delta)) = 0$ for $(h, \delta) \in U_0$. If, in addition, $P'(h_0) \neq 0$, then we can suppose that $\delta_0$ and $\sigma_0$ are so small that $\zeta'_h(h, \delta) \neq 0$, $(h, \delta) \in U_0$, since $\lim_{\substack{\delta \to 0 \\ h \to h_0}} \zeta'_h(h, \delta) = P'(h_0)$. $\zeta'_h(h, \delta) \neq 0$ implies that for every $\zeta$ near $\zeta_0$, $0 \leq \delta \leq \delta_0$ equation (2.15) has a unique solution with respect to $h \in (h_0 - \sigma_0, h_0 + \sigma_0)$.

The above discussion is valid except in two neighborhoods of $h$: (1) near $h = -\frac{1}{4}$, since $I_0(-\frac{1}{4}) = 0$ so the Implicit Function Theorem is invalid; and (2) near $h = h^*$, since $P'(h^*) = 0$ so the above condition is not satisfied. In the first case, we can use Theorem 3.2.6 instead of the Implicit Function Theorem. In fact, the linearized equation of (2.11) at $(1, 0)$ has a matrix with the same form as (1.31); hence the conditions

$(H_1^*)$, $(H_2^*)$, and $(H_3^*)$ are satisfied (see the proof of Lemma 1.10). In the second case, we consider

$$\begin{cases} F(h, \delta, \zeta) = 0, \\ \dfrac{\partial F}{\partial h}(h, \delta, \zeta) = 0. \end{cases}$$

We have that

$$F(h^*, 0, P(h^*)) = 0,$$

$$\frac{\partial F}{\partial h}(h^*, 0, P(h^*)) = 0,$$

and

$$\left. \frac{\partial\left(F, \dfrac{\partial F}{\partial h}\right)}{\partial(\zeta, h)} \right|_{\substack{\delta=0 \\ h=h^* \\ \zeta=P(h^*)}} = \begin{vmatrix} I_0(h^*) & 0 \\ * & -I_0(h^*)P''(h^*) \end{vmatrix}$$

$$= -I_0^2(h^*)P''(h^*) < 0.$$

Hence there are $\delta^* > 0$ and functions $\zeta = \zeta^*(\delta)$, $h = h^*(\delta)$ defined in $0 \le \delta \le \delta^*$ such that $\zeta^*(0) = P(h^*)$, $h^*(0) = h^*$, and

$$\begin{cases} F(h^*(\delta), \delta, \zeta^*(\delta)) = 0, \\ \dfrac{\partial F}{\partial h}(h^*(\delta), \delta, \zeta^*(\delta)) = 0 \end{cases}$$

for $0 \le \delta \le \delta^*$. Since $\partial^2 F/\partial h^2(h^*, 0, P(h^*)) = -I_0(h^*)P''(h^*) \ne 0$, we can suppose that $\delta^*$ is so small that $\partial^2 F/\partial h^2(h^*(\delta), \delta, \zeta^*(\delta)) \ne 0$ for $0 \le \delta \le \delta^*$, which implies that $\zeta = \zeta^*(\delta)$, $h = h^*(\delta)$ correspond to double limit cycle bifurcation, and the numbers of zeros for equations (2.16) and (2.15) are the same near $h = h^*$.

Hence the number of solutions of (2.26) will determine the number of limit cycles of (2.11) (or, equivalently, of (2.4)$^-$). The relationship between them is: one-to-one for $h > 0$ and one-to-two for $-\frac{1}{4} < h < 0$

because of the symmetry (see Figures 2.7 and 2.2). The solution of (2.26) for $h = 0$ corresponds to the pair of homoclinic orbits of $(2.4)^-$.

It is clear that for $\zeta_0 > 1$, (2.26) has a unique solution $h_1 > 0$ corresponding to a limit cycle around three equilibria (the case of $\zeta_0 = 1$ is discussed in Lemma 2.2); for $\frac{4}{5} < \zeta_0 < 1$, (2.26) has two solutions $-\frac{1}{4} < h_2 < 0$ and $h_3 > 0$, $h_2$ corresponding to two limit cycles around two foci respectively while $h_3$ corresponding to one limit cycle which surrounds the two limit cycles and three equilibria; for $\zeta_0 = 4/5$, (2.26) has two solutions $h = 0$ and $h_4 > 0$, the former corresponding to a pair of symmetric homoclinic loops while the latter corresponding to a limit cycle surrounding the three equilibria and the homoclinic loops; for $c < \zeta_0 < \frac{4}{5}$ ($c = P(h^*) \simeq 0.752$), (2.26) has two solutions $h_5 > 0$ and $h_6 > 0$, which correspond to two limit cycles, one surrounding another and both of them surrounding three equilibria; for $\zeta_0 = c$, (2.26) has a double solution $h^*$ which corresponds to a double limit cycle (it is semistable); for $\zeta_0 < c$, (2.26) has no solution, which means $(2.4)^-$ has no limit cycle.

From the scaling (2.10) we know that $\epsilon_2/\epsilon_1 = \zeta(h, \delta) = \zeta_0 + O(\delta) = \zeta_0 + O(\epsilon_1^{1/2})$, which gives the equations of curves $HL$ and $B$ for $\zeta_0 = \frac{4}{5}$ and $\zeta_0 = c$, respectively. This finishes the proof of Lemma 2.8.    $\square$

**Lemma 2.9.** *System $(2.4)^-$ has no limit cycle in region I and has a unique limit cycle in region II.*

*Proof.* This is the case of $\epsilon_1 < 0$. As in the case $\epsilon_1 > 0$, we take the scaling

$$x \to \delta x, \quad y \to \delta^2 y, \quad \epsilon_1 = -\delta^2, \quad \epsilon_2 = \tilde{a}\delta^2, \quad t \to \frac{t}{\delta}.$$

Thus, $(2.4)^-$ becomes

$$\begin{cases} \dot{x} = y, \\ \dot{y} = -x - x^3 + \delta y(\tilde{a} - x^2) + O(\delta^3). \end{cases} \tag{2.27}$$

For $\delta = 0$, (2.27) is a Hamiltonian system

$$\begin{cases} \dot{x} = y, \\ \dot{y} = -x - x^3, \end{cases} \tag{2.28}$$

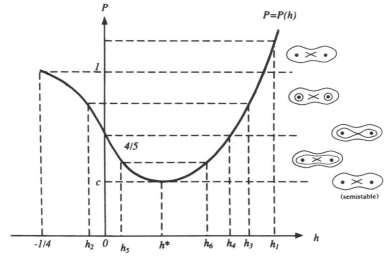

Figure 2.7.

with the first integral

$$H(x, y) = \frac{y^2}{2} + \frac{x^2}{2} + \frac{x^4}{4}.$$     (2.29)

We consider a function

$$P(h) = \begin{cases} \dfrac{I_2(h)}{I_0(h)}, & h > 0, \\ 0, & h = 0, \end{cases}$$

where

$$I_i(h) = \int_{\Gamma_h} x^i y \, dx, \qquad i = 0, 2,$$

and

$$\Gamma_h: H(x, y) = h, \qquad h \geq 0 \text{ (see Figure (2.8))}.$$

$P(h)$ satisfies an equation

$$4h(4h + 1)P'(h) = -5P^2 + 8hP - 4P + 4h,$$

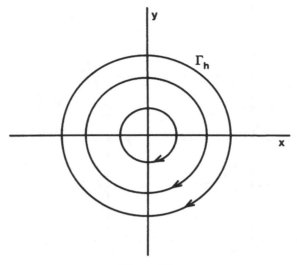

Figure 2.8.

which implies that $P'(h) > 0$ for $h > 0$ and $P'(0) = \frac{1}{2}$ (by a similar analysis as in Lemma 2.7). Then, Lemma 2.9 follows by the same argument as in Lemma 2.8.                                                   □

By Lemmas 2.2, 2.3, 2.8, and 2.9, we have the proof of Theorem 2.1 for the case of $(2.1)^-$. For the case $(2.1)^+$, the difficult part of the proof is to study the existence of limit cycles and their numbers. In a manner similar to the case $(2.4)^-$, one can derive an equation for a similar function $P(h)$ for $(2.4)^+$:

$$4h(4h - 1)P'(h) = -5P^2 + 8hP + 4P - 4h.$$

Corresponding to Figure 2.3, we have Figure 2.8 for the plus case. We leave the details to the readers.

### 4.3 Double Zero Eigenvalue with Symmetry of Order 3

In this section we study the family of vector fields on the plane that are invariant under a rotation through an angle $2\pi/3$. The normal form is (Section 2.11)

$$\dot{z} = \epsilon z + Az|z|^2 + \bar{z}^2, \tag{3.1}$$

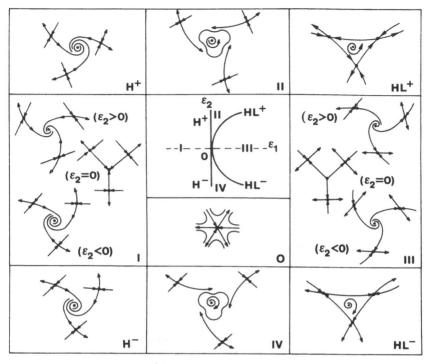

Figure 3.1. The bifurcation diagram and phase portraits of $(3 \cdot 1)\,(a < 0)$.

where

$$z = x + iy, \qquad \epsilon = \epsilon_1 + i\epsilon_2, \quad \text{and} \quad A = a + ib, \qquad a \neq 0.$$

The following theorem belongs to Khorozov [1]. We will use the Picard–Fuchs equation to prove the uniqueness of periodic orbits (see Chow, Li, and Wang [2]).

**Theorem 3.1.** (1) *Equation (3.1) is a versal deformation with symmetry of order* 3 *of the following system*

$$\dot{z} = Az|z|^2 + \bar{z}^2. \tag{3.2}$$

(2) *The bifurcation diagram of (3.1) consists of the origin and the following curves in parameter space:*

(a) $H^+ = \{\epsilon | \epsilon_1 = 0,\ \epsilon_2 > 0\}$,

(b) $H^- = \{\epsilon | \epsilon_1 = 0,\ \epsilon_2 < 0\}$,

(c) $HL^+ = \{\epsilon | \epsilon_1 = -(a/2)\epsilon_2^2 + O(\epsilon_2^3),\ \epsilon_2 > 0\}$,

(d) $HL^- = \{\epsilon | \epsilon_1 = -(a/2)\epsilon_2^2 + O(\epsilon_2^3),\ \epsilon_2 < 0\}$.

(3) *The phase portraits of (3.1) are shown in Figure 3.1.*

We first consider a deformation of (3.2) with symmetry of order 3

$$\dot{z} = Az|z|^2 + \bar{z}^2 + w(z, \bar{z}, \mu), \qquad (3.3)$$

where

$$w \in C^\infty \quad \text{and} \quad w|_{\mu=0} = 0, \qquad \mu \in \mathbb{C}.$$

Equation (3.3) can be written in the form

$$\dot{z} = \sum_{i+j=0}^{3} a_{ij}(\mu) z^i \bar{z}^j + O(|z|^4), \qquad a_{ij}(\mu) \in \mathbb{C}. \qquad (3.4)$$

Since (3.4) is invariant under a rotation through $2\pi/3$, we have in polar coordinates that

$$\dot{r}(r, \theta) = \dot{r}\left(r, \theta + \frac{2\pi}{3}\right), \qquad \dot{\theta}(r, \theta) = \dot{\theta}\left(r, \theta + \frac{2\pi}{3}\right) \qquad (r \neq 0).$$

Using the formulas

$$\dot{r} = \frac{\bar{z}\dot{z} + z\dot{\bar{z}}}{2r}, \qquad \dot{\theta} = \frac{\bar{z}\dot{z} - z\dot{\bar{z}}}{2r^2 i},$$

we obtain in (3.4)

$$a_{00} = a_{01} = a_{20} = a_{11} = a_{30} = a_{12} = a_{03} = 0.$$

Hence, (3.4) has the form (after a linear transformation)

$$\dot{z} = f(\mu)z + \bar{A}(\mu)z|z|^2 + \bar{z}^2 + O(|z|^4), \qquad (3.5)$$

where $f(\mu) = s\mu + O(|\mu|^2)$. If $s \neq 0$, then let $\epsilon = f(\mu)$. Equation (3.5) becomes

$$\dot{z} = \epsilon z + A(\epsilon)z|z|^2 + \bar{z}^2 + O(|z|^4), \qquad (3.6)$$

where $A(\epsilon) = \bar{A}(\mu(\epsilon))$, $A(0) = \bar{A}(0) = A$. We will prove that the topological structure of solutions of (3.6) is independent of the function $A(\epsilon)$ and the higher-order terms $O(|z|^4)$ as long as Re $A(0) \neq 0$. By a transformation

$$\begin{cases} x = \sqrt{2\rho} \cos \phi, \\ y = \sqrt{2\rho} \sin \phi. \end{cases} \qquad (3.7)$$

(3.6) takes the form

$$
\begin{cases}
\dot{\rho} = \epsilon_1(2\rho) + a(\epsilon)(2\rho)^2 + (2\rho)^{3/2}\cos 3\phi + (2\rho)^{5/2}F_1(\rho, 3\phi, \epsilon), \\
\dot{\phi} = \epsilon_2 + b(\epsilon)(2\rho) - (2\rho)^{1/2}\sin 3\phi + (2\rho)^{3/2}F_2(\rho, 3\phi, \epsilon),
\end{cases}
$$

$$(3.8)$$

where $F_j(\rho, 3\phi, \epsilon)$ is $2\pi$-periodic with respect to $\phi$, $F_j|_{\epsilon=0} = 0$ $(j = 1, 2)$, and $a(\epsilon) + ib(\epsilon) = A(\epsilon)$.

We suppose that $a(0) < 0$. The case $a(0) > 0$ can be obtained through the transformation $t \to -t$, $z \to -z$.

In order to determine the number and property of the nonzero equilibria, we consider

$$
\begin{cases}
\epsilon_1 + a(\epsilon)2\rho + (2\rho)^{1/2}\cos 3\phi + (2\rho)^{3/2}F_1 = 0, \\
\epsilon_2 + b(\epsilon)2\rho - (2\rho)^{1/2}\sin 3\phi + (2\rho)^{3/2}F_2 = 0
\end{cases}
$$

$$(3.9)$$

in a small neighborhood of the origin in phase space. Let $\epsilon = \sqrt{\alpha}\, e^{i\psi}$, $\alpha\tilde{\rho} = \rho$. Then (3.9) becomes

$$
\begin{cases}
\sqrt{\alpha}\, G_1(\alpha, \tilde{\rho}, \phi, \psi) = 0, \\
\sqrt{\alpha}\, G_2(\alpha, \tilde{\rho}, \phi, \psi) = 0,
\end{cases}
$$

$$(3.10)$$

where

$$
G_1 = \cos\psi + a(\mu)\sqrt{\alpha}\,(2\tilde{\rho}) + (2\tilde{\rho})^{1/2}\cos 3\phi + \alpha(2\tilde{\rho})^{3/2}F_1,
$$

$$
G_2 = \sin\psi + b(\mu)\sqrt{\alpha}\,(2\tilde{\rho}) - (2\tilde{\rho})^{1/2}\sin 3\phi + \alpha(2\tilde{\rho})^{3/2}F_2.
$$

For $\alpha \neq 0$, (3.10) is equivalent to

$$
\begin{cases}
G_1 = 0, \\
G_2 = 0.
\end{cases}
$$

$$(3.11)$$

For $\alpha = 0$, (3.11) becomes

$$
\begin{cases}
\cos\psi + (2\tilde{\rho})^{1/2}\cos 3\phi = 0, \\
\sin\psi - (2\tilde{\rho})^{1/2}\sin 3\phi = 0.
\end{cases}
$$

Hence it is equivalent to

$$\bar{\rho} = \tfrac{1}{2}, \quad \sin 3\phi = \sin \psi, \quad \cos 3\phi = -\cos \psi.$$

In other words (3.11) has solutions (for $\alpha = 0$)

$$\begin{cases} \bar{\rho} = \tfrac{1}{2}, \\ \phi_k = \tfrac{1}{3}(\pi + 2k\pi - \psi), \quad k = 0,1,2. \end{cases}$$

Since $\partial(G_1, G_2)/\partial(\bar{\rho}, \phi)|_{\alpha=0} = -3$, the Implicit Function Theorem implies that for $\alpha \neq 0$ and $|\alpha|$ small, (3.9) has solutions

$$\begin{cases} \rho = \rho^*(\alpha, \psi), \\ \phi_k = \phi_k^*(\alpha, \psi), \quad k = 0,1,2. \end{cases}$$

Let $z^* = \sqrt{2\rho^*}\, e^{i\phi_k^*}$ be one of the equilibria. We have for small $|\epsilon|$

$$2\rho^* = 2\alpha\bar{\rho} = |\epsilon|^2 2\bar{\rho} = |\epsilon|^2 + o(|\epsilon|^2),$$

$$|z^*| = \sqrt{2\rho^*} = |\epsilon| + o(|\epsilon|).$$

Let $\zeta = z - z^*$. Then (3.6) becomes

$$\dot{\zeta} = P\zeta + Q\bar{\zeta} + O(|\epsilon|^2),$$

where

$$P = \epsilon + 2A(\epsilon)|z^*|^2 + O(|z^*|^4),$$

$$Q = 2z^* + A(\epsilon)z^{*2} + O(|z^*|^4).$$

For sufficiently small $|\epsilon|$, we have

$$|P| \leq |\epsilon| + o(|\epsilon|) < \tfrac{5}{4}|\epsilon|,$$

$$|Q| \geq 2|\epsilon| - o(|\epsilon|) > \tfrac{7}{4}|\epsilon|.$$

The following lemma is useful in determining the property of an equilibrium.

**Lemma 3.2.** (*Arnold* [4]) *Consider a planar linear equation*

$$\dot\zeta = P\zeta + Q\bar\zeta, \qquad P, Q, \zeta \in \mathbb{C}.$$

*If* $|P| < |Q|$, *then the origin is a saddle point. If* $|P| > |Q|$, *then the origin is a focus* (*center*) *or node which is stable for* Re $P < 0$ *and unstable for* Re $P > 0$.

*Proof.* Let $\zeta = \zeta_1 + i\zeta_2$, $P = p_1 + ip_2$, and $Q = q_1 + iq_2$. Then

$$\begin{bmatrix} \dot\zeta_1 \\ \dot\zeta_2 \end{bmatrix} = M \begin{bmatrix} \zeta_1 \\ \zeta_2 \end{bmatrix}, \qquad M = \begin{bmatrix} p_1 + q_1 & -p_2 + q_2 \\ p_2 + q_2 & p_1 - q_1 \end{bmatrix}.$$

Hence as det $M = |P|^2 - |Q|^2$ and tr $M = 2p_1$, the desired result follows.   □

**Lemma 3.3.** *Suppose that* $|\epsilon| \neq 0$. *Then in a small neighborhood of the origin in phase space* (3.8) *has four equilibria.* $\rho = 0$ *is a focus or node which is stable for* $\epsilon_1 \leq 0$ *and unstable for* $\epsilon_1 > 0$. *The other three equilibria are saddle points. The curve* $H^\pm = \{\epsilon | \epsilon_1 = 0, \epsilon_2 \neq 0\}$ *is a Hopf bifurcation curve.*

*Proof.* The behavior of the equilibrium $\rho = 0$ and the curve $H^\pm$ is easy to obtain from (3.8).

The behavior of $(\rho^*, \phi_k^*)$ is obtained from Lemma 3.2 and from the fact that the saddle is structurally stable.   □

Now we turn to the discussion of the existence and the number of limit cycles. The basic idea is the same as in Sections 4.1 and 4.2, but we will use a specific technique.

Let

$$\epsilon_1 = -\frac{1}{a}\delta^2\beta, \qquad \epsilon_2 = \frac{1}{a}\delta, \qquad \rho \to \frac{\delta^2}{a^2}\rho,$$

$$b \to -ab, \qquad t \to -\frac{at}{\delta}, \tag{3.12}$$

where $a = a(\epsilon) < 0$ and $b = b(\epsilon)$. Then (3.8) becomes

$$
\begin{cases}
\dot\rho = \delta\beta(2\rho) - \delta(2\rho)^2 + (2\rho)^{3/2} \\
\qquad \cos 3\phi + \delta^2(2\rho)^{5/2}\overline{F}_1, \\
\dot\phi = 1 + b\delta(2\rho) - (2\rho)^{1/2} \\
\qquad \sin 3\phi + \delta^2(2\rho)^{5/2}\overline{F}_2.
\end{cases}
\tag{3.13}
$$

Suppose that $\rho_0(\delta, \beta)$, $\phi_0(\delta, \beta)$ are coordinates of a nonzero equilibrium of (3.13). Let

$$
\begin{cases}
r = \dfrac{\rho}{2\rho_0}, \\[2mm]
\theta = \dfrac{\pi}{6} + \phi - \phi_0 = \phi - \psi(\beta, \delta).
\end{cases}
\tag{3.14}
$$

Then (3.13) takes the form

$$
\begin{cases}
\dot r = \delta\beta(2r) - \delta(2\rho_0)(2r)^2 \\
\quad + (2\rho_0)^{1/2}(2r)^{3/2}\cos 3(\theta + \psi) + \delta^2(2r)^{5/2}\overline{F}_1, \\
\dot\theta = 1 + \delta(2\rho_0)b(2r) - (2\rho_0)^{1/2}(2r)^{1/2}\sin 3(\theta + \psi) \\
\quad + \delta^2(2r)^{3/2}\overline{F}_2 = \tilde{H}(\delta, r, \theta).
\end{cases}
\tag{3.15}
$$

The coordinates of the equilibria of (3.15) are independent of $\delta$ and $\beta$: $r = 0$ and $r_k = \frac{1}{2}$, $\theta_k = \pi/6 + 2k\pi/3$ ($k = 0, 1, 2$).
For $\delta = 0$, (3.15) is a Hamiltonian system

$$
\begin{cases}
\dot r = (2r)^{3/2}\cos 3\theta, \\
\dot\theta = 1 - (2r)^{1/2}\sin 3\theta,
\end{cases}
\tag{3.16}
$$

with the first integral

$$
H = r - \tfrac{1}{3}(2r)^{3/2}\sin 3\theta = h.
\tag{3.17}
$$

The level curves of $H = h$ are shown in Figure 3.2, where $0 \le h \le \frac{1}{6}$. $h = 0$ corresponds to the equilibrium $r = 0$, and $h = 1/6$ corresponds to the three heteroclinic orbits.

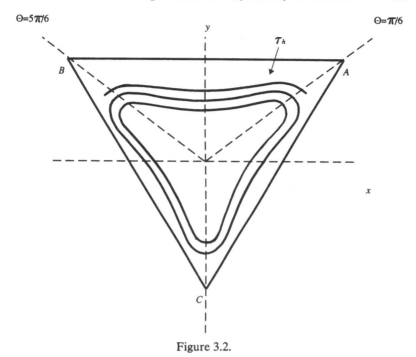

Figure 3.2.

Obviously, any closed orbit of (3.15) must cross the segment $L = \{(r, \theta)|\theta = \pi/6, 0 \le r \le \frac{1}{2}\}$ which could be parameterized by $h$, $H(r, \theta) = h$, $0 \le h \le \frac{1}{6}$.

Let

$$H^\delta(r, \theta) = \int_0^r \tilde{H}(\delta, r, \theta) \, dr \qquad (H^0 \equiv H), \qquad (3.18)$$

where $\tilde{H}$ is defined in (3.15). Then (3.15) can be rewritten in the form

$$\begin{cases} \dot{r} = -\dfrac{\partial H^\delta}{\partial \theta} + 2\delta r \big(\beta - (2\rho_0)(2r) + \delta(2r)^{3/2}\overline{F}\big), \\[2mm] \dot{\theta} = \dfrac{\partial H^\delta}{\partial r}. \end{cases} \qquad (3.19)$$

Let $\gamma(h, \delta, \beta)$ be the part of the orbit of (3.15) from the point on $L$ with $h \in (0, 1/6)$, to the point on the segment $\{(r, \theta)|\theta = 5\pi/6, 0 \le r \le 1/2\}$.

**Lemma 3.4.** *The orbit through* $\gamma = \gamma(h, \delta, \beta)$ *is a closed orbit of (3.15) if and only if*

$$\int_{\gamma(h,\delta,\beta)} \left(\frac{dH^{\delta}}{dt}\right)\bigg|_{(3.19)} dt = 0. \tag{3.20}$$

*Proof.* This proof is similar to that of Lemma 1.4.                          □

The calculation shows that

$$\frac{dH^{\delta}}{dt}\bigg|_{(3.19)} dt = 2\delta r\left(\beta - (2\rho_0)2r + \delta(2r)^{3/2}\bar{F}\right) d\theta.$$

Let

$$\Phi(h, \delta, \beta) = \int_{\gamma(h,\delta,\beta)} r\left[\beta - (2\rho_0)2r + \delta(2r)^{3/2}\bar{F}\right] d\theta. \tag{3.21}$$

Then (3.20) is equivalent to $\Phi(\delta, h, \beta) = 0$ (for $\delta \neq 0$). For $\delta = 0$, we have $2\rho_0 = 1$ and

$$\Phi(h, 0, \beta) = \int_{\Gamma_h} r(\beta - 2r)\, d\theta = \beta I_1 - 2I_2, \tag{3.22}$$

where

$$I_i(h) = \int_{\Gamma_h} r^i\, d\theta \qquad (i = 1, 2),$$

and $\Gamma_h$ is the level curve $H = h$ $(\pi/6 \leq \theta \leq 5\pi/6)$ of (3.16). Obviously, $I_1(h) > 0$ for $h > 0$; and $\lim_{h \to 0} I_2(h)/I_1(h) = 0$. Let

$$P(h) = \frac{I_2(h)}{I_1(h)}, 0 < h \leq \tfrac{1}{6}, \quad \text{and} \quad P(0) = 0. \tag{3.23}$$

Then $\Phi(h, 0, \beta) = 0$ is equivalent to

$$\beta = 2P(h). \tag{3.24}$$

We will prove that $P'(h) > 0$ for $0 \le h \le \frac{1}{6}$. This will give the uniqueness of the periodic orbits of (3.15).

**Lemma 3.5.** $P(h)$ satisfies an equation $(0 < h < \frac{1}{6})$

$$9h(6h - 1)P'(h) = -12P^2 + (28h + 9 - \phi(h))P$$

$$+ 48h^2 - 18h + 6h\phi(h), \qquad (3.25)$$

where

$$\phi(h) = 6h^2(6h - 1)\frac{I_1''(h)}{I_1(h)}. \qquad (3.26)$$

*Proof.* From (3.17), along $\Gamma_h$

$$\frac{\partial r}{\partial h} = \frac{1}{1 - \sqrt{2r}\,\sin 3\theta} = \frac{2r}{3h - r} > 0 \qquad (0 < h < \frac{1}{6}). \quad (3.27)$$

Hence

$$I_k' = 2k\int_{\Gamma_k} \frac{r^k}{3h - r}\, d\theta, \qquad (3.28)$$

where

$$I_k = \int_{\Gamma_h} r^k\, d\theta, \qquad k = 1, 2, 3, \ldots. \qquad (3.29)$$

From (3.29) and (3.28) we obtain

$$I_k = \int_{\Gamma_h} \frac{r^k(3h - r)}{3h - r}\, d\theta = \frac{3h}{2k}I_k' - \frac{1}{2(k + 1)}I_{k+1}'.$$

In particular

$$\begin{cases} I_1 = \frac{3}{2}hI_1' - \frac{1}{4}I_2', \\ I_2 = \frac{3}{4}hI_2' - \frac{1}{6}I_3'. \end{cases} \qquad (3.30)$$

On the other hand,

$$I_3(h) = \int_{\Gamma_h} r^3 \, d\theta = \frac{9}{8} \int_{\Gamma_h} \frac{(r-h)^2}{\sin^2 3\theta} \, d\theta = -\frac{3}{8} \int_{\partial = \pi/6}^{\theta = 5\pi/6} (r-h)^2 \, d(\cot 3\theta)$$

$$= -\frac{3}{8}(r-h)^2 \cot 3\theta \Big|_{\theta = \pi/6}^{\theta = 5\pi/6} + \frac{3}{4} \int_{\Gamma_h} (r-h)\cot 3\theta \, dr.$$

Using (3.16), (3.17), and (3.28), we have

$$I_3 = \frac{3}{4} \int_{\Gamma_h} \frac{(r-h)}{\sin 3\theta} \frac{(2r)^{3/2} \cos^2 3\theta}{(1 - \sqrt{2r} \sin 3\theta)} \, d\theta$$

$$= \frac{1}{2} \int_{\Gamma_h} \frac{r\left[(2r)^3 - 9(r-h)^2\right]}{3h - r} \, d\theta$$

$$= -4I_3 + \left(\frac{9}{2} - 12h\right)I_2 + \left(\frac{9}{2}h - 36h^2\right)I_1 + (54h^3 - 9h^2)I_1'.$$

Hence

$$I_3 = \left(\frac{9}{10} - \frac{12}{5}h\right)I_2 + \left(\frac{9}{10}h - \frac{36}{5}h^2\right)I_1 + \left(\frac{54}{5}h^3 - \frac{9}{5}h^2\right)I_1'.$$

$$(3.31)$$

Substituting (3.31) into (3.30), we have

$$\begin{cases} 4I_1 = 6hI_1' - I_2', \\ 48I_2 = (18h - 48h^2)I_1' + (-9 + 44h)I_2' - 24h^2(6h - 1)I_1''. \end{cases}$$

$$(3.32)$$

Equation (3.22) is equivalent to

$$\begin{cases} 9h(6h - 1)I_1' = (-9 + 44h)I_1 + 12I_2 + 6h^2(6h - 1)I_1'', \\ 9h(6h - 1)I_2' = (-18h + 48h^2)I_1 + 72I_2 + 36h^3(6h - 1)I_1''. \end{cases}$$

$$(3.33)$$

where $0 < h < \frac{1}{6}$. Equation (3.33) and $P'(h) = (I_2'I_1 - I_1'I_2)/I_1^2$ imply (3.25)     □

**Lemma 3.6.** $\lim_{h \to 0} P'(h) = 1$.

*Proof.* Equation (3.27) shows that $\partial r/\partial h \to 1$ as $h \to 0$. Hence $r = O(h)$, $3h - r = O(h)$, and $r - h = O(h^{3/2})$ (see (3.17)). Therefore, as $h \to 0$, $I_1 = O(h)$, $I_2 = O(h^2)$, $P(h) = O(h)$, and

$$I_1'' = 6 \int_{\Gamma_h} \frac{r(r - h)}{(3h - r)^3} \, d\theta = O(h^{-1/2}),$$

$$I_1''' = 12 \int_{\Gamma_h} \frac{r(r - h)(5r - 6h)}{(3h - r)^5} \, d\theta = O(h^{-3/2}).$$

These imply that $\phi(h) = O(h^{1/2})$ and $\phi'(h) = O(h^{-1/2})$ as $h \to 0$. Using the above estimations and L'Hospital's rule, from (3.25) we obtain $\lim_{h \to 0} P'(h) = 1$.     □

**Lemma 3.7.** $P(1/6) = 1/4$.

*Proof.* Let $x = \sqrt{2r} \cos \theta$ and $y = \sqrt{2r} \sin \theta$. Then $\Gamma_{1/6}$ is a line segment $\{(x, y)| y = 1/2, - \sqrt{3}/2 \le x \le \sqrt{3}/2\}$, that is, $\sqrt{2r} \sin \theta = 1/2$, $\pi/6 \le \theta \le 5\pi/6$. Hence

$$I_1(1/6) = \int_{\Gamma_{1/6}} r \, d\theta = \frac{1}{8} \int_{\pi/6}^{5\pi/6} \frac{d\theta}{\sin^2 \theta} = \frac{\sqrt{3}}{4},$$

$$I_2(1/6) = \int_{\Gamma_{1/6}} r^2 \, d\theta = \frac{1}{8^2} \int_{\frac{\pi}{6}}^{\frac{5\pi}{6}} \frac{d\theta}{\sin^4 \theta} = \frac{\sqrt{3}}{16},$$

Therefore,

$$P(1/6) = \frac{I_2(1/6)}{I_1(1/6)} = \frac{1}{4}$$     □

**Lemma 3.8.** $P'(h) > 0$ *for* $0 < h < 1/6$.

*Proof.* We shall prove that if there exists $h_0 \in (0, \frac{1}{6})$ such that $P'(h_0)$ $= 0$, then $P''(h_0) > 0$, which is impossible, since $P(0) = 0$, $P(h) > 0$ for $h > 0$, and $P'(0) = 1$ (Lemma 3.6).

Let $P'(h_0) = 0$ and

$$Q(h) = \frac{I_2'(h)}{I_1'(h)}, \qquad 0 < h < 1/6$$

Then $P(h_0) = Q(h_0)$ and

$$P''(h_0) = \frac{I_1'(h_0)}{I_1(h_0)} Q'(h_0). \tag{3.34}$$

From the first equation of (3.32) we have

$$6hI_1'' = I_2' - 2I_1'.$$

This implies that

$$\frac{I_1''}{I_1'} = \frac{1}{6h} \left( \frac{I_2''}{I_2'} Q(h) - 2 \right).$$

Hence

$$Q'(h) = \frac{1}{I_1'^2} (I_2'' I_1' - I_2' I_1'') = Q \left( \frac{I_2''}{I_2'} - \frac{I_1''}{I_1'} \right)$$

$$= \frac{1}{6h} Q \left[ (6h - Q) \frac{I_2''}{I_2'} + 2 \right].$$

The first equation of (3.32) implies $6h - Q = 4I_1/I_1'$. Thus when $0 < h < 1/6$

$$Q' = \frac{Q}{3h} \left( \frac{2I_1 I_2''}{I_1' I_2'} + 1 \right).$$

Since $P'(h_0) = 0$ $(0 < h_0 < 1/6)$, $Q(h_0) = P(h_0) > 0$ and

$$\frac{I_1(h_0)}{I_1'(h_0)} = \frac{I_2(h_0)}{I_2'(h_0)}.$$

It is easy to see that

$$I_2(h_0) = \int_{\Gamma_{h_0}} r^2 \, d\theta > 0,$$

$$I_2'(h_0) = \int_{\Gamma_{h_0}} \frac{4r^2}{3h_0 - r} \, d\theta > 0,$$

$$I_2''(h_0) = \int_{\Gamma_{h_0}} \frac{4r^2(3h_0 + r)}{(3h_0 - r)^3} \, d\theta > 0.$$

We obtain finally

$$Q'(h_0) = \frac{Q(h_0)}{3h_0} \left( \frac{2I_2(h_0) I_2''(h_0)}{I_2'^2(h_0)} + 1 \right) > 0.$$

By using (3.34) we have $P''(h_0) > 0$. This finishes the proof of Lemma 3.8. $\qquad\square$

**Lemma 3.9.** *Suppose that* Re $A(0) < 0$. *Then system (3.6) has the bifurcation diagram and phase portraits as shown in Figure 3.1, which are topologically independent of the function $A(\epsilon)$ and the higher-order terms $O(|z|^4)$, up to a factor in the equation of the curve HL.*

*Proof.* Equation (3.6) is transformed into (3.15) and the closed orbit of (3.15) is determined by the zero point of (3.21) (Lemma 3.4). Similarly to Theorem 1.2 and Lemma 2.8, we can consider the equation (3.22) instead of (3.21). Lemmas 3.6–3.8 show that for every $\beta_0$, if $0 < \beta_0 < 1/2$ (i.e., $2P(0) < \beta_0 < 2P(1/6)$), then (3.22) has exactly one zero point with respect to $h(0 < h < 1/6)$; if $\beta_0 < 0$ or $\beta_0 > 1/2$, then (3.22) has no zero point.

From (3.12) we have

$$\frac{\epsilon_1}{\epsilon_2^2} = -a\beta = -a\beta_0(1 + O(\delta)) = -a\beta_0(1 + O(\epsilon_2)),$$

where $a = \operatorname{Re} A(0) < 0$. Hence the equation of $HL$ is

$$\epsilon_1 = -\frac{a}{2}\epsilon_2^2 + O(|\epsilon_2|^3).$$

We note that all the above discussions are independent of the higher-order terms $O(|z|^4)$ and the behavior of the function $A(\epsilon)$ as long as $\operatorname{Re} A(0) \neq 0$. $\qquad\square$

### 4.4 Double Zero Eigenvalue with Symmetry of Order 4

In this section we consider the family of vector fields on the plane that are invariant under a rotation through an angle $2\pi/4$. The normal form equation is (see Section 2.11)

$$\dot{z} = \epsilon z + Mz|z|^2 + \bar{z}^3, \qquad z \in \mathbb{C}, \tag{4.1}$$

where $z = x + iy$, $M = a + ib$, and $\epsilon = \epsilon_1 + i\epsilon_2$.

Since the two nonlinear terms in (4.1) are both of order 3, the discussion of unfoldings of (4.1) is more complicated. The problem of versal unfolding has not been solved completely. Some results in this section have been obtained or discussed by Arnold [4], Wan [1], Neishtadt [1], Berezovskaia and Knibnik [1], and Wang [1].

Without loss of generality, we assume in (4.1) that $a \leq 0$ and $b \leq 0$. In fact, if $a > 0$, by a change of variables and parameter $\epsilon$:

$$z \to ze^{-\pi i/4}, \qquad t \to -t, \qquad \epsilon \to -\epsilon,$$

equation (4.1) becomes

$$\dot{z} = \epsilon z - \bar{M}z|z|^2 + \bar{z}^3.$$

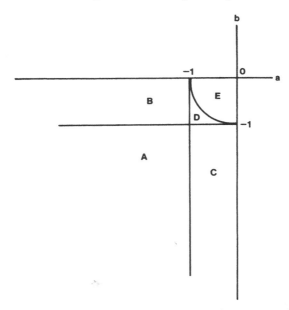

Figure 4.1. The partition of the $ab$-plane ($a < 0, b < 0$).

If $b > 0$, let $v = \bar{z}$. Then equation (4.1) becomes

$$\dot{v} = \epsilon v + \bar{M}v|v|^2 + \bar{v}^3,$$

which keeps the sign of Re $M$ and changes the sign of Im $M$.

First, we present some theorems and give a conjecture; then we introduce the proofs of the theorems.

**Theorem 4.1.** *The third quadrant of the ab-plane is divided into the following regions (see Figure 4.1):*

*Region A:* $\{(a, b)|a < -1, b < -1\}$,
*Region B:* $\{(a, b)|a < -1, -1 < b < 0\}$,
*Region C:* $\{(a, b)| -1 < a < 0, b < -1\}$,
*Region D:* $\{(a, b)| -1 < a < 0, -1 < b < 0, a^2 + b^2 > 1\}$,
*Region E:* $\{(a, b)|a < 0, b < 0, a^2 + b^2 < 1\}$.

*If $(a, b) \in$ region E, then equation (4.1), for any $\epsilon$ (except zero), has four nonzero equilibria which are saddle points. If $(a, b) \in$ one of the regions A–D, then there exists two semi-straight lines $l_1$ and $l_2$ which have*

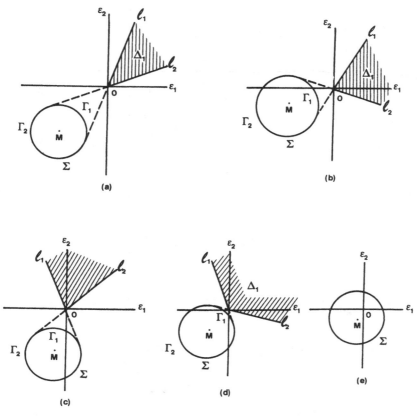

Figure 4.2. (a) $(a, b) \in$ region $A$. (b) $(a, b) \in$ region $B$. (c) $(a, b) \in$ region $C$. (d) $(a, b) \in$ region $D$. (e) $(a, b) \in$ region $E$.

*a common end at the origin in the $\epsilon$-plane, and divide the $\epsilon$-plane into two open angular regions $\Delta_1$ and $\Delta_2$ in a small neighborhood of the origin ($\Delta_1$ has a smaller angle than $\Delta_2$), such that when $\epsilon \in \Delta_1$, equation (4.1) has eight nonzero equilibria (four saddle points and four nodes or foci); when $\epsilon \in \Delta_2$, (4.1) has no nonzero equilibrium; and when $\epsilon \in l_1 \cup l_2$ (4.1) has four saddle-nodes. The positions of $l_1$ and $l_2$ are shown in Figure 4.2.*

**Theorem 4.2.** *If $\epsilon_1 \le 0$, then equation (4.1) has no periodic orbit. For any fixed $\epsilon_2 \ne 0$, as $\epsilon_1$ changes its sign from negative to positive, the Hopf bifurcation occurs at the origin.*

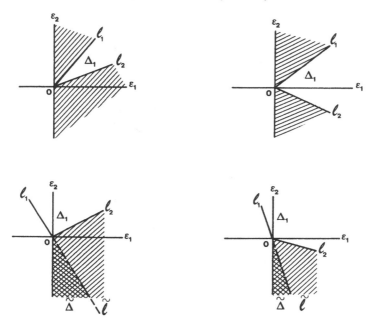

Figure 4.3. The regions of uniqueness of limit cycle in Theorem 4.3.

**Theorem 4.3.** *Equation (4.1) has a unique limit cycle which surrounds the origin and is stable, if one set of the following conditions is satisfied:*

$$\epsilon_1 > 0, \, a^2 + b^2 > 1, \quad \text{and} \quad |b\epsilon_1 - a\epsilon_2| > \sqrt{\epsilon_1^2 + \epsilon_2^2}, \quad (4.2)$$

*or*

$$\epsilon_1 > 0, \, a^2 + b^2 > 1, \, -1 < a < 0, \quad \text{and} \quad \epsilon_2 \leq -\frac{ab + \sqrt{a^2 + b^2 - 1}}{1 - a^2} \epsilon_1.$$

$$(4.3)$$

**Remark 4.4.** *The shaded regions in Figure 4.3 are covered by conditions (4.2) and (4.3), where the curves $l_1$ and $l_2$ are the same as in Figure 4.2. For more details, see the proof of Lemma 4.12.*

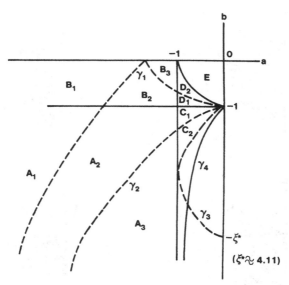

Figure 4.4.

**Theorem 4.5.** *If $a < -1$ (i.e., $(a, b) \in$ region $A \cup B$), then equation (4.1) has at most one limit cycle. If the limit cycle exists, then it is stable.*

**Theorem 4.6.** *For equation (4.1), the Hopf bifurcation occurs at the nonzero foci if and only if $(a, b) \in$ region $C$ and the point $(a, b)$ is below the curve $\gamma_4$ that is given by*

$$\gamma_4: \left\{ (a, b) \middle| b = -\frac{1 + a^2}{\sqrt{1 - a^2}}, \, -1 < a < 0 \right\}.$$

*Moreover, the bifurcating limit cycle is unstable.*

**Theorem 4.7.** *In the ab-plane there are curves $\gamma_1$, $\gamma_2$, and $\gamma_3$ which divide the regions $A$–$D$ into some subregions (Figure 4.4). The asymptotic behavior of these curves are:*

$\gamma_1: b \approx -1 + 0.47a^2$ *as* $b \to -1$, $\quad b \approx -0.35a^2$ *as* $b \to -\infty$;

$\gamma_2: b \approx -1 - 0.13a^2$ *as* $b \to -1$, $\quad b \approx -0.352a^2$ *as* $b \to -\infty$;

$\gamma_3: b \approx -1 - 0.45a^2$ *as* $b \to -1$, $\quad b \approx -4.11 + 0.84a^2$ *as* $b \to -4.11$.

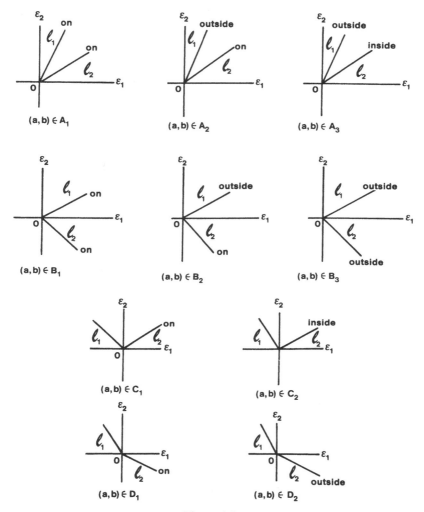

Figure 4.5.

The curve $\gamma_1$(correspondingly $\gamma_2$) is the boundary between regions in which saddle-nodes of equation (4.1) appear in different locations on the phase plane: outside (correspondingly inside) the central cycle and on this central cycle. For more details, see Figure 4.5, where $l_1$ and $l_2$ are the same as in Theorem 4.1. The notation "on" ("outside" or "inside") related to the curve $l_1$ (or $l_2$) means that the saddle-nodes appear on (outside or inside) the central cycle when $(\epsilon_1, \epsilon_2) \in l_1$ (or $l_2$). The curve

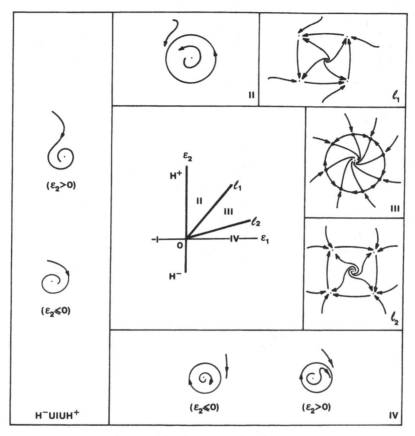

Figure 4.6. The case of $(a, b) \in A_1$.

$\gamma_3$ *is a boundary between regions in which the stabilities of the heteroclinic loop are different.*

Now we can state the main results of this section.

**Theorem 4.8.** *If* $(a, b) \in$ *subregion* $A_i$ *or* $B_j$ $(i, j = 1, 2, 3)$, *then the bifurcation diagram and phase portraits of equation (4.1) are shown in Figures 4.6–4.11, respectively. They are similar to the case of weak resonances (see Section 4.5), but the nonzero equilibria could appear outside, on, or inside the invariant cycle.*

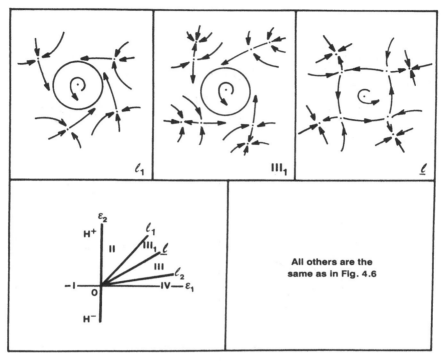

Figure 4.7. The case of $(a, b) \in A_2$.

**Conjecture 4.9.** *If $(a, b) \in$ region $D_i$ $(i = 1, 2)$, or region $E$, then the bifurcation diagram and the phase portraits of equation (4.1) are shown in Figure 4.12, Figure 4.13, and Figure 4.14,, respectively. The last case is similar to the case of $1:3$ resonance (see Section 4.3).*

**Theorem 4.10.** *If $(a, b) \in$ region $C$, and $|a|$ is sufficiently small, then the bifurcation diagram and phase portraits of equation (4.1) are shown in Figure 4.15 and Figure 4.16 for $b \leq -\xi^*$ and $-\xi^* < b < -1$, respectively, where*

$$\xi^* = (3 + \cos \theta^*)/(1 - \cos \theta^*),$$

*and $\theta^*$ is the unique root of the equation*

$$tg\theta - \theta = \pi \quad for \ \theta \in \left(0, \frac{\pi}{2}\right);$$

$\theta^* \approx 1.352$ *and* $\xi^* \approx 4.11$.

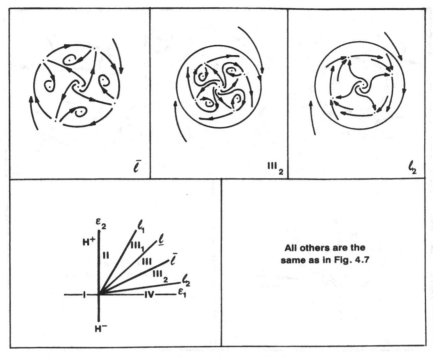

Figure 4.8. The case of $(a, b) \in A_3$.

The aforementioned theorems show that the result for $(a, b) \in A \cup B$ is complete, for $(a, b) \in D \cup E$ is almost complete (except the uniqueness of the central limit cycle), and for $(a, b) \in C$ is far from complete. In fact we need a condition $|a| \ll 1$ in Theorem 4.10 in order to transform equation (4.1) into an equation which is near a Hamiltonian system.

*Proof of Theorem 4.1.* Let $z = re^{i\theta}$ be a nonzero equilibrium of equation (4.1). Then

$$\frac{-\epsilon}{r^2} = M + N, \qquad (4.4)$$

where $N = e^{-4i\theta}$. The point $M + N$ lies on the circle $\Sigma$ centered at $M$ with radius 1 (see Figure 4.2).

If $|M| = a^2 + b^2 < 1$, then the origin lies inside $\Sigma$ (see Figure 4.2(e)). To solve equation (4.4) for $r$ and $\theta$, we must choose $\theta$ so that $\epsilon$ and $M + N \in \Sigma$ have opposite directions. This is always possible. Hence for $\epsilon \neq 0$, equation (4.1) has four nonzero equilibria.

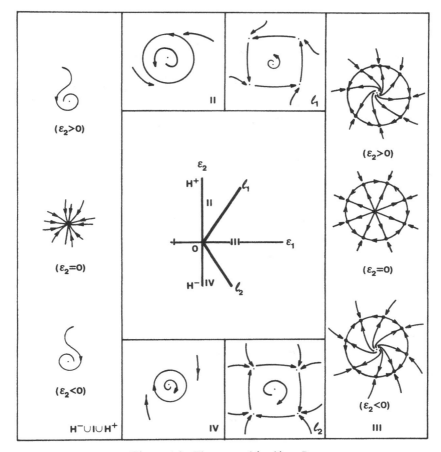

Figure 4.9. The case of $(a, b) \in B_1$.

If $|M| = a^2 + b^2 > 1$, then equation (4.4) is solvable with respect to $(r, \theta)$ if and only if $\epsilon \in l_1 \cup \Delta_1 \cup l_2$, where $\Delta_1$ is an open angular region formed by the semi-lines $l_1$ and $l_2$ which are opposite extensions of the tangent lines from the origin to $\Sigma$ (see Figure 4.2(a)–(d)). Hence, for each $\epsilon \in \Delta_1$, there are two points on the circle $\Sigma$ with directions opposite to $\epsilon$. This means that equation (4.1) has eight nonzero equilibria. For $\epsilon \in l_1 \cup l_2$, equation (4.1) has obviously four nonzero equilibria.

Now let us check the types of the equilibria. Suppose that $z_0 = re^{i\theta}$ is a nonzero equilibrium. In order to use Lemma 3.2, we linearize the equation (4.1) at the point $z_0$. Substituting $z = z_0 + \xi$ into equation (4.1) and retaining the first-order terms in $\xi, \bar{\xi}$ on the right-hand side,

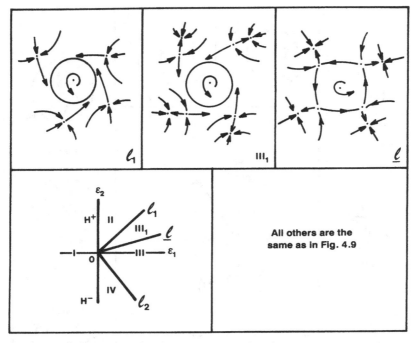

Figure 4.10. The case of $(a, b) \in B_2$.

we obtain

$$\dot{\xi} = P\xi + Q\bar{\xi}, \tag{4.5}$$

where

$$P = \epsilon + 2z_0\bar{z}_0M = R^2(M - N) \ (\text{using } (4.4)),$$

$$Q = Mz_0^2 + 3\bar{z}_0^2 = r^2e^{2i\theta}(M + 3N).$$

By Lemma 3.2, the saddle points appear if $|M - N| < |M + 3N|$. We consider the points $M - N$ and $M + 3N$. These points are symmetric with respect to the point $M + N$, and the straight line connecting them goes through the point $M$ (see Figure 4.17). Let $l$ be the tangent line to circle $\Sigma$ at the point $M + N$. If the origin and the point $M - N$ lie on the same side of the tangent $l$, then $|M - N| < |M + 3N|$. Otherwise, $|M - N| > |M + 3N|$. It is clear now that if $|M| < 1$, then $|M - N| < |M + 3N|$ (see Figure 4.17(a)) and equilibria are saddle points. If

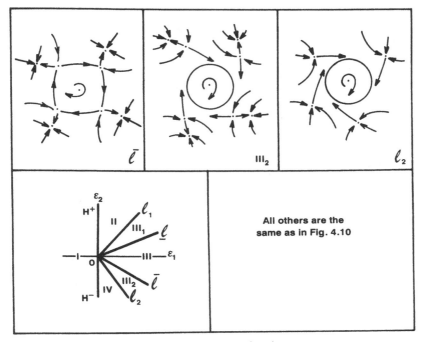

Figure 4.11. The case of $(a, b) \in B_3$.

$|M| > 1$, there are two cases. Let $\Gamma_1$ and $\Gamma_2$ be the two arcs bounded by the tangents to $\Sigma$ from the origin, with the arc $\Gamma_1$ closer to the origin. If $M + N \in \Gamma_2$, then $|M - N| < |M + 3N|$ (see Figure 4.17(b)) and the corresponding four equilibria are saddle points, and if $M + N \in \Gamma_1$, then $|M - N| > |M + 3N|$ and the corresponding four equilibria are foci or nodes.                                                       $\square$

*Proof of Theorem 4.2.* In $(x, y)$ coordinates, equation (4.1) takes the following form:

$$\begin{cases} \dot{x} = \epsilon_1 x - \epsilon_2 y + (x^2 + y^2)(ax - by) + x^3 - 3xy^2 = f(x, y), \\ \dot{y} = \epsilon_2 x + \epsilon_1 y + (x^2 + y^2)(bx + ay) - 3x^2 y + y^3 = g(x, y). \end{cases}$$

$$(4.6)$$

It is easy to see that the divergence of the right-hand side of equation

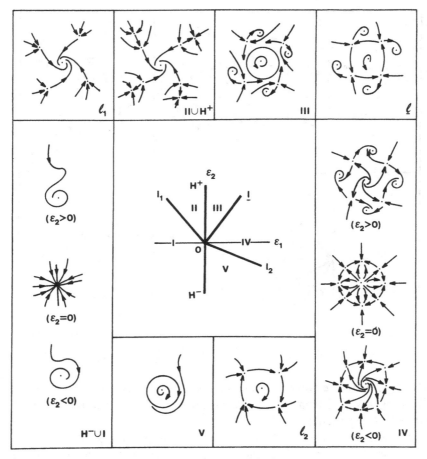

Figure 4.12. The case of $(a, b) \in D_1$.

(4.6) is

$$\mathrm{div}(f, g) = 2\big(\epsilon_1 + 2a(x^2 + y^2)\big). \tag{4.7}$$

Since $a < 0$, $\mathrm{div}(f, g)$ keeps a negative sign throughout $\mathbb{C} \setminus \{0\}$ for $\epsilon_1 \leq 0$. Thus, by a theorem of Bendixson, equation (4.6) has no periodic orbits in $\mathbb{C}$. Next, for any fixed $\epsilon_1 = \mu$ and $\epsilon_2 \neq 0$, it is easy to check that equation (4.6) satisfies the conditions of Theorem 3.2.4. Hence, a Hopf bifurcation occurs at $\epsilon_1 = 0$.      $\square$

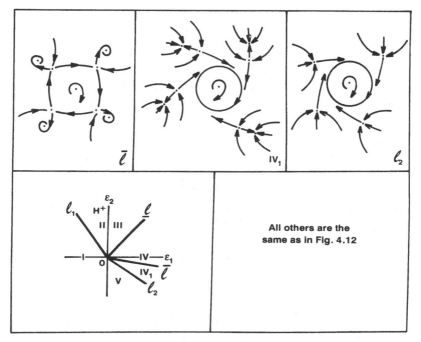

Figure 4.13. The case of $(a, b) \in D_2$.

In polar coordinates $(r, \theta)$, the equation (4.1) takes the following form:

$$\begin{cases} \dot{r} = \epsilon_1 r + r^3 \operatorname{Re}(M + N), \\ \dot{\theta} = \epsilon_2 + r^2 \operatorname{Im}(M + N), \end{cases} \tag{4.8}$$

where $M = a + ib$, $N = e^{-4i\theta}$. In order to prove Theorem 4.3, the following lemmas are needed.

**Lemma 4.11.** *Let*

$$H = |M|^2 \frac{z^2 \bar{z}^2}{2} - \frac{M}{2} \bar{z}^4 \quad \text{and} \quad E = \operatorname{Re} H,$$

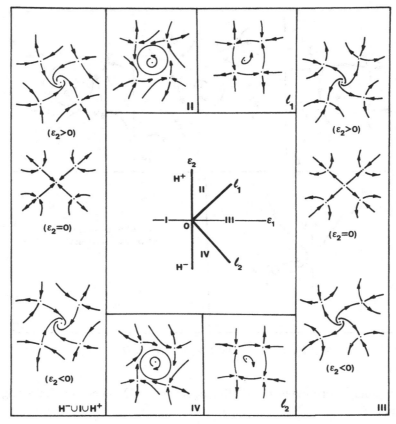

Figure 4.14. The case of $(a, b) \in E$.

where $M = a + ib$ and $z = re^{i\theta}$ satisfies (4.1). Then

$$\frac{d}{dt}(E) = 2\big[\epsilon_1 |M|^2 - \mathrm{Re}(\bar{\epsilon}MN)\big]r^4 + 2a(|M|^2 - 1)r^6. \quad (4.9)$$

*Proof.*

$$\frac{d}{dt}(E) = \mathrm{Re}\,\frac{dH}{dt} = \mathrm{Re}\left\{\frac{\partial H}{\partial z}\dot{z} + \frac{\partial H}{\partial \bar{z}}\dot{\bar{z}}\right\}$$

$$= \mathrm{Re}\left\{|M|^2\bar{z}^2z(\epsilon z + Mz^2\bar{z}^2 + \bar{z}^3)\right.$$

$$\left. + \big(|M|^2z^2\bar{z} - 2M\bar{z}^3\big) \cdot \big(\bar{\epsilon}\bar{z} + \bar{M}\bar{z}^2z + z^3\big)\right\}. \quad (4.10)$$

After simplifying the above expressions, we get the formula (4.9).  □

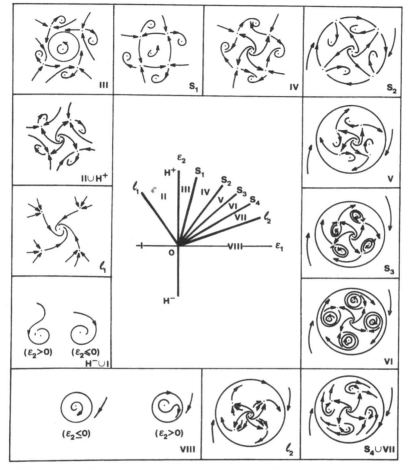

Figure 4.15. The case of $(a, b) \in C$ and $b < -\xi^*$ ($\xi^* \approx 4.11$), $|a| \ll 1$.

**Lemma 4.12.** *If conditions (4.2) or (4.3) is satisfied, then the origin is the only equilibrium point of equation (4.1).*

*Proof.* Direct calculations show that the condition (4.3) implies $(a, b) \in$ region $C$ or $D$ (see Theorem 4.1 and Figure 4.1) and $(\epsilon_1, \epsilon_2) \in \tilde{\Delta} \cup \tilde{l}$, where $\tilde{l}$ is the semi-line opposite to $l_1$ and $\tilde{\Delta}$ is the angular region formed by $\tilde{l}$ and the negative $\epsilon_2$-axis. The condition (4.2) implies

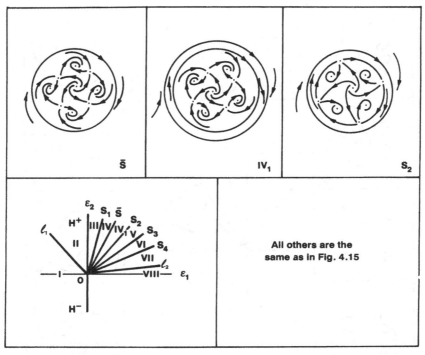

Figure 4.16. The case of $(a, b) \in C$ and $-\xi^* < b < -1$, $|a| \ll 1$.

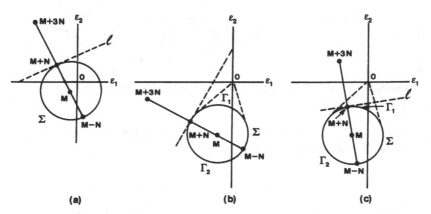

Figure 4.17. (a) The case of $(a, b) \in$ region $E$. (b), (c). The case of $(a, b) \in$ one of regions $A$, $B$, $C$, and $D$.

(i) $(a, b) \in$ region $A$ or $B$ and $(\epsilon_1, \epsilon_2) \in \{(\epsilon_1, \epsilon_2)|\epsilon_1 > 0\} \setminus \{l_1 \cup \Delta_1 \cup l_2\}$, or (ii) $(a, b) \in$ region $C$ or $D$ and $(\epsilon_1, \epsilon_2) \in \{(\epsilon_1, \epsilon_2)|\epsilon_1 > 0\} \setminus \{\Delta_1 \cup l_2 \cup \tilde{l} \cup \tilde{\Delta}\}$. Therefore, by Theorem 4.1, equation (4.1), under condition (4.2), or (4.3), has no nonzero equilibria.      □

**Lemma 4.13.** *Suppose that condition (4.2) is satisfied, and $(\rho(t), \theta(t))$ is a periodic orbit of equation (4.1). Then:*
(i)  *$\dot{\theta}(t) \neq 0$ at any $t$. Therefore, the periodic orbit takes the form $\rho(t) = r(\theta(t))$.*
(ii) *sgn $\dot{\theta}(t) =$ sgn $Im(\bar{e}R) =$ sgn $Im(\bar{e}M)$, where $R = M + N$.*
(iii) *If we set $y = r/\rho - 1$ and consider $y \in (-1, +\infty)$, then in $(y, \theta)$ coordinates the solutions of equation (4.1) satisfy the equation*

$$\dot{y} = -\frac{2\rho^2 \ Im(\bar{e}R)}{\epsilon_2 + \rho^2 \ Im \ R} y + \rho^2(3y^2 + y^3)\left(Re \ R - \frac{1}{\rho}\frac{d\rho}{d\theta} \ Im \ R\right). \quad (4.11)$$

*Proof.* (i) Let

$$S = \{(r, \theta)|\epsilon_2 + r^2 \ Im(M + N) = 0\}.$$

We need to show that $\{(\rho(t), \theta(t))\} \cap S = \varnothing$ (see (4.8)). If $\epsilon_2 = 0$, then $S$ consists of finitely many straight lines through the origin, and $S$ is an invariant set of equation (4.1). Obviously, $\{(\rho(t), \theta(t))\} \cap S = \varnothing$ in this case. If $\epsilon_2 \neq 0$, then $\{0\} \notin S$, and

$$S = \left\{(r, \theta)|r^2 = \frac{-\epsilon_2}{Im(M + N)}\right\}.$$

Each branch of $S$ is a curve $\Gamma$: $r = r(\theta)$ which bounds a closed unbounded region $U = \{re^{i\theta}|r \geq r(\theta)\}$. By Lemma 4.12, the origin is the only equilibrium of equation (4.1). Hence, $\dot{r} \neq 0$ along $\Gamma$. This means that the region $U$ is a positive or negative invariant set of equation (4.1). Therefore, the periodic orbit $\{(\rho(t), \theta(t))\}$ cannot meet the curve $\Gamma$, that is,

$$\{(\rho(t), \theta(t))\} \cap S = \varnothing.$$

(ii) From (4.1) we have

$$\bar{\epsilon}\frac{\dot{z}}{z} = \epsilon\bar{\epsilon} + r^2\bar{\epsilon}R, \tag{4.12}$$

where $R = M + N$. On the other hand, it is easy to see that

$$\frac{\dot{z}}{z} = \frac{\dot{r}}{r} + i\dot{\theta}. \tag{4.13}$$

Equations (4.12) and (4.13) imply

$$\bar{\epsilon}\frac{\dot{r}}{r} + \bar{\epsilon}i\dot{\theta} = \epsilon\bar{\epsilon} + r^2\bar{\epsilon}R. \tag{4.14}$$

Let $(r, \theta)$ in (4.14) be the periodic solution of equation (4.1). Integrating (4.14) over a period $T$ $(T > 0)$, and then taking the imaginary parts, we have

$$2\pi\epsilon_1 \operatorname{sgn}(\dot{\theta}(t)) = \int_0^T r^2 \operatorname{Im}(\bar{\epsilon}R)\, dt. \tag{4.15}$$

Noting $\bar{\epsilon}R = \bar{\epsilon}M + \bar{\epsilon}N$ and $|\operatorname{Im}(\bar{\epsilon}M)| > |\bar{\epsilon}|$ (condition (4.2)), we have $\operatorname{sgn}\operatorname{Im}(\bar{\epsilon}R) = \operatorname{sgn}\operatorname{Im}(\bar{\epsilon}M)$ which is independent of $t$. Therefore, from (4.15) we obtain

$$\operatorname{sgn}\dot{\theta}(t) = \operatorname{sgn}\operatorname{Im}(\bar{\epsilon}M) = \operatorname{sgn}\operatorname{Im}(\bar{\epsilon}R).$$

(iii) Since $y = r/\rho - 1$, $\dot{y} = (1/\rho^2)(\rho\dot{r} - r\dot{\rho}) = (1/\rho^2)(\rho\dot{r} - r\,d\rho/d\theta\dot{\theta})$. Using $r = \rho(1 + y)$, (4.8), and

$$\frac{d\rho}{d\theta} = \frac{\rho\epsilon_1 + \rho^3\operatorname{Re}R}{\epsilon_2 + \rho^2\operatorname{Im}R},$$

we have

$$\dot{y} = \frac{1}{\rho^2} \Bigg\{ \rho \Big[ \epsilon_1 \rho (1 + y) + \rho^3 (1 + y)^3 \operatorname{Re} R \Big]$$

$$-\rho(1 + y) \frac{\rho \epsilon_1 + \rho^3 \operatorname{Re} R}{\epsilon_2 + \rho^2 \operatorname{Im} R} \Big( \epsilon_2 + \rho^2 (1 + y)^2 \operatorname{Im} R \Big) \Bigg\}$$

$$= 2\rho^2 \Bigg( \operatorname{Re} R - \frac{\epsilon_1 + \rho^2 \operatorname{Re} R}{\epsilon_2 + \rho^2 \operatorname{Im} R} \operatorname{Im} R \Bigg) y$$

$$+ \rho^2 (3y^2 + y^3) \Bigg( \operatorname{Re} R - \frac{\epsilon_1 + \rho^2 \operatorname{Re} R}{\epsilon_2 + \rho^2 \operatorname{Im} R} \operatorname{Im} R \Bigg)$$

$$= -\frac{2\rho^2 \operatorname{Im}(\bar{\epsilon} R)}{\epsilon_2 + \rho^2 \operatorname{Im} R} y + \rho^2 (3y^2 + y^3) \Bigg( \operatorname{Re} R - \frac{1}{\rho} \frac{d\rho}{d\theta} \operatorname{Im} R \Bigg). \quad \square$$

*Proof of Theorem 4.3.* From (4.2) and (4.3), we have $|M| > 1$. Hence, the function $E > 0$ on $\mathbb{C} \setminus \{0\}$ and $E(z) \to \infty$ if and only if $|z| \to \infty$, where $E = \operatorname{Re} H$ is defined in Lemma 4.11. By this lemma, one can find an $e > 0$ such that $dE/dt(z) < 0$ on the set $\{z | E(z) \geq e\}$. Consequently, the set $\{z | E(z) \leq e\}$ is compact and positive invariant. Any solution $z(t)$ of equation (4.1) exists for all $t \geq 0$ and its $\omega$-limit set $\Omega \subset \{z | E(z) \leq e\}$. The origin is the only equilibrium of equation (4.1) (Lemma 4.12), and it is a source for $\epsilon_1 > 0$. Thus, by the Poincaré–Bendixson Theorem, the $\omega$-limit set of any nontrivial solution, which lies in the compact set $\{z | E(z) \leq e\}$, is a closed orbit.

First we consider the case of condition (4.2). On any closed orbit, by (i) and (ii) of Lemma 4.13, $\operatorname{Im}(\bar{\epsilon} R)/(\epsilon_2 + \rho^2 \operatorname{Im} R) \geq d > 0$ for some constant $d$. Hence the conclusion (iii) of Lemma 4.13 shows that any closed orbit is hyperbolic and stable. Hence it is unique.

Next, we consider the case of condition (4.3). By the same arguments used in the proof of Lemma 4.13(i) and the condition $\epsilon_2 < 0$ (see (4.3)), one can see that any closed orbit must be located in the region $V = \{re^{i\theta} | \dot{\theta} < 0\}$. Since the right-hand side of equation (4.1) is a polynomial with respect to $z$ and $\bar{z}$, and the only equilibrium, the origin, is a source, any closed orbit is isolated (a limit cycle). Suppose that $\Gamma_1 \subset \Gamma_2$ are two limit cycles in the region $V$, and they have the expressions

$$\Gamma_1: \begin{cases} \rho = \rho_1(t), \\ \theta = \theta_1(t) \end{cases}$$

and

$$\Gamma_2: \begin{cases} \rho = \rho_2(t), \\ \theta = \theta_2(t). \end{cases}$$

From (4.7) and (4.8) we have

$$\oint_{\Gamma_1} \operatorname{div}(f, g)\, dt - \oint_{\Gamma_2} \operatorname{div}(f, g)\, dt$$

$$= 2\int_0^{-2\pi} \frac{\epsilon_1 + 2a\rho_1^2}{\dot\theta_1}\, d\theta_1 - 2\int_0^{-2\pi} \frac{\epsilon_1 + 2a\rho_2^2}{\dot\theta_2}\, d\theta_2$$

$$= 2\int_0^{-2\pi} \frac{[\epsilon_1(b - \sin 4\theta) - 2a\epsilon_2](\rho_2^2 - \rho_1^2)}{\dot\theta_1\dot\theta_2}\, d\theta. \quad (4.16)$$

By condition (4.3), we have

$$\epsilon_1(b - \sin 4\theta) - 2a\epsilon_2 \le \left[ b - \sin 4\theta + 2a\frac{ab + \sqrt{a^2 + b^2 - 1}}{1 - a^2} \right]\epsilon_1.$$

$$(4.17)$$

If $b < -1$, then the right-hand side of (4.17) is obviously negative. If $-1 < b < 0$, then by using $a^2 + b^2 > 1$ and $-1 < a < 0$, we have

$$b + 2a\frac{ab}{1 - a^2} < b + \frac{2a^2b}{b^2} = \frac{2a^2 + b^2}{b} < \frac{a^2 + b^2}{b} < -1.$$

Hence, the right-hand side of (4.17) is also negative. Thus

$$\epsilon_1(b - \sin 4\theta) - 2a\epsilon_2 < 0. \quad (4.18)$$

Noting $\rho_2 > \rho_1$ and $\dot{\theta}_i < 0$, $i = 1, 2$, from (4.16) and (4.18) we have

$$\oint_{\Gamma_1} \text{div}(f, g)\, dt > \oint_{\Gamma_2} \text{div}(f, g)\, dt. \qquad (4.19)$$

Since the origin is a source, let $\Gamma_1$ be the limit cycle that is the closest to the origin. Then $\Gamma_1$ is stable inside. If $\Gamma_1$ is stable, then $\oint_{\Gamma_1} \text{div}(f, g)\, dt \leq 0$. Therefore, $\oint_{\Gamma_2} \text{div}(f, g)\, dt < 0$ (see (4.19)), which implies $\Gamma_2$ is stable. This is impossible. If $\Gamma_1$ is semistable, since the vector field is a rotated vector field with respect to $\epsilon_1$ (with fixed $\epsilon_2$), change $\epsilon_1$ to $\epsilon_1 + \delta$, where $\delta$ is sufficiently small with suitable sign. Then $\Gamma_1$ will become at least two limit cycles, and the inner most one is stable. For more detail, see Section 3 of Chapter 4 in Zhang et al. [1]. By the same argument as above, this is impossible. Therefore, the limit cycle is unique. $\qquad \square$

*Proof of Theorem 4.5.* Theorem 4.2 shows that if equation (4.1) has a limit cycle, then $\epsilon_1 > 0$. Let $z = re^{i\theta}$, and consider

$$r_* = \min_{r \neq 0} r|_{\dot{r}=0}.$$

From (4.8) it is easy to see that

$$r_*^2 = \min_{\theta} \frac{\epsilon_1}{-a - \cos 4\theta} > \frac{\epsilon_1}{-2a} \quad \text{for } a < -1.$$

This means that if $a < -1$, then inside the circle $r = \sqrt{\epsilon_1/(-2a)}$ one has $\dot{r} > 0$. Hence any limit cycle must surround the circle $r = \sqrt{\epsilon_1/(-2a)}$. By using (4.7), along any limit cycle $\Gamma$ we have

$$\text{div}(f, g) = 2(\epsilon_1 + 2ar^2) < 0,$$

which implies

$$\oint_{\Gamma} \text{div}(f, g)\, dt < 0.$$

Therefore, a limit cycle must be stable. Hence, it is unique. $\qquad \square$

*Proof of Theorem 4.6.* Theorem 4.1 shows that if $(a, b) \in$ one of the regions $A$, $B$, $C$ or $D$, and $\epsilon$ varies from $l_1$ into $\Delta_1$ (see Figure 4.2), then each of the four nonzero saddle-nodes of equation (4.1) becomes a saddle point and a node (or focus). Since the center manifold at a saddle node is one-dimensional, there is a connection orbit between a saddle point and an node. Hence, if the node (or focus) does not change its stability when $\epsilon$ varies in $\Delta_1$, then there is no closed orbit surrounding the node or focus.

Let us study the stability of the nonzero node or focus. We recall the analysis in the proof of Theorem 4.1 and keep the same notation. Suppose that $z_0 = re^{i\theta}$ is one of the nonzero nodes or focus. By linearization of equation (4.1) at the point $z_0$, we obtain equation (4.5). By Lemma 3.2, the stability of equilibrium $z_0$ is determined by the sign of Re $p = r^2 \operatorname{Re}(M - N)$ while $M + N \in \Gamma_1$. From Figure 4.17 and Figure 4.2 it is clear that if $(a, b) \in$ one of the regions $A$, $B$, or $D$, then $\operatorname{Re}(M - N) < 0$. Hence the equilibrium $z_0$ is always stable. Suppose that $(a, b) \in$ region $C$. A change of stability of the equilibrium $z_0$ takes place if the point $M - N$ crosses the imaginary axis as the point $M + N$ varies along $\Gamma_1$. The boundary separating the points $\{M\}$ for which such a phenomenon takes place is determined by the following condition: The diameter drawn through the point of intersection of the circle $\Sigma$ with the imaginary axis is perpendicular to a tangent from the origin. The calculation shows that the boundary curve $\gamma_4$ has the following form:

$$\gamma_4 = \left\{ (a, b) \Big| b = -\frac{1 + a^2}{\sqrt{1 - a^2}}, -1 < a < 0 \right\}. \tag{4.20}$$

Summing up the above discussion, we see that if $(a, b)$ is in the third quadrant and above the curve $\gamma_4$, then the nonzero nodes (or foci) are always stable. And there is no periodic orbit around them; if $(a, b)$ is below the curve $\gamma_4$, then there is a $\phi_0 = \tan^{-1}(\epsilon_2^0/\epsilon_1^0) \in (0, \pi/2)$ such that the nonzero focus (or node) is unstable for $\tan^{-1}(\epsilon_2/\epsilon_1) < \phi_0$ and stable for $\tan^{-1}(\epsilon_2/\epsilon_1) > \phi_0$ (see Figure 4.18).

We will determine the stability of the nonzero focus (or node) $z_0 = re^{i\theta}$ for $\tan^{-1}(\epsilon_2^0/\epsilon_1^0) = \phi_0$ by using formula (3.2.36). It is easy to calculate that $M + N = 2a + i(b + \sqrt{1 - a^2})$. Hence, from (4.8) and $\dot{r}|_{z_0} = \dot{\theta}|_{z_0} = 0$, we obtain

$$\frac{\epsilon_2^0}{\epsilon_1^0} = \tan \phi_0 = \frac{\operatorname{Im}(M + N)}{\operatorname{Re}(M + N)} = \frac{b + \sqrt{1 - z_0^2}}{2a}, \tag{4.21}$$

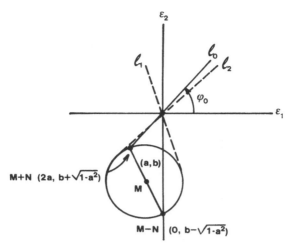

Figure 4.18.

and

$$|z_0|^2 = r^2 = -\frac{\epsilon_1^0}{\mathrm{Re}(M+N)} = -\frac{1}{2a}\epsilon_1^0, \qquad (4.22)$$

where $\epsilon_1^0 > 0$, $\epsilon_2^0 > 0$.

Let $z = z_0 w$. Then equation (4.1) becomes

$$\dot{w} = \epsilon(w - \overline{w}^3) + r^2 M(|w|^2 w - \overline{w}^3), \qquad (4.23)$$

and $w = 1$ is a focus (or node) of equation (4.23).

Let $w = x + iy$. Then equation (4.23) takes the following form:

$$\begin{cases} \dot{x} = \epsilon_1 x - \epsilon_2 y - \epsilon_1 x^3 - (3\epsilon_2 + 4br^2)x^2 y \\ \qquad + (3\epsilon_1 + 4ar^2)xy^2 + \epsilon_2 y^3, \\ \dot{y} = \epsilon_2 x + \epsilon_1 y - \epsilon_2 x^3 + (3\epsilon_1 + 4ar^2)x^2 y \\ \qquad + (3\epsilon_2 + 4br^2)xy^2 - \epsilon_1 y^3, \end{cases} \qquad (4.24)$$

where $\epsilon_2 = \epsilon_1^0$, $\epsilon_2 = \epsilon_2^0$. By substituting (4.21) and (4.22) into (4.24), and by rescaling time $t \to t/\epsilon_1^0$ ($\epsilon_1^0 > 0$), we have for equation (4.24)

$$\begin{cases} \dot{x} = x - \xi y - x^3 - \eta x^2 y + xy^2 + \xi y^3, \\ \dot{y} = \xi x + y - \xi x^3 + x^2 y + \eta xy^2 - y^3, \end{cases} \qquad (4.25)$$

where

$$\xi = \frac{b + \sqrt{1 - a^2}}{2a}, \qquad \eta = \frac{-b + 3\sqrt{1 - a^2}}{2a}. \qquad (4.26)$$

The matrix of the linear part of (4.25) at $(1, 0)$ is

$$\begin{bmatrix} \alpha & \beta \\ \gamma & \delta \end{bmatrix} = \begin{bmatrix} -2 & -\dfrac{2\sqrt{1 - a^2}}{a} \\ -\dfrac{b + \sqrt{1 - a^2}}{a} & 2 \end{bmatrix}. \qquad (4.27)$$

Denote $\omega^2 = \det\begin{bmatrix} \alpha & \beta \\ \gamma & \delta \end{bmatrix}$. Then

$$\omega^2 = -\frac{2(a^2 + 1 + b\sqrt{1 - a^2})}{a^2}. \qquad (4.28)$$

Using the formula (3.2.36) we have (up to a positive factor)

$$\mathrm{Re}(C_1) = -4\beta \left\{ \beta - \gamma + \frac{2}{\omega^2} \left[ \delta(\gamma + 5\beta) - 3\xi\beta^2 + (\omega^2 + 3\delta^2)\eta \right] \right\}. \qquad (4.29)$$

Substituting (4.26), (4.27), and (4.28) into (4.29), we obtain

$$\mathrm{Re}(C_1) = \frac{-32\beta(b + \sqrt{1 - a^2})}{a(a^2 + 1 + b\sqrt{1 - a^2})}.$$

Since $a < 0$, $\beta > 0$ (see (4.27)), $b + \sqrt{1 - a^2} < 0$, $a^2 + 1 + b\sqrt{1 - a^2} < 0$ (see (4.20)), and $(a, b)$ is below the curve $\gamma_4$, it follows that $\mathrm{Re}(C_1) > 0$. Thus, the proof of Theorem 4.6 is complete. $\square$

For a proof of Theorem 4.7, we refer to the paper of Berezovskaia and Knibnik [1], where both the analytic and numerical methods were used.

The proof of Theorem 4.8 follows from Theorems 4.1, 4.2, 4.3, 4.5, and 4.7.

*Proof of Theorem 4.10.* By the transformation

$$\rho = \frac{1}{2}(x^2 + y^2), \qquad \phi = \tan^{-1}\frac{y}{x},$$

equation (4.1) is transformed into the following form:

$$\begin{cases} \dot{\rho} = 2\rho(\epsilon_1 + 2\rho(a + \cos 4\phi)), \\ \dot{\phi} = \epsilon_2 + 2\rho(b - \sin 4\phi), \end{cases} \qquad (4.30)$$

where $-1 < a < 0$, $|a| \ll 1$, $b < -1$, and $\epsilon_1$ and $\epsilon_2$ are small parameters. By Theorems 4.1, 4.2, and 4.3, and Remark 4.4 and Lemma 4.12, we only need to consider the case of $\epsilon_1 > 0$ and $\epsilon_2 > 0$.

Let

$$\epsilon_2 = \delta, \quad \epsilon_1 = \alpha\delta^2, \quad a = -\beta\delta, \quad \rho \to \delta\rho, \quad t \to \frac{t}{\delta}, \quad (4.31)$$

where $\delta > 0$, $\alpha > 0$, $\beta > 0$; $\alpha$ and $\beta$ are new parameters. Equation (4.30) becomes

$$\begin{cases} \dot{\rho} = 4\rho^2 \cos 4\phi + \delta(2\alpha\rho - 4\beta\rho^2), \\ \dot{\phi} = 1 + 2\rho(b - \sin 4\phi). \end{cases} \qquad (4.32)$$

For $\delta = 0$, (4.32) becomes a Hamiltonian system

$$\begin{cases} \dot{\rho} = 4\rho^2 \cos 4\phi, \\ \dot{\phi} = 1 + 2\rho(b - \sin 4\phi), \end{cases} \qquad (4.33)$$

with Hamiltonian function $-H(\rho, \phi)$, where $H$ is defined by

$$H(\rho, \phi) = \rho + \rho^2(b - \sin 4\phi). \qquad (4.34)$$

The closed level curves $\Gamma_h = \{(\rho, \phi)|H(\rho, \phi) = h\}$ are shown in Figure 4.19, and

$$\begin{cases} 0 \leq h < h_s & \text{for region } G_1, \\ -\infty < h < h_s & \text{for region } G_2, \\ h_s < h \leq h_c & \text{for region } G_3^{(k)}, \quad k = 1, 2, 3, 4, \end{cases} \qquad (4.35)$$

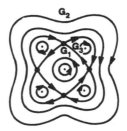

Figure 4.19.

where $h_s = 1/(4(1 - b))$ and $h_c = -1/(4(1 + b))$. When $h \to h_s$, $\Gamma_h$ tends to the separatrix connections between four saddle points which have the coordinates $\rho = \rho_s = 1/(2(1 - b))$ and $\phi = \pi/8 + k\pi/2$ ($k = 1, 2, 3, 4$); when $h \to h_c$, the four symmetric branches of $\Gamma_h$ shrink to the four nonzero centers which have the coordinates $\rho = \rho_c = -1/(2(1 + b))$ and $\phi = 3\pi/8 + k\pi/2$ ($k = 1, 2, 3, 4$); and when $h \to 0$, $\Gamma_h$ shrinks to the center $\rho = 0$ (in the region $G_1$).

As in previous sections, we will study equation (4.32) as a perturbation of equation (4.33) for some small $\delta > 0$. Since

$$\dot{H}|_{(4.32)} = \frac{\partial H}{\partial \rho}\dot{\rho} + \frac{\partial H}{\partial \phi}\dot{\phi} = \delta(2\alpha\rho - 4\beta\rho^2)\dot{\phi},$$

where $H = H(\rho, \phi)$ is defined in (4.34), the bifurcation function for periodic orbits of equation (4.32) is defined, for $\delta = 0$, by

$$G(h, \delta, \alpha, \beta)|_{\delta=0} = \int_{\Gamma_h} (2\alpha\rho - 4\beta\rho^2)\, d\phi = 2\alpha I_1(h) - 4\beta I_2(h),$$

$$(4.36)$$

where

$$I_1(h) = \int_{\Gamma_h} \rho\, d\phi, \; I_2(h) = \int_{\Gamma_h} \rho^2\, d\phi, \qquad (4.37)$$

$\Gamma_h$ is the level curve of $H(\rho, \phi) = h$, and $h$ satisfies (4.35).

Obviously, for regions $G_2$ and $G_3$, $I_1(h) > 0$ and $I_2(h) > 0$; for region $G_1$, $I_1(h) > 0$ and $I_2(h) > 0$ if $h > 0$, and $I_1(0) = I_2(0) = 0$.

Moreover,

$$\lim_{h \to 0^+} \frac{I_2(h)}{I_1(h)} = 0.$$

Hence, similarly to Sections 4.1–4.3, we define

$$P_i(h) = \frac{I_2(h)}{I_1(h)}, \tag{4.38}$$

where $h$ satisfies (4.35) for region $G_i$ ($i = 1, 2, 3$), and

$$P_1(0) = 0.$$

Equations (4.36) and (4.38) show that $G|_{\delta=0} = 0$ is equivalent to

$$\frac{\alpha}{\beta} = 2P_i(h), \qquad i = 1, 2, 3. \tag{4.39}$$

Thus, for a given $\alpha/\beta$ (or equivalently, for a given $\epsilon_2/\epsilon_1$; see (4.31)), the number of periodic orbits of equation (4.32) (or equation (4.1)) is equal to the number of solutions of (4.39) with respect to $h$ satisfying (4.35).

By a similar discussion as in Sections 4.1–4.3, from the following theorem we can obtain the conclusion of Theorem 4.10.

**Theorem 4.14.** *Suppose $\xi^*$ is defined as in Theorem* 4.10, *$P_i(h)$ is defined as* (4.38), *and* $i = 1, 2, 3$.
*(I) If $b \le -\xi^*$, then*
  *(1) $P_1'(h) > 0$, $P_2'(h) < 0$, and $P_3'(h) > 0$;*
  *(2) $\lim_{h \to -\infty} P_2(h) = +\infty$;*
  *(3) $P_1(h_s) < P_2(h_s) < P_3(h_s)$.*
*(II) If $-\xi^* < b < -1$, then*
  *(1) $\exists h_m < h_s$ such that $P_2'(h_m) = 0$, $P_2''(h_m) > 0$, and $P_2'(h) \ne 0$ for $h \ne h_m$;*
  *(2) $P_2(h_m) > P_1(h_s)$;*
  *(3) the behavior of each $P_i(h)$ ($i = 1, 2, 3$) is the same as in Case (I).*
*The behaviors of $P_i(h)$ ($i = 1, 2, 3$) are shown in Figure* 4.20. *Theorem* 4.14 *follows from the following lemmas.*

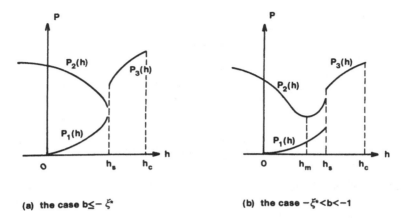

(a) the case $b \leq -\xi^*$                       (b) the case $-\xi^* < b < -1$

Figure 4.20.

**Lemma 4.15.** $P_3'(h) > 0$.

*Proof.* In region $G_3$ (we consider $G_3^{(1)}$, i.e., $\pi/8 < \phi < 5\pi/8$), along $\Gamma_h$ we have from (4.34)

$$\rho_{1,2} = \frac{1 \pm \sqrt{v}}{2u} \quad (\rho_1 \leq \rho_2), \tag{4.40}$$

where

$$u = -b + \sin 4\phi > 0, \qquad v = 1 - 4hu \geq 0. \tag{4.41}$$

Hence,

$$I_1 = \int_{\Gamma_h} \rho \, d\phi = \int_{\phi_+}^{\phi_-} (\rho_2 - \rho_1) \, d\phi = \int_{\phi_+}^{\phi_-} \frac{\sqrt{v}}{u} \, d\phi,$$

$$\tag{4.42}$$

$$I_2 = \int_{\Gamma_h} \rho^2 \, d\phi = \int_{\phi_+}^{\phi_-} (\rho_2^2 - \rho_1^2) \, d\phi = \int_{\phi_+}^{\phi_-} \frac{\sqrt{v}}{u^2} \, d\phi,$$

where $\phi_\pm$ are the limits of the variation of $\phi$ on $\Gamma_h$, $\phi_- < \phi_+$.

Equations (4.10), (4.41), and (4.42) give

$$\begin{cases} I_1' = \int_{\phi_+}^{\phi_-} (\rho_2' - \rho_2') \, d\phi = \int_{\phi_-}^{\phi_+} \dfrac{2}{\sqrt{v}} \, d\phi, \\[4mm] I_2' = \int_{\phi_+}^{\phi_-} 2(\rho_2\rho_2' - \rho_1\rho_1') \, d\phi = 2\int_{\phi_-}^{\phi_+} \dfrac{1}{u\sqrt{v}} \, d\phi, \end{cases} \tag{4.43}$$

where the prime denotes the derivative with respect to $h$. From (4.38), (4.42), and (4.43) we have

$$\begin{aligned} P_3'(h) &= \frac{1}{I_1^2(h)} \left[ I_2'(h) I_1(h) - I_1'(h) I_2(h) \right] \\[3mm] &= -\frac{2}{I_1^2(h)} \left[ \int_{\phi_-}^{\phi_+} \frac{d\phi}{u\sqrt{v}} \int_{\phi_-}^{\phi_+} \frac{\sqrt{v}}{u} \, d\phi - \int_{\phi_-}^{\phi_+} \frac{d\phi}{\sqrt{v}} \int_{\phi_-}^{\phi_+} \frac{\sqrt{v}}{u^2} \, d\phi \right] \\[3mm] &= -\frac{1}{I_1^2(h)} \int_{\phi_-}^{\phi_+} \int_{\phi_-}^{\phi_+} \left[ \frac{\sqrt{v_1}}{u_1 u_2 \sqrt{v_2}} + \frac{\sqrt{v_2}}{u_2 u_1 \sqrt{v_1}} \right. \\[3mm] &\qquad\qquad\qquad \left. - \frac{\sqrt{v_1}}{u_1^2\sqrt{v_2}} - \frac{\sqrt{v_2}}{u_2^2\sqrt{v_1}} \right] d\phi_1 \, d\phi_2, \end{aligned}$$

where $u_i = u(\phi_i)$, $v_i = v(\phi_i)$, $i = 1, 2$. The calculation shows that the integrand is equal to (using (4.41))

$$\frac{(u_1 - u_2)(u_2 v_1 - u_2 v_2)}{u_1^2 u_2^2 \sqrt{v_1 v_2}} = -\frac{(u_1 - u_2)^2}{u_1^2 u_2^2 \sqrt{v_1 v_2}} \leq 0.$$

Hence, $P_3'(h) > 0$.  $\square$

**Lemma 4.16.** $P_1'(h) > 0$.

*Proof.* In region $G_1$, from (4.34) along $\Gamma_h$ we have

$$\rho = \frac{1 - \sqrt{v}}{2u}, \tag{4.44}$$

where $u$ and $v$ are the same as in (4.41).

Since $P_1(0) = 0$ and $P_1(h) > 0$ for $0 < h < h_s$, in order to show $P_1'(h) > 0$, we only need to show that if there exists $h_0 \in (0, h_s)$ such that $P_1'(h_0) = 0$, then $P_1''(h_0) > 0$.

Similarly to the proof of Lemma 3.8, we define

$$Q(h) = \frac{I_2'(h)}{I_1'(h)}. \tag{4.45}$$

If $P_1'(h_0) = 0$, then $P_1(h_0) = Q(h_0)$ and

$$P_1''(h_0) = \frac{I_1'(h_0)}{I_1(h_0)} Q'(h_0). \tag{4.46}$$

From (4.44), (4.41), and (4.37), we have

$$
\begin{cases}
I_1'(h) = \int_0^{2\pi} \rho_h' \, d\phi = \int_0^{2\pi} \dfrac{d\phi}{\sqrt{v}}, \\[2mm]
I_1''(h) = \int_0^{2\pi} \rho_h'' \, d\phi = \int_0^{2\pi} \dfrac{2u}{v^{3/2}} \, d\phi, \\[2mm]
I_2'(h) = \int_0^{2\pi} 2\rho\rho_h' \, d\phi = \int_0^{2\pi} \left( -\dfrac{1}{u} + \dfrac{1}{u\sqrt{v}} \right) d\phi, \\[2mm]
I_2''(h) = \int_0^{2\pi} \dfrac{2}{v^{3/2}} \, d\phi.
\end{cases} \tag{4.47}
$$

From (4.45) and (4.47), we have

$$
\begin{aligned}
Q'(h) &= \frac{1}{(I'(h))^2} [I_2'' I_1' - I_1'' I_2'] \\[2mm]
&= \frac{2}{(I_1')^2} \left[ \int_0^{2\pi} \frac{d\phi}{v^{3/2}} \int_0^{2\pi} \frac{d\phi}{v^{1/2}} + \int_0^{2\pi} \frac{d\phi}{u} \int_0^{2\pi} \frac{u}{v^{3/2}} \, d\phi \right. \\[2mm]
&\qquad\qquad \left. - \int_0^{2\pi} \frac{d\phi}{u\sqrt{v}} \int_0^{2\pi} \frac{u}{v^{3/2}} \, d\phi \right] \\[2mm]
&= \frac{1}{(I_1')^2} \int_0^{2\pi} \int_0^{2\pi} \left[ \frac{1}{v_1^{3/2} v_2^{1/2}} + \frac{1}{v_2^{3/2} v_1^{1/2}} + \frac{u_2}{u_1 v_2^{3/2}} + \frac{u_1}{u_2 v_1^{3/2}} \right. \\[2mm]
&\qquad\qquad \left. - \frac{u_2}{u_1 v_1^{1/2} v_2^{3/2}} - \frac{u_1}{u_2 v_2^{1/2} v_1^{3/2}} \right] d\phi_1 \, d\phi_2,
\end{aligned}
$$

where $u_i = u(\phi_i)$, $v_i = v(\phi_i)$, $i = 1, 2$.

Since $u_2 v_1 - u_1 v_2 = u_2 - u_1$ (see (4.41)), we can transform the integrand to the following form:

$$\frac{A(\phi_1, \phi_2)}{(v_1 v_2)^{3/2}},$$

where

$$A = \frac{-(u_1 - u_2)^2 + u_2^2 v_1^{3/2} + u_1^2 v_2^{3/2}}{u_1 u_2}$$

$$= 2 + \frac{u_2}{u_1}\left(v_1^{3/2} - 1\right) + \frac{u_1}{u_2}\left(v_2^{3/2} - 1\right)$$

$$= 2 + \frac{u_2}{u_1}\frac{\left(v_1 - 1\right)\left(1 + v_1^{1/2} + v_1\right)}{1 + v_1^{1/2}} + \frac{u_1}{u_2}\frac{\left(v_2 - 1\right)\left(1 + v_2^{1/2} + v_2\right)}{1 + v_2^{1/2}}.$$

Since $(v_1 - 1)u_2 = (v_2 - 1)u_1$ (see (4.41)),

$$A = 2 + \frac{\left(v_2 - 1\right)\left(1 + v_1^{1/2} + v_1\right)}{1 + v_1^{1/2}} + \frac{\left(v_1 - 1\right)\left(1 + v_2^{1/2} + v_2\right)}{1 + v_2^{1/2}}$$

$$= \frac{\left(\sqrt{v_2} + \sqrt{v_1}\right)\left(\sqrt{v_2} - \sqrt{v_1}\right)^2}{\left(1 + \sqrt{v_1}\right)\left(1 + \sqrt{v_2}\right)} + v_1\sqrt{v_2} + v_2\sqrt{v_1} \geq 0.$$

Hence $Q'(h) > 0$. This implies $P_1'(h) > 0$ for $h \in (0, h_s]$.     □

**Lemma 4.17.** *We have:*
(1) *If $b \leq -\xi^*$, then $P_2'(h) < 0$;*
(2) *If $-\xi^* < b < -1$, then $\exists h_m < h_s$ such that $P_2'(h_m) = 0$, $P_2''(h_m) > 0$, and $P_2'(h) \neq 0$ if $h \neq h_m$;*
(3) $\lim_{h \to -\infty} P_2(h) = +\infty$.

*Proof.* In the region $G_2$, we have

$$\rho = \frac{1 + \sqrt{v}}{2u}, \qquad (4.48)$$

where $u$ and $v$ are the same as in (4.41).

It is easy to see that for $h < 0$ and $\phi \in (0, 2\pi]$,

$$\frac{1 + \sqrt{1 - 4h(-b-1)}}{2(-b+1)} < \rho < \frac{1 + \sqrt{1 - 4h(-b+1)}}{2(-b-1)}.$$

Hence

$$\lim_{h \to -\infty} P_2(h) = \lim_{h \to -\infty} \frac{\int_0^{-2\pi} \rho^2 \, d\phi}{\int_0^{-2\pi} \rho \, d\phi} = +\infty.$$

Next, in the same manner as in the proof of Lemma 4.16, we can obtain that if there exists an $h_0 \in (-\infty, h_s)$ such that $P_2'(h_0) = 0$, then $P_2''(h_0) > 0$. Obviously, such an $h_0$, if it exists, is unique, and its existence depends on the sign of $P_2'(h_s)$. If $P_2'(h_s) < 0$, then such an $h_0$ does not exist and $P_2'(h) < 0$ for $-\infty < h < h_s$ (using $\lim_{h \to -\infty} P_2(h) = +\infty$). If $P_2'(h_s) > 0$, then such an $h_0$ exists, and it is unique (denote it by $h_m$).

Now we consider the sign of $P_2'(h_s)$. From (4.48) and (4.42) we have

$$I_1'(h) = \int_0^{-2\pi} \left( -\frac{1}{\sqrt{v}} \right) d\phi,$$

$$I_2'(h) = \int_0^{-2\pi} 2\rho \left( -\frac{1}{\sqrt{v}} \right) d\phi.$$

Thus,

$$P_2'(h_s) = \lim_{h \to h_s^-} \frac{1}{I_1^2(h)} \left[ I_2'(h) I_1(h) - I_1'(h) I_2(h) \right]$$

$$= \lim_{h \to h_s^-} \frac{1}{I_1^2(h_s)} \int_0^{2\pi} \left[ 2\rho I_1(h_s) - I_2(h_s) \right] \frac{1}{\sqrt{v}} \, d\phi.$$

At the saddle points $(h = h_s, \; \phi = \pi/8 + k\pi/2, \; k = 1, 2, 3, 4)$, the integral has singularities, and

$$v \to 1 - \sin 4\phi = O\left(\left|\phi - \left(\frac{\pi}{8} + \frac{k\pi}{2}\right)\right|^2\right)$$

$$\text{as } h \to h_s - , \quad \phi \to \frac{\pi}{8} + \frac{k\pi}{2}.$$

Hence

$$P_2'(h_s) = -\infty \quad \text{if } 2\rho_s I_1(h_s) - I_2(h_s) < 0,$$

$$P_2'(h_s) = +\infty \quad \text{if } 2\rho_s I_1(h_s) - I_2(h_s) > 0.$$

If $2\rho_s I_1(h_s) - I_2(h_s) = 0$, then $P_2'(h_s)$ is a finite negative number.

Finally, we need to show that $2\rho_s I_1(h_s) - I_2(h_s)$ is negative, positive, or zero if and only if $b$ is less than, greater than, or equal to $-\xi^*$.

From (4.40) and (4.41) we have

$$\rho_s = \frac{1}{2(-b + 1)}$$

and

$$\rho_{1,2}(h_s, \phi) = \frac{1 \mp \sqrt{\dfrac{1 - \sin 4\phi}{-b + 1}}}{2(-b + \sin 4\phi)}$$

$$= \frac{1}{2\sqrt{-b+1}} \frac{\sqrt{-b+1} \mp \sqrt{1 - \sin 4\phi}}{(-b+1) - (1 - \sin 4\phi)}$$

$$= \frac{1}{2\sqrt{-b+1}} \frac{1}{\sqrt{-b+1} \pm \sqrt{1 - \sin 4\phi}}.$$

Hence

$$\rho_{1,2}(h_s, \phi) = \bar\rho_{1,2}(\phi) = \frac{\rho_s}{1 \pm \eta \left|\sin\left(2\phi - \dfrac{\pi}{4}\right)\right|}, \qquad (4.49)$$

where

$$\eta = \sqrt{\frac{2}{-b+1}} < 1.$$

In this case

$$\rho(h_s, \phi) = \bar{\rho}_2(\phi) = \frac{\rho_s}{1 - \eta \left| \sin\left(2\phi - \frac{\pi}{4}\right)\right|}.$$

Substituting the above expression into the following equality

$$2\rho_s I_1(h_s) - I_2(h_s) = 0, \qquad (4.50)$$

we have

$$\int_0^\pi \left[ \frac{2}{1 - \eta \sin \psi} - \frac{1}{(1 - \eta \sin \psi)^2} \right] d\psi = 0. \qquad (4.51)$$

By a substitution $\tan(\psi/2) = s$ in the above integral, (4.51) is reduced to

$$\frac{1 - 2\eta^2}{(1 - \eta^2)\sqrt{1 - \eta^2}} \left[ \arctan \sqrt{\frac{1 - \eta}{1 + \eta}} + \arctan \frac{\eta}{\sqrt{1 - \eta^2}} \right]$$

$$- \frac{\eta}{2(1 - \eta^2)} = 0. \qquad (4.52)$$

Let $\theta = 2 \arcsin \eta$, $\theta \in (0, \pi)$. Then (4.52) takes the following form:

$$\tan \theta - \theta = \pi. \qquad (4.53)$$

If $\theta^*$ is the root of (4.53), and $\eta = \sin(\theta^*/2)$, then

$$-b = \frac{2 - \eta^2}{\eta^2} = (3 + \cos \theta^*)(1 - \cos \theta^*) = \xi^*.$$

This completes the proof of Lemma 4.17.                    □

**Lemma 4.18.** $P_1(h_s) < P_2(h_s) < P_3(h_s)$ and $P_1(h_s) < P_2(h_m)$.

*Proof.* In order to avoid confusion, we denote

$$P_i(h) = \frac{I_{2,i}(h)}{I_{1,i}(h)},$$

where $h$ satisfies (4.35) for region $G_i$, $i = 1, 2, 3$. Then

$P_2(h_s) - P_1(h_s)$

$$= \frac{I_{2,2}(h_s)}{I_{1,2}(h_s)} - \frac{I_{2,1}(h_s)}{I_{1,1}(h_s)}$$

$$= \frac{\int_0^{-2\pi} \bar{\rho}_2^2(\phi) \, d\phi \int_0^{2\pi} \bar{\rho}_1(\psi) \, d\psi - \int_0^{2\pi} \bar{\rho}_1^2(\psi) \, d\psi \int_0^{-2\pi} \bar{\rho}_2(\phi) \, d\phi}{\int_0^{-2\pi} \bar{\rho}_2(\phi) \, d\phi \int_0^{2\pi} \bar{\rho}_1(\psi) \, d\psi}$$

$$= \frac{\int_{\pi/8}^{5\pi/8} \int_{\pi/8}^{5\pi/8} \bar{\rho}_1(\psi) \bar{\rho}_2(\phi) (\bar{\rho}_2(\phi) - \bar{\rho}_1(\psi)) \, d\phi \, d\psi}{\int_{\pi/8}^{5\pi/8} \int_{\pi/8}^{5\pi/8} \bar{\rho}_1(\psi) \bar{\rho}_2(\phi) \, d\phi \, d\psi}.$$

From (4.49), we see

$$\bar{\rho}_1(\psi) < \rho_s < \bar{\rho}_2(\phi),$$

for any $\phi \neq \pi/8 + k\pi/2$ and $\psi \neq \pi/8 + k\pi/2$, $k = 1, 2, 3, 4$. Hence

$$P_2(h_s) - P_1(h_s) > 0.$$

Noting $\rho_2(h_m, \phi) > \rho_2(h_s, \phi) = \bar{\rho}_2(\phi)$ for all $\phi$, we can prove $P_2(h_m) > P_1(h_s)$ by the same method.

Finally, we consider

$$P_3(h_s) - P_2(h_s)$$

$$= \frac{I_{2,3}(h_s)}{I_{1,3}(h_s)} - \frac{I_{2,2}(h_s)}{I_{1,2}(h_s)}$$

$$= \frac{\int_0^{2\pi} \bar{\rho}_1^2(\phi)\, d\phi + \int_0^{-2\pi} \bar{\rho}_2^2(\phi)\, d\phi}{\int_0^{2\pi} \bar{\rho}_1(\phi)\, d\phi + \int_0^{-2\pi} \bar{\rho}_2(\phi)\, d\phi} - \frac{\int_0^{-2\pi} \bar{\rho}_2^2(\phi)\, d\phi}{\int_0^{-2\pi} \bar{\rho}_2(\phi)\, d\phi}$$

$$= \frac{\int_{\pi/8}^{5\pi/8} \int_{\pi/8}^{5\pi/8} \bar{\rho}_2(\psi)(\bar{\rho}_2(\phi) - \bar{\rho}_1(\phi))(\bar{\rho}_2(\phi) + \bar{\rho}_1(\phi) - \bar{\rho}_2(\psi))\, d\phi\, d\psi}{\int_{\pi/8}^{5\pi/8} \int_{\pi/8}^{5\pi/8} (\bar{\rho}_2(\phi) - \bar{\rho}_1(\phi))\bar{\rho}_2(\psi)\, d\psi\, d\psi}.$$

From (4.49), we have

$$\bar{\rho}_2(\phi) + \bar{\rho}_1(\phi) - \bar{\rho}_2(\psi) = \frac{\rho_s(1 + \eta^2 \sin^2 \bar{\phi} - 2\eta \sin \bar{\psi})}{(1 - \eta^2 \sin^2 \bar{\phi})(1 - \eta \sin \bar{\psi})},$$

where $\bar{\phi} = 2\phi - \pi/4$, $\bar{\psi} = 2\psi - \pi/4$.
Since for some constant $M$,

$$0 < (1 - \eta^2 \sin \bar{\phi})(1 - \eta \sin \bar{\psi}) < M,$$

we obtain that

$$P_3(h_s) - P_2(h_s) > \frac{\rho_s}{M} \int_0^\pi \int_0^\pi (1 + \eta^2 \sin \bar{\phi} - 2\eta \sin \bar{\psi})\, d\bar{\phi}\, d\bar{\psi} > 0.$$

$$\square$$

## 4.5 Double Zero Eigenvalue with Symmetry of Order $\geq 5$

A family of vector fields on the plane that is invariant under a rotation through $2\pi/q$ ($q \geq 3$) takes the form

$$\dot{z} = \epsilon z + C_1 z^2 \bar{z} + C_2 z^3 \bar{z}^2 + \cdots + C_m z^{m+1} \bar{z}^m + A \bar{z}^{q-1} + O(|z|^q),$$

$$(5.1)$$

where $m = [(q - 1)/2]$, $z, \epsilon, C_j, A \in \mathbb{C}$, $\mathrm{Re}\, C_1 \neq 0$, and $A \neq 0$.

In Sections 1–4, we discussed the cases $q \leq 4$, which are called strong resonances. In this section we consider the cases $q \geq 5$, which are called weak resonances. For $q \geq 5$, the term $A\bar{z}^{q-1}$ is smaller than the term $C_1 z^2 \bar{z}$. Hence the behavior of (5.1) is governed mainly by the term $C_1 z^2 \bar{z}$ (see (5.4) below). Therefore, we expect that the discussion of weak resonance is simpler than the case of strong resonance.

By a scaling on $z$ and $t$, we suppose $A = 1$ and $\mathrm{Re}(C_1) = -1$. Let $\epsilon = \epsilon_1 + i\epsilon_2$, $C_j = a_j + ib_j$ $(j = 1, 2, \ldots, m, \ m = [(q - 1)/2])$. We transform (5.1) into polar coordinates:

$$
\begin{cases}
\dot{r} = \epsilon_1 r - r^3 + a_2 r^5 + \cdots + a_m r^{2m+1} + r^{q-1} \cos q\theta + O(r^q), \\
\dot{\theta} = \epsilon_2 + b_1 r^2 + b_2 r^4 + \cdots + b_m r^{2m} - r^{q-2} \sin q\theta + O(r^{q-1}).
\end{cases}
$$

$$(5.2)$$

Obviously, (5.2) has a Hopf bifurcation at $\epsilon_1 = 0$ and $\epsilon_2 \neq 0$. When $\epsilon_1 < 0$, the first equation in system (5.2) shows that in a small neighborhood of $r = 0$, every flow tends to the unique attractive equilibrium $r = 0$ as $t \to +\infty$. When $\epsilon_1 > 0$, we make a scaling

$$
\epsilon_1 = \delta^2, \quad \epsilon_2 = \zeta\delta^2, \quad r = \delta\rho, \quad t \to \frac{t}{\delta^2} \quad (\delta > 0). \quad (5.3)
$$

Thus (5.2) becomes

$$
\begin{cases}
\dot{\rho} = \rho(1 - \rho^2) + \delta f(\delta, \rho, \theta), \\
\dot{\theta} = \zeta + b_1 \rho^2 + \delta g(\delta, \rho, \theta),
\end{cases}
\quad (5.4)
$$

where

$$
f(\delta, \rho, \theta) = a_2 \delta\rho^5 + \cdots + a_l \delta^{2l-3} \rho^{2l+1}
$$

$$
+ \delta^{q-5} \rho^{q-1} \cos q\theta + O(\delta^{q-4} \rho^q),
$$

$$
g(\delta, \rho, \theta) = b_2 \delta\rho^4 + \cdots + b_l \delta^{2l-3} \rho^{2l}
$$

$$
- \delta^{q-5} \rho^{q-2} \sin q\theta + O(\delta^{q-4} \rho^{q-1}).
$$

For $\delta = 0$, (5.4) takes the following form:

$$\begin{cases} \dot{\rho} = \rho(1 - \rho^2), \\ \dot{\theta} = \zeta + b_1\rho^2. \end{cases} \tag{5.5}$$

Equation (5.5) has a unique attractive invariant circle $\Sigma = \{(\rho, \theta)|\rho = 1\}$ which is a limit cycle for $\zeta + b_1 \neq 0$, and consists of only equilibria for $\zeta + b_1 = 0$. Denote the time-1 mapping along the flow of (5.4) by $\Phi_\delta$. Then there exists $\bar{\delta} > 0$ (correspondingly $\bar{\epsilon}_1 > 0$) such that for every $0 < \delta < \bar{\delta}$ (correspondingly $0 < \epsilon_1 < \bar{\epsilon}_1$), the sequence of manifolds $\{\Phi_\delta^n(\Sigma)\}_{n=0}^\infty$ converges to an invariant manifold $\Sigma_\delta^*$ which is an attractive circle of (5.4) (see Ruelle and Takens [1] for details). Hence, the phase portrait of (5.4) is completely determined by the behavior of (5.4) restricted to these invariant circles $\{\Sigma_\delta^*\}$. If $\zeta + b_1 \neq 0$ (equivalently, if $(\epsilon_1, \epsilon_2)$ is not on the line $\mathscr{L}$: $\epsilon_2 = -b_1\epsilon_1$; see (5.3)), then for sufficiently small $\delta$, there are no equilibria on $\Sigma_\delta^*$. This means that, in addition to the Hopf bifurcation on $\epsilon_1 = 0$ and $\epsilon_2 \neq 0$, all other possible bifurcations must take place near the line $\mathscr{L}$. In fact, we will prove that there exist two bifurcation curves $SN_1$ and $SN_2$ which are tangent to $\mathscr{L}$ and form a cusp region $\Omega$ (see Figure 5.1).

The bifurcation of equilibria on $\Sigma_\delta^*$ occurs in the following way: When $(\epsilon_1, \epsilon_2) \in$ region $A$, there are no equilibria on $\Sigma_\delta^*$, and $q$ saddle-nodes appear on $\Sigma_\delta^*$ if $(\epsilon_1, \epsilon_2) \in SN_1$. As $(\epsilon_1, \epsilon_2)$ goes from $SN_1$ to $\Omega$, every saddle-node becomes a saddle point and a node. $q$ new saddle-nodes appear when $(\epsilon_1, \epsilon_2) \in SN_2$. See Figure 5.4, p. 332, for more detail.

We now explain why the equilibrium $(\rho_0, \theta_0)$ on $\Sigma_\delta^*$ of (5.4) must be a saddle point or node, or a saddle-node. If $(\epsilon_1, \epsilon_2) \in SN_1$ or $SN_2$, then $\sin(q\theta_0) = \pm 1$. Hence, $\dot{\theta}$ does not change its sign when $\theta$ passes through $\theta_0$ (see the second equation of (5.2)). Moreover, $\Sigma_\delta^*$ is attractive on both sides. Therefore, $(\rho_0, \theta_0)$ is a saddle-node (see Figure 5.2).

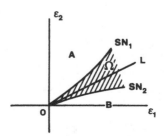

Figure 5.1.

Figure 5.2.

Figure 5.3.

If $(\epsilon_1, \epsilon_2) \in \Omega$, then $|\sin(q\theta_0)| < 1$. Hence $\dot\theta$ changes its sign when $\theta$ passes through $\theta_0$. Therefore, $(\rho_0, \theta_0)$ is a saddle point or a node. Moreover, the saddle points and nodes appear alternately (see Figure 5.3).

**Theorem 5.1.** *The bifurcation diagram of (5.2) consists of the origin* $(\epsilon_1 = \epsilon_2 = 0)$ *and the curves* $H^{\pm} = \{\epsilon_1 = 0, \ \epsilon_2 \neq 0\}$, $SN_1$, *and* $SN_2$. $SN_j$ *is defined by an equation with parameter s in the* $\epsilon_1\epsilon_2$-plane: $SN_j(s) = (s^2, h_j(s))$, $j = 1, 2$, $s > 0$, *with the property*

$$\lim_{s \to 0+} \frac{h_2(s) - h_1(s)}{s^{q-2}} \neq 0.$$

*Along* $SN_1$ *and* $SN_2$, $q$ *saddle-nodes on the invariant circle are created or annihilated. The phase portraits of (5.2) are shown in Figure 5.4.*

Now we prove the existence of the curves $SN_1$ and $SN_2$. We consider (5.4) near $(\rho, \zeta, \delta) = (1, -b_1, 0)$. A calculation shows that

$$\det \begin{bmatrix} \dfrac{\partial R}{\partial \rho} & \dfrac{\partial R}{\partial \theta} \\[2mm] \dfrac{\partial \Theta}{\partial \rho} & \dfrac{\rho \Theta}{\partial \theta} \end{bmatrix} := q\delta^{q-4} \cdot D(\rho, \theta, \delta), \qquad (5.6)$$

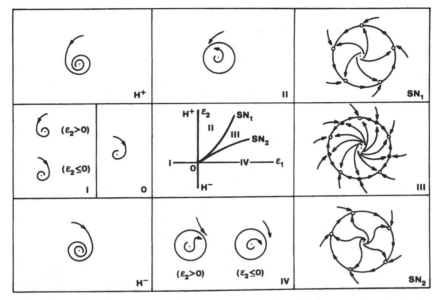

Figure 5.4. The bifurcation diagram and phase portraits of (5.2) ($q = 5$, $b_1 < 0$).

where $R$ and $\Theta$ are defined to be the right-hand side of (5.4). Note that

$$D(\rho, \theta, \delta) = (3\rho^2 - 1)\cos(q\theta) + 2b_1\rho\sin(q\theta) + O(\delta). \quad (5.7)$$

**Lemma 5.2.** (*Takens* [2]) *There is a neighborhood* $U$ *of* $W = \{(\cdot, \theta, -b_1, 0)|0 \le \theta < 2\pi\}$ *in the* $\rho\theta\zeta\delta$-*space such that the intersection of* $U$ *with* $\{(\rho, \theta, \zeta, \delta)|D = 0, R = 0, \Theta = 0\}$ *consists of* $2q$ *curves. The projection of these curves on the* $\zeta\delta$-*plane consists of two curves which are of the form* $\{M_j(\delta), \delta\}$ $(j = 0, 1)$ *with* $M_0(\delta) - M_1(\delta) = \xi\delta^{q-4} + O(\delta^{q-3})$, *where* $\xi$ *is a nonzero constant.*

*Proof.* It is easy to obtain from the first equation of (5.4) that in a neighborhood of $W$ the set $\{(\rho, \theta, \zeta, \delta)|R = 0\}$ has the form $\{\bar{\rho}(\theta, \zeta, \delta), \theta, \zeta, \delta\}$, where $\bar{\rho}$ is a smooth function satisfying $\bar{\rho} = 1 + O(\delta)$.

Next, $D|_{R=\delta=0} = 2\cos(q\theta) + 2b_1\sin(q\theta)$ and $D$ has $2q$ zeros in $W$. Hence, the set $\{(\rho, \theta, \zeta, \delta)|R = 0, D = 0\}$, in a neighborhood of $W$, has the form

$$\bigcup_{j=0}^{2q-1} \left\{\bar{\rho}_j(\zeta, \delta), \bar{\theta}_j(\zeta, \delta), \zeta, \delta\right\},$$

where $\bar{\theta}_j$ and $\bar{\rho}_j$ satisfy the following conditions:

(1) $\bar{\theta}_j(\zeta, \delta) = \beta_0 + j\pi/q + O(\delta)$, where $\beta_0$ satisfies (see (5.7))

$$\cos(q\beta_0) + b_1\sin(q\beta_0) = 0; \tag{5.8}$$

(2) $\bar{\rho}_{j+2} = \bar{\rho}_j$ and $\bar{\theta}_{j+2} - \bar{\theta}_j = 2\pi/q$ (since (5.4) is invariant under a rotation through $2\pi/q$);

(3) $\bar{\rho}_j = 1 + O(\delta)$.

Now we make an estimate of $\bar{\rho}_1 - \bar{\rho}_0$. From $\dot{\rho}(\bar{\rho}_j, \bar{\theta}_j, \zeta, \delta) = 0$, it follows that $\bar{\rho}_1 - \bar{\rho}_0 = O(\delta^{q-4})$. Since $\dot{\rho}(\bar{\rho}_0, \bar{\theta}_0, \zeta, \delta) = \dot{\rho}(\bar{\rho}_1, \bar{\theta}_1, \zeta, \delta) = 0$,

$$\bar{\rho}_0 - \bar{\rho}_0^3 + \delta^{q-4}\cos q\beta_0 = \bar{\rho}_1 - \bar{\rho}_1^3 + \delta^{q-4}\cos(\beta_0 q + \pi) + O(\delta^{q-3}),$$

that is,

$$(\bar{\rho}_1 - \bar{\rho}_0)\left[1 - \left(\bar{\rho}_1^2 + \bar{\rho}_1\bar{\rho}_0 + \bar{\rho}_0^2\right)\right] = 2\delta^{q-4}\cos q\beta_0 + O(\delta^{q-3}),$$

which implies

$$\bar{\rho}_1 - \bar{\rho}_0 = -\delta^{q-4}\cos q\beta_0 + O(\delta^{q-3}). \tag{5.9}$$

Finally, we consider the function $\Theta$ on the set $\{(\rho, \theta, \zeta, \delta)|R = 0, D = 0\}$. This is given by

$$\Theta\left(\bar{\rho}_j(\zeta, \delta), \bar{\theta}_j(\zeta, \delta), \zeta, \delta\right) = \zeta + b_1\rho_j^2 + O(\delta).$$

Hence, there are functions $M_j(\delta)$, $j = 0, 1, \ldots, 2q - 1$, such that

$$\Theta\left(\bar{\rho}_j(M_j(\delta), \delta), \bar{\theta}_j(M_j(\delta), \delta), M_j(\delta), \delta\right) = 0,$$

where $\delta$ is near zero. By symmetry, $M_{j+2} = M_j$. We make an estimate of $M_1(\delta) - M_0(\delta)$. The equation

$$\Theta\big(\bar{\rho}_1(M_1(\delta),\delta),\bar{\theta}_1(M_1(\delta),\delta),M_1(\delta),\delta\big)$$

$$= \Theta\big(\bar{\rho}_0(M_0(\delta),\delta),\bar{\theta}_0(M_0(\delta),\delta),M_0(\delta),\delta\big)$$

gives

$$M_1(\delta) + b_1\big(\rho_1(M_1(\delta),\delta)\big)^2 - \delta^{q-4}\sin(q\beta_0 + \pi)$$

$$= M_0(\delta) + b_1\big(\rho_0(M_0(\delta),\delta)\big)^2 - \delta^{q-4}\sin(q\beta_0) + O(\delta^{q-3}),$$

which implies, by (5.9), that

$$M_1(\delta) - M_0(\delta) = 2\delta^{q-4}\big[b_1\cos(q\beta_0) - \sin(q\beta_0)\big] + O(\delta^{q-3}).$$

By using (5.8), we have

$$\xi \equiv 2\big[b_1\cos(q\beta_0) - \sin(q\beta_0)\big] = -2\big[(b_1^2 + 1)\sin q\beta_0\big] \neq 0.$$

This completes the proof.                                                  □


*Proof of Theorem 5.1.* As mentioned above, we only need to consider equilibria bifurcation on $\Sigma_\delta^*$. This can only occur at the points where $R = \Theta = D = 0$. For $\zeta = M_0(\delta)$, these points are $(\bar{\rho}_j(M_0(\delta),\delta),$ $\bar{\theta}_j(M_0(\delta),\delta))$, where $j = 0, 2, \ldots, 2q - 2$ and $\bar{\theta}_{j+2} - \bar{\theta}_j = 2\pi/q$, $\rho_{j+2} = \rho_j$. For $\zeta = M_1(\delta)$ the situation is similar, but $j = 1, 3, \ldots, 2q - 1$. On the other hand, from the second equation of (5.4), equilibria bifurcation must take place at the values of $(\zeta,\delta)$ where $R = 0$, $\Theta = 0$, and $\sin(q\theta) = \pm 1$, which just correspond to the above two cases. Finally, by (5.3) we can change $(\zeta,\delta)$ back to $(\epsilon_1,\epsilon_2)$; the curves $(M_0(\delta),\delta)$ and $(M_1(\delta),\delta)$ in the $\zeta\delta$-plane become curves $(\delta^2, \delta^2 M_0(\delta))$ and $(\delta^2, \delta^2 M_1(\delta))$ in the $\epsilon_1\epsilon_2$-plane. Denote $h_1(\delta) = \delta^2 M_0(\delta)$, $h_2(\delta) = \delta^2 M_1(\delta)$. Then $h_1(\delta) - h_2(\delta) = \delta^2(M_0(\delta) - M_1(\delta)) = \xi\delta^{q-2} + O(\delta^{q-1})$, $\xi \neq 0$ (see Lemma 5.2). This completes the proof of the theorem.                                                  □


**Remark 5.3.** Figure 5.4 is drawn for $b_1 < 0$. The case $b_1 \geq 0$ is similar.

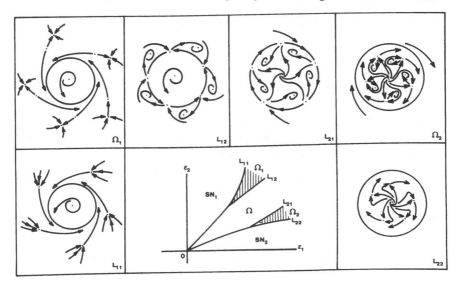

Figure 5.5. The additional bifurcation diagram and phase portraits of (5.2) $(q = 5, b_1 < 0)$.

**Remark 5.4.** The above results are established in a small neighborhood of $\epsilon_1 = \epsilon_2 = 0$. Otherwise, some additional bifurcation curves $L_{ij}$ $(i, j = 1, 2)$ will appear. The system (5.2) will have some additional phase portraits (see Figure 5.5 which is obtained numerically).

### 4.6 A Purely Imaginary Pair of Eigenvalues and a Simple Zero Eigenvalue

As we mentioned in the beginning of this chapter, we need the truncated normal form equations for types $A_2$ and $A_3$, and they have respectively the following forms (see Section 2.11):

$$\begin{cases} \dot{x} = \epsilon_1 x + axy + \bar{d}_1 x^3 + \bar{d}_2 xy^2, \\ \dot{y} = \epsilon_2 + bx^2 + cy^2 + \bar{d}_3 x^2 y + \bar{d}_4 y^3, \end{cases} \tag{$A_2$}$$

where $\epsilon_1, \epsilon_2$ are small parameters, $a, b, c, \bar{d}_j \in \mathbb{R}^1$, and $abc \neq 0$, and

$$\begin{cases} \dot{x} = x\left(\epsilon_1 + p_1 x^2 + p_2 y^2 + q_1 x^4 + q_2 x^2 y^2 + q_3 y^4\right), \\ \dot{y} = y\left(\epsilon_2 + p_3 x^2 + p_4 y^2 + q_4 x^4 + q_5 x^2 y^2 + q_6 y^4\right), \end{cases} \quad (A_3)$$

where $x \geq 0$, $y \geq 0$, $p_j, q_j \in \mathbb{R}^1$, and $\epsilon_1, \epsilon_2$ are small parameters.

We will discuss the dynamical behavior of equation $A_2$ in this section, and equation $A_3$ in the next section. Żoładek [1, 2] obtained the complete results for these two cases. We will use a simpler method to prove the uniqueness of periodic orbits for equation $A_2$.

Since equation $A_2$ is symmetric with respect to the $y$-axis, we only need to consider the half plane $x \geq 0$. To obtain a simpler form, we let

$$x \to |c|^{1/2} x, \quad y \to |b|^{1/2} y, \quad t \to -\frac{t}{c|b|^{1/2}}, \quad (6.1)$$

$$\epsilon_1 \to -c|b|^{1/2}\epsilon_1, \quad \epsilon_2 \to -c|b|\epsilon_2.$$

Thus, $A_2$ becomes

$$\begin{cases} \dot{x} = \epsilon_1 x + Bxy + d_1 x^3 + d_2 xy^2, \\ \dot{y} = \epsilon_2 + \eta x^2 - y^2 + d_3 x^2 y + d_4 y^3, \end{cases} \quad (6.2)$$

where $B = -a/c \neq 0$, $\eta = -\text{sgn}(bc)$, $d_j \in \mathbb{R}^1$.

If we assume

$$K_3 = \eta\left(\frac{2}{B} + 2\right)d_1 + \frac{2}{B}d_2 + \eta d_3 + 3d_4 \neq 0,$$

then the qualitative behavior of (6.2) near $(0, 0)$ with small $\epsilon_1$ and $\epsilon_2$ is the same as that of the equation

$$\begin{cases} \dot{x} = \epsilon_1 x + Bxy + xy^2, \\ \dot{y} = \epsilon_2 + \eta x^2 - y^2 \end{cases} \quad (6.3)$$

(see Remark 6.7). The main result of this section is as follows.

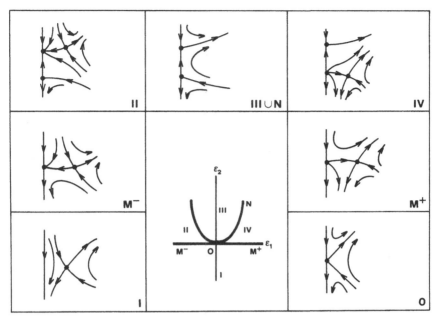

Figure 6.1. Case (I): $\eta = 1$, $B > 0$.

**Theorem 6.1.** *For equations (6.3) there are four cases: (I) $\eta = 1$, $B > 0$; (II) $\eta = 1$, $B < 0$; (III) $\eta = -1$, $B > 0$; and (IV) $\eta = -1$, $B < 0$. We have:*

*(1) In cases (I) and (IV), the bifurcation diagram of (6.3) consists of the origin and the following curves in parameter space:*

$$M = \{(\epsilon_1, \epsilon_2) | \epsilon_2 = 0, \ \epsilon_1 \neq 0\},$$

$$N = \left\{(\epsilon_1, \epsilon_2) \Big| \epsilon_2 = \frac{\epsilon_1^2}{B^2} + O(\epsilon_1^3), \ \epsilon_1 \neq 0\right\}.$$

*Along $M$ and $N$, saddle-node bifurcation and pitchfork bifurcation occur respectively. Equation (6.3) has no periodic orbits, and the phase portraits are shown in Figure 6.1 and Figure 6.4 for cases (I) and (IV), respectively.*

*(2) In case (III), the bifurcation diagram of (6.3) consists of the origin, the curves $M$, $N$, and the following curves:*

$$H = \{(\epsilon_1, \epsilon_2) | \epsilon_1 = 0, \ \epsilon_2 > 0\},$$

*and*

$$S = \left\{ (\epsilon_1, \epsilon_2) | \epsilon_1 = -\frac{B}{3B+2}\epsilon_2 + O\big(|\epsilon_2|^{3/2}\big), \epsilon_2 > 0 \right\}.$$

*Along M and N, we have exactly the same bifurcations as in (1). Along H and S, Hopf bifurcation and heteroclinic bifurcation occur respectively. If $(\epsilon_1, \epsilon_2)$ lies between the curves H and S, then (6.3) has a unique limit cycle which is unstable and becomes a heteroclinic orbit when $(\epsilon_1, \epsilon_2) \in S$. Phase portraits in case (III) for different $(\epsilon_1, \epsilon_2)$ are shown in Figure 6.3.*

*(3) In case (II), the bifurcation diagram of (6.3) consists of the origin, the curves M, N, and the following curve:*

$$\overline{H} = \{ (\epsilon_1, \epsilon_2) | \epsilon_1 = 0, \epsilon_2 < 0 \}.$$

*The bifurcations along M and N are the same as in (1) and (2). Along $\overline{H}$, Hopf bifurcation occurs. If we localize equation (6.3) by restricting $0 < x < \beta, |\epsilon_1|^{1/2} + |\epsilon_2|^{1/2} < \beta, 0 < \beta \ll 1$, then there exists a curve*

$$\overline{S} = \left\{ (\epsilon_1, \epsilon_2) | \epsilon_1 = \phi(\beta, \epsilon_2)\epsilon_2 + O\big(|\epsilon_2|^{3/2}\big), \epsilon_2 < 0 \right\},$$

*where*

$$\phi(\beta, \epsilon_2) = \frac{\iint_{\Omega(\beta, \epsilon_2)} x^q y^2 \, dx \, dy}{\iint_{\Omega(\beta, \epsilon_2)} x^q \, dx \, dy},$$

*with $\Omega(\beta, \epsilon_2)$ the bounded region surrounded by the closed curve*

$$x^{2/B}\left( 1 - \frac{x^2}{B+1} + y^2 \right) = \frac{\beta^{2/B}}{|\epsilon_2|^{1/2}}\left( 1 - \frac{\beta^2}{|\epsilon_2|(B+1)} \right), \qquad if\ B+1 \neq 0,$$

*or*

$$\frac{1+y^2}{2x^2} + \ln x = \frac{|\epsilon_2|}{2\beta^2} + \ln\frac{\beta}{|\epsilon_2|^{1/2}}, \qquad if\ B+1 = 0.$$

*If $(\epsilon_1, \epsilon_2)$ lies between the curves $\overline{H}$ and $\overline{S}$, then (6.3) has a unique limit cycle which is located entirely in the strip $0 < x < \beta$ and touches the line*

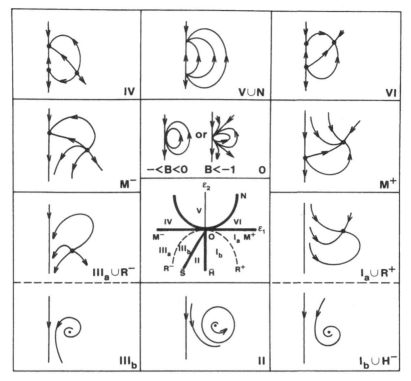

Figure 6.2.  Case (II): $\eta = 1$, $B < 0$.

$x = \beta$ as $(\epsilon_1, \epsilon_2) \to \bar{S}$. *Phase portraits of (6.3) in case (II) are shown in Figure 6.2.*

**Proof.** The proof of Theorem 6.1 is given in the following steps.

*Step 1. Bifurcations of equilibria*  Clearly, (6.3) has an invariant line $x = 0$ and is symmetric with respect to the $y$-axis. Hence, we will only consider the half plane $x \geq 0$.

On the $y$-axis $x = 0$, there are two equilibria $(0, \pm \sqrt{\epsilon_2})$ if $\epsilon_2 > 0$, and the linearized system at $(0, \pm \sqrt{\epsilon_2})$ is given by the matrix

$$A_{\pm} = \begin{bmatrix} \epsilon_1 \pm B\sqrt{\epsilon_2} + \epsilon_2 & 0 \\ 0 & \mp 2\sqrt{\epsilon_2} \end{bmatrix}. \tag{6.4}$$

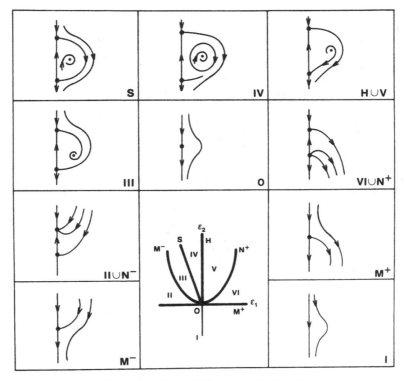

Figure 6.3. Case (III): $\eta = -1$, $B > 0$.

Note that $B \cdot (\det A_+) < 0$ if $\epsilon_2 > \epsilon_1^2/B^2 + O(\epsilon_1^3)$ (i.e., $(\epsilon_1, \epsilon_2)$ is above the curve $N$), and $\epsilon_1 \cdot (\det A_+) < 0$ if $0 < \epsilon_2 < \epsilon_1^2/B^2 + O(\epsilon_1^3)$ (i.e., $(\epsilon_1, \epsilon_2)$ is between curves $N$ and $M$, see Figure 6.5). This means that if $(\epsilon_1, \epsilon_2)$ is above $N$, then both equilibria $(0, \pm \sqrt{\epsilon_2})$ are saddle points in cases (I) and (III) and are nodes in cases (II) and (IV). If $(\epsilon_1, \epsilon_2)$ is between $M$ and $N$, then one of $(0, \pm \sqrt{\epsilon_2})$ is a saddle point and the other one is a node for cases (I)–(IV), and these two equilibria form a saddle-node when $\epsilon_2 = 0$ ($\epsilon_1 \neq 0$).

In the open half plane $x > 0$, equation (6.3) has two equilibria:

$$ x = \left[ \eta \left( y^2 - \epsilon_2 \right) \right]^{1/2}, \qquad y = \frac{1}{2} \left[ -B \pm \left( B^2 - 4\epsilon_1 \right)^{1/2} \right]. $$

Obviously, only one of them exists in a small neighborhood of the origin if $|\epsilon_1|$ and $|\epsilon_2|$ are sufficiently small. If we denote this equilibrium by

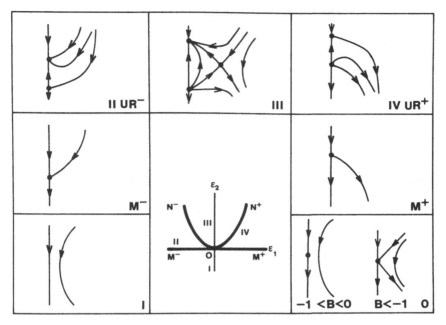

Figure 6.4.  Case (IV): $\eta = -1$, $B < 0$.

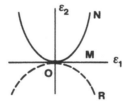

Figure 6.5.

$(x_3, y_3)$, then

$$
\begin{cases}
x_3 = \left[ \eta\left(\dfrac{\epsilon_1^2}{B^2}\right) - \epsilon_2 + O(\epsilon_1^3) \right]^{1/2}, \\
y_3 = -\dfrac{\epsilon_1}{B} + O(\epsilon_1^2),
\end{cases}
\qquad \text{as } \epsilon_1 \to 0, \qquad (6.5)
$$

provided $\eta(\epsilon_1^2/B^2 - \epsilon_2 + O(\epsilon_1^3)) > 0$. This means that $(x_3, y_3)$ exists in cases (I) and (II) if $(\epsilon_1, \epsilon_2)$ is below the curve $N$, and in cases (III) and

(IV) if $(\epsilon_1, \epsilon_2)$ is above the curve $N$. If $(\epsilon_1, \epsilon_2) \in N$ (i.e., $x_3 = 0$), then $(x_3, y_3)$ merges into one of the equilibria $(0, \pm \sqrt{\epsilon_2})$. If $(\epsilon_1, \epsilon_2) = (0, 0)$, then all equilibria merge into the origin $(x, y) = (0, 0)$.

The linearized system at $(x_3, y_3)$ is given by the matrix:

$$A = \begin{bmatrix} 0 & Bx_3\left(1 + \dfrac{2}{B}y_3\right) \\ 2\eta x_3 & -2y_3 \end{bmatrix}. \tag{6.6}$$

The determinant of $A$ has the same sign as $-\eta B$ (if $x_3 \neq 0$). Hence $(x_3, y_3)$ is a saddle point in cases (I) and (IV), and is an equilibrium which may be a focus, node, or center in cases (II) and (III). It can be shown that:

(1) In case (II) if $(\epsilon_1, \epsilon_2)$ lies between the curves $N$ and $R$, where

$$R: \epsilon_2 = \frac{2B + 1}{2B^3}\epsilon_1^2 + O(\epsilon_1^3),$$

then $(x_3, y_3)$ is a node.

(2) If $(\epsilon_1, \epsilon_2)$ is below $R$ in case (II) or above $N$ in case (III), then $(x_3, y_3)$ is a focus which changes its stability when $(\epsilon_1, \epsilon_2)$ crosses the curve $\overline{H}$ in case (II) or $(\epsilon_1, \epsilon_2)$ crosses the curve $H$ in case (III).

Summing up the above discussion, we conclude that in cases (I) and (IV) the system (6.3) has no periodic orbits. In fact, the $y$-axis is an invariant line. Hence, periodic orbits in the half plane $x > 0$, if they exist, must surround the equilibrium $(x_3, y_3)$. But this is impossible, since $(x_3, y_3)$ is a saddle in cases (I) and (IV). In these cases, the bifurcation diagram consists of curves $N$ and $M$. Saddle-node bifurcations occur on $M$ while pitchfork bifurcations occur on $N$. In cases (II) and (III), the bifurcations on $M$ and $N$ are similar to the cases (I) and (IV). However, a Hopf bifurcation occurs on $\overline{H}$ and $H$, respectively.

*Step 2. The stability of $(x_3, y_3)$ when $(\epsilon_1, \epsilon_2) \in \overline{H}$ in case (II) and $(\epsilon_1, \epsilon_2) \in H$ in case (III)* From (6.5) and (6.6), it is easy to determine the stability of $(x_3, y_3)$ when $(\epsilon_1, \epsilon_2)$ is in the two regions divided by $H$ (or $\overline{H}$), and the stability is opposite in these regions. Hence $H(\overline{H})$ is a Hopf bifurcation curve. It is necessary to determine the stability of $(x_3, y_3)$ when $(\epsilon_1, \epsilon_2) \in \overline{H}$ (or $H$). By using the formula (3.2.34) we have

$$\text{Re}(C_1) = a\epsilon_2 \qquad a > 0 \text{ constant.}$$

Hence, $(x_3, y_3)$ is an unstable focus when $(\epsilon_1, \epsilon_2) \in H$ (case (III)), and a stable focus when $(\epsilon_1, \epsilon_2) \in \bar{H}$ (case (II)). Therefore, in both cases a limit cycle appears when $(\epsilon_1, \epsilon_2)$ moves across $H$ or $\bar{H}$ from the right to the left, and the bifurcating limit cycle is stable in case (II) and unstable in case (III).

*Step 3. Uniqueness of periodic orbits* As shown in Step 2, (6.3) can have periodic orbits only if $\epsilon_2 < 0$ in case (II) and $\epsilon_2 > 0$ in case (III). We choose a small parameter $\delta > 0$ and let

$$x \to \delta x, \quad y \to \delta y, \quad dt \to \frac{x^q}{\delta} \, dt, \quad \epsilon_1 = a\delta^2, \quad \epsilon_2 = -\eta\delta^2,$$

where $q + 1 = 2/B$ and $\eta = -\text{sgn } B$. Then (6.3) is transformed into the form

$$\begin{cases} \dot{x} = x^q (Bxy + \delta(\alpha x + xy^2)), \\ \dot{y} = x^q (-\eta + \eta x^2 - y^2), \end{cases} \tag{6.7}$$

where $\eta = 1$, $B < 0$ (case (II)) or $\eta = -1$, $B > 0$ (case (III)). When $\delta = 0$, (6.7) becomes a Hamiltonian system

$$\begin{cases} \dot{x} = x^q \cdot Bxy, \\ \dot{y} = x^q (-\eta + \eta x^2 - y^2), \end{cases} \tag{6.8}$$

with a first integral $H(x, y) = h$, where

$$H(x, y) = \frac{B}{2} x^{q+1} \left( -\eta - y^2 + \eta \frac{x^2}{B+1} \right), \qquad \text{if } B + 1 \neq 0, \quad (6.9)$$

and

$$H(x, y) = \frac{1 + y^2}{2x^2} + \ln x, \qquad \text{if } B + 1 = 0. \tag{6.10}$$

Note that $H(x, y)$ is the Hamiltonian function of (6.8).

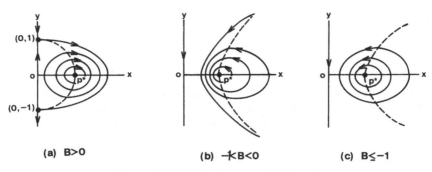

(a) B>0                (b) −1<B<0                (c) B≤−1

Figure 6.6. The level curves of $H(x, y)$.

The closed curves $\Gamma_h = \{(x, y)|H(x, y) = h\}$ are shown in Figure 6.6 for cases (a) $\eta = -1$, $B > 0$, (b) $\eta = 1$, $-1 < B < 0$, and (c) $\eta = 1$, $B \le -1$, respectively (see Remark 6.5), where $h \in J_i$ ($i = 1, 2, 3, 4$) and $J_i$ corresponds to different values of $B$:

$$\begin{cases} J_1 = (0, h_1^*), \; h_1^* = \dfrac{B^2}{2(B + 1)}, & \text{if } B > 0, \\[3mm] J_2 = (h_2^*, 0), \; h_2^* = \dfrac{-B^2}{2(B + 1)} < 0, & \text{if } -1 < B < 0, \\[3mm] J_3 = (h_3^*, +\infty), \; h_3^* = \dfrac{1}{2}, & \text{if } B = -1, \\[3mm] J_4 = (h_4^*, +\infty), \; h_4^* = -\dfrac{B^2}{2(B + 1)} > 0, & \text{if } B < -1. \end{cases} \qquad (6.11)$$

When $h \to h_i^*$ ($i = 1, 2, 3, 4$), the level curve $\Gamma_h$ shrinks to the equilibrium $(x^*, y^*) = (1, 0)$ of (6.8); and when $h \to 0$, $\Gamma_h$ expands to the heteroclinic orbit in the case $B > 0$. Since

$$\left.\frac{dH}{dt}\right|_{(6.7)} = \delta x^q(\alpha x + xy^2)\frac{dy}{dt},$$

as in Sections 4.1–4.3, to study the periodic orbits of (6.7), we obtain a bifurcation function $\Phi(h, \delta, \alpha, B)$ which for $\delta = 0$ is given by

$$\Phi|_{\delta=0} = \int_{\Gamma_h} x^q(\alpha x + xy^2)\, dy = -(q + 1)\int_{\Gamma_h} x^q\left(\alpha y + \frac{1}{3}y^3\right)\, dx,$$

$$(6.12)$$

where $h \in J_i$ and the orientation of $\Gamma_h$ is defined by the direction of the vector field (6.8). The condition $\Phi|_{\delta=0} = 0$ is equivalent to

$$\alpha = -\frac{1}{3}P(h, B),$$

where

$$P(h, B) = \frac{I_3(h, B)}{I_1(h, B)}, \quad I_j(h, B) = \int_{\Gamma_h} x^q y^j \, dx, \quad j = 1, 3. \quad (6.13)$$

In a manner similar to Sections 4.1–4.3, the uniqueness of periodic orbits of (6.7) is equivalent to the monotonicity of $P(h, B)$ with respect to $h$. The following three lemmas give the monotonic property of $P(h, B)$.

**Lemma 6.2.** *If* $h \in J_i$, *then* $P(h, B) > 0$ *and*

$$\lim_{h \to h_i^*} P(h, B) = 0, \quad i = 1, 2, 3, 4.$$

*Proof.* For simplicity, we denote $P(h, B)$ and $I_i(h, B)$ by $P(h)$ and $I_i(h)$, respectively. Using Green's Theorem, we have from (6.13) that

$$P(h) = \frac{3 \iint_{\Omega_h} x^q y^2 \, dx \, dy}{\iint_{\Omega_h} x^q \, dx \, dy} > 0 \quad \text{for } h \in J_i,$$

where $\Omega_h$ is the compact region surrounded by $\Gamma_h$, and it is contained in the open right half plane. From the above expression of $P(h)$, it is easy to see that

$$\lim_{h \to h^*} P(h) = \lim_{y \to y^*} 3y^2 = 0,$$

since $\Omega(h)$ shrinks to the point $(x^*, y^*)$ as $h \to h^*$, and $(x^*, y^*) = (1, 0)$. $\qquad \square$

**Lemma 6.3.** *If there exists $h_0 \in J_i$ such that $P'(h_0) = 0$, then $P(h) - P(h_0) > 0$ for $0 < |h - h_0| \ll 1$.*

*Proof.* Suppose $y = y_1(x, h)$ and $y = y_2(x, h)$ are defined by $H(x, y) = h$ for $h \in J_i$ and $a_1(h) < x < a_2(h)$, where $a_1(h)$ and $a_2(h)$ are the intersection points between $\Gamma_h$ and the $x$-axis. Obviously, $0 < a_1(h) < 1 < a_2(h)$. We will use $y(x, h)$, or simply use $y$, to denote $y_1(x, h)$ or $y_2(x, h)$ if there is no confusion. Using (6.9), (6.10), and (6.13), we have

$$\frac{\partial y}{\partial h} = -\frac{1}{Bx^{q+1}y} \qquad (\text{along } \Gamma_h) \qquad (6.14)$$

and

$$I_1(h) = \int_{\Gamma_h} x^q y \, dx = 2(\operatorname{sgn} B) \int_{a_1(h)}^{a_2(h)} x^q y_1(x, h) \, dx. \qquad (6.15)$$

Since $0 < a_1(h) < 1 < a_2(h)$, for fixed $h \in J_i$ we have that

$$\lim_{\substack{x \to a_i(h) \\ (x, y) \in \Gamma_h}} \frac{y^2}{x - a_i(h)} = \lim_{\substack{x \to a_i(h) \\ (x, y) \in \Gamma_h}} 2y \frac{dy}{dx} = \frac{2\eta(a_i^2(h) - 1)}{Ba_i(h)} \neq 0.$$

This implies that

$$|y| = O\left(|x - a_i(h)|^{1/2}\right) \qquad \text{as } x \to a_i(h),$$

$$(x, y) \in \Gamma_h, \qquad i = 1, 2. \quad (6.16)$$

Noting $y_1(a_1(h), h) = y_2(a_2(h), h) = 0$, from (6.14) and (6.15) we have

$$I_1'(h) = -\frac{1}{B} \int_{\Gamma_h} \frac{1}{xy} \, dx. \qquad (6.17)$$

Here the integral is convergent because of (6.16). From (6.13) and (6.14) it is not hard to obtain that

$$I_3'(h) = -\frac{3}{B} \int_{\Gamma_h} \frac{y}{x} \, dx. \qquad (6.18)$$

On the other hand, from (6.13) we have

$$P(h) - P(h_0) = \frac{I_3(h)}{I_1(h)} - \frac{I_3(h_0)}{I_1(h_0)} = \frac{\xi(h) - \xi(h_0)}{I_1(h)I_1(h_0)},$$

where

$$\xi(h) = I_1(h_0)I_3(h) - I_3(h_0)I_1(h).$$

Hence

$$P(h) - P(h_0) = \frac{\xi'(\theta)(h - h_0)}{I_1(h)I_1(h_0)} \equiv \frac{h - h_0}{I_1(h)} Q(\theta), \qquad (6.19)$$

where $\theta$ is in between $h$ and $h_0$, and

$$Q(h) = I_3'(h) - P(h_0)I_1'(h). \qquad (6.20)$$

From (6.20), (6.13), and the condition $P'(h_0) = 0$, we have

$$Q(h_0) = 0. \qquad (6.21)$$

We consider two cases separately:

(i) $\eta = -1$, $B > 0$ (case (III)). The direction of the vector field (6.8) on $\Gamma_h$ is clockwise. Substituting (6.17) and (6.18) into (6.20), we have

$$Q(h) = \int_{\Gamma_h} \frac{P(h_0) - 3y^2}{Bxy} \, dx. \qquad (6.22)$$

Equations (6.21) and (6.22) imply

$$Q(h) = Q(h) - Q(h_0) = \int_{\Gamma_h} \frac{P(h_0) - 3y^2}{Bxy} \, dx - \int_{\Gamma_{h_0}} \frac{P(h_0) - 3y^2}{Bxy} \, dx.$$

$$(6.23)$$

If $h > h_0$ then $\Gamma_h \subset \Omega_{h_0}$, where $\Omega_{h_0}$ is the compact region surrounded

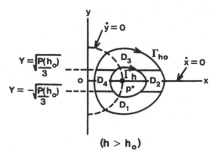

Figure 6.7.

by $\Gamma_{h_0}$. From (6.23) we obtain

$$Q(h) = \frac{1}{B}\int_{\partial D+} \frac{P(h_0) - 3y^2}{xy}\, dx$$

$$= \frac{1}{B}\left[\int_{\partial D_1^+ \cup \partial D_3^+} \frac{P(h_0) - 3y^2}{xy}\, dx - \int_{\partial D_2^+ \cup \partial D_4^+} \frac{P(h_0) - 3y^2}{1 - x^2 - y^2}\, dy\right]$$

$$= \frac{1}{B}\left[\iint_{D_1 \cup D_3} \frac{P(h_0) + 3y^2}{xy^2}\, dx\, dy\right.$$

$$\left. + \iint_{D_2 \cup D_4} \frac{2x(P(h_0) - 3y^2)}{(1 - x^2 - y^2)^2}\, dx\, dy\right] > 0, \qquad (6.24)$$

where $D = \cup_{i=1}^4 D_i$ is the annular domain bounded by $\Gamma_h$ and $\Gamma_{h_0}$, and $D_1$, $D_2$, $D_3$, and $D_4$ are formed by the truncation lines $y = \pm(P(h_0)/3)^{1/2}$. We note that they are mutually disjoint and satisfy the following properties (see Figure 6.7):

$$\{(x, y)|x^2 + y^2 = 1\} \cap D \subset D_1 \cup D_3,$$

$$\{(x, y)|y = 0\} \cap D \subset D_2 \cup D_4.$$

Because $P(h_0) > 0$ (Lemma 6.2) and $Q(h_0) = 0$, the lines $y = \pm(P(h_0)/3)^{1/2}$ must intersect $\Gamma_{h_0}$ (see (6.22)). Hence the above partition of $D$ into $D_1, \ldots, D_4$ is always possible if $0 < |h - h_0| \ll 1$. The orientation of $\partial D^+$ (or $\partial D_i^+$) is defined in the usual way: The region $D$

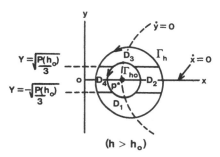

Figure 6.8.

(or $D_i$) is always on the left side if one goes along $\partial D$ (or $\partial D_i$). In (6.24) the integrand along $\partial D_2^+ \cup \partial D_4^+$ is transformed by using (6.8) and the integrand along every part of the truncation lines $y = \pm (P(h_0)/3)^{1/2}$ is zero. If $h < h_0$, then $\Gamma_{h_0} \subset \Omega_h$. We obtain $Q(h) < 0$ in the same way. Hence $Q(h)(h - h_0) > 0$ for $0 < |h - h_0| \ll 1$. By using (6.19), we have $P(h) - P(h_0) > 0$ for $0 < |h - h_0| \ll 1$ since $I_1(h) > 0$.

(ii) $\eta = 1$, $B < 0$ (case (II)). This case is similar to case (i). We will only indicate the main differences between the two cases. The direction of the vector field (6.8) on $\Gamma_h$ is counterclockwise. If $h > h_0$ (the case $h < h_0$ is similar), then $\Gamma_{h_0} \subset \Omega_h$, and (6.23) gives

$$Q(h) = \frac{1}{B} \left[ \int_{\partial D_1^+ \cup \partial D_3^+} \frac{P(h_0) - 3y^2}{xy} \, dx + \int_{\partial D_2^+ \cup \partial D_4^+} \frac{P(h_0) - 3y^2}{-1 + x^2 - y^2} \, dy \right]$$

$$= \frac{1}{B} \left[ \iint_{D_1 \cup D_3} \frac{P(h_0) + 3y^2}{xy^2} \, dx \, dy \right.$$

$$\left. - \iint_{D_2 \cup D_4} \frac{2x(P(h_0) - 3y^2)}{(-1 + x^2 - y^2)^2} \, dx \, dy \right] < 0,$$

where $D$ is the annular domain bounded by $\Gamma_h$ and $\Gamma_{h_0}$, and $D_1, \ldots, D_4$ are formed by truncation lines $y = \pm (P(h_0)/3)^{1/2}$. They are mutually disjoint and satisfy the following properties (see Figure 6.8):

$$\{(x, y) | x^2 - y^2 = 1\} \cap D \subset D_1 \cup D_3,$$

$$\{(x, y) | y = 0\} \cap D \subset D_2 \cup D_4.$$

Therefore, $P(h) - P(h_0) > 0$ for $0 < |h - h_0| \ll 1$ since $I_1(h) < 0$. $\quad\square$

**Lemma 6.4.** $P(h)$ is monotone in $h \in J_i$, $i = 1, 2, 3, 4$.

*Proof.* By (6.13), (6.17), and (6.18), we have that $P(h) \in C^1(J_i)$. Suppose there is an $h_0 \in J_i$ such that $P'(h_0) = 0$. Lemma 6.3 implies that such an $h_0$ is unique. Hence $P(h) - P(h_0) > 0$ for all $h \in J_i$, $h \neq h_0$. Therefore, by using Lemma 6.3 and the first part of Lemma 6.2 we obtain that $P(h) > P(h_0) > 0$ for $h \in J_i$, $h \neq h_0$. This implies

$$\lim_{h \to h_i^*} P(h) \geq P(h_0) > 0,$$

which contradicts the second part of Lemma 6.2, and the desired result follows.    □

**Remark 6.5.** Instead of equation (6.8), we consider

$$\begin{cases} \dot{x} = Bxy, \\ \dot{y} = -\eta + \eta x^2 - y^2, \end{cases} \tag{6.25}$$

where $B \neq 0$, $\eta = \pm 1$, $\eta B < 0$. It is easy to see that equation (6.25) has an equilibrium $P^*(1,0)$, and trace $(A(P^*)) = 0$, $\det(A(P^*)) = -2B\eta > 0$, where $A(P^*)$ is the matrix of the linear part of (6.25) at point $P^*$. On the other hand, equations (6.25) are symmetric with respect to the $x$-axis. Hence, $P^*$ is a center. Using (6.9) and (6.10) we obtain Figure 6.6.

**Remark 6.6.** In order to give the equation of the heteroclinic bifurcation curve $S$ in case (III), we need the value $P(0, B)$ for $B > 0$. In this case

$$P(0, B) = \frac{\displaystyle\int_{\Gamma_0} x^q y^3 \, dx}{\displaystyle\int_{\Gamma_0} x^q y \, dx} = \frac{\displaystyle\int_0^1 x^q (1 - x^2)^{3/2} \, dx}{\displaystyle\int_0^1 x^q (1 - x^2)^{1/2} \, dx}$$

$$= \frac{\displaystyle\int_0^1 u^{(q-1)/2} (1 - u)^{3/2} \, du}{\displaystyle\int_0^1 u^{(q-1)/2} (1 - u)^{1/2} \, du} = \frac{B\left(\dfrac{q+1}{2}, \dfrac{5}{2}\right)}{B\left(\dfrac{q+1}{2}, \dfrac{3}{2}\right)},$$

where $B(\alpha, \beta)$ is the usual beta function. By using the properties:

$$B(\alpha, \beta) = \frac{\Gamma(\alpha)\Gamma(\beta)}{\Gamma(\alpha + \beta)}$$

and $\Gamma(a) = (a - 1)\Gamma(a - 1)$, we obtain

$$P(0, B) = \frac{3}{q + 4} = \frac{3B}{3B + 2}.$$

Noting

$$\alpha = -\frac{1}{3}P(h, B) \quad \text{and} \quad \frac{\epsilon_2}{\epsilon_1} = -\frac{\eta}{\alpha},$$

we obtain

$$\frac{\epsilon_1}{\epsilon_2} = -\frac{\alpha}{\eta} = \frac{P(h, B)}{3\eta} + O(\delta). \tag{6.26}$$

In our case $\eta = -1$, $h = 0$, and $\delta \sim |\epsilon_2|^{1/2}$ as $\epsilon_2 \to 0$. Hence, the equation of the curve $S$ is given by

$$\epsilon_1 = -\frac{B}{3B + 2}\epsilon_2 + O\left(|\epsilon_2|^{3/2}\right).$$

The equation of the curve $\bar{S}$ in case (II) is easily determined by using (6.26), where $\eta = 1$, $h = H(\xi, 0)$ (for the definition of $H(x, y)$, see (6.9) and (6.10)), $\xi = \beta/|\epsilon_2|^{1/2} > 1$, and $\delta \sim |\epsilon_2|^{1/2}$ as $\epsilon_2 \to 0$.

**Remark 6.7.** We now show why we can consider equation (6.3) instead of equation (6.2). In fact, if (6.2) has no periodic orbits, then its qualitative behavior does not depend on the third-order terms. Thus, we only need to consider (6.2) for $\epsilon_2 < 0$ in case (II), and for $\epsilon_2 > 0$ in case (III). In these cases, we can make the same scaling for (6.2) as for (6.3), and obtain the same Hamiltonian system (6.8) if $\delta = 0$. The bifurcation function, similarly to (6.12), is

$$\Phi|_{\delta=0} = \int_{\Gamma_h} x^q (\alpha x + d_1 x^3 + d_2 xy^2)\, dy - x^q (d_3 x^2 y + d_4 y^3)\, dx$$

$$= -(q + 1)\alpha I_1(h) - \left(\frac{q + 1}{3} d_2 + d_4\right) I_3(h)$$

$$- [(q + 3)d_1 + d_3] \int_{\Gamma_h} x^{q+2} y\, dx,$$

where $I_1(h)$ and $I_3(h)$ are the same as in (6.13).

Along $\Gamma_h$ we have

$$x^q (-\eta + \eta x^2 - y^2)\, dx - x^q Bxy\, dy = 0.$$

Hence

$$\eta \int_{\Gamma_h} x^{q+2} y\, dx = \int_{\Gamma_h} x^q (\eta + y^2) y\, dx + B \int_{\Gamma_h} x^{q+1} y^2\, dy$$

$$= \eta I_1 + \left(1 - \frac{q + 1}{3} B\right) I_3 = \eta I_1 + \frac{1}{3} I_3.$$

Thus,

$$\Phi|_{\delta=0} = -(q + 1)\alpha I_1 + K_3 I_3 + K_1 I_1,$$

where

$$K_3 = -\frac{1}{3} [\eta(q + 3)d_1 + (q + 1)d_2 + \eta d_3 + 3d_4],$$

$$K_1 = -[(q + 3)d_1 + d_3].$$

This means that as long as $K_3 \neq 0$ we can choose any values for $d_1$, $d_2$, $d_3$, and $d_4$ without changing the existence and uniqueness of periodic orbits of system (6.2). For system (6.3), $d_1 = d_3 = d_4 = 0$, $d_2 = 1$, and

hence

$$K_3 = -\frac{1}{3}(q + 1) = -\frac{2}{3B} \neq 0.$$

### 4.7 Two Purely Imaginary Pairs of Eigenvalues

As we mentioned at the beginning of Section 4.6, in this case the truncated normal form equation is

$$\begin{cases} \dot{x} = x\left(\epsilon_1 + p_1 x^2 + p_2 y^2 + \bar{q}_1 x^4 + \bar{q}_2 x^2 y^2 + \bar{q}_3 y^4\right), \\ \dot{y} = y\left(\epsilon_2 + p_3 x^2 + p_4 y^2 + \bar{q}_4 x^4 + \bar{q}_5 x^2 y^2 + \bar{q}_6 y^4\right), \end{cases} \quad (7.1)$$

where $x \geq 0$, $y \geq 0$, $p_i, \bar{q}_j \in \mathbb{R}^1$, $i = 1, 2, 3, 4$, $j = 1, 2, \ldots, 6$, and $\epsilon_1$ and $\epsilon_2$ are small parameters.

By changing $(x^2, y^2) \to (x, y)$ and $t \to \frac{1}{2}t$, equation (7.1) becomes

$$\begin{cases} \dot{x} = x\left(\epsilon_1 + p_1 x + p_2 y + \bar{q}_1 x^2 + \bar{q}_2 xy + \bar{q}_3 y^2\right), \\ \dot{y} = y\left(\epsilon_2 + p_3 x + p_4 y + \bar{q}_4 x^2 + \bar{q}_5 xy + \bar{q}_6 y^2\right). \end{cases} \quad (7.2)$$

We suppose that

$$p_i \neq 0 \quad (i = 1, 2, 3, 4) \quad \text{and} \quad \begin{vmatrix} p_1 & p_2 \\ p_3 & p_4 \end{vmatrix} \neq 0. \quad (7.3)$$

Let

$$x \to \frac{x}{|p_1|}, \quad y \to \frac{y}{|p_2|}, \quad t \to -(\operatorname{sgn} p_2)t. \quad (7.4)$$

Then (7.2) takes the form

$$\begin{cases} \dot{x} = x\left(\mu_1 + \eta x - y + q_1 x^2 + q_2 xy + q_3 y^2\right), \\ \dot{y} = y\left(\mu_2 - \frac{\alpha + 1}{\beta}\eta x + \frac{\alpha}{\beta + 1}y + q_4 x^2 + q_5 xy + q_6 y^2\right), \end{cases} \quad (7.5)$$

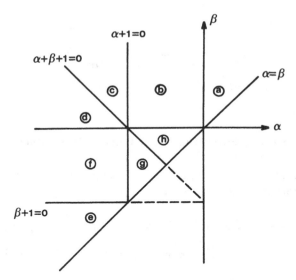

Figure 7.1. The partition of half plane $\alpha \le \beta$.

where $\eta = 1$ if $p_1 p_2 < 0$ and $\eta = -1$ if $p_1 p_2 > 0$,

$$\frac{\alpha + 1}{\beta} = -\frac{p_3}{p_1}, \qquad \frac{\alpha}{\beta + 1} = -\frac{p_4}{p_2},$$

and condition (7.3) becomes

$$\alpha \beta (\alpha + 1)(\beta + 1)(\alpha + \beta + 1) \ne 0. \qquad (7.6)$$

We only need to consider the case $\alpha \le \beta$. Otherwise, let $(x, y) \to (y, x)$. Then by using the following change of variables: $x = |(\beta + 1)/\alpha|\bar{x}$, $y = |\beta/(\alpha + 1)|\bar{y}$, and $t = \eta(\text{sgn}(\beta/(\alpha + 1)))\bar{t}$, $\alpha$ and $\beta$ in equation (7.5) are interchanged $(\bar{\mu}_1 = \mu_2(\text{sgn}(\beta/(\alpha + 1)))\eta$, $\bar{\mu}_2 = \mu_1(\text{sgn}(\beta/(\alpha + 1)))\eta$, $\bar{\eta} = \eta(\text{sgn}(\beta(\beta + 1)/\alpha(\alpha + 1)))$.

The half plane $\alpha \le \beta$ is divided into eight regions $(a, b, \ldots, h)$ by five lines: $\alpha = 0$, $\beta = 0$, $\alpha + 1 = 0$, $\beta + 1 = 0$, and $\alpha + \beta + 1 = 0$ (see Figure 7.1).

We denote (7.5) with $(\alpha, \beta) \in$ region $a$ and $\eta = 1$ $(\eta = -1)$ by $a_+(a_-)$. The notations $b_+, \ldots, h_+$ have the same meanings.

Although the total number of $a_+, a_-, \ldots, h_+, h_-$ is sixteen, the total number of unfoldings for system (7.5) with different behavior is thirteen. In fact, we will show that $b_- \sim f_-$, $c_- \sim g_-$, and $d_- \sim h_-$.

In order to state the main theorem, we need the following definition.

**Definition 7.1.** Equation (7.5) is called nondegenerate if

(1) $\alpha\beta(\alpha + 1)(\beta + 1)(\alpha + \beta + 1) \neq 0$;

(2) $\tilde{\tilde{q}}_6 \neq 0$ for cases $a_-$, $d_+$, and $h_+$,

where

$$\tilde{\tilde{q}}_6 = \frac{\alpha\beta^2}{(\alpha + 1)(\beta + 1)(\beta + 2)}\left(\frac{\alpha + 2}{\beta}q_1 + q_4\right)$$

$$+ \frac{\eta\alpha\beta}{(\alpha + 1)(\beta + 2)}\left(\frac{\alpha + 1}{\beta + 1}q_2 + q_5\right)$$

$$+ \left(\frac{\alpha}{\beta + 2}q_3 + q_6\right). \tag{7.7}$$

In Lemma 7.10 we will show that if (7.5) is nondegenerate, then instead of (7.5), we only need to consider equation

$$\begin{cases} \dot{x} = x(\mu_1 + \eta x - y), \\ \dot{y} = y\left(\mu_2 - \frac{\alpha + 1}{\beta}\eta x + \frac{\alpha}{\beta + 1}y + \nu g(\mu_1 + \eta x, y)\right), \end{cases} \tag{7.8}$$

where $\nu = \text{sgn}\,\tilde{\tilde{q}}_6$ for cases $a_-$, $d_+$, and $h_+$, $\nu = 0$ for other cases, and

$$g(x, y) = \frac{x^2}{\beta} - \frac{2xy}{\beta + 1} + \frac{y^2}{\beta + 2}. \tag{7.9}$$

**Theorem 7.2.** (1) *For different values of* $\alpha, \beta$ $(\alpha \leq \beta)$ *and* $\eta = \pm 1$, *the nondegenerate system (7.5) has thirteen different types of bifurcation. The bifurcation diagrams and phase portraits are shown in Figures 7.2–7.14, respectively.*

(2) *In every case, except* $a_-$, $d_+$ *and* $h_+$, *the bifurcation diagram consists of*

$$\Phi = K^+ \cup K^- \cup L^+ \cup L^- \cup M \cup N \cup \{0\}, \tag{7.10}$$

*where*

$$K^+: \mu_1 = 0, \quad \mu_2 > 0,$$

$$K^-: \mu_1 = 0, \quad \mu_2 < 0,$$

$$L^+: \mu_2 = 0, \quad \mu_1 > 0,$$

$$L^-: \mu_2 = 0, \quad \mu_1 < 0,$$

$$M: \mu_2 = -\frac{\alpha}{\beta + 1}\mu_1 + O(\mu_1^2), \quad \mu_1 > 0,$$

$$N: \mu_2 = -\frac{\alpha}{\beta}\mu_1 + O(\mu_1^2), \quad \mu_1 > 0$$

*for $\eta = -1$ and $\mu_1 < 0$ for $\eta = 1$.*

*Along curves $K^\pm, L^\pm, M, N$, saddle-node bifurcations occur. In these cases, system (7.5) has no limit cycle.*

(3) *In case $a_-$, the bifurcation diagram consists of*

$$H \cup HL \cup \Phi \qquad (\text{see } (7.10)),$$

*where*

$$H: \mu_2 = -\frac{\alpha}{\beta}\mu_1 + O(\mu_1^3), \qquad \mu_1 > 0,$$

$$HL: \mu_2 = -\frac{\alpha}{\beta}\mu_1 - \frac{\nu(\beta + 1)}{\beta^2(\alpha + \beta + 1)(\alpha + \beta + 2)}\mu_1^2 + O(\mu_1^3),$$

$$\mu_1 > 0.$$

*Along the curve $H$, Hopf bifurcation occurs. When $(\mu_1, \mu_2)$ is between curves $H$ and $HL$, (7.5) has a unique limit cycle which forms a heteroclinic loop if $(\mu_1, \mu_2) \in HL$.*

(4) *In cases $d_+$ and $h_+$, the bifurcation diagram consists of $\Phi \cup H$, where*

$$H: \mu_2 = -\frac{\alpha}{\beta}\mu_1 + O(\mu_1^3), \qquad \mu_1 > 0 \text{ for case } h_+$$

*and $\mu_1 < 0$ for case $d_+$.*

*Along curve $H$, Hopf bifurcation occurs. If we localize equation (7.5) by restricting $0 < x < \epsilon$ and $|\mu_1| + |\mu_2| < -\epsilon\xi\alpha/(\alpha + \beta + 1)$, $\epsilon > 0$, then there exists a curve $S$,*

$$S: \mu_2 = -\frac{\alpha}{\beta}\mu_1 - \nu\phi(\epsilon, \mu_1)\mu_1^2 + O(|\mu_1|^3),$$

*where $\mu_1 > 0$ in case $h_+$ and $\mu_1 < 0$ in case $d_+$,*

$$\phi(\epsilon, \mu_1) = \frac{\iint_{\Omega_{\epsilon, \mu_1}} x^{\alpha-1} y^{\beta-1} (\xi + x - y)^2 \, dx \, dy}{\beta \iint_{\Omega_{\epsilon, \mu_1}} x^{\alpha-1} y^{\beta-1} \, dx \, dy},$$

$\Omega_{\epsilon, \mu_1}$ *is the region bounded by the closed curve*

$$x^{\alpha} y^{\beta} \left( \frac{\xi + x}{\beta} - \frac{y}{\beta + 1} \right) = \frac{\left( \dfrac{\epsilon}{|\mu_1|} \right)^{\alpha} \left( \xi + \dfrac{\epsilon}{|\mu_1|} \right)^{\beta+1}}{\beta(\beta + 1)},$$

$\xi = 1$ *in case $h_+$ and $\xi = -1$ in case $d_+$, and $\epsilon / |\mu_1| > -\xi \alpha / (\alpha + \beta + 1) > 0$. When $(\mu_1, \mu_2)$ is between curves $H$ and $S$, (7.5) has a unique limit cycle which is located entirely in the strip $0 < x < \epsilon$ and touches the line $x = \epsilon$ if $(\mu_1, \mu_2) \in S$.*

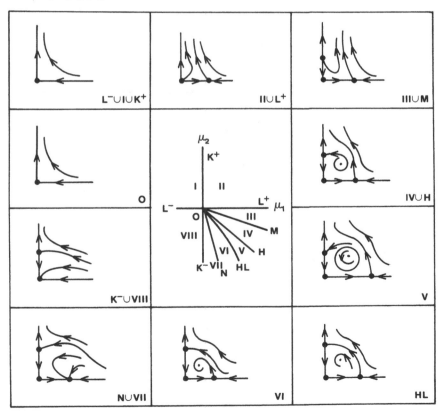

Figure 7.2. The unfoldings for case $a_-$.

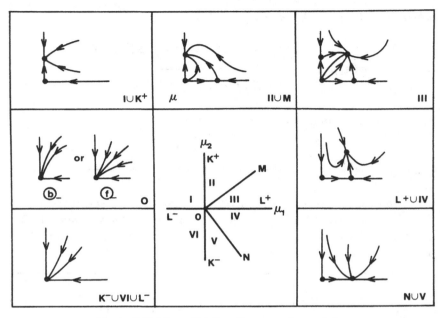

Figure 7.3.  The unfoldings for cases $b_-$ and $f_-$.

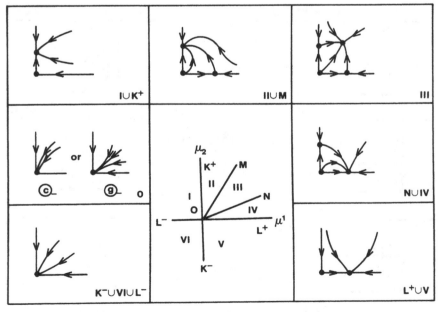

Figure 7.4.  The unfoldings for cases $c_-$ and $g_-$.

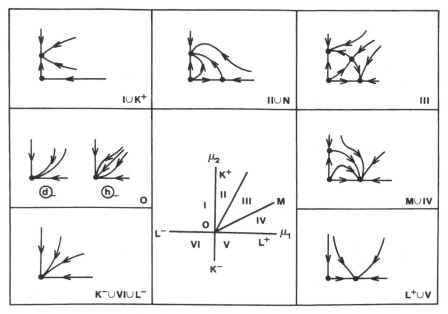

Figure 7.5. The unfoldings for cases $d_-$ and $h_-$.

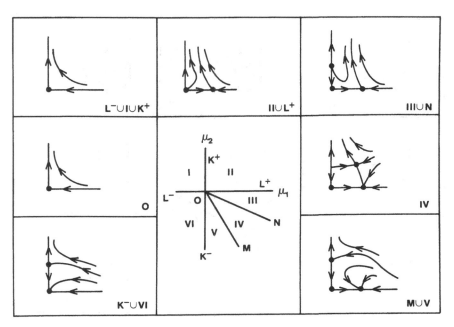

Figure 7.6. The unfoldings for case $e_-$.

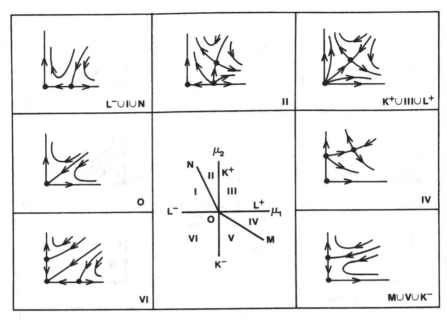

Figure 7.7. The unfoldings for case $a_+$.

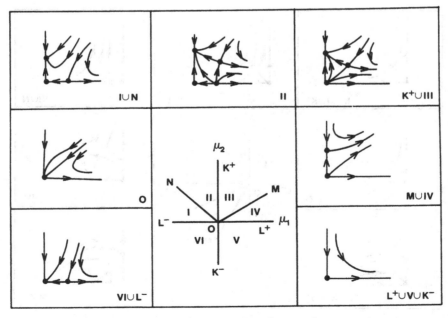

Figure 7.8. The unfoldings for case $b_+$.

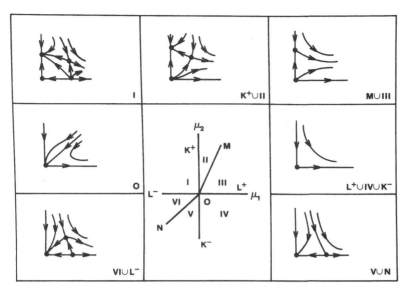

Figure 7.9. The unfoldings for case $c_+$.

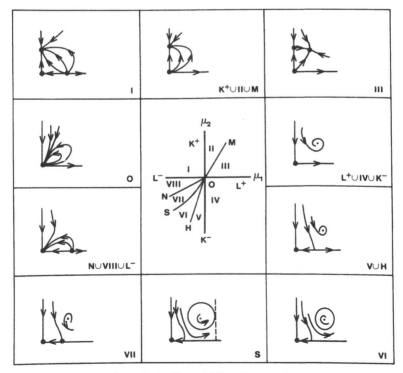

Figure 7.10. The unfoldings for case $d_+$.

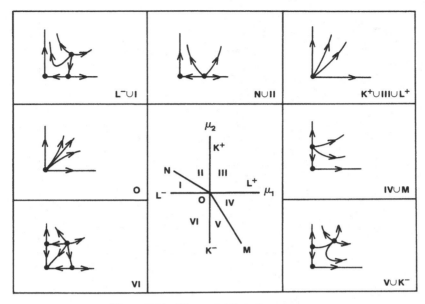

Figure 7.11. The unfoldings for case $e_+$.

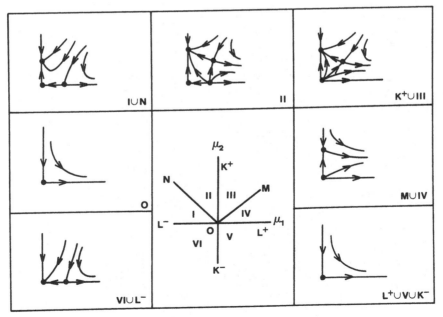

Figure 7.12. The unfoldings for case $f_+$.

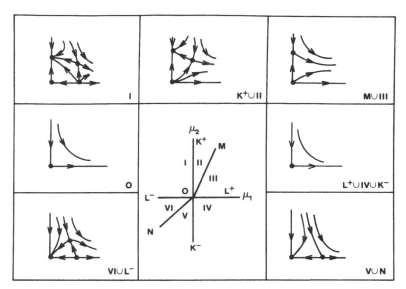

Figure 7.13. The unfoldings for case $g_+$.

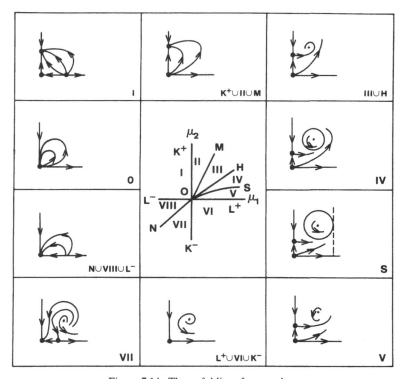

Figure 7.14. The unfoldings for case $h_+$.

*Proof.* We divide the proof into four steps.

*Step 1. The phase portraits for $\mu_1 = \mu_2 = 0$* We consider first the equation

$$\begin{cases} \dot{x} = x(\mu_1 + \eta x - y), \\ \dot{y} = y\left(\mu_2 - \dfrac{\alpha + 1}{\beta}\eta x + \dfrac{\alpha}{\beta + 1}y\right), \end{cases} \qquad (7.11)$$

where $x \geq 0$, $y \geq 0$, $\eta = \pm 1$, $\alpha \leq \beta$, and $\alpha\beta(\alpha + 1)(\beta + 1)(\alpha + \beta + 1) \neq 0$.

Let

$$x = r\cos\theta, \qquad y = r\sin\theta, \qquad dt \to \dfrac{dt}{r},$$

where $r > 0$ and $0 \leq \theta \leq \pi/2$. Equation (7.11) with $\mu_1 = \mu_2 = 0$ becomes

$$\begin{cases} \dot{r} = r\left(\eta\cos^3\theta - \cos^2\theta\sin\theta - \dfrac{\alpha + 1}{\beta}\eta\cos\theta\sin^2\theta + \dfrac{\alpha}{\beta + 1}\sin^3\theta\right), \\ \dot{\theta} = (\alpha + \beta + 1)\cos\theta\sin\theta\left(-\dfrac{\eta}{\beta}\cos\theta + \dfrac{1}{\beta + 1}\sin\theta\right). \end{cases}$$

$$(7.12)$$

Equation (7.12) has the following equilibria whose linear parts are given by the following matrices:

$$\begin{cases} (0,0), & \begin{bmatrix} \eta & 0 \\ 0 & -\dfrac{\alpha + \beta + 1}{\beta}\eta \end{bmatrix}; \\[3em] (0,\bar{\theta}), & \left(1 + \left(\dfrac{\beta}{\beta + 1}\right)^2\right)\sin^3\bar{\theta}\begin{bmatrix} -\dfrac{1}{\beta + 1} & 0 \\ 0 & \dfrac{\alpha + \beta + 1}{\beta + 1} \end{bmatrix}; \quad (7.13) \\[3em] \left(0,\dfrac{\pi}{2}\right), & \begin{bmatrix} \dfrac{\alpha}{\beta + 1} & 0 \\ 0 & -\dfrac{\alpha + \beta + 1}{\beta + 1} \end{bmatrix}. \end{cases}$$

The equilibria $(0, 0)$ and $(0, \pi/2)$ always exist, but the equilibrium $(0, \bar{\theta})$ exists only if $(\beta + 1)\eta/\beta > 0$. We note that under this condition $0 < \bar{\theta} < \pi/2$.

From (7.13) we can obtain the phase portraits of (7.12) and (7.11) ($\mu_1 = \mu_2 = 0$) in the $r\theta$-plane ($r \geq 0$, $0 \leq \theta \leq \pi/2$) and in the $xy$-plane ($x \geq 0$, $y \geq 0$) for every case of $a_\pm, \ldots, h_\pm$; see Figure 7.15 and Figure 7.16.

It is not difficult to see that the phase portraits of (7.5) for the case $\mu_1 = \mu_2 = 0$ are the same as for (7.11).


*Step 2. The phase portraits for $\mu_1^2 + \mu_2^2 \neq 0$* In a small neighborhood of $x = y = 0$ in the phase space, (7.11) has the following equilibria and their linear parts:

$$
\left\{
\begin{array}{lll}
p_1(0,0), & A(p_1) = \begin{bmatrix} \mu_1 & 0 \\ 0 & \mu_2 \end{bmatrix}, & \\[3em]
p_2\left(0, -\dfrac{\beta + 1}{\alpha}\mu_2\right), & A(p_2) = \begin{bmatrix} \mu_1 + \dfrac{\beta + 1}{\alpha}\mu_2 & 0 \\ * & -\mu_2 \end{bmatrix}, & \text{if } -\dfrac{\beta + 1}{\alpha}\mu_2 \geq 0, \\[3em]
p_3(-\eta\mu_1, 0), & A(p_3) = \begin{bmatrix} -\mu_1 & * \\ 0 & \dfrac{\alpha + 1}{\beta}\mu_1 + \mu_2 \end{bmatrix}, & \text{if } -\eta\mu_1 \geq 0, \\[3em]
p_4(x_4, y_4), & A(p_4) = \begin{bmatrix} \eta x_4 & -x_4 \\ -\dfrac{\alpha + 1}{\beta}\eta y_4 & \dfrac{\alpha}{\beta + 1}y_4 \end{bmatrix}, & \text{if } x_4 \geq 0, y_4 \geq 0,
\end{array}
\right.
$$

$$(7.14)$$

where

$$
\left\{
\begin{array}{l}
x_4 = \dfrac{\beta(\beta + 1)\eta}{\alpha + \beta + 1}\left(\dfrac{\alpha}{\beta + 1}\mu_1 + \mu_2\right), \\[2em]
y_4 = \dfrac{\beta(\beta + 1)}{\alpha + \beta + 1}\left(\dfrac{\alpha + 1}{\beta}\mu_1 + \mu_2\right).
\end{array}
\right.
$$

$$(7.15)$$

From (7.14) and (7.15) it is not difficult to obtain the following properties for equation (7.11):

(1) $p_2(p_3)$ appears or disappears when $(\mu_1, \mu_2)$ moves across the curves $L^\pm(K^\pm)$, and $p_2(p_3)$ merges into $p_1(0,0)$ when $(\mu_1, \mu_2) \in$

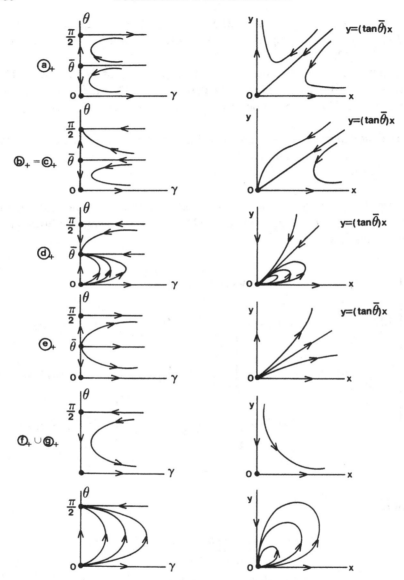

Figure 7.15. The phase portraits near $x = y = 0$ for the case $\mu_1 = \mu_2 = 0$ and $\eta = 1$.

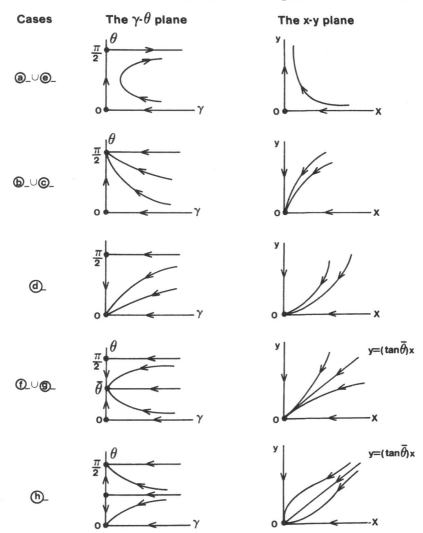

Figure 7.16. The phase portraits near $x = y = 0$ for the case $\mu_1 = \mu_2 = 0$ and $\eta = -1$.

$L^{\pm}(K^{\pm})$. $p_2(p_3)$ changes from a saddle point to a node when $(\mu_1, \mu_2)$ moves across the curve $M(N)$, and $p_4$ merges into $p_2(p_3)$ when $(\mu_1, \mu_2) \in M(N)$.

(2) $p_4$ exists if and only if one of the following conditions is satisfied:
(i) $(\mu_1, \mu_2)$ is between the curves $M$ and $N$ ($\mu_1 > 0$) for the case $\eta = -1$ (i.e., cases $a_-, b_-, \ldots, h_-$).
(ii) $(\mu_1, \mu_2)$ is above the curves $M$ and $N$ for the cases $a_+, b_+, c_+, f_+$, and $g_+$.
(iii) $(\mu_1, \mu_2)$ is below the curves $M$ and $N$ for the cases $d_+, e_+$, and $g_+$.

Moreover, the property of equilibrium $p_4$ is determined by

$$\begin{cases} \det(A(p_4)) = -\eta \dfrac{\alpha + \beta + 1}{\beta(\beta + 1)} x_4 y_4, \\ \operatorname{trace}(A(p_4)) = \alpha\mu_1 + \beta\mu_2. \end{cases} \tag{7.16}$$

If $p_4$ is a focus keeping its stability when $(\mu_1, \mu_2)$ varies in some subset of a small neighborhood of $(0, 0)$, then in a small neighborhood of $(x, y) = (0, 0)$ the qualitative behavior of (7.5) is the same as that of (7.11) (see Bautin [2]).

Suppose now that $p_4$ is a focus. The condition

$$\det(A(p_4)) > 0 \text{ and } \operatorname{trace}(A(p_4)) = 0 \tag{7.17}$$

gives a possibility for Hopf bifurcation. Equations (7.16) and (7.17) give $\mu_2 = -(\alpha/\beta)\mu_1$ and

$$\frac{\alpha}{\beta}(\alpha + \beta + 1) > 0. \tag{7.18}$$

Noting $\alpha \le \beta$, $x_4 > 0$, we obtain only the following three cases which satisfy (7.18):
(1) $\alpha > 0$, $\beta > 0$, $\eta = -1$, and $\mu_1 > 0$ (i.e., $a_-$ with $\mu_1 > 0$);
(2) $\beta < 0$, $-1 < \alpha + \beta < 0$, $\eta = 1$, and $\mu_1 > 0$ (i.e., $h_+$ with $\mu_1 > 0$);
(3) $\beta > 0$, $\alpha + \beta < -1$, $\eta = 1$, and $\mu_1 < 0$ (i.e., $d_+$ with $\mu_1 < 0$).

$p_1$, $p_2$, and $p_3$ are always on the invariant lines $x = 0$ and $y = 0$. Hence, if (7.5) has a limit cycle, then it surrounds $p_4$. In every case except $a_-$, $h_+$, and $d_+$, $p_4$ is a saddle point or a node which arises from a saddle-node bifurcation, keeps its stability, and moves into another saddle-node. This means that (7.5) has no periodic orbits for every case except $a_-$, $d_+$, and $h_+$. Thus, the bifurcation diagrams and phase portraits for (7.5) (also for (7.11)) with $\mu_1^2 + \mu_2^2 \neq 0$ in these cases are determined completely by the signs of $(\beta + 1)/\alpha$, $(\alpha + 1)/\beta$, and $(\alpha + \beta + 1)/(\beta(\beta + 1))$ (see (7.14), (7.15), and (7.16)). Hence, the behaviors of $b_-$ and $f_-$ are the same, and there is a similar situation for $c_-$ and $g_-$, $d_-$ and $h_-$, $b_+$ and $f_+$, and $c_+$ and $g_+$. On the other hand, as we discussed in Step 1, the phase portraits of (7.5) with $\mu_1 = \mu_2 = 0$ are topologically equivalent for cases $b_-$, $c_-$, $d_-$, $f_-$, $g_-$, and $h_-$, but are different for cases $b_+$ and $f_+$, and for $c_+$ and $g_+$ (see Figure 7.15 and Figure 7.16). Therefore, the total number of partitions of the half plane $\alpha \leq \beta$ for (7.5) with different dynamical behavior is thirteen (see Figures 7.2–7.14).

*Step 3. The uniqueness of periodic orbits of (7.5) for cases $a_-$, $d_+$, and $h_+$* Instead of equation (7.5), we consider equation (7.8) in three cases: $a_-$ with $\mu_1 > 0$, $d_+$ with $\mu_1 < 0$, and $h_+$ with $\mu_1 > 0$ (we will give the reason in Step 4). Let

$$\mu_1 = \xi\delta, \qquad \mu_2 = -\xi\frac{\alpha}{\beta}\delta - \sigma\delta^2, \qquad x \to \delta x,$$

$$y \to \delta y, \qquad dt \to \frac{x^{\alpha-1}y^{\beta-1}}{\delta} \, dt, \qquad (7.19)$$

where $\delta > 0$, $\delta$ and $\sigma$ are new parameters, $\xi = 1$ for cases $a_-$ and $h_+$, and $\xi = -1$ for case $d_+$.

Equation (7.8) is transformed into

$$\begin{cases} \dot{x} = x^\alpha y^{\beta-1}(\xi + \eta x - y), \\ \dot{y} = x^{\alpha-1}y^\beta\left[-\xi\dfrac{\alpha}{\beta} - \dfrac{\alpha+1}{\beta}\eta x + \dfrac{\alpha}{\beta+1}y \right. \\ \qquad\qquad \left. +\delta(-\sigma + g(\xi + \eta x, y))\right], \end{cases} \qquad (7.20)$$

where $\xi = 1$, $\eta = -1$ for $a_-$, $\xi = -1$, $\eta = 1$ for $d_+$, and $\xi = 1$, $\eta = 1$ for $h_+$.

When $\delta = 0$, (7.20) becomes a Hamiltonian system

$$\begin{cases} \dot{x} = x^\alpha y^{\beta-1}(\xi + \eta x - y), \\ \dot{y} = x^{\alpha-1} y^\beta \left( -\xi \dfrac{\alpha}{\beta} - \dfrac{\alpha+1}{\beta} \eta x + \dfrac{\alpha}{\beta+1} y \right), \end{cases} \tag{7.21}$$

with a first integral

$$H(x, y) = x^\alpha y^\beta \left( \frac{\xi + \eta x}{\beta} - \frac{y}{\beta+1} \right). \tag{7.22}$$

The closed curves $\{H = h\}$ are shown in Figure 7.17. The values of $h$ that correspond to the closed level curves are

$$\begin{cases} 0 < h < h_1^* & \text{for cases } a_- \text{ and } d_+, \\ h_2^* < h < 0 & \text{for case } h_+ \end{cases} \tag{7.23}$$

where $H = h_i^*$ corresponds to the equilibrium $(x_4, y_4)$, and $H = 0$ corresponds to three straight lines (they form three heteroclinic orbits in case $a_-$).

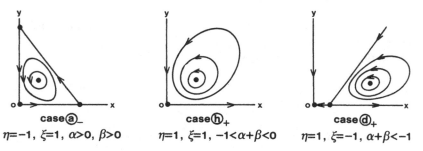

$$\underset{\substack{\text{case}\,\textcircled{a}_- \\ \eta=-1,\ \xi=1,\ \alpha>0,\ \beta>0}}{} \qquad \underset{\substack{\text{case}\,\textcircled{h}_+ \\ \eta=1,\ \xi=1,\ -1<\alpha+\beta<0}}{} \qquad \underset{\substack{\text{case}\,\textcircled{d}_+ \\ \eta=1,\ \xi=-1,\ \alpha+\beta<-1}}{}$$

Figure 7.17.  The closed level curves of equation (7.21).

Similarly to Section 4.6, for the study of periodic orbits of (7.8), we use the bifurcation function $\Phi(\sigma, \delta, h, \alpha, \beta)$ which for $\delta = 0$ is given by

$$\Phi|_{\delta=0} = \int_{\Gamma_h} x^{\alpha-1} y^{\beta}(-\sigma + \nu g(\xi + \eta x, y))\, dx, \qquad (7.24)$$

where $\Gamma_h$ is the level curve of $H = h$. The orientation of $\Gamma_h$ is defined by the direction of the vector field (7.21). We note that

$$\Phi|_{\delta=0} = 0 \quad \text{is equivalent to} \quad \sigma = \nu P(h),$$

where

$$P(h) = \frac{I_2(h)}{I_1(h)},$$

$$(7.25)$$

$$I_1(h) = \int_{\Gamma_h} x^{\alpha-1} y^{\beta}\, dx, \qquad I_2(h) = \int_{\Gamma_h} x^{\alpha-1} y^{\beta} g(\xi + \eta x, y)\, dx.$$

To prove the uniqueness of periodic orbits of (7.8), we only need to show that $P'(h) \neq 0$ for $h$ satisfying (7.23).

From (7.22) we have that along $\Gamma_h$

$$\frac{\partial y}{\partial h} = \frac{1}{x^{\alpha} y^{\beta-1}(\xi + \eta x - y)}.$$

Hence

$$I_1'(h) = \beta \int_{\Gamma_h} \frac{1}{x(\xi + \eta x - y)}\, dx, \qquad (7.26)$$

$$I_2'(h) = \int_{\Gamma_h} \frac{\xi + \eta x - y}{x}\, dx. \qquad (7.27)$$

We define

$$G(h) = \beta P(h). \qquad (7.28)$$

**Theorem 7.3.** $G'(h) \neq 0$ *for h satisfying (7.23).*

From (7.28), (7.25), (7.26), and (7.27), it is easy to see that $G'(h) \neq 0$ is equivalent to

$$I_2'(h) - P(h)I_1'(h) = \int_{\Gamma_h} \frac{(\xi + \eta x - y)^2 - G(h)}{x(\xi + \eta x - y)}\, dx \neq 0. \quad (7.29)$$

Let $z = \xi + \eta x - y$ and $z_i = z_i(x) = \xi + \eta x - y_i(x)$ $(i = 1, 2)$, where $\{y = y_i(x)\}$ are branches of $\Gamma_h$ lying below and above the line $z = 0$, whence $z_1(x) > 0 > z_2(x)$ for $x_1 < x < x_2$. $x_1$ and $x_2$ are $x$-components of the intersection of the line $z = 0$ with $\Gamma_h$. Thus, to show (7.29), we only need to prove that

$$\int_{x_1}^{x_2}\left[ z_1 - z_2 - G(h)\left(\frac{1}{z_1} - \frac{1}{z_2}\right)\right]\frac{dx}{x} > 0. \quad (7.30)$$

The following lemmas of Żołądek [2] are needed.

**Lemma 7.4.** *If Theorem 7.3 holds for $\beta \geq 1$, then in the other cases it is also true.*

**Lemma 7.5.** *Let $x_0 \in (x_1, x_2)$ be such a point that the function $(z_1 z_2)(x) < 0$ takes its minimal value at $x_0$. Then there exist $\gamma > 0$ and $x_5 > x_0$ such that*

$$z_1(x) \geq a_1\left[(x_5^\gamma - x_1^\gamma)^2 - (x_5^\gamma - x^\gamma)^2\right]^{1/2} = a_1 u(x),$$
$$z_2(x) \leq a_2 u(x), \quad (7.31)$$

*for $x_1 \leq x \leq x_0$, where $a_i = z_i(x_0)/u(x_0)$, $i = 1, 2$.*

**Lemma 7.6.** *There exist $\delta > 0$ and $x_6 > x_2$ such that*

$$z_1(x) \geq \tilde{a}_1\left[(x_6^\delta - x^\delta)^2 - (x_6^\delta - x_2^\delta)^2\right]^{1/2} = \tilde{a}_1 w(x),$$
$$z_2(x) \leq \tilde{a}_2 w(x), \quad (7.32)$$

*for $x_0 < x < x_2$, where $\tilde{a}_i = z_i(x_0)/w(x_0)$, $i = 1, 2$.*

**Lemma 7.7.** *We have the inequality*

$$0 < -\frac{G(h)}{(z_1 z_2)(x_0)} \le \frac{1}{3} \tag{7.33}$$

*for h satisfying (7.23) and* $\beta \ge 1$.

**Lemma 7.8.** *We have the inequality*

$$\int_{x_1}^{x_0} \left[ u(x) - \frac{u^2(x_0)}{3u(x)} \right] \frac{dx}{x} > 0. \tag{7.34}$$

**Lemma 7.9.** *We have the inequality*

$$\int_{x_0}^{x_2} \left[ w(x) - \frac{w^2(x_0)}{3w(x)} \right] \frac{dx}{x} > 0. \tag{7.35}$$

*Proof of Theorem 7.3.* By (7.31), (7.33), and (7.34) we have

$$\int_{x_1}^{x_0} \left[ z_1 - z_2 - G(h) \left( \frac{1}{z_1} - \frac{1}{z_2} \right) \right] \frac{dx}{x}$$

$$\ge \int_{x_1}^{x_0} \left[ (a_1 - a_2) u(x) - G(h) \left( \frac{1}{a_1} - \frac{1}{a_2} \right) \frac{1}{u(x)} \right] \frac{dx}{x}$$

$$= (a_1 - a_2) \int_{x_1}^{x_0} \left[ u(x) + \frac{G(h) u^2(x_0)}{(z_1 z_2)(x_0) u(x)} \right] \frac{dx}{x}$$

$$\ge (a_1 - a_2) \int_{x_1}^{x_0} \left[ u(x) - \frac{u^2(x_0)}{3u(x)} \right] \frac{dx}{x} > 0.$$

Similarly, by (7.32), (7.33), and (7.35) we have

$$\int_{x_0}^{x_2}\left[z_1 - z_2 - G(h)\left(\frac{1}{z_1} - \frac{1}{z_2}\right)\right]\frac{dx}{x} > 0.$$

This gives (7.30), and Theorem 7.3 is proved.        □

*Step 4. Reasons for using equation (7.8) instead of equation (7.5)* Let us consider equation (7.5). For cases $a_-$, $d_+$, and $h_+$, we take the same transformation as (7.19). Then (7.5) becomes

$$\begin{cases} \dot{x} = x^\alpha y^{\beta-1}\left[\xi + \eta x - y + \delta\left(q_1 x^2 + q_2 xy + q_3 y^2\right)\right], \\ \dot{y} = x^{\alpha-1}y^\beta\left[-\xi\frac{\alpha}{\beta} - \frac{\alpha+1}{\beta}\eta x + \frac{\alpha}{\beta+1}y \right. \\ \left. \qquad\qquad + \delta\left(-\sigma + q_4 x^2 + q_5 xy + q_6 y^2\right)\right]. \end{cases} \tag{7.36}$$

For $\delta = 0$, (7.36) becomes (7.21) with the first integral (7.22). Hence, to study the periodic orbits of (7.5), we obtain a bifurcation function $G(\sigma, \delta, h, \alpha, \beta, \{q_i\})$ which, for $\delta = 0$, is given by

$$G|_{\delta=0} = \int_{\Gamma_h} x^\alpha y^{\beta-1}\left(-q_1 x^2 - q_2 xy - q_3 y^2\right) dy$$

$$+ x^{\alpha-1}y^\beta\left(-\sigma + q_4 x^2 + q_5 xy + q_6 y^2\right) dx$$

$$= -\sigma I_{\alpha-1,\beta} + \tilde{q}_4 I_{\alpha+1,\beta} + \tilde{q}_5 I_{\alpha,\beta+1} + \tilde{q}_6 I_{\alpha-1,\beta+2}, \tag{7.37}$$

where

$$I_{a,b} = \int_{\Gamma_h} x^a y^b \, dx,$$

$\Gamma_h$ is the level curve of $H = h$, and

$$\tilde{q}_4 = q_4 + \frac{\alpha + 2}{\beta}q_1, \qquad \tilde{q}_5 = q_5 + \frac{\alpha + 1}{\beta + 1}q_2, \qquad \tilde{q}_6 = q_6 + \frac{\alpha}{\beta + 2}q_3.$$

$$(7.38)$$

Along $\Gamma_h$ we have

$$x^{\alpha - 1}y^{\beta}\left(-\xi\frac{\alpha}{\beta} - \frac{\alpha + 1}{\beta}\eta x + \frac{\alpha}{\beta + 1}y\right) dx$$

$$- x^{\alpha}y^{\beta - 1}(\xi + \eta x - y) \, dy = 0. \qquad (7.39)$$

Multiplying (7.39) by $x$ and then integrating it along $\Gamma_h$, we have

$$\frac{\xi}{\beta}I_{\alpha, \beta} + \frac{\eta}{\beta}I_{\alpha + 1, \beta} - \frac{1}{\beta + 1}I_{\alpha, \beta + 1} = 0. \qquad (7.40)$$

Multiplying (7.39) by $y$ and then integrating it along $\Gamma_h$, we have

$$-\frac{\alpha\xi}{\beta}I_{\alpha - 1, \beta + 1} - \frac{\eta(\alpha + 1)}{\beta}I_{\alpha, \beta + 1} + \frac{\alpha}{\beta + 2}I_{\alpha - 1, \beta + 2} = 0. \quad (7.41)$$

Equations (7.40) and (7.41) give

$$\left\{ \begin{aligned} I_{\alpha, \beta + 1} &= -\frac{\xi\eta\alpha}{\alpha + 1}I_{\alpha - 1, \beta + 1} + \frac{\eta\alpha\beta}{(\alpha + 1)(\beta + 2)}I_{\alpha - 1, \beta + 2}, \\ I_{\alpha + 1, \beta} &= -\xi\eta I_{\alpha, \beta} - \frac{\xi\alpha\beta}{(\alpha + 1)(\beta + 1)}I_{\alpha - 1, \beta + 1} \\ &\quad + \frac{\alpha\beta^2}{(\alpha + 1)(\beta + 1)(\beta + 2)}I_{\alpha - 1, \beta + 2}. \end{aligned} \right. \qquad (7.42)$$

Substituting (7.42) into (7.37), we obtain

$$G|_{\delta = 0} = -\sigma I_{\alpha - 1, \beta} + \gamma_1 I_{\alpha, \beta} + \gamma_2 I_{\alpha - 1, \beta + 1} + \tilde{\tilde{q}}_6 I_{\alpha - 1, \beta + 2}, \quad (7.43)$$

where

$$\gamma_1 = -\xi\eta\bar{q}_4, \qquad \gamma_2 = \xi\left[-\frac{\alpha\beta}{(\alpha+1)(\beta+1)}\bar{q}_4 - \frac{\eta\alpha}{\alpha+1}\bar{q}_5\right],$$

$$\bar{\bar{q}}_6 = \frac{\alpha\beta^2}{(\alpha+1)(\beta+1)(\beta+2)}\bar{q}_4 + \frac{\eta\alpha\beta}{(\alpha+1)(\beta+2)}\bar{q}_5 + \bar{q}_6. \quad (7.44)$$

We remark here that (7.44) and (7.38) give (7.7).                    □

**Lemma 7.10.** *Suppose that equation (7.5) is nondegenerate. Then there exist a small neighborhood $\mathscr{N}$ of the origin $(x, y) = (0, 0)$ and a small neighborhood $\Delta$ of $(\mu_1, \mu_2) = (0, 0)$ such that equation (7.5) has at most one periodic orbit in $\mathscr{N}$ for all $(\mu_1, \mu_2) \in \Delta$ if and only if the same property holds for the following equation*

$$\begin{cases} \dot{x} = x(\mu_1 + \eta x - y), \\ \dot{y} = y(\mu_2 + \eta ax + by + \nu g(\mu_1 + \eta x, y)), \end{cases} \quad (7.45)$$

*where*

$$a = -\frac{\alpha+1}{\beta}, \qquad b = \frac{\alpha}{\beta+1}, \qquad g(x, y) = \frac{x^2}{\beta} - \frac{2xy}{\beta+1} + \frac{y^2}{\beta+2},$$

*and $\nu$ is a constant determined in the following manner: $\nu = 0$ if one of the conditions $a_-$, $d_+$, or $h_+$ is not satisfied; $\nu = \mathrm{sgn}(\bar{\bar{q}}_6)$ if one of the conditions $a_-$, $d_+$, or $h_+$ is satisfied.*

*Proof.* In this proof, we may have to choose different neighborhoods $\mathscr{N}$ and $\Delta$ for different equations. We will always take the intersections of such neighborhoods and continue to denote them by $\mathscr{N}$ and $\Delta$. This will not cause any confusion.

From our earlier discussions, we only need to consider the case that one of the conditions $a_-$, $d_+$, or $h_+$ is satisfied. Thus, we assume that

$\nu \neq 0$. Moreover, for the uniqueness of periodic orbits of equation (7.5) in $\mathcal{N}$ for $(\mu_1, \mu_2) \in \Delta$, we only need to show the uniqueness of periodic orbits of equation (7.36) in the $xy$-plane. Furthermore, we have shown that the uniqueness of periodic orbits of equation (7.36) is equivalent to the unique solvability of the bifurcation equation:

$$G|_{\delta=0} = 0, \tag{7.46}$$

where $G|_{\delta=0}$ is given in (7.43).

On the other hand, consider the following equation:

$$\begin{cases} \dot{x} = x(\mu_1 + \eta x - y), \\ \dot{y} = y(\mu_2 + (\eta a + \xi \gamma_1 \mu_1)x + (b + \xi \gamma_2 \mu_1)y + \tilde{\tilde{q}}_6 y^2) \end{cases} \tag{7.47}$$

in the neighborhood $\mathcal{N}$ for $(\mu_1, \mu_2) \in \Delta$, where $\xi = \pm 1$ is chosen according to the conditions $a_-$, $d_+$, and $h_+$ (see the change of variables (7.19)). By a scaling of (7.9), equation (7.47) is transformed to

$$\begin{cases} \dot{x} = x^\alpha y^{\beta-1}[\xi + \eta x - y], \\ \dot{y} = x^{\alpha-1} y^\beta \left[ -\xi \dfrac{\alpha}{\beta} - \dfrac{\alpha+1}{\beta}\eta x + \dfrac{\alpha}{\beta+1}y \right. \\ \qquad\qquad \left. + \delta\left(-\sigma + \gamma_1 x + \gamma_2 y + \tilde{\tilde{q}}_6 y^2\right) \right]. \end{cases} \tag{7.48}$$

Thus, the uniqueness of periodic orbits of equation (7.47) in $\mathcal{N}$ for $(\mu_1, \mu_2) \in \Delta$ is equivalent to the uniqueness of periodic orbits of equation (7.48) in the $xy$-plane. If $\delta = 0$, then (7.48) is reduced to the Hamiltonian system (7.21). Furthermore, the uniqueness of periodic orbits of equation (7.48) in the $xy$-plane is equivalent to the unique solvability of the bifurcation equation (7.46). This says that equation (7.36) has at most one periodic orbit in the $xy$-plane if and only if equation (7.48) does. In other words, equation (7.5) has at most one periodic orbit in $\mathcal{N}$ for $(\mu_1, \mu_2) \in \Delta$ if and only if equation (7.47) does. We will use the notation $(7.5) \simeq (7.47)$ to mean this kind of equivalence relation between equations.

Since equation (7.5) is nondegenerate, $\tilde{\tilde{q}}_6 \neq 0$. Thus, let

$$x \rightarrow \frac{1}{|\tilde{\tilde{q}}_6|}x, \qquad y \rightarrow \frac{1}{|\tilde{\tilde{q}}_6|}y, \qquad t \rightarrow |\tilde{\tilde{q}}_6|t.$$

This transforms equation (7.47) into the following form:

$$\begin{cases} \dot{x} = x(\mu_1 + \eta x - y), \\ \dot{y} = y[\mu_2 + (\eta a + \xi\gamma_1\mu_1)x + (b + \xi\gamma_2\mu_1)y + \nu y^2]. \end{cases} \quad (7.49)$$

Hence, $(7.47) \simeq (7.49)$.

Now, consider equation (7.45). By the definition of $g$, we write equation (7.45) as follows:

$$\begin{cases} \dot{x} = x(\mu_1^* + \eta x - y), \\ \dot{y} = y(\mu_2^* + \eta a^* x + b^* y + \nu g^*(x, y)), \end{cases} \quad (7.50)$$

where

$$\mu_1^* = \mu_1, \qquad \mu_2^* = \mu_2 + \frac{\nu\mu_1^2}{\beta}, \qquad a^* = a + \frac{2\nu\eta}{\beta}\mu_1,$$

$$b^* = b - \frac{2\nu}{\beta + 1}\mu_1, \qquad \nu = \text{sgn}(\tilde{\tilde{q}}_6),$$

$$g^*(x, y) = \frac{x^2}{\beta} - \frac{2\eta xy}{\beta + 1} + \frac{y^2}{\beta + 2}.$$

Thus, equation (7.50) is a special case of equation (7.5) with $q_1 = q_2 = q_3 = 0$ but $a^*$ and $b^*$ are dependent on $\mu_1^* = \mu_1$. By repeating the arguments as for equation (7.5), we obtain that equation (7.50) $\simeq$ the following equation:

$$\begin{cases} \dot{x} = x(\mu_1 + \eta x - y), \\ \dot{y} = y[\mu_2^* + (\eta a + \xi\gamma_1^*\mu_1)x + (b + \xi\gamma_2^*\mu_1)y + \tilde{\tilde{q}}_6 y^2], \end{cases} \quad (7.51)$$

where

$$\mu_2^* = \mu_2 + \frac{\nu}{\beta}\mu_1^2, \qquad \gamma_1^* = \frac{\nu\xi\eta}{\beta}, \qquad \gamma_2^* = -\frac{\xi\nu(\alpha + 2)}{(\alpha + 1)(\beta + 1)},$$

$\xi$ is the same as in (7.47), and

$$\tilde{\bar{q}}_6^* = \frac{\nu(\alpha + \beta + 1)}{(\alpha + 1)(\beta + 1)(\beta + 2)}.$$

Since

$$\nu = \text{sgn}(\tilde{\bar{q}}_6) \quad \text{and} \quad \frac{\alpha + \beta + 1}{(\alpha + 1)(\beta + 1)(\beta + 2)} > 0$$

(see conditions $a_-$, $d_+$, and $h_+$ with $\alpha \le \beta$), $\tilde{\bar{q}}_6^*$ and $\tilde{\bar{q}}_6$ have the same sign.

Let

$$x \to \frac{1}{|\tilde{\bar{q}}_6^*|}x, \qquad y \to \frac{1}{|\tilde{\bar{q}}_6^*|}y, \qquad t \to |\tilde{\bar{q}}_6^*|t.$$

Then equation (7.51) is transformed into the following form:

$$\begin{cases} \dot{x} = x(\mu_1 + \eta x - y), \\ \dot{y} = y[\mu_2^* + (\eta a + \xi\gamma_1^*\mu_1)x + (b + \xi\gamma_2^*\mu_1)y + \nu y^2]. \end{cases} \tag{7.52}$$

We have proved that $(7.5) \simeq (7.47) \simeq (7.49)$ and $(7.45) \simeq (7.51) \simeq (7.52)$. It is clear that $(7.49) \simeq (7.52)$. Therefore, $(7.5) \simeq (7.45)$. $\square$

**Remark 7.11.** Since $P(h^*) = 0$ and $\sigma = \nu P(h)$, the equation of Hopf bifurcation curve $H$ is obtained from (7.19), that is, $\mu_2 = -(\alpha/\beta)\mu_1 + O(\mu_1^3)$. In order to obtain the equation of the heteroclinic bifurcation curve, we need $P(0)$ in case $a_-$. The calculation shows

$$P(0) = \frac{\dfrac{1}{\beta^2(\beta + 2)} \displaystyle\int_0^1 x^{\alpha-1}(1 - x)^{\beta+2}\, dx}{\displaystyle\int_0^1 x^{\alpha-1}(1 - x)^{\beta}\, dx} = \frac{1}{\beta^2(\beta + 2)}\frac{B(\alpha, \beta + 3)}{B(\alpha, \beta + 1)}$$

$$= \frac{\beta + 1}{\beta^2(\alpha + \beta + 1)(\alpha + \beta + 2)}.$$

Finally, in order to obtain the equation of curve $S$ for cases $d_+$ and $h_+$ ($\eta = 1$), we note that the coordinates of the center $(x_4, y_4)$ of equation (7.21) are

$$\begin{cases} \bar{x}_4 = -\dfrac{\xi\alpha}{\alpha + \beta + 1}, \\ \bar{y}_4 = \xi + \bar{x}_4. \end{cases}$$

If we take $\epsilon/|\mu_1| > \bar{x}_4$, $h_{\epsilon,\mu_1} = H(\epsilon/|\mu_1|, \xi + \epsilon/|\mu_1|)$, then the closed level curve $H(x, y) = h_{\epsilon,\mu}$ is tangent to the line $x/|\mu_1| = \epsilon$. Returning to equation (7.5) by using (7.9), we see that if $(\mu_1,\mu_2) \in S$, $S$: $\mu_2 = -(\alpha/\beta)\mu_1 - \phi(\epsilon, \mu_1)\nu\mu_1^2 + O(|\mu_1|^3)$, where $\phi(\epsilon, \mu_1) = P(h_{\epsilon,\mu_1})$, then the limit cycle touches the line $x = \epsilon$. Of course, the limit cycle could be larger when $(\mu_1, \mu_2)$ crosses the curve $S$ and is still in a small neighborhood of $(0, 0)$. But the behavior of the limit cycle will depend on the global property of the vector field.

### 4.8 Bibliographical Notes

The families of vector fields on the plane that are invariant under a rotation $2\pi/q$, $q = 1, 2, \ldots$, have been investigated by many authors; see, for example, Afraimovich, et al. [1] ($q = 1, 2, 3, 4$), Arnold [4] ($q = 1, 2, 3, 4$), Berezovskaia and Knibnik [1] ($q = 4$), Bogdanov [1–3] ($q = 1$), Carr [1] ($q = 2$), Chow and Hale [1] ($q = 1$), Chow, Li, and Wang [2] ($q = 3$), Guckenheimer [3] ($q = 1, 2$), Guckenheimer and Holmes [1] ($q = 1, 2$), Khorozov [1] ($q = 2, 3$), Neishtadt [1] ($q = 4$), Takens [1] ($q = 1, 2$ and $q \geq 5$), Wan [1] ($q = 4$), Wang [1] ($q = 4$), and so forth.

Theorem 1.13 and Lemmas 1.5, 1.20, 1.21, and 1.22 are due to Bogdanov [1, 2]. Lemmas 1.6 and 1.7 are essentially due to Cushman and Sanders [1] and Dumortier, Roussarie, and Sotomayor [1]. In Li, Rousseau, and Wang [1] there is an alternate proof on the uniqueness of periodic orbits for the Bogdanov–Takens system (1.5), and the result is global. Lemma 2.2 belongs to Khorozov [1]. Lemma 2.6 is taken from

Carr [1]. There is also an alternate study on the number of limit cycles for equation $(2.1)^{\pm}$ in Li and Rousseau [2]. In the first part of Section 3, we follow Khorozov [1], Lemma 3.2 belongs to Arnold [2, 4], and the proof on uniqueness of periodic orbits for $q = 3$ is taken from Chow, Li, and Wang [2]. The case $q = 4$ has not been solved completely. It was conjectured by Arnold [5] that this probably is the only case in 1: $q$ resonance problems that is beyond the techniques of modern mathematics. Proofs of Theorems 4.1, 4.5, and 4.10 can be found in Arnold [4]. Theorem 4.6 is taken from Wang [1]. Theorem 4.7 belongs to Berezovskaia and Knibnik [1]. Lemmas 4.11 and 4.13 are due to Wan [1]. Theorem 4.14 is obtained by Neishtadt [1]. Lemma 5.2 is taken from Takens [2].

If the linear part of the (unperturbed) vector fields has a purely imaginary pair and a simple zero eigenvalue, then the normal form equation can be reduced to a planar system which is invariant under a dihedral group generated by reflection with respect to a line. Theorem 6.1 was first obtained by Żołądek [1]. Carr, Chow, and Hale [1], Chow and Hale [1], Chow, Li, and Wang [1], Gavrilov [1], Guckenheimer [1], Guckenheimer and Holmes [1], Langford [1], and van Gils [1] also investigated this case. Figure 6.6 can be obtained from Li [2].

If the linear part of the unperturbed vector field has two different pairs of purely imaginary eigenvalues which satisfy some nonresonance conditions, then the normal form equation can be reduced to a planar system which is invariant under a dihedral group generated by two reflections with respect to two orthogonal axes. This case has been considered by Chow and Hale [1], Gavrilov [2], Guckenheimer and Holmes [1], and Holmes [1], Iooss and Langford [1]. Żołądek [2] gave in this case the first complete codimension-two bifurcation result. Carr, van Gils, and Sanders [1], van Gils and Horozov [1], and van Gils and Sanders [1], gave simpler proofs on the uniqueness of periodic orbits for some special cases.

In the discussion of unfoldings of singular vector fields, an important problem is the study of the number of periodic orbits. It is related to determining the numbers of zeros of Abelian integrals. Khovansky [1], Il'yashenko [1, 2], Petrov [1, 2], and Varchenko [1] gave estimates on the numbers of zeros for certain Abelian integrals. By using these results Li and Huang [1] gave an example of a planar cubic system which has at least eleven limit cycles. See also Chow and Sanders [1], Chow and Wang [1], and Wang [6].

For the study of the existence, the stability, and the number of periodic orbits of planar vector fields, we refer to Andronov et al. [1], Sotomayor [2], Takens [3], Ye [1], and Zhang et al. [1].

We refer to Dumortier, Roussarie, Sotomayor, and Żoladek [2], Guckenheimer [1, 3, 4], Hale [1–4], and Holmes [2], for additional results on codimension-two bifurcation.

# 5

# Bifurcations with Codimension Higher than Two

In the first two sections of this chapter we will introduce Hopf bifurcation and homoclinic bifurcation with higher codimension. We will introduce codimension 3 and codimension 4 results concerning the Bogdanov–Takens system in the last two sections.

## 5.1 Hopf Bifurcation of Higher Order

As in Section 3.1, the classical Hopf Bifurcation Theorem is a local result which deals with the occurrence (or annihilation) of a periodic orbit at an equilibrium point of a system

$$\begin{cases} \dot{x} = f(x, y, \mu), \\ \dot{y} = g(x, y, \mu), \end{cases} \quad x, y \in \mathbb{R}^1, \mu \in \mathbb{R}^n, f, g \in C^\infty, \quad (1.1)$$

when two eigenvalues $\lambda_{1,2} = \alpha(\mu) \pm i\beta(\mu)$ of the linear part of (1.1) cross the imaginary axis. Namely, we suppose:

$$\alpha(0) = 0, \beta(0) = \beta_0 \neq 0, \quad (\text{H}_1)$$

$$\alpha'(0) \neq 0 \quad (\text{if } \mu \in \mathbb{R}^1), \text{ and} \quad (\text{H}_2)$$

$$\text{Re } C_1 \neq 0, \quad (\text{H}_3)$$

where $C_1$ is the first coefficient of nonlinear terms of the normal form equation obtained from $(1.1)_{\mu=0}$. This normal form equation has the

383

following form (see Section 2.11 or Lemma 1.1 below):

$$\dot{w} = i\beta_0 w + C_1 w^2 \overline{w} + C_2 w^3 \overline{w}^2 + \cdots + C_k w^{k+1} \overline{w}^k + O(|w|^{2k+3}).$$

$$(1.2)$$

In this section, we show that a Hopf bifurcation may occur when either condition $(H_2)$ or $(H_3)$ fails. This is called a degenerate Hopf bifurcation. Many authors considered this problem (see Section 5.5). The proof of Theorem 1.3 in this section belongs to Rousseau and Schlomiuk [1].

**Lemma 1.1.** *Suppose that the linear part of system (1.1) at $(x, y) = (0, 0)$ has eigenvalues $\lambda_{1,2} = \alpha(\mu) \pm i\beta(\mu)$ satisfying condition $(H_1)$. Then for any integer $k > 0$, there are $\sigma > 0$ and a polynomial change of variables depending smoothly on the parameter $\mu$ for $|\mu| < \sigma$ such that in complex coordinates system (1.1) can be transformed into the following form*:

$$\dot{w} = (\alpha(\mu) + i\beta(\mu))w + C_1(\mu)w^2\overline{w} + C_2(\mu)w^3\overline{w}^2$$

$$+ \cdots + C_k(\mu)w^{k+1}\overline{w}^k + O(|w|^{2k+3}), \qquad (1.3)$$

*where $C_k(\cdot) \in C^\infty$ and $C_k(0) = C_k$, with $C_k$ the coefficient in (1.2).*

*Proof.* As in Section 2.1, we let $m = (m_1, m_2)$, $m_j \geq 0$ ($j = 1, 2$) are integers, $\mathscr{M}_k = \{m | 2 \leq m_1 + m_2 \leq 2k + 2\}$, $\lambda(\mu) = (\lambda_1(\mu), \lambda_2(\mu))$, and

$$(m, \lambda(\mu)) = m_1\lambda_1(\mu) + m_2\lambda_2(\mu).$$

It follows from $(H_1)$ that

$$\lambda_1(0) = (m, \lambda(0))$$

gives a resonance of order $\leq 2k + 2$ if and only if

$$m \in \mathscr{M}^* = \{m | m_1 = m_2 + 1, m_2 = 1, 2, \ldots, k\} \subset \mathscr{M}_k.$$

Then by Theorem 2.1.5, there is a polynomial change of variables that transforms system (1.1) for $\mu = 0$ into its normal form (1.2) up to order $2k + 2$.

Since $\lambda_j(\mu)$ is a smooth function $(j = 1, 2)$, $\exists \sigma > 0$ such that for $|\mu| < \sigma$ we have that

$$\lambda_1(\mu) \neq (m, \lambda(\mu)) \qquad \text{if } m \in \mathscr{M}_k \backslash \mathscr{M}^*.$$

Hence, we can use the same arguments as in the proof of Theorem 2.1.5 to find a polynomial change of variables, depending smoothly on $\mu$, to get rid of all terms $w^{m_1}\bar{w}^{m_2}$ with $(m_1, m_2) \in \mathscr{M}_k \backslash \mathscr{M}^*$. Hence, system (1.1) for $0 < |\mu| < \sigma$ is transformed into the form (1.3). ☐

**Definition 1.2.** We say that (1.1) has a Hopf bifurcation of order $k$ $(k \geq 1)$ at the origin if $\alpha(0) = 0$, $\beta(0) = \beta_0 \neq 0$, and

$$\mathrm{Re}(C_1) = \mathrm{Re}(C_2) = \cdots = \mathrm{Re}(C_{k-1}) = 0, \qquad \mathrm{Re}(C_k) \neq 0, \quad (1.4)$$

where $C_1, \ldots, C_k$ are the coefficients of (1.2) which is the normal form equation of $(1.1)_{\mu=0}$. In this case, we also say that the origin is a weak focus of order $k$ for equation $(1.1)_{\mu=0}$.

**Theorem 1.3.** *Let*

$$\begin{cases} \dot{x} = f(x, y), \\ \dot{y} = g(x, y) \end{cases} \qquad (x, y \in \mathbb{R}^1) \qquad (1.5)$$

*be a $C^\infty$ system with an equilibrium $(0,0)$ that is a weak focus of order $k$. Then*
*(1) if $n \geq k$ and $(1.1)_{\mu=0} \equiv (1.5)$, then there are $\sigma > 0$ and a neighborhood $\Delta$ of $(x, y) = (0,0)$ such that for $|\mu| < \sigma$, (1.1) has at most $k$ limit cycles in $\Delta$;*
*(2) for any integer $j$, $1 \leq j \leq k$, and a neighborhood $\Delta^* \subset \Delta$ of $(x, y) = (0,0)$, there exists a system of the form $(1.1)_\mu$ with $(1.1)_{\mu=0} = (1.5)$, and a number $\sigma^* > 0$ such that $(1.1)_\mu$ has exactly $j$ limit cycles in $\Delta^*$ for $\mu \in S$, where $S$ is an open subset of $\{\mu | 0 < |\mu| < \sigma^*\}$ and $0 \in \bar{S}$.*

*Proof.* (1) Suppose that (1.2) is a normal form equation of (1.5). Then by condition (1.4) it can be transformed into the following form in polar coordinates:

$$\begin{cases} \dot{r} = \mathrm{Re}(C_k)r^{2k+1} + O(r^{2k+3}), \\ \dot{\theta} = \beta_0 + O(r^2). \end{cases} \tag{1.6}$$

Noting $\beta_0 \neq 0$, in a small neighborhood of $r = 0$, we obtain from (1.6) the following equation:

$$\frac{dr}{d\theta} = \frac{\mathrm{Re}(C_k)}{\beta_0} r^{2k+1} + O(r^{2k+3}). \tag{1.7}$$

By Lemma 1.1, we can transform (1.1) into the following form:

$$\frac{dr}{d\theta} = \frac{\alpha(\mu)}{\beta(\mu)} r + h_1(\theta, \mu)r^3 + h_2(\theta, \mu)r^5$$

$$+ \cdots + h_k(\theta, \mu)r^{2k+1} + O(r^{2k+3}), \tag{1.8}$$

where $h_j(\theta, \mu) \in C^\infty$ in $\theta \in [0, 2\pi]$ and $\mu$ near 0, and $\alpha(0) = h_1(\theta, 0) = \cdots = h_{k-1}(\theta, 0) = 0$, $h_k(\theta, 0) = \mathrm{Re}(C_k)/\beta_0 \neq 0$.
Suppose that

$$R(r_0, \theta, \mu) = u_1(\theta, \mu)r_0 + u_2(\theta, \mu)r_0^2$$

$$+ \cdots + u_{2k+1}(\theta, \mu)r_0^{2k+1} + \cdots \tag{1.9}$$

is the solution of (1.8) satisfying the initial condition $R(r_0, 0, \mu) = r_0$, and

$$\psi(r_0, \theta) \equiv R(r_0, \theta, 0) \tag{1.10}$$

is the solution of (1.7) satisfying $\psi(r_0, 0) = r_0$. This implies that

$$\frac{\partial}{\partial \theta} \frac{\partial^i}{\partial r_0^i} \bigg|_{r_0=0} \psi(r_0, \theta)$$

$$= \begin{cases} 0 & \text{for } 1 \leq i < 2k + 1, \\ [(2k + 1)!] \dfrac{\mathrm{Re}\, C_k}{\beta_0} \neq 0, & \text{for } i = 2k + 1. \end{cases}$$

Hence, $\partial^i/\partial r_0^i|_{r_0=0}\psi(r_0, \theta)$ is a constant for $i < 2k + 1$ and is equal to $[(2k + 1)!](\operatorname{Re} C_k/\beta_0)\theta$ for $i = 2k + 1$. Therefore, by using $\psi(r_0, 0) = r_0$ we obtain

$$\left.\frac{\partial^i}{\partial r_0^i}\right|_{r_0=0}\psi(r_0, 2\pi)$$

$$= \begin{cases} \left.\dfrac{\partial^i}{\partial r_0^i}\right|_{r_0=0}\psi(r_0, 0) = \begin{cases} 1, & \text{for } i = 1, \\ 0, & \text{for } 1 < i < 2k + 1, \end{cases} \\ 2\pi[(2k + 1)!]\dfrac{\operatorname{Re} C_k}{\beta_0}, & \text{for } i = 2k + 1. \end{cases} \quad (1.11)$$

As in the proof of Theorem 3.2.4, we define the Poincaré map $P(x, \mu)$ for system (1.8) along the $x$-axis near $x = 0$ and $\mu = 0$, and let

$$V(x, \mu) = P(x, \mu) - x.$$

The number of periodic orbits of (1.8) near $x = 0$ for small $|\mu|$ is determined by the number of zeros of function $V(x, \mu)$ near $(0, 0)$ for $x > 0$. When $x > 0$, we have that

$$V(x, \mu) = R(x, 2\pi, \mu) - x,$$

and

$$V(x, 0) = \psi(x, 2\pi) - x.$$

Clearly, we have that

$$\begin{cases} \dfrac{\partial V}{\partial x}(0, 0) = \dfrac{\partial P}{\partial x}(0, 0) - 1 = \left.\dfrac{\partial}{\partial r_0}\right|_{r_0=0}\psi(r, 2\pi) - 1, \\ \dfrac{\partial^i V}{\partial x^i}(0, 0) = \dfrac{\partial^i P}{\partial x^i}(0, 0) = \left.\dfrac{\partial^i}{\partial r_0^i}\right|_{r_0=0}\psi(r, 2\pi), \quad \text{for } i > 1. \end{cases} \quad (1.12)$$

Equations (1.12) and (1.11) give

$$\frac{\partial^i V}{\partial x^i}(0,0) = \begin{cases} 0, & \text{for } 1 \leq i < 2k+1, \\ 2\pi[(2k+1)!]\dfrac{\operatorname{Re} C_k}{\beta_0}, & \text{for } i = 2k+1. \end{cases} \qquad (1.13)$$

Thus, by using the Malgrange Preparation Theorem (see Theorem 3.1.10), we have $V(x,\mu) = Q(x,\mu) \cdot \eta(x,\mu)$, where $Q$ is a polynomial of degree $2k+1$ with respect to $x$, and $\eta(x,\mu)$ is invertible in a neighborhood of $(x,\mu) = (0,0)$. We remark that $Q$ is divisible by $x$ (since $r = 0$ is the equilibrium of (1.8)), and that other roots of $Q$ appear in pairs: one positive and one negative (since any periodic orbit, surrounding the origin, must cross the positive and negative $x$-axes, respectively). Therefore, there are $\sigma > 0$ and a neighborhood $\Delta$ of $r = 0$ such that (1.8) has at most $k$ limit cycles in $\Delta$.

(2) Suppose that the origin is a weak focus of order $k$ of system (1.5). Then (1.5) has the following normal form equation:

$$\dot{z} = i\beta_0 z + C_1 Z^2 \bar{Z} + \cdots + C_k z^{k+1}\bar{z}^k + O(|z|^{2k+3}) \equiv F(z,\bar{z}). \tag{1.14}$$

For a fixed $j$, $1 \leq j \leq k$, we take a perturbation of (1.14) in the form

$$\dot{z} = F(z,\bar{z}) + \mu_{k-j} z^{k-j+1}\bar{z}^{k-j} + \cdots + \mu_{k-1} z^k \bar{z}^{k-1}, \tag{1.15}$$

where $\mu_l \in \mathbb{R}$, $k - j \leq l \leq k - 1$. In polar coordinates (1.15) gives

$$\dot{r} = \mu_{k-j} r^{2(k-j)+1} + \cdots + \mu_{k-1} r^{2k-1} + \operatorname{Re}(C_k) r^{2k+1} + O(r^{2k+3}) \tag{1.16}$$

$$\equiv G(\mu_{k-j}, \ldots, \mu_{k-2}, \mu_{k-1}; r).$$

In order to obtain $j$ limit cycles for (1.16), we take $\mu_{k-1}, \ldots, \mu_{k-j}$ successively in the following way. Suppose $\operatorname{Re}(C_k) > 0$ (the discussion for $\operatorname{Re}(C_k) < 0$ is similar). We choose $0 < r_k < 1$ so that

$$G(0, \ldots, 0, 0; r_k) > 0.$$

$\mu_{k-1}$ is chosen negative with $|\mu_{k-1}| \ll \operatorname{Re}(C_k)$ so that

$$G(0, \ldots, 0, \mu_{k-1}; r_k) > 0.$$

Then there is $r_{k-1} \in (0, r_k)$ such that

$$G(0, \ldots, 0, \mu_{k-1}; r_{k-1}) < 0.$$

Next, $\mu_{k-2}$ is chosen positive with $|\mu_{k-2}| \ll |\mu_{k-1}|$ so that

$$G(0, \ldots, \mu_{k-2}, \mu_{k-1}; r_k) > 0,$$

$$G(0, \ldots, \mu_{k-2}, \mu_{k-1}; r_{k-1}) < 0.$$

Then there is $r_{k-2} \in (0, r_{k-1})$ such that

$$G(0, \ldots, \mu_{k-2}, \mu_{k-1}; r_{k-2}) > 0.$$

$\mu_{k-3}, r_{k-3}, \ldots, \mu_{k-j}, r_{k-j}$ are chosen similarly, where $\mathrm{Re}(C_k)$, $\mu_{k-1}, \ldots, \mu_{k-j}$ have alternating signs, $0 < |\mu_{k-j}| \ll \cdots \ll |\mu_{k-1}| \ll \mathrm{Re}(C_k)$, and $0 < r_{k-j} < \cdots < r_{k-1} < r_k$. Thus we have finally:

$$\dot{r} > 0 \text{ on } r = r_k, r_{k-2}, \ldots,$$

$$\dot{r} < 0 \text{ on } r = r_{k-1}, r_{k-3}, \ldots.$$

This gives $j$ Poincaré–Bendixson domains; hence there are at least $j$ limit cycles. We claim that there are $\sigma^* > 0$ and a neighborhood $\Delta^*$ of $r = 0$ such that by the choices of $\mu_{k-1}, \ldots, \mu_{k-j}$ described above, (1.15) has exactly $j$ limit cycles in $\Delta^*$ for $|\mu| < \sigma^*$. Otherwise, for any choices of $\Delta$ and $\sigma$, we can find a system of form (1.15) ($|\mu| < \sigma$) which has more than $j$ limit cycles in $\Delta$. Then we can choose also $\mu_{k-j-1}, \ldots, \mu_1$ successively to obtain $(k - j)$ other limit cycles in $\Delta$. Since the total number of limit cycles will be more than $k$, this contradicts the conclusion (1).                                                                          □

**Remark 1.4.** For applications, it is important to determine the order of the weak focus of equation $(1.1)_{\mu=0}$, that is, to find the first nonzero coefficient $\mathrm{Re}(C_k)$ in (1.2). Here we introduce briefly another method, the method of Lyapunov coefficients, which is more convenient in practical calculations.

Suppose that $(1.1)_{\mu=0}$ has the following form:

$$\begin{cases} \dot{x} = -\beta_0 y + p(x, y), \\ \dot{y} = \beta_0 x + q(x, y), \end{cases} \qquad (1.17)$$

where $p, q = O(|x, y|^2)$, $\beta_0 \neq 0$. It is known (see, for example, Zhang et al. [1] or Blows and Lloyd [1]) that for any $m \geq 1$ there exists a smooth function

$$F(x, y) = \frac{\beta_0}{2}(x^2 + y^2) + O(|x, y|^3)$$

such that

$$\left. \frac{d}{dt} F \right|_{(1.17)} = \sum_{i=1}^{m} V_i (x^2 + y^2)^{i+1} + O\left((x^2 + y^2)^{m+1}\right). \qquad (1.18)$$

The coefficients $\{V_j\}$ are called the Lyapunov coefficients of (1.17). We note that $F$ in general is not unique. However, the sign and the position $k$ of the first nonzero coefficient $V_k$ in (1.18) is the same for any $F$. Lyapunov coefficients are equivalent to the coefficients of Hopf bifurcation of order $k$ in the following sense.

**Theorem 1.5.** (Bonin and Legault [1]) The first $(k-1)$ Lyapunov coefficients are zero and the kth coefficient is positive (respectively, negative) if and only if the same is true for the first k coefficients $\{\operatorname{Re}(C_i)|i = 1, 2, \ldots, k\}$.

**Example 1.6.** (Li [1, 3]) For a quadratic system

$$\begin{cases} \dot{x} = -y + a_{20}x^2 + a_{11}xy + a_{02}y^2, \\ \dot{y} = x + b_{20}x^2 + b_{11}xy + b_{02}y^2, \end{cases} \qquad (1.19)$$

let

$$A = a_{20} + a_{02}, \qquad B = b_{20} + b_{02},$$

$$\alpha = a_{11} + 2b_{02}, \qquad \beta = b_{11} + 2a_{20},$$

$$\gamma = b_{20}A^3 - (a_{20} - b_{11})A^2B + (b_{02} - a_{11})AB^2 - a_{02}B^3,$$

$$\delta = a_{02}^2 + b_{20}^2 + a_{02}A + b_{20}B.$$

Then we have (up to a positive factor)
(i)   $V_1 = A\alpha - B\beta$,
(ii)  $V_2 = [\beta(5A - \beta) + \alpha(5B - \alpha)]\gamma$, if $V_1 = 0$,
(iii) $V_3 = (A\beta + B\alpha)\gamma\delta$, if $V_1 = V_2 = 0$,
(iv)  $V_k = 0$ for $k > 3$ if $V_1 = V_2 = V_3 = 0$. In this case, (1.19) is inte-
      grable. (This says that the highest order of weak focus for a
      quadratic system is 3).
(v)   Around a weak focus of order 3, (1.19) has no limit cycle globally.

**Example 1.7.** (Sibirskii [1]) Consider a cubic system without quadratic
terms

$$
\begin{cases}
\dot{x} = -y + \displaystyle\sum_{j+k=3} B_{jk} x^j y^k, \\
\dot{y} = x + \displaystyle\sum_{j+k=3} C_{jk} x^j y^k.
\end{cases}
\tag{1.20}
$$

If $B_{21} + C_{12} = 0$ (this is always possible by a rotation of axes), then we
can rewrite (1.20) in the following form:

$$
\begin{cases}
\dot{x} = -y + (a - \omega - \theta)x^3 + (3\mu - \eta)x^2 y \\
\quad + (3\theta + \xi - 3\omega - 2a)xy^2 + (\nu - \mu)y^3, \\
\dot{y} = x + (\mu + \nu)x^3 + (3\omega + 3\theta + 2a)x^2 y + (\eta - 3\mu)xy^2 \\
\quad + (\omega - \theta - a)y^3.
\end{cases}
\tag{1.21}
$$

We have that (up to a positive factor)
(i)   $V_1 = \xi$,
(ii)  $V_2 = -a\nu$, if $V_1 = 0$,
(iii) $V_3 = a\theta\omega$, if $V_1 = V_2 = 0$,
(iv)  $V_4 = a^2\theta\eta$, if $V_1 = V_2 = V_3 = 0$,
(v)   $V_5 = -a^2\theta[4(\mu^2 + \theta^2) - a^2]$, if $V_1 = V_2 = V_3 = V_4 = 0$,
(vi)  $V_k = 0$, $k \geq 6$, if $V_i = 0$, $i = 1, 2, 3, 4, 5$. In this case equation (1.21)
      is integrable. (For system (1.21), the highest order of weak focus
      is 5.)

**Example 1.8.** Consider Hopf bifurcation for the equation

$$\begin{cases} \dot{x} = y, \\ \dot{y} = -1 + x^2 + \mu_1 y + \mu_2 xy + \mu_3 x^3 y + \mu_4 x^4 y. \end{cases} \quad (1.22)$$

Since (1.22) has two equilibria $(\pm 1, 0)$ and $(1, 0)$ is always a saddle point, we only need to consider $(-1, 0)$. If we make a change of variable $X = x + 1$, (1.22) becomes

$$\begin{cases} \dot{X} = y, \\ \dot{y} = 2X + y(\mu_1 - \mu_2 - \mu_3 + \mu_4) + X^2 + (\mu_2 + 3\mu_3 - 4\mu_4)Xy \\ \quad + (-3\mu_3 + 6\mu_4)X^2 y + (\mu_3 - 4\mu_4)X^3 y + \mu_4 X^4 y. \end{cases}$$

$$(1.23)$$

The linear part of (1.23) at $(0, 0)$ has a pair of purely imaginary eigenvalues if and only if

$$\mu_1 - \mu_2 - \mu_3 + \mu_4 = 0. \quad (1.24)$$

Under condition (1.24), let $y = -\sqrt{2}\, Y$ so (1.23) becomes

$$\begin{cases} \dot{X} = -\sqrt{2}\, Y, \\ \dot{Y} = \sqrt{2}\, X - \dfrac{1}{\sqrt{2}} X^2 + (\mu_2 + 3\mu_3 - 4\mu_4)XY + (-3\mu_3 + 6\mu_4)X^2 Y \\ \quad + (\mu_3 - 4\mu_4)X^3 Y + \mu_4 X^4 Y. \end{cases}$$

$$(1.25)$$

By using the method of Lyapunov coefficients (Remark 1.4), we obtain

$$V_1 = \frac{1}{16}(\mu_2 - 3\mu_3 + 8\mu_4),$$

$$V_2 = \frac{1}{96\sqrt{2}}(5\mu_3 - 14\mu_4), \qquad \text{if } V_1 = 0,$$

$$V_3 = \frac{14}{5}\mu_4, \qquad \text{if } V_1 = V_2 = 0.$$

Therefore, if $\mu_4 \neq 0$, then a Hopf bifurcation of order 3 takes place at

$$\mu_1 = \frac{11}{5}\mu_4, \qquad \mu_2 = \frac{2}{5}\mu_4, \qquad \mu_3 = \frac{14}{5}\mu_4.$$

System (1.22) has three limit cycles if $(\mu_1, \mu_2, \mu_3)$ is located inside the region

$$\mu_4(5\mu_3 - 14\mu_4) < 0, \qquad \mu_4(\mu_2 - 3\mu_3 + 8\mu_4) > 0,$$

$$\mu_4(\mu_1 - \mu_2 - \mu_3 + \mu_4) < 0$$

and $0 < |\mu_1 - \mu_2 - \mu_3 + \mu_4| \ll |V_1| \ll |V_2| \ll |V_3|$.

If $\mu_4 = 0$, $\mu_3 \neq 0$, then a Hopf bifurcation of order 2 takes place at

$$\mu_1 = 4\mu_3, \qquad \mu_2 = 3\mu_3.$$

System (1.22) has two limit cycles if $(\mu_1, \mu_2, \mu_3)$ is located inside the region

$$\mu_3(\mu_2 - 3\mu_3) < 0, \qquad \mu_3(\mu_1 - \mu_2 - \mu_3) > 0$$

and $0 < |\mu_1 - \mu_2 - \mu_3| \ll |V_1| \ll |V_2| \ll 1$.

## 5.2 Homoclinic Bifurcation of Higher Order

Consider

$$\begin{cases} \dot{x} = f(x, y, \mu), \\ \dot{y} = g(x, y, \mu), \end{cases} \tag{2.1}$$

where $x, y \in \mathbb{R}^1$, parameter $\mu \in \mathbb{R}^n$, and $f, g \in C^{2(n+1)}$.

Suppose that $(x, y) = (0, 0)$ is a hyperbolic saddle point of (2.1). Then we can transform (2.1) by a linear change of variables such that the matrix of its linear part at $(0, 0)$ becomes

$$A(\mu) = \begin{bmatrix} a_0(\mu) & b_0(\mu) \\ b_0(\mu) & a_0(\mu) \end{bmatrix}.$$

Thus $A(\mu)$ has eigenvalues $\lambda_{1,2} = a_0(\mu) \pm jb_0(\mu)$, where $j^2 = 1$ and $a_0^2(\mu) - b_0^2(\mu) < 0$. The stable and unstable manifolds of the linear part at $(0, 0)$ are lines $x \pm jy = 0$.

Since trace $A(\mu) = 2a_0(\mu)$, the condition $a_0(0) \neq 0$ is needed for the homoclinic bifurcation of codimension one (see Section 3.2). Now we suppose that

$$a_0(0) = 0, \qquad b_0(0) \neq 0. \tag{2.2}$$

When $\mu = 0$, we have that $\lambda_1 + \lambda_2 = 0$ and hence $A(0)$ has the resonances $\lambda_i = k(\lambda_1 + \lambda_2) + \lambda_i$, $i = 1, 2$, and $k \geq 1$. If $\mu$ is close to zero, the orders of the other resonances of $A(\mu)$ are greater than $2n + 1$. Let

$$\begin{cases} w = x + jy, \\ \overline{w} = x - jy \end{cases} \qquad (j^2 = 1).$$

Then by an argument similar to Lemma 1.1, we can make a polynomial change of variables to transform the system (2.1) into the form

$$\begin{cases} \dot{w} = (a_0(\mu) + jb_0(\mu))w + (a_1(\mu) + jb_1(\mu))w^2\overline{w} + \cdots \\ \qquad + (a_n(\mu) + jb_n(\mu))w^{n+1}\overline{w}^n + A(w, \overline{w}), \\ \dot{\overline{w}} = (a_0(\mu) - jb_0(\mu))\overline{w} + (a_1(\mu) - jb_1(\mu))\overline{w}^2 w + \cdots \\ \qquad + (a_n(\mu) - jb_n(\mu))\overline{w}^{n+1}w^n + \overline{A}(w, \overline{w}), \end{cases} \tag{2.3}$$

where $A(w, \overline{w}) = O(|w, \overline{w}|^{2n+2})$.

**Remark 2.1.** Since $j^2 = 1$, both $w$ and $\overline{w}$ are real. The "conjugate" $\overline{w}$ of $w$, introduced by Joyal [1], is different from the usual complex conjugate, but it is convenient for the discussion in this section.

Before any further discussion, let us present briefly the problem, the method, and the result in this section.

Suppose that for $\mu = 0$ (2.1) has a homoclinic loop $\Gamma$. We are interested in the occurrence of periodic orbits in a small neighborhood of $\Gamma$ by perturbation of the system, that is, for $0 < |\mu| \ll 1$. We will try to find the expression of the Poincaré map of the flows on a transversal segment to $\Gamma$ and near $\Gamma$, and then to study the number of fixed points of this map for small $\mu$.

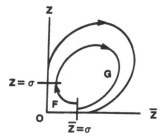

Figure 2.1.

We will find a coordinate $z = w +$ higher-order terms such that the stable and unstable manifolds coincide locally (near the saddle point) with the $\bar{z}$- and $z$-axes, respectively. We define a map $F_\mu(\rho_0)$ from the transversal segment $\bar{z} = \sigma$ to another transversal segment $z = \sigma$ by the flow ($\sigma > 0$ small), where $\rho_0 = 0$ corresponds to the intersection point between $\Gamma$ and $\bar{z} = \sigma$. Then we define a map $G_\mu(\rho_0)$ from $z = \sigma$ to $\bar{z} = \sigma$ by the flow (see Figure 2.1). Thus, the Poincaré map is $P_\mu = G_\mu \circ F_\mu$, and a fixed point of the map $P_\mu$ corresponds to a periodic orbit of the system.

Suppose that $G_\mu(\rho_0)$ has the Taylor expansion

$$G_\mu(\rho_0) = \rho_0 + \beta_0(\mu) + \beta_1(\mu)\rho_0 + \beta_2(\mu)\rho_0^2$$

$$+ \cdots + \beta_n(\mu)\rho_0^n + \phi(\rho_0, \mu), \qquad -1 < \beta_1(\mu), \quad (2.4)$$

where $\phi(\rho_0, \mu) = O(\rho_0^{n+1})$. Then we have a sequence of numbers:

$$\beta_0(\mu), a_0(\mu), \beta_1(\mu), a_1(\mu), \ldots, \beta_k(\mu), a_k(\mu), \ldots, \quad (2.5)$$

where $a_i(\mu)$ and $\beta_i(\mu)$ are coefficients in (2.3) and (2.4), respectively.

The main result in this section can be described roughly as follows: If $\beta_k(0)$ is the first nonzero coefficient in the list (2.5) for $\mu = 0$, then

$$P_0(\rho_0) - \rho_0 \sim \beta_k(0)\rho_0^k,$$

and the system can have at most $2k$ limit cycles near the loop $\Gamma$ for small $\mu$; if $a_k(0)$ is the first one, then

$$P_0(\rho_0) - \rho_0 \sim a_k(0)\rho_0^{k+1}\ln x,$$

and the system can have at most $2k + 1$ limit cycles near $\Gamma$ for small $\mu$.

To prove the above result, we need to find an expression of map $F_\mu(\rho_0)$ (the singular part of $P$). To our knowledge, this problem has been solved by Roussarie [1] and Joyal [1] independently and Leontovich [1] announced a result much earlier. Most results in this section, as well as some notation, belong to Joyal [1].

**Lemma 2.2.** *(Joyal [1]) There exists a $C^{2n+2}$ change of coordinates which transforms (2.3) into the following form*:

$$
\begin{cases}
\dot{w} = (a_0(\mu) + jb_0(\mu))w + (a_1(\mu) + jb_1(\mu))w^2\overline{w} + \cdots \\
\qquad + (a_n(\mu) + jb_n(\mu))w^{n+1}\overline{w}^n + w^{n+1}\overline{w}^n E(w,\overline{w}), \\
\dot{\overline{w}} = (a_0(\mu) - jb_0(\mu))\overline{w} + (a_1(\mu) - jb_1(\mu))\overline{w}^2 w + \cdots \\
\qquad + (a_n(\mu) - jb_n(\mu))\overline{w}^{n+1}w^n + \overline{w}^{n+1}w^n \overline{E(w,\overline{w})},
\end{cases}
\tag{2.6}
$$

*where $E(w,\overline{w}) \in C^0$ and $E(w,\overline{w}) \to 0$ as $(w,\overline{w}) \to (0,0)$.*

Since $a_0(0) = 0$, $b_0(0) \neq 0$ (see (2.2)), we suppose that $jb_0(0) > 0$. Then $\alpha_0(\mu) = a_0(\mu) + jb_0(\mu) > 0$, $\overline{\alpha}_0(\mu) = a_0(\mu) - jb_0(\mu) < 0$ for small $\mu$. It is easy to see from (2.6[2ybn]) that in a small neighborhood $\Omega$ of the origin, the two lines $w = 0$ and $\overline{w} = 0$ are the stable and unstable manifolds of (2.6), respectively. Without loss of generality, we suppose that the region $\{(w,\overline{w})|w > 0, \overline{w} > 0\} \cap \Omega$ is inside the homoclinic loop $\Gamma$ which exists for $\mu = 0$. Thus, to establish the transition map $F_\mu(p)$, we only need to consider $w > 0$ and $\overline{w} > 0$. Let

$$
\begin{cases}
\rho = w\overline{w}, \\
\phi = \ln w.
\end{cases}
\tag{2.7}
$$

Then (2.6) becomes

$$
\begin{cases}
\dot{\rho} = 2(a_0(\mu)\rho + a_1(\mu)\rho^2 + \cdots + a_n(\mu)\rho^{n+1} + \rho^{n+1}A(\rho,\phi)), \\
\dot{\phi} = \alpha_0(\mu) + \alpha_1(\mu)\rho + \cdots + \alpha_n(\mu)\rho^n + \rho^n B(\rho,\phi),
\end{cases}
$$

$$
\tag{2.8}
$$

where $\alpha_i(\mu) = a_i(\mu) + jb_i(\mu)$, $\rho^{n+1}A \in C^{2n+1}$, $\rho^n B \in C^{2n}$ in $\rho \geq 0$, $A, B$ are continuous, and $A, B \to 0$ as $(\rho, \phi)$ tends to the saddle point.

Since $\alpha_0(0) = jb_0(0) > 0$, the right-hand side of the second equation in (2.8) is positive for small $\rho$ and small $\mu$. Hence, we obtain from (2.8)

$$\frac{d\rho}{d\phi} = c_0(\mu)\rho + c_1(\mu)\rho^2 + \cdots + c_n(\mu)\rho^{n+1} + \rho^{n+1}\tilde{A}(\rho, \phi), \quad (2.9)$$

where $\rho^{n+1}\tilde{A}(\rho, \phi) \in C^{2n+1}$ and $\tilde{A}(\rho, \phi) \to 0$ as $(w, \bar{w}) \to 0 (\rho \to 0, \phi \to -\infty)$. We have the following relation among the $a_i(\mu)$ and $c_i(\mu)$:

$$a_0 = a_1 = \cdots = a_{k-1} = 0, \qquad a_k \neq 0$$

$$\Leftrightarrow c_0 = c_1 = \cdots = c_{k-1} = 0, \qquad c_k \neq 0, \qquad (2.10)$$

and $\text{sgn}(a_k) = \text{sgn}(c_k b_0)$.

Suppose the solution of (2.9), $\rho = \rho(\phi; \rho_0, \mu)$ with $\rho(\phi_0; \rho_0, \mu) = \rho_0$, has the form

$$\rho = h_1(\phi)\rho_0 + h_2(\phi)\rho_0^2 + \cdots + h_{n+1}(\phi)\rho_0^{n+1} + H(\rho_0, \phi), \quad (2.11)$$

where $H(\rho_0, \phi)$ satisfies

$$\lim_{\rho_0 \to 0} \frac{H(\rho_0, \phi)}{\rho_0^{n+1}} = 0.$$

Then the $h_i(\phi)$ satisfy the following equations:

$$\begin{cases} h_1' = c_0 h_1, \\ h_2' = c_0 h_2 + c_1 h_1^2, \\ h_3' = c_0 h_3 + 2c_1 h_1 h_2 + c_2 h_1^3, \\ \quad \vdots \\ h_{n+1}' = c_0 h_{n+1} + c_1 \left( \sum_{i_1+i_2=n+1} h_{i_1} h_{i_2} \right) + \cdots \\ \qquad + c_{n-1} \left( \sum_{i_1+\cdots+i_n=n+1} h_{i_1} \cdots h_{i_n} \right) + c_n h_1^{n+1}, \end{cases} \quad (2.12)$$

where $h_1(\phi_0) = 1$ and $h_i(\phi_0) = 0$ for $i \geq 2$.

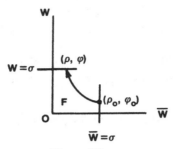

Figure 2.2.

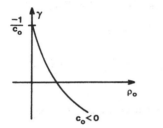 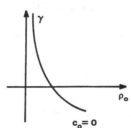 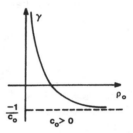

Figure 2.3.

If we consider a transition map $\rho = F_\mu(\rho_0)$ which is defined by the flow from a point $(\rho_0, \phi_0) \in \{\overline{w} = \sigma\}$ to point $(\rho, \phi) \in \{w = \sigma\}$ (Figure 2.2), then $\phi_0 = \ln \rho_0 - \ln \sigma$ and $\phi = \ln \sigma$. If we make a rescaling $w \to \sigma w$, then we can suppose that the flow is defined from a point $(\rho_0, \phi_0)$ to point $(\rho, \phi)$ satisfying $\phi_0 = \ln \rho_0$ and $\phi = 0$.

Let $\exp(c_0(\phi - \phi_0)) = c_0\gamma + 1$. Then (2.12) gives

$$\begin{cases} h_1 = c_0\gamma + 1, \\ h_{k+1} = (c_0\gamma + 1)p_k(\gamma, c_0, c_1, \ldots, c_k), \quad 1 \le k \le n, \end{cases} \quad (2.13)$$

where $p_k$ is a polynomial in $\gamma$ of degree $k$, vanishing at $\gamma = 0$, and $c_k\gamma$ is the linear term in $p_k$. $\gamma = (\exp(c_0(\phi - \phi_0)) - 1)/c_0 = (\rho_0^{-c_0} - 1)/c_0$ for $c_0 \ne 0$ and $\gamma = \phi - \phi_0$ for $c_0 = 0$. For small $\mu$, $\gamma$ is always positive in a neighborhood of the origin $(w, \overline{w}) = 0$ (see Figure 2.3). Thus, substituting (2.13) into (2.11), we have

$$F_\mu(\rho_0) = \rho = \rho_0 + c_0[\rho_0\gamma + \tilde{p}_0(\rho_0, \gamma, c_0, \ldots, c_n)]$$

$$+ c_1[\rho_0^2\gamma + \tilde{p}_1(\rho_0, \gamma, c_0, \ldots, c_n)] \quad (2.14)$$

$$+ \cdots + c_n[\rho_0^{n+1}\gamma + \widetilde{p_n}(\rho_0, \mu)],$$

where $\tilde{p}_i$ $(i < n)$ is a polynomial and $\widetilde{p_n} = H/c_n$ ($H$ is defined in (2.11)), with $\tilde{p}_k = o(\rho_0^{k+1}\gamma)$, $k = 0, 1, \ldots, n$.

Now suppose the map $G_\mu$: $\{w = \sigma\} \to \{\bar{w} = \sigma\}$ has the expression (2.4). Then

$$P_\mu(\rho_0) = G_\mu \circ F_\mu(\rho_0) = \beta_0 + (1 + \beta_1)[\rho_0 + c_0(\rho_0 + \gamma) + \cdots]$$

$$+ \beta_2[\rho_0 + c_0(\rho_0 + \gamma) + \cdots]^2$$

$$+ \cdots + \beta_n[\rho_0 + c_0(\rho_0 + \gamma) + \cdots]^n$$

$$+ R(\rho_0, \mu).$$

We expand $P_\mu(\rho_0) - \rho_0$ in two different cases corresponding to the manner of $\gamma$ at $\rho_0 = 0$ (see Figure 2.3):

(a) $c_0 \geq 0$, $\rho_0^k = o(\rho_0^k\gamma)$:

$$P_\mu(\rho_0) - \rho_0$$

$$= \beta_0 + c_0[(1 + \beta_1)\rho_0\gamma + q_0(\rho_0, \gamma, c_0, \ldots, c_n, \beta_0, \ldots, \beta_n)]$$

$$+ \beta_1\rho_0 + \cdots + c_{n-1}[(1 + \beta_1)\rho_0^n\gamma + q_{n-1}(\rho_0, \gamma, c_{n-1}, c_n, \beta_n)]$$

$$+ \beta_n\rho_0^n + c_n[(1 + \beta_1)\rho_0^{n+1}\gamma + q_n(\rho_0, \mu)]; \qquad (2.15a)$$

(b) $c_0 < 0$, $[\rho_0(c_0\gamma + 1)]^k = o(\rho_0^k\gamma)$:

$$P_\mu(\rho_0) - \rho_0 = \beta_0 + c_0(\rho_0\gamma + \tilde{q}_0) + \beta_1\rho_0(c_0\gamma + 1) + \cdots$$

$$+ c_{n-1}[(1 + \beta_1)\rho_0^n\gamma + \tilde{q}_{n-1}] + \beta_n[\rho_0(c_0\gamma + 1)]^n$$

$$+ c_n[(1 + \beta_1)\rho_0^{n+1}\gamma + \tilde{q}_n(\rho_0, \mu)], \qquad (2.15b)$$

where $q_k, \tilde{q}_k = o(\rho_0^{k+1}\gamma)$ for $0 \leq k \leq n - 1$. Let

$$\begin{cases} \xi_{2i} = \beta_i, & i = 0, 1, 2, \ldots, n, \\ \xi_{2i+1} = c_i, & i = 0, 1, 2, \ldots, n, \end{cases} \qquad (2.16)$$

where $\beta_i$ and $c_i$ are the coefficients in (2.4) and (2.14), respectively.

**Definition 2.3.** Suppose the linear part of system (2.1) at the origin has the eigenvalues $a_0(\mu) \pm jb_0(\mu)$ with $a_0^2 - b_0^2 < 0$ $(j^2 = 1)$. The system $(2.1)_{\mu=0}$ is said to have a homoclinic bifurcation of order $m$ if for $\mu = 0$ (2.1) has a homoclinic loop and

$$\xi_0(0) = \cdots = \xi_{m-1}(0) = 0 \quad \text{and} \quad \xi_m(0) \neq 0.$$

**Theorem 2.4.** If $(2.1)_{\mu=0}$ has a homoclinic bifurcation of order $m$ $(m \leq 2n + 1)$, then (1) in a sufficiently small neighborhood of $\mu = 0$, any $(2.1)_\mu$ has at most $m$ limit cycles near the loop;

(2) for any $k$, $0 \leq k \leq m$, there exists a perturbation system $(2.1)_\mu$ and a neighborhood $U$ of the loop such that the system has exactly $k$ limit cycles in $U$.

In order to prove Theorem 2.4, we need the following notation and lemma.

**Notation 2.5.** Let $f_i(x, \mu) \in C^n$ for $x > 0$ and be continuous at $x = 0$ $(1 \leq i \leq n)$. We define the following functions:

$$\begin{cases} \cdots \ f_{i_1 i_2} = \dfrac{f'_{i_2}}{f'_{i_1}}, \\[2mm] f_{i_1 i_2 \cdots i_k} = \dfrac{f'_{i_1 \cdots i_{k-2} i_k}}{f'_{i_1 \cdots i_{k-1}}} \quad \text{for } 3 \leq k \leq n, \end{cases}$$

where $'$ means $\partial/\partial x$ and $1 \leq i_1 < i_2 < \cdots < i_k \leq n$.

If $f_{i_1}(0, \mu) = 0$ and $f_{i_2} = o(f_{i_1})$ as $x \to 0$, then $\lim_{x \to 0} f_{i_1 i_2} = 0$.

**Lemma 2.6.** Let $f_1(x, \mu), \ldots, f_n(x, \mu)$ be functions continuous at $x = 0$ and of class $C^n$ for $\mu \in R^m$ and $x > 0$. If $f_i(0, \mu) = 0$, $f_{i_2} = o(f_{i_1})$ for $i_2 > i_1$, and $f'_{i_1 \cdots i_k} > 0$ for $0 < x < \epsilon$, $\mu \in \mathbb{R}^m$, and $1 \leq k \leq n$, then the function

$$P(x, \mu) = c_0(\mu) + c_1(\mu)f_1(x, \mu) + \cdots + c_n(\mu)f_n(x, \mu)$$

has at most $n$ zeros for $0 \leq x \leq \epsilon$, where $c_i(\mu) \in \mathbb{R}$, $i = 0, \ldots, n$, and $c_n(\mu) \neq 0$.

*Proof.* Suppose that $P$ has more than $n$ zeros in $[0, \epsilon]$. Then

$$\frac{\partial P}{\partial x} = f_1'(c_1 + c_2 f_{12} + \cdots + c_n f_{1n}) := f_1' P_1$$

has at least $n$ zeros in $(0, \epsilon)$. Since $f_1' > 0$, $P_1$ has at least $n$ zeros in $(0, \epsilon)$. Repeating the above argument, we see that $\partial P_1/\partial x$ has at least $n - 1$ zeros in $(0, \epsilon)$ and as many as

$$P_2 = c_2(\mu) + c_3(\mu) f_{123} + \cdots + c_n(\mu) f_{12n}.$$

By induction, $\partial P_n/\partial x = c_n f_{12 \cdots n}'$ has at least one zero in $(0, \epsilon)$, which contradicts the assumptions of the lemma.          □

*Proof of Theorem of 2.4.* (1) We will prove that (2.15a) and (2.15b) have at most $m$ zeros for $0 \leq \rho_0 \leq \epsilon$. In order to use Lemma 2.6, we only need to show that the sequence

$$\rho_0 \gamma, \rho_0, \rho_0^2 \gamma, \rho_0^2, \ldots, \rho_0^n \gamma, \rho_0^n \qquad (c_0 \geq 0) \qquad (2.17a)$$

or

$$\rho_0 \gamma, \rho_0(c_0 \gamma + 1), \ldots, \rho_0^n \gamma, [\rho_0(c_0 \gamma + 1)]^n \qquad (c_0 < 0) \quad (2.17b)$$

satisfies the properties of the sequence $\{f_i\}$ in Lemma 2.6.

From (2.10) we have that $c_0 = 0 \Leftrightarrow a_0 = 0$. In the following we suppose that $|c_0|$ is sufficiently small. We denote (2.17a) or (2.17b) by $\{f_i(\rho_0, \mu)\}$. Since for $c_0 \neq 0$, $c_0 \gamma + 1 = \rho_0^{-c_0}$, we consider the following sequence instead of (2.17a) or (2.17b):

$$x^{k_1} \gamma^{m_1}, \ldots, x^{k_n} \gamma^{m_n}, \qquad (2.18)$$

where $x = \rho_0$, $\gamma = (x^{-c_0} - 1)/c_0$, and $k_i$ and $m_i \in \mathbb{R}$. Equation (2.18) satisfies the following conditions:

(i) when $c_0 \geq 0$, we have $k_i \leq k_j$ for $i < j$; if $k_i = k_j$ then $m_i > m_j$; when $c_0 < 0$ ($|c_0|$ small), we have $k_i < k_j$ for $i < j$.

(ii) $k_i > 0$ and $m_i \geq 0$.

It follows that $f_{i_2} = o(f_{i_1})$ as $x \to 0$ for $i_1 < i_2$, and

$$(x^k \gamma^m)' = (k - c_0 m) x^{k-1} \gamma^m + o(x^{k-1} \gamma^m),$$

where $k - c_0 m > 0$ for small $|c_0|$. Hence.

$$f_{i_1 i_2} = \frac{f'_{i_2}}{f'_{i_1}} = Mx^{k_2 - k_1} \gamma^{m_2 - m_1} + o(x^{k_2 - k_1} \gamma^{m_2 - m_1}),$$

where $M > 0$. Thus $f'_{i_1 i_2} > 0$ in $(0, \epsilon)$ for a small $\epsilon > 0$. By induction, we have that $f'_{i_1 \cdots i_k} > 0$ in $(0, \epsilon)$ for a small fixed $\epsilon > 0$. Thus the conditions of Lemma 2.6 are satisfied.

(2) We first suppose $k = m$. Without loss of generality, suppose $m = 2n + 1$. Then

$$\xi_0(0) = \xi_1(0) = \cdots = \xi_{2n}(0) = 0, \qquad \xi_{2n+1}(0) \neq 0.$$

We will find a perturbation system $(2.1)_\mu$ such that it has exactly $2n + 1$ limit cycles near the loop.

By using (2.16) and (2.9), we know that $(2.1)_{\mu = 0}$ can be transformed into the following form

$$\frac{d\rho}{d\phi} = c_n(0)\rho^{n+1} + \rho^{n+1}\tilde{A}(\rho, \phi) \equiv S(\rho, \phi). \tag{2.19}$$

We will construct a system in a small tubular neighborhood $T$ of the loop. Let

$$\begin{cases} \dot{x} = \tilde{u}, \\ \dot{y} = \tilde{v}, \end{cases} \tag{2.20}$$

where

$$\tilde{u} = u + (\tilde{u}_1 - u)\omega_1, \qquad \tilde{v} = v + (\tilde{v}_1 - v)\omega_1,$$

$$\tilde{u}_1 = u_2 + (\tilde{u}_2 - u_2)\omega_2, \qquad \tilde{v}_1 = v_2 + (\tilde{v}_2 - v_2)\omega_2,$$

$$\tilde{u}_2 = u_3 + (\tilde{u}_3 - u_3)\omega_3, \qquad \tilde{v}_2 = v_3 + (\tilde{v}_3 - v_3)\omega_3,$$

where $(u, v)$ is the original system $(2.1)_{\mu = 0}$ with the form (2.19) in $(\rho, \phi)$ coordinates. The systems $(u_2, v_2)$, $(u_3, v_3)$, and $(\tilde{u}_3, \tilde{v}_3)$ expressed in the $(\rho, \phi)$ coordinates are respectively:

$$\frac{d\rho}{d\phi} = \mu_2 \rho + \mu_4 \rho^2 + \cdots + \mu_{2n} \rho^n + S(\rho, \phi), \tag{2.21}$$

$$\frac{d\rho}{d\phi} = S(\rho, \phi), \tag{2.22}$$

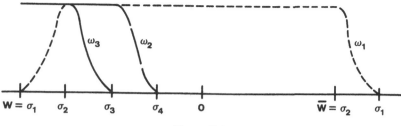

Figure 2.4.

and

$$\frac{d\rho}{d\phi} = \mu_3\rho + \mu_5\rho^2 + \cdots + \mu_{2n+1}\rho^n + S(\rho,\phi). \qquad (2.23)$$

Here $S(\rho,\phi)$ is the same as in (2.19) and $\mu_{2n+1}, \mu_{2n}, \ldots, \mu_3, \mu_2$ are small parameters satisfying

$$\mu_i\mu_{i+1} < 0, \qquad 0 < |\mu_2| \ll |\mu_3| \ll \cdots \ll |\mu_{2n+1}| \ll |c_n(0)|. \qquad (2.24)$$

Finally, the $C^\infty$ functions $\omega_1$, $\omega_2$, and $\omega_3$ are defined as follows:

(i) $\omega_1(x, y) \equiv 0$ in a strip of $T$. This strip going from $w = \sigma_1$ to $\bar{w} = \sigma_1$ does not contain the origin. $\omega_1(x, y) \equiv 1$ from $\bar{w} = \sigma_2$ to $w = \sigma_2$ ($0 < \sigma_2 < \sigma_1$). Elsewhere, $0 < \omega_1 < 1$ and $\omega_1$ is a constant on $w = \sigma$ (i.e., on $\phi = \ln \sigma$) for $\sigma_2 < \sigma < \sigma_1$.

(ii) $\omega_2(x, y) \equiv 0$ from $\bar{w} = \sigma_1$ to $w = \sigma_4$, $\omega_2(x, y) \equiv 1$ from $w = \sigma_3$ to $w = \sigma_1$ ($0 < \sigma_4 < \sigma_3 < \sigma_2 < \sigma_1$), and $0 < \omega_2(x, y) < 1$ and $\omega_2$ is a constant on $w = \sigma$ for $\sigma_4 < \sigma < \sigma_3$. Elsewhere, it does not matter.

(iii) $\omega_3(x, y) \equiv 0$ from $\bar{w} = \sigma_1$ to $w = \sigma_3$, $0 < \omega_3 < 1$ and $\omega_3$ is a constant on $w = \sigma$ for $\sigma_3 < \sigma < \sigma_2$, and $\omega_3(x, y) \equiv 1$ from $w = \sigma_2$ to $w = \sigma_1$. Elsewhere, it does not matter (see Figure 2.4).

Obviously, (2.20) is a perturbation system of (2.19). The equations

$$\frac{d\rho}{d\phi} = \left(\mu_2\rho + \mu_4\rho^2 + \cdots + \mu_{2n}\rho^n\right)\omega_1(\rho) + S(\rho,\phi)$$

(from $\bar{w} = \sigma_1$ to $w = \sigma_4$).

$$\frac{d\rho}{d\phi} = \left(\mu_2\rho + \mu_4\rho^2 + \cdots + \mu_{2n}\rho^n\right)\left(1 - \omega_2(\rho)\right) + S(\rho,\phi)$$

(from $w = \sigma_4$ to $w = \sigma_3$)

give the map $F: \{\bar{w} = \sigma_1\} \to \{w = \sigma_3\}$ which has an expansion of the form (2.14), with $\phi - \phi_0 = -\ln \rho_0 + \ln \sigma_3 - \ln \sigma_1 = -\ln(\sigma_1/\rho_0\sigma_3)$ and the $c_i$ having the same sign as that of $\mu_{i+2}$. The map $E: \{w = \sigma_3\} \to \{\bar{w} = \sigma_1\}$ is defined by

$$\frac{d\rho}{d\phi} = \left(\mu_3\rho + \mu_5\rho^2 + \cdots + \mu_{2n+1}\rho^n\right)\omega_3(\rho) + S(\rho, \phi)$$

$$(\text{from } w = \sigma_3 \text{ to } w = \sigma_2),$$

$$\frac{d\rho}{d\phi} = \left(\mu_3\rho + \mu_5\rho^2 + \cdots + \mu_{2n+1}\rho^n\right)\omega_1(\rho) + S(\rho, \phi)$$

$$(\text{from } w = \sigma_2 \text{ to } \bar{w} = \sigma_1).$$

$E$ has an expansion of the form (2.14), with $\phi - \phi_0 = -\ln \sigma_3/\sigma_1 > 0$ (finite). On the other hand, replacing the terms, $E$ has an expansion of the form (2.4) with $\beta_0 = 0$ (this means that the homoclinic loop exists). According to (2.13), the $\beta_i$ of the expansion satisfy

$$\beta_1 = \mu_3\gamma,$$

$$\beta_2 = (\mu_3\gamma + 1)\mu_5\left(\frac{\mu_3}{2}\gamma^2 + \gamma\right),$$

$$\beta_n = (\mu_3\gamma + 1)p_{n-1}(\gamma, \mu_3, \mu_5, \ldots, \mu_{2n+1}),$$

where $p_j$ $(j \geq 1)$ is a polynomial in $\gamma$ of degree $j$ and having $\mu_{2j+1}\gamma$ as the first-degree term. Hence $\beta_i$ has the same sign as that of $\mu_{2i+1}$ whenever $|\mu_{2j-1}| \ll |\mu_{2j+1}|(2 \leq j \leq n)$.

Since the $\mu_i(2 \leq i \leq 2n + 1)$ can be chosen independently, and they are continuous functions of the coefficients of the Taylor expansion of the system, condition (2.24) implies that system (2.20) has at least $2n$ limit cycles inside the annulus region $T$.

Up to now, we still have the condition $\beta_0 = 0$, that is, system (2.20) has a homoclinic loop. $\mu_2 \neq 0$ implies $c_0 \neq 0$ $(a_0 \neq 0)$, that is, the trace at the saddle point of (2.20) is nonzero. Hence, as a perturbation of (2.20), the following system

$$\begin{cases} \dot{x} = \bar{u} - \mu_1\bar{v} \\ \dot{y} = \mu_1\bar{u} + \bar{v}. \end{cases} \tag{2.25}$$

has one more limit cycle if $|\mu_1| \ll |\mu_2|$ and sgn $\mu_1$ is well chosen (see Example 3.3.9).

From conclusion (1) of Theorem 2.4, we know that system (2.25), as a perturbation of $(2.1)_{\mu=0}$, has at most $2n + 1$ limit cycles. Therefore, (2.25) has exactly $2n + 1 = m$ limit cycles.

For the case $k < m$, we can use the same argument to prove part (2) of Theorem 1.3. We use the same perturbed systems (2.25) and (2.20), but take the first $(m - k)$ elements of the list $\mu_m, \mu_{m-1}, \ldots, \mu_1$ to be zero. Then we get a perturbed system having at least $k$ limit cycles. We can obtain a system in this way which has exactly $k$ limit cycles. Otherwise we can make a perturbation with $\mu_m, \mu_{m-1}, \ldots, \mu_{m-k}$ satisfying (2.24) to get other $(m - k)$ limit cycles, and the total number of limit cycles will be more than $m$, which contradicts conclusion (1).     □

**Remark 2.7.** For applications, it is important to determine the first nonzero coefficient in (2.5) for $\mu = 0$, that is, the order of homoclinic bifurcation (see (2.16) and Definition 2.3). We will introduce a method which is called the method of dual Lyapunov constants to determine the first nonzero coefficient among $a_1(0), a_2(0), \ldots$ if $a_0(0) = 0$ and $b_0(0) = 1$ (see (2.5)). Then we will give some results and examples without a detailed discussion.

If $a_0(0) = 0$ (i.e., the trace $A(0) = 0$), then system $(2.1)_{\mu=0}$ with a saddle point at the origin can be transformed into the following form:

$$\begin{cases} \dot{x} = y + p(x, y), \\ \dot{y} = x + q(x, y). \end{cases} \tag{2.26}$$

We try to find a function

$$F(x, y) = (x^2 - y^2) + F_3(x, y) + \cdots + F_k(x, y) + \cdots,$$

where $F_k(x, y)$ is a homogeneous polynomial of order $k$, such that

$$\left. \frac{dF}{dt} \right|_{(2.26)} = v_1^*(x^2 - y^2)^2 + v_2^*(x^2 - y^2)^3$$

$$+ \cdots + v_k^*(x^2 - y^2)^{k+1} + \cdots.$$

**Theorem 2.8.** *(Joyal and Rousseau [1]) Consider the system (2.1) with* $a_0(0) = 0$, $b_0(0) = 1$, *and the sequence (2.5). Then at* $\mu = 0$, $a_1 = a_2 = \cdots = a_{k-1} = 0$ *and* $a_k \neq 0$ *if and only if* $v_1^* = v_2^* = \cdots = v_{k-1}^* = 0$ *and* $v_k^* \neq 0$. *Moreover* $\mathrm{sgn}(a_k) = \mathrm{sgn}(v_k^*)$.

**Definition 2.9.** The origin is called a weak saddle point of order $k$ if $v_1^* = \cdots = v_{k-1}^* = 0$ and $v_k^* \neq 0$ for system (2.1) with $a_0(0) = 0$ and $b_0(0) = 1$. $v_k^*$ is called a $k$th saddle quantity.

**Example 2.10.** For system (2.26), the first saddle quantity is given by

$$v_1^* = \left(f_{xxx} - f_{xyy} + g_{xxy} - g_{yyy}\right)$$

$$+ \left[f_{xy}(f_{yy} - f_{xx}) + g_{xy}(g_{yy} - g_{xx}) - f_{xx}g_{xx} + f_{yy}g_{yy}\right]. \quad (2.27)$$

**Example 2.11.** (Cai [1] and Zhang and Cai [1]) For a quadratic system

$$\begin{cases} \dot{x} = x + Ax^2 + Bxy + Cy^2, \\ \dot{y} = -y - Kx^2 - Lxy - My^2: \end{cases} \quad (2.28)$$

(i)   The first three saddle quantities are

$$v_1^* = LM - AB,$$

$$v_2^* = KB(2M - B)(M + 2B) - CL(2A - L)(A + 2L),$$

$$\text{if } v_1^* = 0,$$

$$v_3^* = (CK - LB)\left[ACL(2A - L) - BKM(2M - B)\right],$$

$$\text{if } v_1^* = v_2^* = 0,$$

and

$$v_k^* = 0 \quad \text{for all} \quad k > 3 \text{ if } v_1^* = v_2^* = v_3^* = 0.$$

(ii) If $v_k^* = 0$ for all $k$, and if the quadratic system has a homoclinic loop (or compound homoclinic cycle) through the saddle point(s), then the system is integrable in the interior of the loop (cycle).

(iii) The nonintegrable quadratic systems satisfying $v_1^* = 0$ have no limit cycle or homoclinic loop.

From Theorem 2.4 and Example 2.11 (iii), one has immediately the following theorem.

**Theorem 2.12.** *There are at most three limit cycles which may arise from a homoclinic loop bifurcation in a nonintegrable quadratic system.*

**Example 2.13.** (Joyal and Rousseau [1]) Consider the homoclinic bifurcation of the system

$$\begin{cases} \dot{x} = y = F, \\ \dot{y} = -1 + x^2 + \delta\left(v_1 y + v_2 xy + v_3 x^3 y + v_4 x^4 y\right) = G, \end{cases} \quad (2.29)$$

with a saddle point at $(1, 0)$. For $\delta = 0$, (2.29) becomes a Hamiltonian system (see Figure 4.1.2)

$$\begin{cases} \dot{x} = y, \\ \dot{y} = -1 + x^2, \end{cases} \quad (2.30)$$

with Hamiltonian function

$$H(x, y) = \frac{y^2}{2} + x - \frac{x^3}{3}. \quad (2.31)$$

$\{H(x, y) = h, \ -\frac{2}{3} \le h \le \frac{2}{3}\}$ are closed level curves. $H = -2/3$ corresponds to the equilibrium $(-1, 0)$ and $H = 2/3$ corresponds to the homoclinic loop.

We consider (2.29) as a small perturbation of (2.30). From the discussion in Section 4.1 (Lemma 4.1.4), we have that the fixed points of the Poincaré map correspond to the zeros of the following bifurcation function

$$M(h) = \int_{H=h} \left(v_1 y + v_2 xy + v_3 x^3 y + v_4 x^4 y\right) dx. \quad (2.32)$$

Therefore, we have that

(i)  System (2.29) has a homoclinic loop bifurcation (HLB) if $M(2/3)$ = 0.

(ii)  The HLB is of order 1 if $M(2/3) = 0$ and $M'(2/3)$ is infinite. The latter is equivalent to $\operatorname{div}(F, G)|_{(1,0)} \neq 0$.

(iii)  The HLB is of order 2 if $M(2/3) = \operatorname{div}(F, G)|_{(1,0)} = 0$ and $M'(2/3) \neq 0$.

(iv)  The HLB is of order 3 if $M(2/3) = \operatorname{div}(F, G)|_{(1,0)} = M'(2/3) = 0$ and $M''(2/3)$ is infinite. The last condition is equivalent to $v_1^* \neq 0$.

By calculation, we have

$$M(2/3) = 4\sqrt{2/3} \int_{-2}^{1} (v_1 + v_2 x + v_3 x^3 + v_4 x^4)(1 + x)\sqrt{x + 2}\, dx.$$

Hence, $M(2/3) = 0$ is equivalent to

$$v_1 - \frac{5}{7}v_2 - \frac{103}{77}v_3 + \frac{187}{91}v_4 = 0. \qquad (2.33)$$

From (2.32) and (2.31) we have

$M'(2/3)$

$$= 2\int_{-2}^{1} \frac{1}{y}(v_1 + v_2 x + v_3 x^3 + v_4 x^4)\, dx$$

$$= (3/2)^{1/2} \int_{-2}^{1} \left[ \frac{v_1 + v_2 + v_3 + v_4}{(1 - x)(x + 2)^{1/2}} \right.$$

$$\left. - \frac{v_2 + v_3(x^2 + x + 1) + v_4(x^2 + 1)(x + 1)}{(x + 2)^{1/2}} \right] dx.$$

It is clear that if $M'(2/3)$ is finite, then it is equivalent to

$$\frac{1}{\delta} \operatorname{div}(F, G)|_{(1,0)} = v_1 + v_2 + v_3 + v_4 = 0. \qquad (2.34)$$

When $\operatorname{div}(F, G)|_{(1,0)} = 0$, we have that $M'(2/3) = 0$ is equivalent to

$$v_2 + \frac{9}{5}v_3 - \frac{8}{7}v_4 = 0. \qquad (2.35)$$

Finally, transforming system (2.29) into the form (2.26) and using the formula (2.27), we obtain

$$v_1^* = \frac{\delta}{2}(-v_2 + 3v_3 + 8v_4).$$

It is easy to see that $v_1^* = c\delta v_4$, where $c > 0$ is a constant, if (2.33), (2.34), and (2.35) are satisfied.

Thus, if $v_4 \neq 0$ and $0 < |\delta| \ll 1$, then (2.29) has a homoclinic loop bifurcation of order 3 when (2.33), (2.34), and (2.35) are satisfied. If $v_4 = 0$, $v_3 \neq 0$, and $0 < |\delta| \ll 1$, then (2.29) has a homoclinic loop bifurcation of order 2 when (2.33) and (2.34) are satisfied.

## 5.3 A Codimension 3 Bifurcation: Cusp of Order 3

In Section 4.1 it is shown that the Bogdanov–Takens system

$$\begin{cases} \dot{x} = y, \\ \dot{y} = \epsilon_1 + \epsilon_2 y + x^2 \pm xy \end{cases} \tag{3.1}$$

is a versal deformation of the vector field

$$\begin{cases} \dot{x} = y, \\ \dot{y} = ax^2 + bxy, \end{cases} \tag{3.2}$$

where $a$ and $b$ are constants satisfying $ab \neq 0$.

If $a \neq 0$ (without loss of generality, suppose $a = 1$), then the phase portrait of (3.2) is shown in Figure 3.1. We note that the phase portrait is the same for all values of $b$, including $b = 0$. Since there is a cusp at the origin, the singularity is said to be a cusp type. If $b \neq 0$, it is a cusp of codimension 2; if $b = 0$, it is a cusp of higher codimension. In particular, the families of vector fields with cusps of codimension 3 and 4 are respectively

$$\begin{cases} \dot{x} = y, \\ \dot{y} = \epsilon_1 + \epsilon_2 y + \epsilon_3 xy + x^2 \pm x^3 y \end{cases} \tag{3.3}^{\pm}$$

Figure 3.1. The flow near a cusp-type equilibrium.

and

$$\begin{cases} \dot{x} = y, \\ \dot{y} = \epsilon_1 + \epsilon_2 y + \epsilon_3 xy + \epsilon_4 x^3 y + x^2 \pm x^4 y, \end{cases} \qquad (3.4)^{\pm}$$

where $\epsilon_i$ $(i = 1, 2, 3, 4)$ are small parameters. In this section, we study $(3.3)^{\pm}$; the results are due to Dumortier, Roussarie, and Sotomayor [1].

The following lemma gives an explanation why the $x^2 y$ term is not considered in the second equations of $(3.3)^{\pm}$ and $(3.4)^{\pm}$.

**Lemma 3.1.** *In a small neighborhood of the origin, the following two vector fields*

$$\begin{cases} \dot{x} = y, \\ \dot{y} = x^2 + y(\alpha x^2 + \beta x^3) + o\big((|x| + |y|)^4\big) \end{cases} \qquad (3.5)$$

*and*

$$\begin{cases} \dot{x} = y, \\ \dot{y} = x^2 + \beta x^3 y + o\big((|x| + |y|)^4\big), \end{cases} \qquad (3.6)$$

*are $C^\infty$-equivalent.*

*Proof.* Let

$$H(x, y) = \frac{1}{2}y^2 - \frac{1}{3}x^3.$$

Then

$$dH = y\,dy - x^2\,dx. \tag{3.7}$$

Hence

$$yx^2\,dx = y^2\,dy - y\,dH. \tag{3.8}$$

Equation (3.5) is equivalent to

$$dH - \left[y(\alpha x^2 + \beta x^3) + o\big((|x| + |y|)^4\big)\right]dx = 0. \tag{3.9}$$

Substituting (3.8) into (3.9), we have

$$dH - \frac{\alpha y^2}{1 + \alpha y}\,dy - \frac{\beta yx^3 + o\big((|x| + |y|)^4\big)}{1 + \alpha y}\,dx = 0. \tag{3.10}$$

It is not difficult to see that, in a small neighborhood of $(x, y) = (0, 0)$, there exists a coordinate change of the form

$$\begin{cases} \bar{x} = x, \\ \bar{y} = y + O\big(|(x, y)|^2\big) \end{cases}$$

such that $\bar{y}\,d\bar{y} = (y - \alpha y^2/(1 + \alpha y))\,dy$. Thus, this change of coordinates transforms (3.10) into the form

$$d\bar{H} - \left[\beta \bar{y}\bar{x}^3 + o\big((|\bar{x}| + |\bar{y}|)^4\big)\right]d\bar{x} = 0, \tag{3.11}$$

where $\bar{H} = \frac{1}{2}\bar{y}^2 - \frac{1}{3}\bar{x}^3$.
Since (3.11) is equivalent to equation (3.6), the lemma is proved          □

By using the method of Section 4.1, it can be shown that the family of vector fields (3.3) is a versal deformation of the vector field

$$\begin{cases} \dot{x} = y, \\ \dot{y} = x^2 \pm yx^3. \end{cases} \tag{3.12}$$

It is easy to see that by a change $(x, y, t, \epsilon_1, \epsilon_2, \epsilon_3) \to (x, -y, -t, \epsilon_1, -\epsilon_2, -\epsilon_3)$, equation $(3.3)^-$ is transformed into the form $(3.3)^+$. Hence we only need to consider the following equation:

$$\begin{cases} \dot{x} = y, \\ \dot{y} = \epsilon_1 + \epsilon_2 y + \epsilon_3 xy + x^2 + x^3 y. \end{cases} \tag{3.13}$$

We first present the main result (Theorem 3.2), then give the proof in detail.

It is obvious that the equilibria of (3.13) are determined by the equations

$$y = 0, \qquad x^2 + \epsilon_1 = 0.$$

Hence (3.13) has no equilibria for $\epsilon_1 > 0$. The plane $\{\epsilon_1 = $ ⌄ excluding the origin in $\epsilon_1 \epsilon_2 \epsilon_3$-space is a bifurcation surface of sado. -node type: When $\epsilon_1$ decreases from this surface, the saddle-node of (3.13) becomes a saddle point and a node. The other bifurcation surfaces are located in the half space $\{\epsilon_1 < 0\}$. We describe them by their intersection with the half 2-sphere $S_\sigma = \{(\epsilon_1, \epsilon_2, \epsilon_3) | \epsilon_1 < 0, \epsilon_1^2 + \epsilon_2^2 + \epsilon_3^2 = \sigma^2, \sigma > 0 \text{ sufficiently small}\}$. The bifurcation diagram of equation (3.13) is a cone based on this intersection which consists of three curves on $S_\sigma$: a curve $H$ of Hopf bifurcation, a curve $HL$ of homoclinic bifurcation, and a curve $C$ of double limit cycle bifurcation. The points $h_2$ on $H$ and $hl_2$ on $HL$ are the endpoints of the curve $C$ (see Figure 3.2 and Figure 3.3).

On the other hand, both curves $H$ and $HL$ touch $\partial S_\sigma = \{(\epsilon_1, \epsilon_2, \epsilon_3) | \epsilon_1 = 0, \epsilon_2^2 + \epsilon_3^2 = \sigma^2\}$ with a first-order tangency at the points $b_1$ and $b_2$. In some small neighborhoods of $b_1$ and $b_2$ one finds

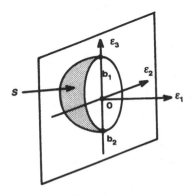

Figure 3.2. The parameter space.

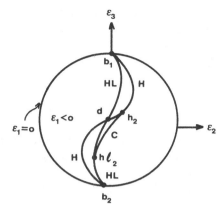

Figure 3.3. The trace of the bifurcation diagram on $S(\epsilon_1 \leq 0)$.

cusp bifurcation of codimension 2 (Bogdanov–Takens bifurcation, see Section 4.1). System (3.13) has a unique unstable limit cycle when $\epsilon$ lies between $H$ and $HL$ and is in a small neighborhood of $b_1$. System (3.13) has a unique stable limit cycle when $\epsilon$ lies between $H$ and $HL$ and is in a small neighborhood of $b_2$.

Along the curve $H$, not including the point $h_2$, a Hopf bifurcation of order 1 occurs. System (3.13) has an unstable limit cycle when $\epsilon$ crosses the arc $\overline{b_1 h_2}$ in $H$ joining $b_1$ and $h_2$ from the right to the left and has a stable limit cycle when $\epsilon$ crosses the arc $\overline{h_2 b_2}$ in $H$ from the left to the right.

The point $h_2$ corresponds to a Hopf bifurcation of order 2.

Along the curve $HL$, excluding the point $hl_2$, a homoclinic bifurcation of order 1 occurs. When $\epsilon$ crosses the arc $\overline{b_1 hl_2}$ in $HL$ from the left to the right, two separatrices of the saddle point change their relative positions and an unstable limit cycle appears. A similar phenomenon happens when $\epsilon$ crosses the arc $\overline{hl_2 b_2}$ in $HL$ from the right to the left, and a stable limit cycle appears.

The point $hl_2$ corresponds to a homoclinic bifurcation of order 2.

The curves $H$ and $HL$ intersect transversally at a unique point $d$ which corresponds to the simultaneous occurrence of a Hopf bifurcation of order 1 and a homoclinic bifurcation of order 1.

If the parameter values are in the curved triangle $dh_2 hl_2$, then system (3.1) has exactly two limit cycles. The inner one is stable and the outer one is unstable.

These two limit cycles coalesce when $\epsilon$ crosses the curve $C$ from the left to the right. On $C$ itself there exists a unique semistable limit cycle.

**Theorem 3.2.** *Let*

$$\overline{\Sigma} = \partial S_\sigma \cup H \cup HL \cup C,$$

*where $\partial S_\sigma$ and curves $H$, $HL$, and $C$ are described above.*
*The bifurcation diagram of equation (3.13) inside the ball*

$$B_\sigma = \{\epsilon | \epsilon_1^2 + \epsilon_2^2 + \epsilon_3^2 \leq \sigma^2\}$$

*is a cone homeomorphic to*

$$\{(\delta^4\bar{\epsilon}_1, \delta^6\bar{\epsilon}_2, \delta^4\bar{\epsilon}_3) | \delta \in [0, \sigma], (\bar{\epsilon}_1, \bar{\epsilon}_2, \bar{\epsilon}_3) \in \overline{\Sigma}\}.$$

*The topological type of the phase portraits of equation (3.13) in a fixed neighborhood of $0 \in \mathbb{R}^2$ is the same in each of six connected components $\{D_i\}$ of the complement of the bifurcation diagram, and is the same in each surface or curve in the bifurcation diagram (there are nine surfaces $\{S_i\}$ and five curves $\{\Gamma_i\}$).*

*The cone regions $D_1, \ldots, D_5$ are based on the open regions, $I, II, \ldots, V$, respectively (see Figure 3.4), and $D_6$ is the half ball $\{\epsilon | \epsilon \in B_\sigma, \epsilon_1 > 0\}$. When $\epsilon \in D_1 \cup \cdots \cup D_5$, the phase portraits of equation (3.13) are shown in Figure 3.4.*

*The surfaces $S_1, \ldots, S_9$ are based on the arcs $\widehat{b_1 \partial S b_2}$ (left), $\widehat{b_1 HLd}, \ldots,$ $\widehat{h_2 Chl_2}$, respectively, and the phase portraits of (3.13) for $\epsilon \in S_1 \cup \cdots \cup S_9$ are shown in Figure 3.5.*

*The curves $\Gamma_1, \ldots, \Gamma_5$ are based on the points $b_1$, $h_2$, $d$, $hl_2$, and $b_2$, respectively, and the phase portraits of (3.13) for $\epsilon \in \Gamma_1 \cup \cdots \cup \Gamma_5$ are shown in Figure 3.6.*

The proof of Theorem 3.2 will be given in the rest of this section. We begin by introducing a blow-up technique for $\epsilon_1 < 0$. Let

$$x \to \delta^2 x, \qquad y \to \delta^3 y, \qquad \epsilon_1 = -\delta^4,$$

$$\epsilon_2 = \delta^6 \nu_1, \qquad \epsilon_3 = \delta^4 \nu_2, \qquad t \to \frac{1}{\delta} t, \qquad (3.14)$$

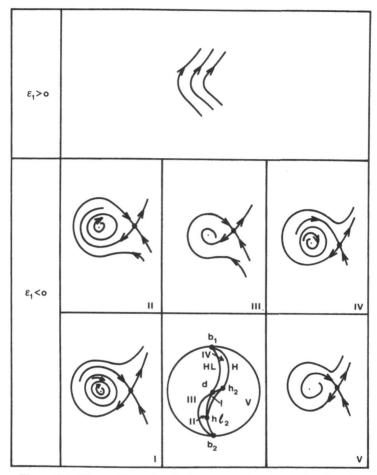

Figure 3.4.  The codimension 0 phase portraits of equation (3.13).

where $\delta > 0$. Equation (3.13) is transformed into the form

$$\begin{cases} \dot{x} = y, \\ \dot{y} = -1 + x^2 + \delta^5\big(\nu_1 + \nu_2 x + x^3\big)y. \end{cases} \qquad (3.15)$$

Let

$$\mu_3 = \delta^5, \qquad \mu_i = \delta^5 \nu_i, \qquad i = 1, 2. \qquad (3.16)$$

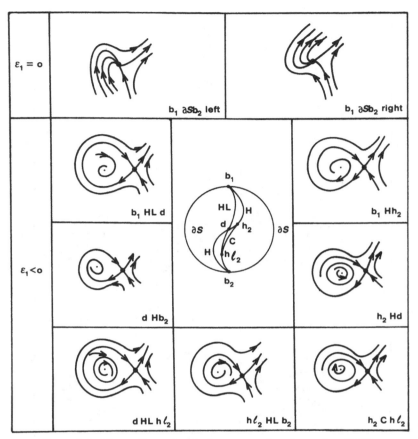

Figure 3.5. The codimension 1 phase portraits of equation (3.13).

Equation (3.15) becomes

$$\begin{cases} \dot{x} = y, \\ \dot{y} = -1 + x^2 + \mu_1 y + \mu_2 xy + \mu_3 x^3 y, \end{cases} \tag{3.17}$$

with condition $\mu_3 > 0$.

Equation (3.17) has equilibria $(-1, 0)$ and $(1, 0)$. The point $(1, 0)$ is always a saddle point whereas the point $(-1, 0)$ is a focus.

*(i) Hopf bifurcation and the homoclinic loop bifurcation* Using the results in Sections 5.1 and 5.2 (Examples 1.8 and 2.13), we have the

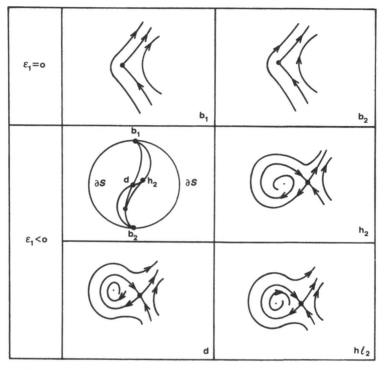

Figure 3.6. The codimension two phase portraits of equation (3.13).

following lemmas:

**Lemma 3.3.** *The equilibrium point $(-1, 0)$ of equation (3.15) is a sink (a source) if $\nu_1 - \nu_2 - 1 < 0 (> 0)$. There is a Hopf bifurcation of order 1 along the line $\tilde{H}$: $\{(\nu_1, \nu_2)|\nu_1 - \nu_2 - 1 = 0\}$, except at the point $\tilde{h}_2$: $(\nu_1, \nu_2) = (4, 3)$ at which a Hopf bifurcation of order 2 takes place. Moreover, there are two limit cycles around the point $(-1, 0)$ if $\nu_2 - 3 < 0, \nu_1 - \nu_2 - 1 > 0$, and $0 \ll |\nu_1 - \nu_2 - 1| \ll |\nu_2 - 3| \ll 1$.*

**Lemma 3.4.** *Equation (3.15) has a homoclinic loop bifurcation of order 1 along the curve $\widetilde{HL}$: $\{(\nu_1, \nu_2)|\nu_1 - \frac{5}{7}\nu_2 - \frac{103}{77} + O(\delta) = 0\}$, except at the point $\widetilde{hl_2}$: $(\nu_1, \nu_2) = (\frac{4}{11} + O(\delta), -\frac{15}{11} + O(\delta))$ at which a homoclinic loop bifurcation of order 2 takes place. The curves $\tilde{H}$ and $\widetilde{HL}$ intersect* *transversely at point* $\tilde{d}$: $(v_1, v_2) = (\frac{24}{11} + O(\delta), \frac{13}{11} + O(\delta))$, *which corresponds to the simultaneous occurrence of a Hopf bifurcation of order* 1 *and a homoclinic loop bifurcation of order* 1.

*(ii) Double limit cycle bifurcation* As in Chapter 4, we consider equation (3.15) with small $\delta$ as a perturbation of a Hamiltonian system. If $\delta = 0$, (3.15) becomes

$$\begin{cases} \dot{x} = y, \\ \dot{y} = -1 + x^2, \end{cases}$$

with a Hamiltonian function $H(x, y)$, where

$$H(x, y) = \frac{y^2}{2} + x - \frac{x^3}{3},$$

and this is exactly the same expression as (4.1.9) of Section 4.1. By Lemma 4.1.4, we can find a bifurcation function for periodic orbits of equation (3.15) as follows:

$$F(h, \delta, v_1, v_2) = \int_{\gamma(h, \delta, v_1, v_2)} (v_1 + v_2 x + x^3) y \, dx, \qquad (3.18)$$

where $\gamma(h, \delta, v_1, v_2)$ is defined as in Section 4.1 and $F(h, \delta, v_1, v_2)$ is well approximated by $M(h) = F|_{\delta=0}$, that is,

$$M(h) = \int_{\Gamma_h} (v_1 + v_2 x + x^3) y \, dx, \qquad (3.19)$$

where $\Gamma_h$ is the level curve of $H(x, y) = h$, $-\frac{2}{3} \le h \le \frac{2}{3}$. $H(x, y) = -\frac{2}{3}$ corresponds to the equilibrium $(-1, 0)$ whereas $H(x, y) = \frac{2}{3}$ corresponds to the homoclinic loop (see Section 4.1 and Figure 4.1.2). The number of periodic orbits of equation (3.15) is the same as the number of solutions of $M(h) = 0$, $-\frac{2}{3} < h < \frac{2}{3}$.

As in Section 4.1, we define

$$I_k(h) = \int_{\Gamma_h} x^k y \, dx, \qquad k = 0, 1, 3, 4. \qquad (3.20)$$

**Lemma 3.5.**

$$I_3 = -\frac{6}{11} h I_0 + \frac{15}{11} I_1,$$

$$I_4 = \frac{21}{13} I_0 - \frac{12}{13} h I_1.$$

*Proof.* Along $\Gamma_h$ we have

$$\frac{y^2}{2} + x - \frac{x^3}{3} = h, \tag{3.21}$$

and

$$y\, dy + (1 - x^2)\, dx = 0. \tag{3.22}$$

From (3.22) and (3.21), we have

$$x^3 y\, dx = xy\, dx + xy^2\, dy = xy\, dx + 2xh\, dy - 2x^2\, dy + \frac{2}{3} x^4\, dy.$$

Integrating the above equality and using integration by parts, we have

$$I_3 = -2h I_0 + 5 I_1 - \frac{8}{3} I_3,$$

which implies the first desired expression. The second one can be obtained in the same way. $\qquad\square$

By Lemma 3.5, we rewrite (3.19) in the form

$$M(h) = \left( \nu_1 - \frac{6}{11} h \right) I_0(h) + \left( \nu_2 + \frac{15}{11} \right) I_1(h). \tag{3.23}$$

As in Section 4.1, we define

$$
P(h) = \begin{cases} -\dfrac{I_1(h)}{I_0(h)}, & -\dfrac{2}{3} < h \le \dfrac{2}{3}, \\[3mm] 1, & h = -\dfrac{2}{3}. \end{cases}
$$

Then $P(h) \in C^0[-2/3, 2/3] \cup C^1[-2/3, 2/3)$ and (3.23) becomes

$$
\overline{M}(h) = \left( \nu_1 - \frac{6}{11}h \right) - \left( \nu_2 + \frac{15}{11} \right) P(h), \tag{3.24}
$$

where

$$
\overline{M}(h) = \frac{M(h)}{I_0(h)}, \qquad -\frac{2}{3} < h \le \frac{2}{3}.
$$

It is obvious that $M(h) = M'(h) = \cdots = M^{(k)}(h) = 0$, $M^{(k+1)}(h) \neq 0$ if and only if $\overline{M}(h) = \overline{M}'(h) = \cdots = \overline{M}^{(k)}(h) = 0$, $\overline{M}^{(k+1)}(h) \neq 0$, where $h \in (-2/3, 2/3]$.

We recall the following properties of the function $P(h)$ (see Lemmas 4.1.6 and 4.1.7 in Section 4.1):

(1) $P([-\frac{2}{3}, \frac{2}{3}]) \subset [\frac{5}{7}, 1]$, $P(-\frac{2}{3}) = 1$, and $P(\frac{2}{3}) = \frac{5}{7}$;

(2) $P'(h) < 0$ for $-\frac{2}{3} < h < \frac{2}{3}$, $P'(-\frac{2}{3}) = -\frac{1}{8}$, and $P'(\frac{2}{3}) = -\infty$;

(3) $P(h)$ satisfies the equation

$$
(9h^2 - 4)P' = 7P^2 + 3hP - 5. \tag{3.25}
$$

We rewrite the last property in the following way:
$(h, P)$ is a solution of the system

$$
\begin{cases} \dfrac{dP}{dt} = -7P^2 - 3hP + 5, \\[3mm] \dfrac{dh}{dt} = 4 - 9h^2, \end{cases} \tag{3.26}
$$

satisfying the condition

$$\lim_{t \to -\infty} h = -\frac{2}{3}, \qquad \lim_{t \to -\infty} P = 1.$$

This system has a saddle point at $(-\frac{2}{3}, 1)$ and an attractive node at $(\frac{2}{3}, \frac{5}{7})$ (Figure 4.1.5). The graph of the function $P(h)$ is the unstable separatrix of $(-\frac{2}{3}, 1)$ of the system (3.26). It joins the point $(-\frac{2}{3}, 1)$ to the point $(\frac{2}{3}, \frac{5}{7})$.

**Lemma 3.6.** $P''(h) < 0$ for $-\frac{2}{3} \le h \le \frac{2}{3}$.

*Proof.* From (3.25) we get $P''(-\frac{2}{3}) = -\frac{55}{1152}$, $P''(\frac{2}{3}) = -\infty$, and

$$(9h^2 - 4)P'' = P'(14P - 15h) + 3P, \tag{3.27}$$

$$(9h^2 - 4)P''' = P''(14P - 33h) + P'(14P' - 12). \tag{3.28}$$

Let us prove that $P''(h) < 0$ for $-2/3 < h < 2/3$. If this is not true, then we let $h^* = \inf\{h | P''(h) = 0, -2/3 < h < 2/3\}$. Then $P''(h^*) = 0$ and $P'''(h^*) \ge 0$ because $P''(-2/3) < 0$.

On the other hand, taking $h = h^*$ on both sides of (3.28) and noting $P''(h^*) = 0$ and $P'(h^*) < 0$ (Lemma 4.1.7), we have $P'''(h^*) < 0$. This contradiction proves the desired result. $\qquad\square$

Now we turn to the problem of the double limit cycle.

The condition for the existence of a multiple limit cycle is given by the equation $\overline{M}(h) = \overline{M}'(h) = 0$ which determines a curve $\tilde{C}$ in the $\nu_1\nu_2$-plane.

**Lemma 3.7.** $\tilde{C}$ *is a convex curve, joins the point* $\bar{h}_2$ *on* $\bar{H}$ *to the point* $\widetilde{hl}_2$ *on* $\overline{HL}$ *(see Lemmas 3.3 and 3.4), and is tangent to* $\bar{H}$ *and* $\overline{HL}$ *at these points. Along* $\tilde{C}$, *a double limit cycle bifurcation for equation (3.15) occurs.*

*Proof.* From (3.24) we have that

$$\overline{M}(h) = \left(\nu_1 - \frac{6}{11}h\right) - \left(\nu_2 + \frac{15}{11}\right)P(h),$$

$$\overline{M}'(h) = -\frac{6}{11} - \left(\nu_2 + \frac{15}{11}\right)P'(h),$$

$$\overline{M}''(h) = -\left(\nu_2 + \frac{15}{11}\right)P''(h).$$

It is clear that if $\overline{M}(h) = \overline{M}'(h) = 0$, then $(\nu_2 + \frac{15}{11}) \neq 0$. By Lemma 3.6, $\overline{M}''(h) \neq 0$. Hence, by the Implicit Function Theorem, we can determine a function $h = h(\nu_2)$ from $\overline{M}'(h) = 0$, and a function $\nu_1 = \nu_1(h, \nu_2) = \nu_1(h(\nu_2), \nu_2)$ from $\overline{M}(h) = 0$. This means that the curve $\tilde{C}$ has an expression $\nu_1 = \nu_1(\nu_2)$. Hence it is a regular curve along which a double limit cycle bifurcation occurs.

Next, from $\overline{M}(h) = \overline{M}'(h) = 0$ we have

$$\begin{cases} \nu_1 = \dfrac{6}{11}h - \dfrac{6}{11}\dfrac{P(h)}{P'(h)}, \\[2mm] \nu_2 = -\dfrac{15}{11} - \dfrac{6}{11}\dfrac{1}{P'(h)}. \end{cases} \qquad (3.29)$$

When $h \to -\frac{2}{3}$, we have that $P(h) \to 1$, $P'(h) \to -\frac{1}{8}$. Hence $(\nu_1, \nu_2) \to (4, 3) = \tilde{h}_2$. When $h \to \frac{2}{3}$, we have that $P(h) \to \frac{5}{7}$, $P'(h) \to -\infty$. Hence $(\nu_1, \nu_2) \to (\frac{4}{11}, -\frac{15}{11}) = \widetilde{hl}_2$ (for convenience, we omit $O(\delta)$ terms; see Section 4.1 for details). From (3.29), we obtain $d\nu_1/d\nu_2 = P(h)$ along the curve $\tilde{C}$: $\nu_1 = \nu_1(\nu_2)$. This implies, by Lemmas 3.3 and 3.4, $\tilde{C}$ is tangent to $\tilde{H}$ and $\widetilde{HL}$ at points $\tilde{h}_2$ and $\widetilde{hl}_2$, respectively.

Finally, we prove the convexity of curve $\tilde{C}$. Since $\tilde{C}$ is defined by $M(h) = M'(h) = 0$, it is the envelope of the family of lines $\{L_h\}$ on the $\nu_1\nu_2$-plane defined by

$$L_h: \nu_1 - P(h)\nu_2 - \frac{6}{11}h - \frac{15}{11}P(h) = 0,$$

where the parameter $h \in [-\frac{2}{3}, \frac{2}{3}]$. Since $P(h)$ is invertible, we can choose its values as a parameter so that the lines $\{L_h\}$ can be parameterized by their slopes $P \in [\frac{5}{7}, 1]$. Thus $L_h$ takes the form

$$L_p: \nu_1 - P\nu_2 - H(P) = 0,$$

where $H(P) = \frac{6}{11}h(P) - \frac{15}{11}P$ and $h(P)$ is the inverse function of $P(h)$. It is easy to see that

$$H''(P) = -\frac{P''(h)}{(P'(h))^3} < 0,$$

since $P'(h) < 0$, $P''(h) < 0$ (Lemma 3.6).

It is known (see, for example, Chapter 1 in Arnold [4]) that the envelope of a family of lines such as $L_h$ ($L_p$), parameterized by their slopes and defined by $\nu_1 = P\nu_2 - (-H)$ with a convex function $-H$, is the graph of a convex function $\nu_1 = \nu_1(\nu_2)$ which is the Legendre transform of the function $-H$. $\quad\square$

*(iii) The number of limit cycles* As in Section 4.1, for a given $(\nu_1, \nu_2)$, the number of limit cycles of equation (3.15) is determined by the number of zeros of equation $\overline{M}(h) = 0$ for $-\frac{2}{3} < h < \frac{2}{3}$.

Suppose $\nu_2 + \frac{15}{11} = 0$. Then $\overline{M}(h) = 0$ if and only if $\nu_1 = \frac{6}{11}h$ (see (3.24)). This means for $-\frac{4}{11} < \nu_1 < \frac{4}{11}$, $\overline{M}(h) = 0$ has a unique root.

We suppose $\nu_2 + \frac{15}{11} \neq 0$, and rewrite (3.24) in the form

$$\overline{M}(h) = \left(\nu_2 + \frac{15}{11}\right)(A(h) - P(h)), \tag{3.30}$$

where

$$A(h) = \left(\nu_2 + \frac{15}{11}\right)^{-1}\left(\nu_1 - \frac{6}{11}h\right)$$

is a linear function of $h$. Obviously, zeros of $\overline{M}(h)$ correspond to intersection points of the straight line $L_A$: $P = A(h)$ and the curve $\Gamma_P$: $P = P(h)$ on the $hP$-plane. Since $P'(h) < 0$ and $P''(h) < 0$ (Lemma 4.1.7 and Lemma 3.6), $\Gamma_P$ is strictly convex. On the other hand, $\Gamma_P$ is independent of $\nu_1$ and $\nu_2$, $P(-2/3) = 1$, and $P(2/3) = 5/7$. The straight line $L_A$ depends on $\nu_1$ and $\nu_2$, but it has the following

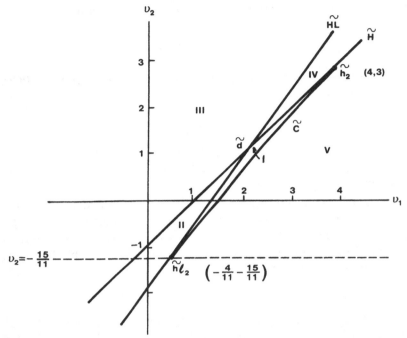

Figure 3.7. Bifurcation diagram of equation (3.15) in the $\nu_1\nu_2$-plane for small $\delta$.

properties for $\nu_2 + \frac{15}{11} > 0$:

($1^0$) $A(-2/3) = 1 \ (> 1 \text{ or } < 1)$ if and only if $(\nu_1, \nu_2) \in \tilde{H}$ (is below or above $\tilde{H}$).

($2^0$) $A(2/3) = 5/7 \ (\geq 5/7 \text{ or } < 5/7)$ if and only if $(\nu_1, \nu_2) \in \widetilde{HL}$ (is below or above $\widetilde{HL}$).

($3^0$) $L_A$ is tangent to curve $\Gamma_P$ if and only if $(\nu_1, \nu_2) \in \tilde{C}$ (i.e., $\overline{M}(h) = \overline{M}'(h) = 0$ for some $h \in (-\frac{2}{3}, \frac{2}{3})$).

In the case $\nu_2 + \frac{15}{11} < 0$, we only need to replace "$>$" (correspondingly "$<$") by "$<$" (correspondingly "$>$") in properties ($1^0$) and ($2^0$).

The $\nu_1\nu_2$-plane is divided into five regions by the curves $\tilde{H}$, $\widetilde{HL}$, and $\tilde{C}$ (see Figure 3.7).

Since $L_A$ is a straight line and $\Gamma_P$ is a convex curve, we show all the different intersection possibilities in Figure 3.8. We note that the dotted lines in Figure 3.8 indicate a different position for $L_A$ determined by the sign of $\nu_2 + \frac{15}{11}$.

Extending the result from $\delta = 0$ to $\delta > 0$ by using the Implicit Function, Hopf Bifurcation, and Homoclinic Bifurcation Theorems (see

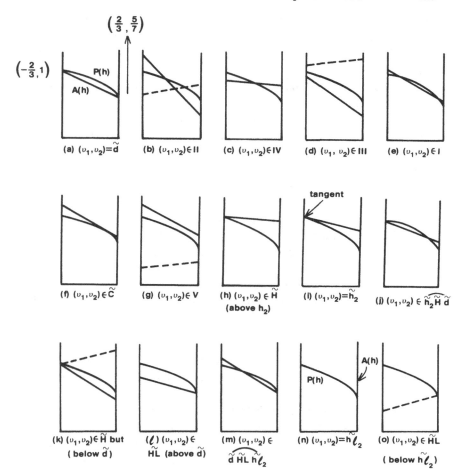

Figure 3.8. The relative positions of curve $p = p(h)$ and straight line $P = A(h)$ (———— for $\nu_2 + \frac{15}{11} > 0$; - - - - - for $\nu_2 + \frac{15}{11} < 0$).

Section 3.2, Section 4.1, Section 5.1, and Section 5.2), we obtain the following lemma.

**Lemma 3.8.** *Let $K$ be a compact neighborhood of the curved triangle $\widetilde{dh_2}\widetilde{hl_2}$ in the $\nu_1\nu_2$-plane and $N$ be a compact neighborhood of the singular disc $\{H(x, y) \leq \frac{2}{3}\} \cap \{x \geq 1\}$ in the xy-plane. There exists a value $\Delta > 0$ such that if $(\delta, \nu_1, \nu_2) \in C(K) := (0, \Delta) \times K \subset \mathbb{R}^3$, then the bifurcation diagram of equation (3.15) consists of three surfaces and* *three curves which can be described as follows, up to a diffeomorphism of C(K) in which the diffeomorphism is the identity at $\delta = 0$:*

$S_H = (0, \Delta) \times (\tilde{H} \setminus \{\tilde{h}_2\})$ *is a surface of Hopf bifurcation of codimension* 1;

$S_{HL} = (0, \Delta) \times (\widetilde{HL} \setminus \{\widetilde{hl}_2\})$ *is a surface of homoclinic loop bifurcation of codimension* 1;

$S_C = (0, \Delta) \times \tilde{C}$ *is a surface of double limit cycle bifurcation;*

$(0, \Delta) \times \{\tilde{h}_2\}$ *is a curve of Hopf bifurcation of codimension* 2;

$(0, \Delta) \times \{\widetilde{hl}_2\}$ *is a curve of homoclinic bifurcation of codimension* 2;

$(0, \Delta) \times \{\tilde{d}\}$ *(i.e.,* $S_H \cap S_{HL}$*) is a curve of Hopf bifurcation and homoclinic loop bifurcation.*

For $\bar{\delta} \in (0, \Delta)$, denote the intersection of the bifurcation diagram of (3.15) and the plane $\delta = \bar{\delta}$ by $\Sigma_{\bar{\delta}}$. Then $\Sigma_{\bar{\delta}}$ has a structure as shown in Figure 3.7. The phase portraits of (3.15) for $(\bar{\delta}, \nu_1, \nu_2)$ in regions I–V are the same as in Figure 3.4 I–V ($\epsilon_1 < 0$), respectively. The phase portraits of (3.15) for $(\bar{\delta}, \nu_1, \nu_2)$ along each part of $\Sigma_{\bar{\delta}}$ are the same as in Figures 3.5 and 3.6 ($\epsilon_1 < 0$) for each corresponding part of $H$, $HL$, and $C$, respectively.

**Remark 3.9.** Since (3.15) is obtained from (3.13), we can obtain a description of the bifurcation diagram for equation (3.13) with $\epsilon_1 < 0$.

In fact, (3.14) gives a transformation $\Phi$: $(\delta, \nu_1, \nu_2) \rightarrow (\epsilon_1, \epsilon_2, \epsilon_3)$. By this transformation $\Phi$, $C(K) = (0, \Delta) \times K \rightarrow C_{\epsilon_1}(K) = \{(-\delta^4, \delta^6 \nu_1, \delta^4 \nu_2) | \delta \in (0, \Delta), (\nu_1, \nu_2) \in K\}$. $C_{\epsilon_1}(K)$ is a cone in $\epsilon_1 \epsilon_2 \epsilon_3$-space around the $\epsilon_1$-axis for $\epsilon_1 < 0$ (see Figure 3.9). The bifurcation diagram of (3.13) in $C_{\epsilon_1}(K)$ is the image by $\Phi$ of those described in Lemma 3.8, and hence homeomorphic to the cones based on $\tilde{H}$, $\widetilde{HL}$, $\tilde{C}$, $\tilde{h}_2$, $\widetilde{hl}_2$, and $\tilde{d}$ with generating curves $\delta \rightarrow (-\delta^4, \delta^6 \nu_1, \delta^4 \nu_2)$, or equivalently $\epsilon_1 \rightarrow (\epsilon_1, (-\epsilon_1)^{3/2} \nu_1, (-\epsilon_1) \nu_2)$ for $\epsilon_1 < 0$.

*(iv) The behavior of (3.13) around the $\epsilon_3$-axis for $\epsilon_1 \leq 0$* Consider a change of coordinates and parameters

$$\begin{cases} \epsilon_1 = \delta^4 \bar{\epsilon}_1, \\ \epsilon_2 = \delta^6 \nu_1, \\ \epsilon_3 = \delta^4 \nu_2, \end{cases} \qquad \begin{cases} x \rightarrow \delta^2 x, \\ y \rightarrow \delta^3 y, \\ t \rightarrow \dfrac{1}{\delta} t. \end{cases} \qquad (3.31)$$

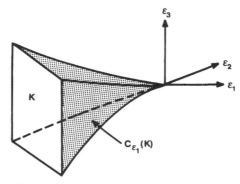

Figure 3.9. The cone $C_{\epsilon_1}(K)$ for $\epsilon_1 < 0$.

In order to consider the parameters in a cone in the neighborhood of the $\epsilon_3$-axis, we fix $\nu_2 = \pm 1$ in (3.31), and let $(\delta, \bar{\epsilon}_1, \nu_1)$ be the new parameters. For $\nu_2 = 1$, we have a cone around the $\epsilon_3$-axis, $\epsilon_3 > 0$, and for $\nu_2 = -1$, a cone around the $\epsilon_3$-axis, $\epsilon_3 < 0$. We consider only $\nu_2 = 1$. The case $\nu_2 = -1$ can be treated in the same way.

By (3.31) with $\nu_2 = 1$, equation (3.13) becomes

$$\begin{cases} \dot{x} = y, \\ \dot{y} = \bar{\epsilon}_1 + x^2 + \delta^5(\nu_1 + x + x^3)y. \end{cases} \tag{3.32}$$

For each fixed $\delta \in (0, T]$, where $T > 0$, we make a second blow up:

$$\begin{cases} \bar{\epsilon}_2 = -\tau^4, \\ \nu_1 = \tau^2 \bar{\nu}_1, \end{cases} \quad \begin{cases} x \to \tau^2 x, \\ y \to \tau^3 y, \\ t \to \dfrac{1}{\tau} t. \end{cases} \tag{3.33}$$

Then (3.32) is transformed into the form

$$\begin{cases} \dot{x} = y, \\ \dot{y} = -1 + x^2 + \delta^5 [\tau(\bar{\nu}_1 + x)y + O(\tau^5)]. \end{cases} \tag{3.34}$$

Comparing (3.34) with equation (4.1.7) and using the method of Section 4.1, we can obtain the following lemma.

**Lemma 3.10.** *In the half plane $\{(\bar{\epsilon}_1, \nu_1, \nu_2) | \nu_2 = 1, \bar{\epsilon}_1 \leq 0\}$ there is a fixed compact subset $B^+$, diffeomorphic to a disk, having a tangency of*

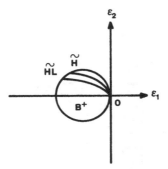

Figure 3.10.

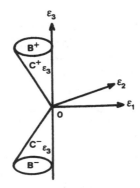

Figure 3.11.

*order 1 at $\bar{b}_1 = (0, 0, 1)$ with the $\nu_1$-axis, such that for equation (3.32) the results of Bogdanov–Takens (see Theorem 4.1.2) are valid for any $(\bar{\epsilon}_1, \nu_1) \in B^+$ and any $\delta \in [0, \tau]$.*

**Remark 3.11.** Similarly to Remark 3.9, (3.31) with $\nu_2 = 1$ gives a mapping $(\delta, \bar{\epsilon}_1, \nu_1) \rightarrow (\epsilon_1, \epsilon_2, \epsilon_3)$ which maps $(0, \tau] \times B^+$ to $C_{\epsilon_3}^+ = \{(\delta^4 \bar{\epsilon}_1, \delta^6 \nu_1, \delta^4) | \delta \in (0, \tau], (\bar{\epsilon}_1, \nu_1) \in B^+\}$. $C_{\epsilon_3}^+$ is a cone in $\epsilon_1 \epsilon_2 \epsilon_3$-space around the $\epsilon_3$-axis and based on $B^+$. The bifurcation diagram of (3.13) in $C_{\epsilon_3}^+$ consists of cones based on $\tilde{H}$, $\widetilde{HL}$, and $\{\bar{b}_1\}$ with generating curves $\delta \rightarrow (\delta^4 \bar{\epsilon}_1, \delta^6 \nu_1, \delta^4)$, where $\tilde{H}$ and $\widetilde{HL}$ are the curves of Hopf and homoclinic bifurcations, respectively (see Figures 3.10 and 3.11). Similarly, taking $\nu_2 = -1$ we can obtain a cone $C_{\epsilon_3}^-$ around the $\epsilon_3$-axis for $\epsilon_3 < 0$.

*Proof of Theorem 3.2.* Let $C_{\epsilon_3}^+$ and $C_{\epsilon_3}^-$ be the two cones obtained in Lemma 3.10 and Remark 3.11. By Lemma 3.8 we can choose a compact set $K$ in the $\nu_1 \nu_2$-plane such that:

(1) $C_{\epsilon_1}(K) \cup C_{\epsilon_3}^+ \cup C_{\epsilon_3}^-$ contains a cone $C(D)$ based on a disk $D$ belonging to the half sphere $S_\sigma = \{(\epsilon_1, \epsilon_2, \epsilon_3) | \epsilon_1^2 + \epsilon_2^2 + \epsilon_3^2 = \sigma^2$, $\epsilon_1 \leq 0$ and $\sigma > 0$ sufficiently small$\}$, where $C_{\epsilon_1}(K)$ is described in Remark 3.9.

(2) $D$ contains the half circle $S_\sigma \cap \{\epsilon_2 = 0\}$. $\partial D$ is tangent to $\partial S_\sigma$ with a tangency of order 1 at the points $b_1$ and $b_2$, where $b_1 = (0, 0, \sigma)$, $b_2 = (0, 0, -\sigma)$.

(3) $D$ contains the curves $H = S_H \cap S_\sigma$, $HL = S_{HL} \cap S_\sigma$, and $C = S_C \cap S_\sigma$, where $S_H$, $S_{HL}$, and $S_C$ are described in Lemma 3.8.

We can obtain condition (3) because the curve of Hopf bifurcation and the curve of homoclinic bifurcation in $D \cap C_{\epsilon_3}^\pm$ are connected with $H$ and $HL$, respectively. To show this, we consider the equations for the curves $H = S_H \cap S_\sigma$ and $HL = S_{HL} \cap S_\sigma$. From Lemma 3.3, we obtain

$$H: \begin{cases} \nu_1 - \nu_2 - 1 = 0, \\ \epsilon_1^2 + \epsilon_2^2 + \epsilon_3^2 = \sigma^2. \end{cases} \tag{3.35}$$

From (3.14) we have

$$\begin{cases} \delta = (-\epsilon)^{1/4}, \\ \nu_1 = \epsilon_2 / (-\epsilon_1)^{3/2}, \\ \nu_2 = \epsilon_3 / (-\epsilon_1), \end{cases} \tag{3.36}$$

where $\epsilon_1 < 0$. Substituting (3.36) into (3.35), we obtain

$$H: \begin{cases} f(\epsilon_1, \epsilon_2, \epsilon_3) = 0, \\ g(\epsilon_1, \epsilon_2, \epsilon_3) = 0, \end{cases} \tag{3.37}$$

where

$$f = \epsilon_2 - (-\epsilon_1)^{1/2} \epsilon_3 - (-\epsilon_1)^{3/2},$$

$$g = \epsilon_1^2 + \epsilon_2^2 + \epsilon_3^2 - \sigma^2.$$

This implies that if $\epsilon_1 \to 0$, the curve $H$ tends to points $b_1$ and $b_2$, respectively. This means that $H$ is connected to the Hopf bifurcation curve in $D \cap C_{\epsilon_3}^{\pm}$ because of the uniqueness of the Hopf bifurcation curve in a neighborhood of $b_1$ or $b_2$ (see Section 4.1).

We can consider the equation for $HL = S_{HL} \cap S_{\sigma}$ in the same way.

Thus, we can choose $K$ and $D$ satisfying the conditions (1), (2), and (3). The conclusion of Theorem 3.2 for $\epsilon_1 < 0$ and $(\epsilon_1, \epsilon_2, \epsilon_3) \in C(D)$ follows from Lemmas 3.8 and 3.10. The conclusion for $\epsilon_1 > 0$ or $\epsilon_1 = 0$ but $(\epsilon_1, \epsilon_2, \epsilon_3) \neq (0, 0, \pm \sigma)$ is obvious. The only remaining case is $\epsilon_1 < 0$ and $(\epsilon_1, \epsilon_2, \epsilon_3) \notin C(D)$. In this case, by using the same method as in the proof of Lemma 4.1.12, it can be shown that equation (3.13) has no periodic orbits.                                              $\square$

### 5.4 A Codimension 4 Bifurcation: Cusp of Order 4

In this section we consider a cusp of codimension 4; the family of vector fields for this case is $(3.4)^{\pm}$. Most results in this section are due to Li and Rousseau [1]. We discuss only the case $(3.4)^+$ because the case $(3.4)^-$ is similar. Thus, we consider the family of vector fields

$$\begin{cases} \dot{x} = y, \\ \dot{y} = \epsilon_1 + \epsilon_2 y + \epsilon_3 xy + \epsilon_4 x^3 y + x^2 + x^4 y. \end{cases} \tag{4.1}$$

Obviously, if $\epsilon_1 > 0$, (4.1) has no equilibrium; if $\epsilon_1 = 0$, we have a saddle-node bifurcation. When $\epsilon_1 < 0$, we make a scaling

$$x \rightarrow \delta^2 x, \qquad y \rightarrow \delta^3 y, \qquad t \rightarrow \frac{t}{\delta}, \qquad \epsilon_1 = -\delta^4,$$

$$\epsilon_2 = \delta^8 \nu_1, \qquad \epsilon_3 = \delta^6 \nu_2, \qquad \epsilon_4 = \delta^2 \nu_3, \tag{4.2}$$

where $\delta > 0$. Then equation (4.1) becomes

$$\begin{cases} \dot{x} = y, \\ \dot{y} = -1 + x^2 + \delta^7 (\nu_1 + \nu_2 x + \nu_3 x^3 + x^4) y. \end{cases} \tag{4.3}$$

We will study first the bifurcation diagram of equation (4.3) in $\nu_1 \nu_2 \nu_3$-space, and then glue it in $\epsilon_1 \epsilon_2 \epsilon_3 \epsilon_4$-space (back to equation (4.1)) with the saddle-node bifurcation on $\epsilon_1 = 0$, with a cusp of order 2 on $\epsilon_1 = \epsilon_2 = 0$, $\epsilon_3 \neq 0$, and with a cusp of order 3 on $\epsilon_1 = \epsilon_2 = \epsilon_3 = 0$,

$\epsilon_4 \neq 0$. Thus, we can obtain the bifurcation diagram of (4.1), as we did for equation (3.13) in the last section.

By using the results in Sections 5.1–5.2 (Examples 1.8 and 2.13), we have the following two lemmas.

**Lemma 4.1.** *Let*

$$\begin{cases} V_0 = \nu_1 - \nu_2 - \nu_3 + 1, \\ V_1 = \nu_2 - 3\nu_3 + 8, \\ V_2 = 5\nu_2 - 14. \end{cases} \qquad (4.4)$$

*Then for each* $k = 1, 2, 3$, *equation (4.3) has a Hopf bifurcation of order* $k$ ($HB_k$) *if the following kth condition is satisfied:*
*(1)* $V_0 = 0$, $\delta V_1 \neq 0$;
*(2)* $V_0 = V_1 = 0$, $\delta V_2 \neq 0$;
*(3)* $V_0 = V_1 = V_2 = 0$, $\delta \neq 0$ (*i.e.,* $(\nu_1, \nu_2, \nu_3) = (\frac{11}{5}, \frac{2}{5}, \frac{14}{5})$, $\delta \neq 0$).
    *System (4.3) has three limit cycles near the focus if* $(\nu_1, \nu_2, \nu_3)$ *is in the region*

$$\begin{cases} V_2 < 0, V_1 > 0, V_0 < 0, \\ 0 < |V_0| \ll |V_1| \ll |V_2| \ll 1. \end{cases} \qquad (4.5)$$

**Lemma 4.2.** *Let*

$$\begin{cases} W_0 = \nu_1 - \dfrac{5}{7}\nu_2 - \dfrac{103}{77}\nu_3 + \dfrac{187}{91}, \\ W_1 = \nu_1 + \nu_2 + \nu_3 + 1, \\ W_2 = -\nu_2 - \dfrac{9}{5}\nu_3 + \dfrac{8}{7}. \end{cases} \qquad (4.6)$$

*Then for each* $k = 1, 2, 3$, *equation (4.3) has a homoclinic loop bifurcation of order* $k$ ($HLB_k$) *if the following kth condition is satisfied:*
*(1)* $W_0 = 0$, $\delta W_1 \neq 0$;
*(2)* $W_0 = W_1 = 0$, $\delta W_2 \neq 0$;
*(3)* $W_0 = W_1 = W_2 = 0$, $\delta \neq 0$.

(Condition (3) is equivalent to $(\nu_1, \nu_2, \nu_3) = (-\frac{107}{91}, -\frac{94}{91}, \frac{110}{91})$, $\delta \neq 0$.)
System (4.3) with $\delta \neq 0$ has three limit cycles near the saddle loop $\Gamma_0$ of
(4.3) with $\delta = 0$ if $(\nu_1, \nu_2, \nu_3)$ is in the region

$$W_2 < 0, \quad W_1 > 0, \quad W_0 > 0, \quad \text{and} \quad 0 < |W_0| \ll |W_1| \ll |W_2| \ll 1.$$

$$(4.7)$$

From the previous two lemmas we have that $HLB_2 \cap HB_1$ occurs at

$$(\nu_1, \nu_2, \nu_3) = \left(-1, -\frac{22}{13}, \frac{22}{13}\right),$$

and (4.3) has three limit cycles when $(\nu_1, \nu_2, \nu_3)$ satisfies

$$V_0 < 0, \quad W_1 > 0, \quad W_0 > 0,$$

$$0 < |V_0| \ll 1, \quad \text{and} \quad 0 < |W_0| \ll |W_1| \ll 1.$$

Also, $HLB_1 \cap HB_2$ occurs at $(\nu_1, \nu_2, \nu_3) = (\frac{31}{65}, -\frac{58}{65}, \frac{154}{65})$, and (4.3)
has three limit cycles when $(\nu_1, \nu_2, \nu_3)$ satisfies

$$V_1 > 0, \quad V_0 < 0, \quad W_0 > 0,$$

$$0 < |V_0| \ll |V_1| \ll 1, \quad \text{and} \quad 0 < |W_0| \ll 1.$$

In order to discuss the periodic orbits of equation (4.3), we consider
(4.3) as a perturbation of the Hamiltonian system

$$\begin{cases} \dot{x} = y, \\ \dot{y} = -1 + x^2, \end{cases} \tag{4.8}$$

with the Hamiltonian function $-H(x, y)$, where

$$H(x, y) = \frac{y^2}{2} + x - \frac{x^3}{3}. \tag{4.9}$$

By Lemma 4.1.4, the condition for the existence of periodic orbits of equation (4.3) is

$$F(h, \delta, \nu_1, \nu_2, \nu_3) = \int_{\gamma(h, \delta, \nu_1, \nu_2, \nu_3)} (\nu_1 + \nu_2 x + \nu_3 x^3 + x^4) y \, dx = 0,$$

$$(4.10)$$

where $\gamma(h, \delta, \nu_1, \nu_2, \nu_3)$ is defined as in Section 4.1 and $F(h, \delta, \nu_1, \nu_2, \nu_3)$ is well approximated by $M(h) = F|_{\delta=0}$. For $\delta = 0$, the condition (4.10) becomes

$$M(h) = \int_{\Gamma_h} (\nu_1 + \nu_2 x + \nu_3 x^3 + x^4) y \, dx = 0, \qquad (4.11)$$

where $\Gamma_h$ is the level curve of $H(x, y) = h, \ -\frac{2}{3} \le h \le \frac{2}{3}. \ H(x, y) = -\frac{2}{3}$ corresponds to the equilibrium $(-1, 0)$ and $H(x, y) = \frac{2}{3}$ corresponds to the homoclinic loop.

We will study the number of solutions of equation (4.11) with respect to $h \in (-\frac{2}{3}, \frac{2}{3})$, for given $(\nu_1, \nu_2, \nu_3)$. This number is just the number of periodic orbits of the system (4.3).

As in Sections 4.1 and 5.3, we define

$$I_k(h) = \int_{\Gamma_h} x^k y \, dx, \qquad k = 0, 1, 2, 3, 4, \qquad (4.12)$$

and then $M(h)$ takes the form

$$M(h) = \nu_1 I_0 + \nu_2 I_1 + \nu_3 I_3 + I_4. \qquad (4.13)$$

By Lemma 3.5, we have

$$\begin{cases} I_3 = -\dfrac{6}{11} h I_0 + \dfrac{15}{11} I_1, \\[2mm] I_4 = \dfrac{21}{13} I_0 - \dfrac{12}{13} h I_1. \end{cases} \qquad (4.14)$$

Substituting (4.14) into (4.13), we obtain

$$M(h) = A(h) I_0(h) - B(h) I_1(h), \qquad (4.15)$$

where

$$\begin{cases} A(h) = v_1 - \dfrac{6}{11}hv_3 + \dfrac{21}{13}, \\ B(h) = v_2 + \dfrac{15}{11}v_3 - \dfrac{12}{13}h. \end{cases} \tag{4.16}$$

As in Section 5.3, we consider $\overline{M}(h) = M(h)/I_0(h)$ instead of $M(h)$, and then from (4.15) we have

$$\overline{M}(h) = A(h) - B(h)P(h), \tag{4.17}$$

where $P(h)$ is the same as in Sections 4.1 and 5.3. We recall the properties of function $P(h)$ in the following lemma (see Lemmas 4.1.6, 4.1.7, and 3.6).

**Lemma 4.3.** *The function $P(h)$ has the following properties:*
*(1) $P([-\frac{2}{3}, \frac{2}{3}]) \subset [\frac{5}{7}, 1]$, $P(-2/3) = 1$, and $P(2/3) = 5/7$;*
*(2) $P'(h) < 0$, $P'(-2/3) = -1/8$, and $P'(2/3) = -\infty$;*
*(3) $P''(h) < 0$ and $P''(-2/3) = -55/1152$;*
*(4) $P = P(h)$ satisfies the following differential equation:*

$$(9h^2 - 4)P' = 7P^2 + 3hP - 5. \tag{4.18}$$

Or, equivalently, $(h, P)$ is a solution of the following system

$$\begin{cases} \dfrac{dP}{dt} = -7P^2 - 3hP + 5, \\ \dfrac{dh}{dt} = 4 - 9h^2, \end{cases} \tag{4.19}$$

satisfying $\lim_{t \to -\infty} h = -2/3$ and $\lim_{t \to -\infty} P = 1$.

Now we consider the bifurcation of multiple limit cycles. The condition for occurrence of multiple limit cycles is given by $\overline{M}(h) = \overline{M}'(h) = 0$, which determines a surface $S$. The points of this surface satisfying $\overline{M}''(h) \neq 0$ correspond to a double limit cycle bifurcation. We show that the surface is regular at these points. We also prove that the points of $S$ satisfying $\overline{M}''(h) = 0$ form a smooth curve $C$, on which $\overline{M}'''(h) \neq 0$. Hence, it corresponds to a triple limit cycle bifurcation. From (4.17) and

(4.16), we have

$$\overline{M}'(h) = -\frac{6}{11}\nu_3 + \frac{12}{13}P(h) - B(h)P'(h), \qquad (4.20)$$

$$\overline{M}''(h) = \frac{24}{13}P'(h) - B(h)P''(h), \qquad (4.21)$$

$$\overline{M}'''(h) = \frac{36}{13}P''(h) - B(h)P'''(h). \qquad (4.22)$$

**Lemma 4.4.** *S is a regular surface at the points where $\overline{M}''(h) \neq 0$.*

*Proof.* From $\overline{M}'(h) = 0$ and $\overline{M}''(h) \neq 0$ we have $h = h(\nu_2, \nu_3)$ by the Implicit Function Theorem. Since $\partial \overline{M}/\partial \nu_1 \neq 0$ (see (4.17) and (4.16)), we have $\nu_1 = \nu_1(h, \nu_2, \nu_3)$ from $\overline{M}(h) = 0$. If we replace $h$ by $h(\nu_2, \nu_3)$, the desired result follows.    □

**Lemma 4.5.** *On the curve $C = \{(\nu_1, \nu_2, \nu_3) | \overline{M}(h) = \overline{M}'(h) = \overline{M}''(h) = 0\}$ we have that $\overline{M}'''(h) \neq 0$, $-2/3 < h < 2/3$.*

*Proof.* The first step is to prove that if $\overline{M}''(h) = 0$ then $\overline{M}'''(h) \neq 0$ is equivalent to

$$3[P''(h)]^2 - 2P'(h)P'''(h) \neq 0. \qquad (4.23)$$

In fact, we will show that the left-hand side of (4.23) is negative for $h \in [-2/3, 2/3)$.

From (4.21) we get that $\overline{M}''(h) = 0$ is equivalent to

$$B(h) = \frac{24}{13}\frac{P'(h)}{P''(h)}, \qquad (4.24)$$

where $P = P(h)$ and $P'' \neq 0$ for $h \in [-2/3, 2/3]$ (see Lemma 4.3).

Substituting (4.24) into (4.22), we have

$$\overline{M}'''(h) = \frac{12}{13P''}\left[3(P'')^2 - 2P'P'''\right].$$

Hence $\overline{M}'''(h) \neq 0$ is equivalent to (4.23).

Now we prove (4.23). From (4.18) we get

$$(9h^2 - 4)P' = 7P^2 + 3hP - 5, \qquad (4.25)$$

$$(9h^2 - 4)P'' = P'(14P - 15h) + 3P, \qquad (4.26)$$

$$(9h^2 - 4)P''' = P''(14P - 33h) + P'(14P' - 12), \qquad (4.27)$$

$$(9h^2 - 4)P^{(4)} = P'''(14P - 51h) + P''(42P' - 45). \qquad (4.28)$$

We denote

$$F(h) \equiv 3(P''(h))^2 - 2P'(h)P'''(h). \qquad (4.29)$$

We have from Lemma 4.3, $P'(-2/3) = -1/8$ and $P''(-2/3) = -55/1152$. From (4.18) it is not difficult to obtain that $P'''(-2/3) = -3685/73728$, whence $F(-2/3) < 0$. We need to show $F(h) < 0$ for $h \in (-2/3, 2/3)$. If this is not true, we let $h^* = \inf\{h|F(h) = 0, h \in (-2/3, 2/3)\}$, and then it is obvious that $F(h^*) = 0$ and $F'(h^*) \geq 0$. But we will show that $F(h^*) = 0$ implies $F'(h^*) < 0$. This contradiction means that such an $h^*$ does not exist.

We suppose now that $F(h^*) = 0$, $h^* \in (-2/3, 2/3)$. From (4.25)–(4.29) we have that

$$\frac{(9h^2 - 4)F'(h)}{2} = (9h^2 - 4)(2P''P''' - P'P^{(4)})$$

$$= 2P''^2(14P - 33h) - P'P'''(14P - 51h)$$

$$+ P'P''(-14P' + 21).$$

$$(4.30)$$

By using $F(h^*) = 0$, from (4.29) and (4.30) we have

$$\frac{(9h^{*2} - 4)F'(h^*)}{2} = \frac{1}{3}P'P'''(14P + 21h) + P'P''(-14P' + 21)\big|_{h=h^*}.$$

$$(4.31)$$

We have already that $P'(h) < 0$ and $P''(h) < 0$ for $h \in (-2/3, 2/3)$. We claim that $14P(h) + 21h > 0$ and $P'''(h) < 0$ for $h \in (-2/3, 2/3)$. Hence (4.31) implies $F'(h^*) < 0$ since $9h^{*2} - 4 < 0$. This yields the desired result.

We need to show finally that the above claim is true.

To show $G(h) \equiv 14P(h) + 21h > 0$ for $h \in (-2/3, 2/3)$, it is sufficient to note that $G(-2/3) = 0$, $G(2/3) > 0$, $G'(-2/3) > 0$, and $G''(h) = 14P''(h) < 0$ for $h \in (-2/3, 2/3)$.

To show $P'''(h) < 0$ for $h \in (-2/3, 2/3)$, we use the same argument as to show $F(h) < 0$. We have that $P'''(-2/3) < 0$. Suppose $\bar{h} = \inf\{h \mid P'''(h) = 0, h \in (-2/3, 2/3)\}$. Then $P^{(4)}(\bar{h}) \geq 0$. From (4.28) and $F'''(\bar{h}) = 0$ we obtain that $(9\bar{h}^2 - 4)P^{(4)}(\bar{h}) = P''(42P' - 45)\big|_{\bar{h}} > 0$. This implies $P^{(4)}(\bar{h}) < 0$. The contradiction means that such an $\bar{h}$ does not exist. □

**Lemma 4.6.** *The curve $C$ corresponds to a triple limit cycle bifurcation, and it is a smooth curve.*

*Proof.* The fact that $C$ corresponds to a triple limit cycle bifurcation follows from the definition of $C$ as the set of $\{(\nu_1, \nu_2, \nu_3)\}$ such that $\overline{M}(h) = \overline{M}'(h) = \overline{M}''(h) = 0$ and from Lemma 4.5 which ensures that $\overline{M}'''(h) \neq 0$.

We prove now the smoothness of $C$. From (4.17), (4.20), (4.21), and (4.16), we have that $\overline{M}(h) = \overline{M}'(h) = \overline{M}''(h) = 0$ is equivalent to

$$\begin{cases} \nu_1 = \dfrac{3}{13}\left(4hP - 8h\dfrac{P'^2}{P''} + 8\dfrac{PP'}{P''} - 7\right), \\[2ex] \nu_2 = \dfrac{3}{13}\left(4h - 10P + 20\dfrac{P'^2}{P''} + 8\dfrac{P'}{P''}\right), \\[2ex] \nu_3 = \dfrac{22}{13}\left(P - \dfrac{2P'^2}{P''}\right). \end{cases} \qquad (4.32)$$

By Lemma 4.5, $\overline{M}'''(h) \neq 0$ is equivalent to (4.23). From the third equation of (4.32) we have

$$\frac{\partial \nu_3}{\partial h} = -\frac{22}{13} \frac{P'(3P''^2 - 2P'P''')}{P''^2},$$

which is different from zero by (4.23). Hence we get $h = h(\nu_3)$ from the third equation of (4.32), and the first two equations of (4.32) give $\nu_1 = \nu_1(h(\nu_3))$ and $\nu_2 = \nu_2(h(\nu_3))$ which are differentiable. Therefore, $C$ is a regular smooth curve.          $\square$

**Lemma 4.7.** *For sufficiently small $\delta$ there is a smooth curve $(HB, S)$ in the parameter space, corresponding to the simultaneous occurrence of an $HB_1$ and a double limit cycle. This curve joins the point $(\nu_1, \nu_2, \nu_3) = (\frac{11}{5}, \frac{2}{5}, \frac{14}{5})$, corresponding to $HB_3$, to the point $(\nu_1, \nu_2, \nu_3) = (-1, -\frac{22}{13}, \frac{22}{13})$, corresponding to $HB_1 \cap HLB_2$ (the coordinates at these points are up to $O(\delta)$). The curve is a convex envelope of the family of lines in the HB plane, given by $\overline{M}(h) = 0$, $h \in (-2/3, 2/3)$.*

*Proof.* We make a change of coordinates $(\nu_1, \nu_2, \nu_3) \to (m_1, m_2, m_3)$, which transforms the two lines $H_2$ and $H \cap HL$ to coordinate axes (see (4.4) and (4.6)):

$$\begin{cases} m_1 = \nu_1 - \nu_2 - \nu_3 + 1, \\ m_2 = \nu_2 - 3\nu_3 + 8, \\ m_3 = \nu_1 - \frac{5}{7}\nu_2 - \frac{103}{77}\nu_3 + \frac{187}{91}. \end{cases} \tag{4.33}$$

The equation $\overline{M}(h) = 0$ (see (4.17) and (4.16)), under the condition $m_1 = 0$ (on the $HB$ plane; see Lemma 4.1), gives

$$\overline{M}(h) = \frac{1}{10}(3h + 14P - 12)m_2 + \frac{7}{20}(-3h - 24P + 22)m_3$$

$$+ \frac{4}{65}(15hP - 21h - 38P + 34) = 0,$$

where $P = P(h)$. The above equation is equivalent to

$$m_2 = \frac{7}{2}Q(h)m_3 + \frac{8}{13}R(h),$$  (4.34)

where

$$Q(h) = \frac{3h + 24P - 22}{3h + 14P - 12},$$  (4.35)

and

$$R(h) = \frac{38P + 21h - 15hP - 34}{3h + 14P - 12}.$$  (4.36)

Since $Q < 0$ and $R > 0$, we will show that $dQ/dh \neq 0$ and $d^2R/dQ^2 < 0$. Hence, $-R$ is a convex function of the slope $Q$. Therefore, by arguments as in Lemma 3.7, the curve $(HB, S)$ is the graph of the Legendre transform of $-R$, that is, a convex curve. From (4.35), we have that

$$\frac{dQ}{dh} = \frac{10[3(1 - P) + (3h + 2)P']}{(3h + 14P - 12)^2} < 0,$$  (4.37)

except at $h = -2/3$, since the numerator is zero at $h = -2/3$ and its derivative $10(3h + 2)P'' < 0$ for $h \in (-2/3, 2/3)$ (see Lemma 4.3).
From (4.36), we get

$$\frac{dR}{dh} = \frac{5[-30 + 72P - 42P^2 + (4 - 9h^2)P']}{(3h + 14P - 12)^2}.$$  (4.38)

Equations (4.37) and (4.38) give

$$2\frac{dR}{dQ} = \frac{-30 + 72P - 42P^2 - (4 - 9h^2)P'}{3(1 - P) + (3h + 2)P'}.$$

Hence,

$$2\frac{d^2R}{dQ\,dh} = \frac{[(1 - P)(3h + 2)P'' + 6(1 - P)P' + 2(3h + 2)P'^2]}{[3(1 - P) + (3h + 2)P']^2}$$

$$\times 3(12 - 3h - 14P).$$

Since $d^2R/dQ^2 = d/dh(dR/dQ) \cdot dh/dQ$, and for $h \in (-2/3, 2/3)$ we have that

$$(12 - 3h - 14P) < 0 \quad \text{and} \quad \frac{dQ}{dh} < 0 \qquad (\text{see } (4.37)),$$

we only need to show

$$G(h) \equiv (1 - P)(3h + 2)P'' + 6(1 - P)P' + 2(3h + 2)P'^2 < 0$$

$$(4.39)$$

instead of $d^2R/dQ^2 < 0$ for $h \in (-2/3, 2/3)$. Since $G(-2/3) = G'(-2/3) = G''(-2/3) = 0$ and $G'''(-2/3) = 3(3P''^2 - 2P'P''')|_{h=-2/3} < 0$ (see the explanation following (4.23)), it is enough to show that $G \neq 0$ for $h \in (-2/3, 2/3)$.

Assume that $G(h)$ has a zero point at some $h \in (-2/3, 2/3)$. We repeat the technique used in the proof of Lemma 4.5 to deduce a contradiction. In fact, let $h^* = \inf\{h|G(h) = 0, -2/3 < h < 2/3\}$; then $G(h^*) = 0$ and $G'(h^*) \geq 0$.

On the other hand, from (4.39) we have

$$G'(h) = 3(3h + 2)P'P'' + 9(1 - P)P'' + (1 - P)(3h + 2)P''',$$

$$(4.40)$$

and $G(h^*) = 0$ implies

$$(3h^* + 2)P'(h^*) = -\frac{1 - P}{2P'}[(3h + 2)P'' + 6P']|_{h=h^*}.$$

Substituting the above expression into (4.40), we obtain

$$G'(h^*) = -\frac{1}{2P'}(1 - P)(3h + 2)(3P''^2 - 2P'P''')|_{h=h^*} < 0,$$

since $-2/3 < h^* < 2/3$, $P(h^*) < 1$, $P'(h^*) < 0$, and $(3P''^2 - 2P'P''') < 0$.

The contradiction proves that $G(h) < 0$ for $h \in [-2/3, 2/3)$.  $\square$

**Lemma 4.8.** *For sufficiently small $\delta$ there is a smooth curve $(HLB, S)$ in the parameter space, corresponding to the simultaneous occurrence of a*

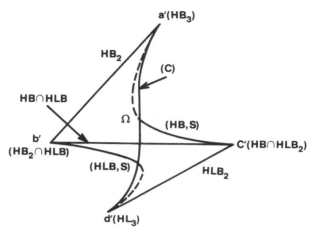

Figure 4.1.

*homoclinic bifurcation of order* 1 *($HLB_1$) and a double limit cycle. This curve joins the point $(\nu_1, \nu_2, \nu_3) = (-\frac{107}{91}, -\frac{94}{91}, \frac{110}{91})$, corresponding to $HLB_3$, to the point $(\nu_1, \nu_2, \nu_3) = (\frac{31}{65}, -\frac{58}{65}, \frac{154}{65})$, corresponding to $HLB_1 \cap HB_2$ (the coordinates at these points are up to $O(\delta)$). The curve is the convex envelope of the family of lines in the HLB plane, given by $\overline{M}(h) = 0$, $h \in (-2/3, 2/3)$.*

As the proof is similar to that of Lemma 4.7, we omit it here.

**Lemma 4.9.** *The parameter region $\Omega$ for which the equation (4.3) has three limit cycles has the form of a "topological 3-simplex" (Figure 4.1).*

*Proof.* We consider the 3-simplex $-2/3 < h_1 < h_2 < h_3 < 2/3$ (Figure 4.2), and the map $F$ from that 3-simplex to the parameter space $\nu = (\nu_1, \nu_2 \nu_3)$, defined by $h = (h_1, h_2, h_3) \to \nu(h)$, where $\nu(h)$ is the solution of $\overline{M}(h_1) = \overline{M}(h_2) = \overline{M}(h_3) = 0$. This solution is unique. In fact, from (4.17) and (4.16) we have

$$\overline{M}(h) = \nu_1 - P(h)\nu_2 + \left(-\frac{6}{11}h - \frac{15}{11}P(h)\right)\nu_3 + \frac{21}{13} + \frac{12}{13}hP(h),$$

and the coefficient determinant of $\overline{M}(h_1) = \overline{M}(h_2) = \overline{M}(h_3) = 0$ with

442        Bifurcations with Codimension Higher than Two

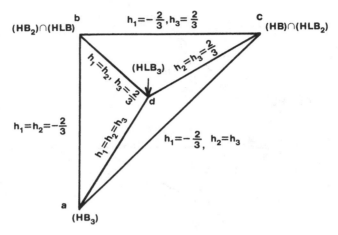

Figure 4.2.

respect to $(\nu_1, \nu_2, \nu_3)$ is

$$
\begin{vmatrix}
1 & -P(h_1) & -\dfrac{6}{11}h_1 - \dfrac{15}{11}P(h_1) \\[2mm]
1 & -P(h_2) & -\dfrac{6}{11}h_2 - \dfrac{15}{11}P(h_2) \\[2mm]
1 & -P(h_3) & -\dfrac{6}{11}h_3 - \dfrac{15}{11}P(h_3)
\end{vmatrix}
$$

$$
= \frac{6}{11}
\begin{vmatrix}
1 & P(h_1) & h_1 \\
1 & P(h_2) & h_2 \\
1 & P(h_3) & h_3
\end{vmatrix}
$$

$$
= (h_2 - h_1)(h_3 - h_2)\left[ \frac{P(h_2) - P(h_1)}{h_2 - h_1} - \frac{p(h_3) - P(h_2)}{h_3 - h_2} \right] > 0,
$$

since $P' < 0$, $P'' < 0$ for $h \in (-2/3, 2/3)$.

The function $F$ is a local diffeomorphism on the 3-simplex in the $h$-space. $F$ is of rank 2 on the faces $h_1 = h_2 \neq h_3$ and $h_1 \neq h_2 = h_3$, and is a local diffeomorphism when restricted to these faces. Similarly $F$ is of rank 1 on the edge $h_1 = h_2 = h_3$, and a local homeomorphism when restricted to this edge. We can conclude that $F$ is a global diffeomorphism on the 3-simplex if we can prove that the system cannot have more than three limit cycles, that is, no four planes $\overline{M}(h_j) = 0$, $h_1 < h_2 < h_3 < h_4$, can intersect. This is shown in the next lemma.  □

**Lemma 4.10.** *For sufficiently small* δ, *the system* (4.3) *has at most three limit cycles for each value of the parameters* $v_1$, $v_2$, *and* $v_3$.

*Proof.* For given $v_1$, $v_2$, and $v_3$ we determine the number of zeros of

$$\overline{M}(h) = A(h) - B(h)P(h)$$

for $h \in (-2/3, 2/3)$, where $A(h)$ and $B(h)$ are linear in $h$ and given by (4.16). Let $h^* \in (-\infty, +\infty)$ such that $B(h^*) = 0$.

Suppose that $\overline{M}(h^*) = 0$. Then we have $A(h^*) = 0$ and $\overline{M}(h) = D(h - h^*)(P - P^*)$ for some constant $D$. Since $P = P(h)$ is monotonic, $\overline{M}(h)$ has at most two zeros.

For the rest of the proof we can suppose that $\overline{M}(h^*) \neq 0$. Then

$$\overline{M}(h) = B(h)(Q(h) - P(h)), \qquad (4.41)$$

where $h \in (-2/3, 2/3)$, $h \neq h^*$, and

$$Q(h) = \frac{A(h)}{B(h)} = \frac{143v_1 - 78hv_3 + 231}{143v_2 + 195v_3 - 132h}. \qquad (4.42)$$

We need to determine the number of intersection points of the curve $P = P(h)$, called $\Gamma_P$, and the curve $P = Q(h)$, called $\Gamma_Q$, for $h \in (-2/3, 2/3) \setminus \{h^*\}$. The curve $\Gamma_Q$ is a hyperbola (Figure 4.3) and

$$Q'(h) = \frac{\alpha}{B^2(h)},$$

where

$$\alpha = -78v_3(143v_2 + 195v_3) + 132(143v_1 + 231).$$

In the case $\alpha \geq 0$, it is obvious that $\Gamma_Q$ and $\Gamma_P$ have at most two intersection points. Hence, we only consider the case $\alpha < 0$.

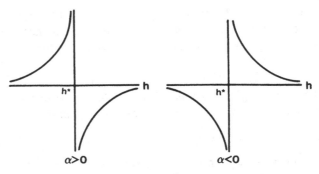

Figure 4.3.

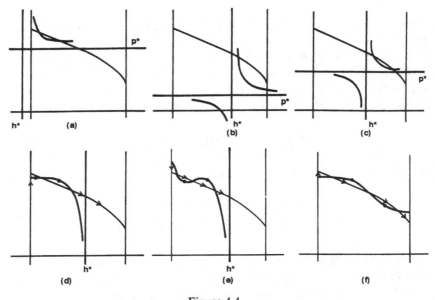

Figure 4.4.

(i) If $h^* \leq -2/3$, then $\Gamma_P$ and $\Gamma_Q$ have, obviously, at most two intersection points (Figure 4.4(a)).

(ii) If $-2/3 < h^* \leq 2/3$, then the right branch $\Gamma_Q^+$ of $\Gamma_Q$ always has at most two intersection points with $\Gamma_P$ (Figure 4.4(b) and (c)). It has exactly one intersection point if and only if $Q(2/3) < 5/7$. In the case where $\Gamma_Q^+$ intersects $\Gamma_P$, the left branch $\Gamma_Q^-$ of $\Gamma_Q$ is then below $P = 5/7$, and hence it has no intersection with $\Gamma_P$. Therefore, we need only consider the case where $Q(2/3) > 5/7$. In this case we study the number of intersection points of $\Gamma_Q^-$ and $\Gamma_P$ (Figure 4.4(d) and (e)).

For this purpose we count the number of contact points of $\Gamma_Q$ with the vector field (4.19), that is, we consider the number of zeros of

$$\dot{G} = \frac{dG}{dt}\bigg|_{(4.19)} = \dot{P} - Q'(h)\dot{h}$$

for $G = P - Q(h)$. A calculation shows that

$$\dot{G} = (-7P^2 - 3hP + 5) - \frac{\alpha}{B^2}(4 - 9h)^2\big|_{P=A/B}$$

$$= \frac{1}{B^2}\left[-7A^2 - 3hAB + 5B^2 - \alpha(4 - 9h^2)\right].$$

The numerator is a polynomial of degree 3 with respect to $h$ (see (4.16)); therefore, it has at most three roots. It is easy to see that the left branch $\Gamma_Q^-$ has at least as many contact points with the vector field (4.19) as it has intersection points with $\Gamma_P$ (Figure 4.4(d) and (e)): Between any two intersection points there is always a contact point, and there is always a contact point on the left of the first intersection point, due to the direction of the vector field at the intersection of $\Gamma_Q^-$ with the line $h = -2/3$. The number of intersection points is therefore at most three.

(iii) The case $h^* > 2/3$ can be discussed by the same arguments as in (ii) (e.g., see Figure 4.4(f)).                                                          □

**Remark 4.11.** From (4.42), Lemma 4.1, and Lemma 4.2, we obtain, similarly to the discussion in Section 5.3, that $Q(-2/3) = 1 = P(-2/3)$ if and only if $(\nu_1, \nu_2, \nu_3) \in HB$ (i.e., $\nu_1 - \nu_2 - \nu_3 + 1 = 0$) and $Q(2/3) = 5/7 = P(2/3)$ if and only if $(\nu_1, \nu_2, \nu_3) \in HLB$ (i.e., $\nu_1 - \frac{5}{7}\nu_2 - \frac{103}{77}\nu_3 + \frac{187}{91} = 0$). Suppose $(\nu_1, \nu_2, \nu_3) \in \Omega$, the topological 3-simplex formed by surfaces $HB, HLB$, two pieces of the double limit cycle bifurcation surface $S$ (see Lemma 4.9), and the curve $C$, and suppose condition (4.5) is satisfied. By Lemma 4.1, equation (4.3) has three limit cycles. In this case, $Q(-2/3) < 1$ and $Q(2/3) > 5/7$ by Lemma 4.10. The relative positions of $\Gamma_P$ and $\Gamma_Q$ is shown in Figure 4.4(f). As $(\nu_1, \nu_2, \nu_3)$ varies inside $\Omega$, the number of intersection points between $\Gamma_P$ and $\Gamma_Q$ is always three. In fact any decreasing of this number corresponds to at least one of the following situations:

(1) $Q(-2/3)$ becomes larger than 1.
(2) $Q(2/3)$ becomes less than $5/7$.

(3) Two intersection points of $\Gamma_P$ and $\Gamma_Q$ become a tangent point, and then disappear.

(4) Three intersection points of $\Gamma_P$ and $\Gamma_Q$ become a tangent point.

These situations mean that $(\nu_1, \nu_2, \nu_3)$ goes to the boundary of $\Omega$ (the surface $HB$, $HLB$, $S$, or $C$, respectively), and then leaves $\Omega$. In this way, we can discuss the phase portrait of equation (4.3) for $(\nu_1, \nu_2, \nu_3)$ as any position in the parameter space.

Summing up the above lemmas and Remark 4.11, we have the following theorem.

**Theorem 4.12.** *For sufficiently small $\delta$, the bifurcation diagram of equation (4.3) is shown in Figure 4.5. It consists of the following:*

(1) *Surfaces (codimension-1 bifurcation): $HB_1$, $HLB_1$, and $S$ ($S$ has two smooth pieces divided by the curve $C$).*

(2) *Curves (codimension-2 bifurcation): $HB_2$, $HB_1 \cap HLB_1$, $HLB_2$, $(HB, S)$, $(HLB, S)$, and $C$.*

(3) *Points (codimension-3 bifurcation): $HB_3$, $HLB_3$, $HB_2 \cap HLB_1$, and $HB_1 \cap HLB_2$.*

*When $\nu = (\nu_1, \nu_2, \nu_3) \in \Omega$, surrounded by surfaces $HB_1$, $HLB_1$, and $S$ and curve $C$, equation (4.3) has exactly three limit cycles; when $\nu$ varies from $\Omega$ through $S$, then two of the three limit cycles merge as a semistable limit cycle, and then disappear; when $\nu$ varies from $\Omega$ through the surface $HB_1$, the most inner limit cycle shrinks into the focus $(x, y) = (-1, 0)$ which changes its stability; and when $\nu$ varies from $\Omega$ through the surface $HLB_1$, the most outer limit cycle expands and forms a homoclinic orbit, and then the connection from the saddle point to itself breaks down and the homoclinic loop disappears.*

Now we return from equation (4.3) to the original equation (4.1). We describe the bifurcation diagram of (4.1) by taking its intersections with a 3-sphere around the origin in the $\epsilon$-space. Equation (4.1) has equilibria only on the closed half 3-sphere $\{(\epsilon_1, \epsilon_2, \epsilon_3) | \epsilon_1^2 + \epsilon_2^2 + \epsilon_3^2 = 1, \epsilon_1 \le 0\}$, which can be transformed into a closed 3-ball (Figure 4.6). The bifurcation diagram inside the ball is similar to the bifurcation diagram of (4.3) in $\nu$-space (Figure 4.5), containing a topological 3-simplex $\Omega$ with exactly three limit cycles. The boundary of the ball (a 2-sphere) corresponds to the saddle-node bifurcation ($\epsilon_1 = 0$, $\epsilon_2 \ne 0$). On it, the Bogdanov–Takens bifurcation appears on a circle ($\epsilon_1 = \epsilon_2 = 0$, $\epsilon_3 \ne 0$). Two points of the circle ($\epsilon_1 = \epsilon_2 = \epsilon_3 = 0$, $\epsilon_4 \ne 0$, one with $\epsilon_4 > 0$, the

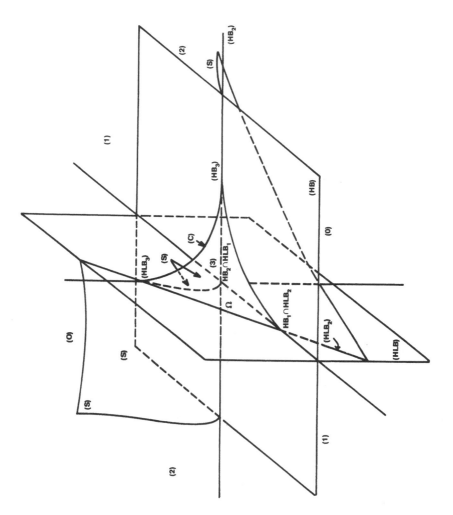

Figure 4.5.

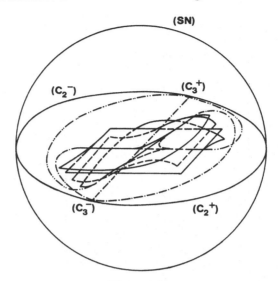

Figure 4.6.

other with $\epsilon_4 < 0$) correspond to the cusps of order 3, and separate the two cases of Bogdanov–Takens bifurcation: $\epsilon_3 < 0$ and $\epsilon_3 > 0$. The Hopf bifurcation surface $(HB)$ and the homoclinic bifurcation surface $(HLB)$ inside the ball branch along the circle of the Bogdanov–Takens bifurcation. Moreover, the different curves of codimension-2 bifurcations inside the ball meet on the boundary of the ball at two cusps of order 3, giving the conic structure described in Section 5.3. In Figure 4.6, for clarity we do not draw completely the Hopf bifurcation surface $(HB)$ and the homoclinic bifurcation surface $(HLB)$. We just draw the continuation of the codimension-2 curves until they meet the boundary of the ball.

To verify the above description, as we did in the last section, we need to construct a union of cones:

(1) The half space $\epsilon_1 > 0$.

(2) A cone $K_4$ constructed around the $\epsilon_4$-axis on a small neighborhood in $\epsilon_1\epsilon_2\epsilon_3$-space.

(3) A cone $K_3$ constructed around the $\epsilon_3$-axis on the product of a small neighborhood in $\epsilon_1\epsilon_2$-space with an arbitrary compact set in $\epsilon_4$-space.

(4) A cone $K_2$ constructed around the $\epsilon_2$-axis on the product of a small neighborhood in $\epsilon_1$-space with an arbitrary compact set in $\epsilon_3\epsilon_4$-space.

(5) A cone $K_1$ constructed around the $\epsilon_1$-axis ($\epsilon_1 < 0$) on an arbitrary compact set in $\epsilon_2\epsilon_3\epsilon_4$-space.

If we choose well the arbitrary compact sets, we will produce a neighborhood of the origin.

The last cone $K_1$ can be obtained from Theorem 4.12 ($(\nu_1, \nu_2, \nu_3) \to (\epsilon_2, \epsilon_3, \epsilon_4)$, see (4.2)). For the other cones $K_j$ we must use the universal unfolding of the cusps of order $j - 1 \le 3$; the cusp of order 1 is just the saddle-node.

## 5.5 Bibliographical Notes

There are at least six different methods used to study degenerate Hopf bifurcation. We list here these methods and some references: the method of Poincaré normal forms, see Arnold [1] and Guckenheimer and Holmes [1]; the method of averaging, see Chow and Hale [1], Guckenheimer and Holmes [1], and Sanders and Verhulst [1]; the method of the succession function, see Andronov, et al. [2]; the method of Lyapunov constants, see Bonin and Legault [1] and Göbber and Williamowski [1]; the method of Lyapunov–Schmidt, see Golubitsky and Langford [1], Golubitsky and Schaeffer [1], and Vanderbauwhede [1]; and the method of intrinsic harmonic balancing, see Allwright [1], Huseyin and Yu [1], and Mees [1]. In the paper of Farr et al. [1] there is a review of these different methods, and there are some explicit formulas of the first three Lyapunov coefficients for degenerate Hopf bifurcation problems of the general case of a differential equation with dimension $n \ge 2$.

The proof of Theorem 1.3 is due to Rousseau and Schlomiuk [1]. They used the Poincaré normal form and the Malgrange Preparation Theorem, which made the proof simpler. Theorem 1.5 was given by Bonin and Legault [1]. The results in Example 1.6 were given by Li [1, 3]. Example 1.6 (i)–(iii) are generalizations of some well-known formulas given by Bautin [2]. Example 1.7 was given by Sibirskii [1].

The degenerate homoclinic bifurcation, however, has been much less studied, although Poincaré [1] and Dulac [1] provided some ideas and approaches many years ago. The first result on this subject, to our knowledge, was presented by Leontovich [1] in an abstract paper, and the complete proof of this result has been given by Roussarie [1] (the first part of Theorem 2.5) and by Joyal [1] (both parts of Theorem 2.5). Most of Section 2 is due to Joyal [1] and Joyal and Rousseau [1].

The finite cyclicity problem for an equilibrium or a singular closed orbit is closely related to the Hilbert 16th problem and Hopf or homoclinic bifurcation; see, for example, Dumortier, Roussarie, and Rousseau [1], Dumortier, Roussarie, Sotomayor, and Żołądek [1], Ecalle [1], Il'yashenko [3], Li and Liu [1], Roussarie [2, 3], and Schlomiuk [1].

Consider the following equation on a plane

$$\begin{cases} \dot{x} = y, \\ \dot{y} = ax^2 + bxy. \end{cases} \tag{5.1}$$

It is well known that if $ab \neq 0$, then the bifurcation of (5.1) is of codimension 2, and the Bogdanov–Takens system is a versal deformation of (5.1) (see Section 4.1). If $b = 0$ and $a \neq 0$ in (5.1), then there is a higher-codimension bifurcation of cusp type. In Sections 3 and 4 we discussed the cusps of codimension 3 and 4, which were obtained by Dumortier, Roussarie, and Sotomayor [1], and Li and Rousseau [1], respectively. Joyal [2] considered cusps of codimension $n$. The maximum number of limit cycles in this case is $(n - 1)$. The proofs of Lemmas 3.6 and 4.5 were suggested by Rousseau, and the idea was stimulated by Drachman, van Gils, and Zhang [1]. See also Dumortier and Fiddelaers [1].

If $a = 0$ and $b \neq 0$ in (5.1), then there are higher-codimension bifurcations of other types: saddle, focus, and elliptic cases. Dumortier, Roussarie, and Sotomayor [2], Dumortier and Rousseau [1], Medved [1], Xiao [1], and Żołądek [3] studied these cases with codimension 3. The results on the focus and elliptic cases are still open, and a conjecture on the bifurcation diagrams for those cases is proposed in Dumortier, Roussaire, and Sotomayor [2].

We have considered codimension 2 bifurcation of the $1:2$ resonance in Section 4.2. The unperturbed system is

$$\begin{cases} \dot{x} = y, \\ \dot{y} = ax^3 + bx^2y, \end{cases} \tag{5.2}$$

where $ab \neq 0$. In the case $b = 0$ and $a \neq 0$, codimension-3 and -4 bifurcations in the $1:2$ resonance were considered by Li and Rousseau [2], Rousseau [2], and Rousseau and Żołądek [1]. In the case $a = 0$ and

$b \neq 0$, codimension-3 bifurcation of the $1:2$ resonance was considered by Dangelmayr, Armbruster, and Neveling [1].

Dangelmayr and Guckenheimer [1] studied a bifurcation problem arising from (5.2) by adding four parameters, and the result was improved by Żoładek [4].

There are many references concerning homoclinic (or heteroclinic) bifurcations in higher-dimensional phase spaces. Silnikov [1, 2] gave an efficient method to study these problems. Chow, Hale, and Mallet-Paret [1], Chow and Lin [1], Deng [3], Li, Li, and Zhang [1], and Schecter [1] studied the case of a homoclinic orbit with a degenerate singular point. See also Chow, Deng, and Fielder [1], Chow, Deng, and Terman [1, 2], Deng [1, 2], Fiedler [1], Kisaka, Kokubu, and Oka [1, 2], and Lin [1].

# Bibliography

Afraimovich, V. S.; Arnold, V. I.; Il'yashenko, Y. S.; and Sil'nikov, L. P.
  [1] *Bifurcation Theory*, Nauka i Tekhnika, VINITI, Moscow, 1986 (in Russian), *Dynamical Systems V*. Springer-Verlag,, New York, 1988 (in English).
Allwright, D. J.
  [1] Harmonic balance and the Hopf bifurcation. *Math. Proc. Camb. Philos. Soc.* **82** (1977): 453–67.
Andronov, A. A.; Leontovich, E. A.; Gordon, I. I.; and Maier, A. G.
  [1] *Qualitative Theory of Second-Order Dynamic Systems*. Nauka, Moscow, 1966.
  [2] *Theory of Bifurcations of Dynamic Systems on a Plane*. Israel Program for Sci. Transl., Wiley, New York, 1973.
Arnold, V. I.
  [1] Lectures on bifurcation in versal families. *Russ. Math. Surv.* **27**, 5 (1972): 54–123.
  [2] On the loss of stability of oscillations near a resonance and deformation of equivariant vector fields. *Funct. Anal. Appl.* **11**, 2 (1977): 1–10.
  [3] *Mathematical Methods of Classical Mechanics*. Springer-Verlag, New York, 1978.
  [4] *Geometric Methods in the Theory of Ordinary Differential Equations*. Springer-Verlag, New York, 1983.
  [5] Bifurcations and singularities in mathematics and mechanics. Lecture on 17th Congress of TCTAM, 1988; *Adv. in Mech.* **19**, 2 (1989): 217–31 (in Chinese).
  [6] On matrices depending on parameters. *Russ. Math. Surv.* **26**, 2 (1971): 29–43.
  [7] Critical points of smooth functions and their normal forms. *Russ. Math. Surv.* **30**, 5 (1975): 1–75.
Arnold, V. I., and Il'yashenko, Y. S.
  [1] *Ordinary Differential Equations, Dynamical Systems I*. Springer-Verlag, New York, 1988.
Baider, A., and Churchill, R. C.
  [1] Uniqueness and nonuniqueness of normal forms of vector fields. *Proc. Roy. Soc. Edinburgh, Sect. A* **108** (1988): 27–33.
Barbancon, M.
  [1] Théorème de Newton pour les fonctions de class $C^r$, preprint, Strashcurg.

452

Bates, P., and Jones, C. K.
  [1] Invariant manifolds for semilinear partial differential equations.
      *Dynamics Reported* **2** (1989): 1–38.
Bautin, N. N.
  [1] On periodic solutions of certain systems of differential equations. *Prikl.
      Mat. Mech.* **18** (1954): 128.
  [2] On the number of limit cycles which appear with variation of the
      coefficients from an equilibrium position of focus or center type. *Math.
      Sb.* **30** (72) (1952): 181–96 (in Russian), *Amer. Math. Soc. Trans. Series*
      **5** (1962): 396–413 (in English).
Belitskii, G. R.
  [1] The equivalence of families of local diffeomorphisms. *Funct. Anal.
      Appl.* **8**, 4 (1974): 338–9.
  [2] The normal forms of local mappings. *Usp. Mat. Nauk.* **30**, 1 (1975): 223
      (in Russian).
  [3] Equivalence and normal forms of germs of smooth mappings. *Russ.
      Math. Surv.* **33**, 1 (1978): 107–77.
  [4] Normal forms in relation to the filtering action of a group. *Trudy
      Moskov Mat. Obsc.* **40** (1979): 3–46.
  [5] *Normal forms, invariant and local mappings.* Naukova Dumka, Kiev,
      1979 (in Russian).
Berezovskaia, F. S., and Knibnik, A. I.
  [1] On bifurcations of separatrices in the problem of loss of stability of
      self-oscillations near the resonance 1:4. *Prikl. Mat. Mekh.* **44**, 5 (1980):
      938–43.
Birkhoff, G. D.
  [1] *Dynamical Systems.* A.M.S. Coll. Publication IX, New York, 1927.
Blows, T. R., and Lloyd, N. G.
  [1] The number of limit cycles of certain polynomial differential equations.
      *Proc. Roy. Soc. Edinburgh, Sect. A* **98** (1984): 215–39.
Bogaevskii, V. N., and Povzner, A. Y.
  [1] A nonlinear generalization of the shearing transformation. *Funct. Anal.
      Appl.* **16** (1982): 192–4.
Bogdanov, R. I.
  [1] Bifurcations of the limit cycle of a family of plane vector fields. *Trudy
      Sem. Petrovsk.* **2** (1976): 23–35 (in Russian); *Sel. Math. Sov.* **1** (1981):
      373–87 (in English).
  [2] Versal deformations of a singular point of a vector field on the plane in
      the case of zero eigenvalues. *Trudy Sem. Petrovsk.* **2** (1976): 37–65 (in
      Russian); *Sel. Math. Sov.* **1** (1981): 389–421 (in English).
  [3] Orbital equivalence of singular points of vector fields. *Funct. Anal.
      Appl.* **10** (1976): 316–17.
Bonin, G., and Legault, J.
  [1] Comparison de la methode des constantes de Lyapunov et la bifurca-
      tion de Hopf. *Canad. Math. Bull.* **31**, 2 (1988): 200–9.
Briot and Bouquet
  [1] Recherches sur les proprietes des fonction definies par des equations
      differentielles. *J. Wecole Impwr. Polytech.* **21**, 36 (1856): 133–98.
Broer, H. W.
  [1] Bifurcations of singularities in volume preserving vector fields. Thesis,
      Univ. of Groningen, Holland, 1979.
  [2] Formal normal form theorems for vector fields and some consequences
      for bifurcations in the volume preserving case. In *Dynamical Systems*

*and Turbulence*, Lect. Notes Math. **898**, Springer, Berlin, Heidelberg, 1981, 54–74.

Broer, H. W., Huitema, G. B., Takens, F., and Braaksma, B. L. J.
  [1] Unfoldings and bifurcations of quasi-periodic tori, *Mem. AMS* **83** 421, *American Mathematical Society*, 1990.

Bruno, A. D.
  [1] The normal forms of differential equations. *Sov. Math. Dokl.* **5** (1964): 1105–8.
  [2] The analytical form of differential equations. *Trans. Moscow Math. Soc.* **25** (1971): 131–288.
  [3] The analytical form of differential equations II. *Trans. Moscow Math. Soc.* **26** (1972): 199–238.
  [4] The normal form of differential equations with a small parameter. *Math. Notes*, **16**, 3 (1974): 832–6.
  [5] The normal form and averaging methods. *Sov. Math. Dokl.* **17** (1976): 1268–73.
  [6] *Local Methods in Nonlinear Differential Equations*. Springer-Verlag, New York, 1989.

Burchard, A.; Deng, B.; and Lu, K.
  [1] Smooth conjugacy of centre manifolds. *Proc. Royal Soc. Edinburg, Sect. A* **120** (1992): 61–77.

Burgoyne, H., and Cushman, R.
  [1] Normal forms for linear Hamiltonian systems. In *1976 NASA Conference on Geometric Control Theory*, ed. C. Martin and R. Hermann, Mathematical Sci. Press, Brookline, MA, 1977, 483–529.

Cai, S.
  [1] Weak saddle and separatrix cycle of quadratic systems. *Acta. Math. Sinica* **30** (1987): 553–9 (*in Chinese*).

Carr, J.
  [1] *Applications of Center Manifold Theory*. Springer-Verlag, New York, 1981.

Carr, J.; Chow, S.-N.; and Hale, J. K.
  [1] Abelian integrals and bifurcation theory. *J. Diff. Eq.* **59** (1985): 413–63.

Carr, J.; van Sanders, J. A.; and Gils, S. A.
  [1] Nonresonant bifurcations with symmetry, *SIAM J. Math. Anal.* **18**, 3 (1987): 579–91.

Cerkas, L. A.
  [1] On the stability of singular cycles. *Diff. Eq.* **4** (1968): 1012–17 (in Russian).

Chen, K. T.
  [1] Equivalence and decomposition of vector fields about an elementary critical point. *Am. J. Math.* **85** (1963): 693–722.

Cherry, T. M.
  [1] On the solution of Hamiltonian systems of differential equations in the neighborhood of a singular point. *Proc. London Math. Soc., Series 2* **27** (1927): 151–70.

Chow, S.-N.; Deng, B.; and Fiedler, B.
  [1] Homoclinic bifurcation at resonant eigenvalues. *Dynamics and Diff. Eq.* **2** (1990): 177–244.

Chow, S.-N.; Deng, B.; and Terman, D.
  [1] The bifurcation of homoclinic orbits from two heteroclinic orbits—An analytic approach. *SIAM J. Math. Anal.* **21** (1990): 179–204.

[2] The bifurcation of homoclinic orbits from two heteroclinic orbits—A topological approach. *Appl. Anal.* **42** (1991): 275–300.

Chow, S.-N.; Drachman, B.; and Wang, D.
[1] Computation of normal forms. *J. Comput. and Appl. Math.* **29** (1990): 129–43.

Chow, S.-N., and Hale, J. K.
[1] *Methods of Bifurcation Theory.* Springer-Verlag, New York, 1982.

Chow, S.-N.; Hale, J. K.; and Mallet-Paret, J.
[1] An example of bifurcation to homoclinic orbits. *J. Diff. Eq.* **37** (1980): 351–73.

Chow, S.-N.; Li, C.; and Wang, D.
[1] Uniqueness of periodic orbits of some vector fields with codimension two singularities. *J. Diff. Eq.* **77** (1989): 231–53.
[2] A simple proof of the uniqueness of periodic orbits in the 1:3 resonance problem. *Proc. A.M.S.* **105** (1989): 1025–32.

Chow, S.-N., and Lin, X.-B.
[1] Bifurcation of a homoclinic orbit with a saddle-node equilibrium. *Diff. Int. Equ.* **3**, 3 (1990): 435–66.

Chow, S.-N.; Lin, X.-B.; and Lu, K.
[1] Smooth invariant foliations in infinite dimensional spaces. *J. Diff. Eq.* **94** (1991): 266–91.

Chow, S.-N., and Lu, K.
[1] $C^k$ center unstable manifolds. *Proc. R. Soc., Edinburgh, Sect. A* **108** (1988): 303–20.
[2] Invariant manifolds for flows in Banach spaces. *J. Diff. Eq.* **74** (1988): 285–317.
[3] Invariant manifolds and foliations for quasiperiodic systems, *J. Diff. Eq.*, to appear.

Chow, S.-N.; Lu, K.; and Sell, G. R.
[1] Smoothness of inertial manifolds. *J. Math. Anal. Appl.* **169** (1992): 283–312.

Chow, S.-N.; Lu, K.; and Shen, Y.
[1] Normal form and linearization for quasiperiodic systems. *Trans. AMS* **331**, 1 (1992): 361–76.

Chow, S.-N., and Mallet-Paret, J.
[1] Integral averaging and bifurcation. *J. Diff. Eq.* **26** (1977): 112–59.

Chow, S.-N., and Sanders, J. A.
[1] On the number of critical points of the period. *J. Diff. Eq.* **64** (1986): 51–66.

Chow, S.-N., and Wang, D.
[1] Normal form of bifurcating periodic orbits. In *Multi-Parameter Bifurcation Theory*, Contemporary Math. Vol. 56, ed. M. Golubitsky and J. Guckenheimer, 9–18. Amer. Math. Soc., Providence, RI, 1986.
[2] On the monotonicity of the period function of some second order equations. *Čas. Pro. Pěs. Mat. Roč.* **111** (1986): 14–25.

Chow, S.-N., and White, R. G.
[1] On the transition from supercritical to subcritical Hopf bifurcation. *Math. Meth. in Appl. Sic.* **4** (1982): 143–63.

Chow, S.-N., and Yamashita, M.
[1] Geometry of the Melnikov vector. In *Nonlinear Equations in the Applied Sciences*, eds. W. F. Ames and C. Rogers, 78–148. Academic, New York, 1991.

Chow, S.-N., and Yi, Y.
  [1] Dynamical systems and singularly perturbed differential equations, preprint, 1991.
Chua, L. O., and Oka, H.
  [1] Normal form for constrained nonlinear differential equations, II: Bifurcation. *IEEE Trans. Circuits Syst.* **36**, 1 (1989): 71–88.
Cushman, R.
  [1] Normal form for Hamiltonian vector fields with periodic flows. In *Diff. Geom. Methods in Math. Phy.*, eds. S. Sternberg et al., Reidel, 125–44, Dordrecht, 1984.
  [2] Reduction of the 1:1 nonsemisimple resonance. *Hadronia J.* **5** (1982): 2109–24.
  [3] Notes on normal forms. *Lectures Notes of Michigan State Univ.*, 1985.
Cushman, R.; Deprit, A.; and Mosak. R.
  [1] Normal forms and representation theory. *J. Math. Phys.* **24** (1983): 2103–16.
Cushman, R., and Rod, D. L.
  [1] Reduction of the semisimple 1:1 resonance. *Physica D* **6** (1982): 105–12.
Cushman, R., and Sanders, J.
  [1] A codimension two bifurcation with a third order Picard–Fuchs equation. *J. Diff. Eq.* **59** (1985): 243–56.
  [2] Nilpotent normal forms and representation theory of $sl(2, R)$. In *Multi-parameter bifurcation theory*, eds. M. Golubitsky and J. Guckenheimer, 31–51. Amer. Math. Soc., Providence, RI, 1986.
  [3] Nilpotent normal form in dimension 4. In *Dynamics of infinite dimensional systems*, eds. S.-N. Chow and J. Hale, 61–6. Springer-Verlag, Berlin, 1987.
  [4] Invariant theory and normal form of Hamiltonian vector fields with nilpotent linear part. *Canadian Math. Soc., Conference Proceedings* **8** (1987): 353–71.
  [5] Splitting algorithm for nilpotent normal forms. *Dynamics and Stability of Systems* **2**, 4 (1988): 235–46.
Cushman, R.; Sanders, J.; and White, N.
  [1] Normal form for the $(2, n)$-nilpotent vector field, using invariant theory. *Physica D* **30** (1988): 399–412.
Dangelmayr, G.; Armbruster, D.; and Neveling, M.
  [1] A codimension three bifurcation for the laser with saturable absorber. *Zeitschrift für physik B—Condensed Matter* **59** (1985): 356–70.
Dangelmayr, G., and Guckenheimer, J.
  [1] On a four parameter family of planar vector fields. *Arch. Rat. Mech. Anal.* **97** (1987): 321–52.
Deng, B.
  [1] The Sil'nikov problem, exponential expansion, strong λ-lemma, $C^1$-linearization, and homoclinic bifurcation. *J. Diff. Eq.* **79** (1989): 189–231.
  [2] Exponential expansion with Sil'nikov's saddle-focus. *J. Diff. Eq.* **82** (1989): 156–73.
  [3] Homoclinic bifurcation with nonhyperbolic equilibria. *SIAM J. Math. Anal.* **21** (1990): 693–726.
Deprit, A.; Henrard, J.; Price, F.; and Rom, A.
  [1] Birkhoff's normalization. *Celest. Mech.* **1** (1969): 225–51.
Drachman, B.; van Gils, S.; and Zhang, Z.
  [1] Abelian integrals for quadratic vector fields. *J. für die Reine und Angewandte Mathematik* **382** (1987): 165–80.

Dulac, H.
[1] Sur les cycles limites. *Bull. Soc. Math. Fr.* **51** (1923): 45–188.
[2] Solution d'un systeme d'equations differentielles dans le voisinage de valeurs singulieres. *Bull. Soc. Math. Fr.* **40** (1912): 324–83.

Dumortier, F., and Fiddelaers, D.
[1] Quadratic models for the generic local 3-parameter bifurcations on the plane. *Trans. AMS* **326** (1991): 101–26.

Dumortier, F., and Roussarie, R.
[1] On the saddle bifurcation. In *Lecture Notes in Math.*, vol. 1455, 43–73. Springer-Verlag, New York, 1989.
[2] Tracking limit cycles escaping from rescaling domain. In *Proc. Inter Conf. on Dynamical Systems and related topics* (1990), ed. K. Shiraiwa, 80–99, World Scientific, 1991.

Dumortier, F.; Roussarie, R.; and Rousseau, C.
[1] Hilbert 16th problem for quadratic vector fields, preprint, 1991.

Dumortier, F.; Roussarie, R.; and Sotomayor, J.
[1] Generic 3-parameter families of vector fields on the plane, unfolding a singularity with nilpotent linear part. The cusp case of codimension 3. *Ergodic Theory and Dynamical Systems* **7** (1987): 375–413.
[2] Generic 3-parameter families of planar vector fields, unfolding of saddle, focus and elliptic singularities with nilpotent linear parts. In *Lecture Notes in Math.*, vol. 1480, 1–164. Springer-Verlag, New York, 1991.

Dumortier, F.; Roussarie, R.; Sotomayor, J.; and Żołądek, H.
[1] Bifurcation of Planar Vector Fields. In *Lecture Notes in Math.*, vol. 1480. Springer-Verlag, New York, 1991.

Dumortier, F., and Rousseau, C.
[1] Cubic Liénard equations with linear damping. *Nonlinearity* **3** (1990): 1015–39.

Ecalle, E. J.
[1] Finitude des cycles limites et accéléro-sommation de l'application de retour. In *Lecture Notes in Mathematics*, vol. 1455, 74–159. Springer-Verlag, New York, 1990.

Elphick, C.
[1] Global aspects of Hamiltonian normal forms. *Physics Letters A* **127** (1988): 418–24.

Elphick, C.; Tirrapegni, E.; Brachat, M. E.; Coullet, D.; and Iooss, G.
[1] A simple global characterization for normal forms of singular vector fields. *Physica D* **29** (1987): 95–117.

Farr, W. W.; Li, C.; Labouriau, I. S.; and Langford, W. F.
[1] Degenerate Hopf bifurcation formulas and Hilbert's 16th problem. *SIAM Math. Anal.* **20** (1989): 13–30.

Feng, B.
[1] Condition for generating limit cycles by bifurcation of the loop of a saddle-point separatrix. *Acta Math. Sinica, New Series* **3**, 4 (1987): 373–82.
[2] On the stability of a heteroclinic cycle for the critical case. *Acta Math. Sinica* **33**, 1 (1990): 114–34 (in Chinese).

Feng, B., and Qian, M.
[1] On the stability of a saddle-point separatrix loop and analytic criterion for its bifurcation limit cycles. *Acta Math. Sinica* **28**, 1 (1985): 53–70 (in Chinese).

Fenichel, N.
[1] Persistence and smoothness of invariant manifolds for flows. *Indiana Univ. Math. J.* **21** (1971): 193–226.
[2] Asymptotic stability with rate conditions. *Indiana Univ. Math. J.* **23** (1974): 1109–37.
[3] Asymptotic stability with rate conditions II. *Indiana Univ. Math. J.* **26** (1977): 87–93.
[4] Geometric singular perturbation theory. *J. Diff. Eq.* **31** (1979): 53–98.

Fiedler, B.
[1] Global path following of homoclinic orbits in two parameter flows, preprint.

Galin, D. M.
[1] Symplectic matrix versal deformations of linear Hamiltonian systems. *Am. Math. Soc. Transl. Ser.* **2** (1982): 1–12.
[2] On real matrices depending on parameters. *Usp. Math. Nauk* **27** (1972): 241–2 (in Russian).

Gavrilov, N. K.
[1] On bifurcation of equilibrium state with one zero and pair of purely imaginary roots. In *Methods of Qualitative Theory of Differential Equations*, 33–40. Gorki University, Gorki, 1978 (in Russian).
[2] Bifurcations of an equilibrium state with two pairs of purely imaginary roots. In *Methods of Qualitative Theory of Differential Equations*, 17–30. Gorki University, Gorki, 1980 (in Russian).

Göbber, F., and Williamowski, K. D.
[1] Liapunov approach to multiple Hopf bifurcation. *J. Math. Anal. Appl.* **71** (1979): 333–50.

Golubitsky, M., and Langford, W. F.
[1] Classification and unfolding of degenerate Hopf bifurcation. *J. Diff. Eq.* **41** (1981): 375–415.

Golubitsky, M., and Schaeffer, D. G.
[1] *Singularities and Groups in Bifurcation Theory*, Vol. I. Springer-Verlag, New York, 1985.
[2] *Singularities and Groups in Bifurcation Theory*, Vol. II. Springer-Verlag, New York, 1988.

Gruendler, J.
[1] The existence of homoclinic orbits and the method of Melnikov for systems in $\mathbb{R}^n$. *SIAM Math. Anal.* **16** (1985): 907–31.

Guckenheimer, J.
[1] Bifurcations of dynamical systems. In *C.I.M.E. Lectures* 115–232. Birkhäuser, Boston, 1980.
[2] On a codimension two bifurcation. In *Dynamical Systems and Turbulence*, eds. D. A. Rand and L.-S. Young. Warwick, 1980. In *Lecture Notes in Math.*, vol. 898, 99–142. Springer-Verlag, New York, 1981.
[3] Multiple bifurcation problems of codimension two. *SIAM J. Math. Anal.* **15**, 1 (1984): 1–49.
[4] A codimension two bifurcation with circular symmetry. In Contemporary Math. 56, *Multi-Parameter Bifurcation Theory*, ed. M. Golubitsky and J. Guckenheimer, AMS, Providence, RI, 1986.

Guckenheimer, J., and Holmes, P.
[1] *Nonlinear Oscillations, Dynamical Systems and Bifurcations of Vector Fields*. Springer-Verlag, New York, 1983.

Gustavson, F. G.
[1] On constructing formal integrals of a Hamiltonian system near an equilibrium point. *Astron. J.* **71** (1966): 670–86.

Hale, J. K.
[1] Bifurcation with several parameters. In *VII Internat. Konf. Nichtlin. Schwing.*, Vol. I, 1, Adh. Akad. Wiss. D. D. R., Akademie-Verlag, Berlin, 1977.
[2] Bifurcation near families of solutions. *Proc. Int. Conf. on Diff. Eqns., Uppsala*, (1977): 91–100, Sympos. Univ. Upsaliensis Ann. Quingentesimum Celebrantis, No. 7, Almqvist and Wiksell, Stockholm, 1977.
[3] Bifurcation from simple eigenvalues for several parameter values. *Nonlin. Anal.* **2** (1978): 491–7.
[4] Topics in dynamic bifurcation theory. In *NSF-CBMS Lectures*, vol. 47. Am. Math. Soc., Providence, RI, 1981.

Hale, J. K., and Lin, X.-B.
[1] Symbolic dynamics and nonlinear flows. *Ann. Mat. Para. Appls.* **14** (1986): 229–60.

Hale, J. K.; Magalhães, L. T.; and Oliva, W. M.
[1] An introduction to infinite dimensional systems—Geometric theory. In *Appl. Math. Sci.*, Vol. 47. Springer-Verlag, New York, 1984.

Hartman, P.
[1] *Ordinary Differential Equations*. Wiley, New York, 1964.

Hassard, B. D.; Kazarinoff, N. D.; and Wan, Y.-H.
[1] *Theory and Applications of the Hopf Bifurcation*. Cambridge University Press, Cambridge, 1980.

Henry, D.
[1] Geometric theory of semilinear parabolic equations. In *Lecture Notes in Math.*, vol. 840. Springer-Verlag, New York, 1981.

Hirsch, M. W.
[1] *Differential Topology*. Springer-Verlag, New York, 1976.

Hirsch, M. W., and Pugh, C.
[1] Stable manifolds and hyperbolic sets. *Bull. A.M.S.* **75** (1969): 149–52.

Hirsch, M. W.; Pugh, C.; and Shub, M.
[1] Invariant manifolds. In *Lecture Notes in Math.*, vol. 583. Springer-Verlag, New York, 1977.

Holmes, P.
[1] Unfolding a degenerate nonlinear oscillator: a codimension two bifurcation. In *Nonlinear Dynamics*, ed. R. Helleman, 437–88. N.Y. Acad. Sci., New York, 1980.
[2] Center manifolds, normal forms, and bifurcation of vector fields. *Physica D* **2** (1981): 449–81.

Hsu, L., and Favretto, L.
[1] Recursive bifurcation formulae for normal form and center manifold theory. *J. Math. Anal. Appl.* **101** (1984): 562–74.

Huseyin, K., and Yu, P.
[1] Bifurcations associated with a simple zero eigenvalue and two pairs of pure imaginary eigenvalues. In *Oscillations, Bifurcations, and Chaos*, C.M.S. Conference Proceedings, eds. F. V. Atkinson, W. F. Langford, A. B. Mingarelli. A.M.S., Providence, RI, 1987.

Il'yasehenko, Yu. S.
[1] Zeros of special abelian integrals in a real domain. *Funct. Anal. Appl.* **11** (1977): 309–11.

[2] The multiplicity of limit cycles arising from perturbations of the form $W = P2/Q1$ of a Hamiltonian equation in the real and complex domain. *Am. Math. Soc. Transl. Ser. 2* **118** (1982): 191–202.

[3] Finiteness theorems for limit cycles. *Russian Math. Surveys* **40** (1990): 143–200.

Iooss, G., and Langford, W. F.

[1] Conjecture on the routes to turbulence via bifurcation. In *Nonlinear Dynamics*, ed. R. Helleman, 498–506. N.Y. Acad. Sci., New York, 1980.

Joyal, P.

[1] Generalized Hopf bifurcation and its dual generalized homoclinic bifurcation. *SIAM J. Appl. Math.* **48** (188): 481–96.

[2] The cusp of order *n*. *J. Diff. Eq.* **88**, 1 (1990): 1–14.

Joyal, P., and Rousseau, C.

[1] Saddle quantities and applications. *J. Diff. Eq.* **78**, 2 (1989): 374–99.

Kelley, A.

[1] The stable, center-stable, center, center-unstable and unstable manifolds. *J. Diff. Eq.* **3** (1967): 546–570.

Khazin, L. G., and Shnol, E. E.

[1] *Stability of Critical Equilibrium States*. Manchester Univ. Press, Manchester, 1991 (Russian original, 1985).

Khorozov, E. I.

[1] Versal deformations of equivariant vector fields in the case of symmetry of order 2 and 3. *Trans. of Petrovski Seminar* **5** (1979): 163–92 (in Russian).

Khovansky, A. G.

[1] Real analytic manifolds with finiteness properties and complex abelian integrals. *Funct. Anal. Appl.* **18** (1984): 119–27.

Kisaka, M.; Kokubu, H.; and Oka, H.

[1] Supplement to homoclinic doubling bifurcation in vector fields. In *Dynamical Systems*, Santiago 1990, eds. R. Bamon et al., *Pitman Research Notes in Math.*, 285 (1993), 92–116. Longman Scientific and Technical.

[2] Bifurcation of *N*-homoclinic orbits and *N*-periodic orbits in vector fields. *J. Dynamics Diff. Eqn.*, **5** (1993): 305–57.

Knobloch, E.

[1] Normal forms for bifurcations at a double zero eigenvalue. *Phys. Lett. A.* **115**, 5 (1986): 199–201.

Koçak, H.

[1] Normal forms and versal deformations of linear Hamiltonian systems. *J. Diff. Eq.* **51** (1984): 359–407.

Langford, W. F.

[1] Periodic and steady mode interactions lead to tori. *SIAM J. Appl. Math.* **37** (1979): 22–48.

Lasalle, M. G.

[1] Une démonstration du théorème de division pour les fonctions différentiables. *Topology* **12** (1973): 41–62.

Leontovich, E. A.

[1] The creation of limit cycles from a separatrix. *Sov. Math. Dokl.* **4** (1951): 641–4.

Li, C.

[1] Two problems of planar quadratic systems. *Scientia Sinica (Series A)* **26** (1983): 471–81.

[2] The quadratic system possessing two centers. *Acta Math. Sinica* **28**, 5 (1985): 644–8 (in Chinese).

[3] Non-existence of limit cycle around a weak focus of order three for any quadratic system. *Chinese Ann. Math.* **7B**, 2 (1986): 174–90.

Li, C., and Rousseau, C.

[1] A sysem with three limit cycles appearing in a Hopf bifurcation and dying in a homoclinic bifurcation, the cusp of order 4. *J. Diff. Eq.* **79** (1989): 132–67.

[2] Codimension 2 symmetric homoclinic bifurcation. *Can. J. Math.* **42** (1990): 191–212.

Li, C.; Rousseau, C.; and Wang, X.

[1] A simple proof for the unicity of the limit cycle in the Bogdanov–Takens system. *Can. Math. Bull.* **33** (1990): 84–92.

Li, J., and Huang, Q.

[1] Bifurcations of limit cycles forming compound eyes in the cubic system. *Chin. Ann. Math.* **8B** (1987): 391–403.

Li, J., and Li, C.

[1] Planar cubic Hamiltonian systems and distributions of limit cycles of $(E_3)$. *Acta Math. Sinica* **28** (1985): 509–21.

Li, J., and Liu, Y.

[1] Theory of values of singular point in complex autonomous differential systems. *Science in China (Series A)* **33**, (1990): 10–23.

Li, W.; Li, C.; and Zhang, Z.

[1] Unfolding the critical homoclinic orbit of a class of degenerate equilibrium points, to appear in *Dynamical Systems*, Nankai Series in Pure, Applied Mathematics and theoretical Physics, vol. 4, eds. S.-T. Liao, T.-R. Ding, and Y.-Q. Ye, World Scientific Publishing, Singapore, 1993.

Lin, X.-B.

[1] Heteroclinic bifurcation and singularly perturbed boundary problems. *J. Diff. Eq.* **84** (1990): 319–82.

Lyapunov, A.

[1] Probleme general de la stabilite du mouvement. In *Ann. Math. Studies*, vol. 17. Princeton Univ. Press, Princeton, NJ, 1947 (reproduction of a French translation dated 1907 of a Russian memoir dated 1892).

Ma, Z., and Wang, E.

[1] Stability of homoclinic loop and condition for bifurcating limit cycles from the loop. *Chinese Ann. Math.* **4A** (1983): 10–110 (in Chinese).

Marsden, J. and McCracken, M.

[1] *The Hopf Bifurcation and Its Applications.* Springer-Verlag, New York, 1976.

Marsden, J. and Scheurle, J.

[1] The construction and smoothness of invariant manifolds by the deformation methods, *SIAM J. Math. An.* **18**, 5 (1987): 1261–74.

Medved, M.

[1] The unfoldings of a germ of vector fields in the plane with a singularity of codimension 3. *Czech. Math. J.* **35**, 110 (1985): 1–41.

Meer, J. C. van der

[1] The Hamiltonian Hopf bifurcation. *Lect. Notes Math.*, **1160**, Springer, 1985.

[2] Nonsemisimple 1:1 resonance at an equilibrium. *Cel. Mech.* **27** (1982): 131–49.

Mees, A. I.
  [1] *Dynamics of Feedback Systems*. Wiley, New York, 1981.
Melnikov, V. K.
  [1] On the stability of the center for time periodic perturbations. *Trans. Moscow Math. Soc.* **12** (1963): 1–57.
Meyer, K. R.
  [1] Normal forms for Hamiltonian systems. *Celestial Mech.* **9** (1974): 517–22.
  [2] The implicit function theorem and analytic differential equations. In *Lecture Notes in Math.*, vol. 468, 191–208. Springer-Verlag, New York, 1975.
  [3] Normal forms for the general equilibrium. *Funkcial. Ekvac.* **27**, 2 (1984): 261–71.
Mielke, A.
  [1] A reduction principle for nonautonomous systems in infinite-dimensional spaces. *J. Diff. Eq.* **65** (1986): 68–88.
Mitionagumo, K.
  [1] On the normal forms of differential equations in the neighborhood of an equilibrium point. *Tamura Takazuiki* 1957: 221–34.
Moser, J.
  [1] *Stable and Random Motions in Dynamical Systems*. Princeton University Press, Princeton, 1973.
  [2] Lectures on Hamiltonian systems. *Mem. AMS* **81** (1968): 1–60.
Nagumo, M., and Ise, K.
  [1] On the normal forms of differential equations in the neighborhood of an equilibrium point. *Osaka J. Math.* **9** (1957): 221–34.
Neishtadt, A. I.
  [1] Bifurcations of phase portrait of certain system arising in the problem of loss of stability of self-oscillations near the resonance 1:4. *Prikl. Mat. Mekh.* **42** (1978): 830–40.
Nikolenko, N. V.
  [1] The method of Poincaré normal forms in problem of integrability of equations of evolution type. *Russ. Math. Surv.* **41**, 5 (1986): 63–114.
Palmer, K.
  [1] On the stability of the center manifold. *J. Appl. Math. Phys.* (*ZAMP*) **38** (1987): 273–8.
  [2] Exponential dichotomies and transversal homoclinic points. *J. Diff. Eq.* **33** (1979): 368–405.
Petrov, G. S.
  [1] On the number of complete elliptic integral. *Funct. Anal. Appl.* **18**, 2 (1984): 74–5.
  [2] Complex zeros of an elliptic integral. *Funct. Anal. Appl.* **21**, 3 (1987): 87–8.
Pliss, V. A.
  [1] Principal reduction in the theory of stability of motion. *Izv. Akad. Nauk SSSR, Mat. Ser.* **28** (1964): 1044–6 (in Russian); *Soviet Math.* **5** (1964): 247–50.
Poenaru, V.
  [1] Singularités $C^\infty$ en présence de symétrie. In Lecture Notes in Math. vol. 510. Springer-Verlag, New York, 1976.
Poincaré, H.
  [1] Sur les courbes definies par les equations differentielles. In *Euvres de Henri Poincaré*, vol. 1, 85–162. Gauthier Villars, Paris, 1951.

[2] Thesis, 1879; also *Oeuvres I*, 59–129, Gauthier Villars, Paris, 1928.

Rand, R. H., and Armbruster, D.
[1] *Perturbation Methods, Bifurcation Theory and Computer Algebra.* Springer-Verlag, New York, 1987.

Rand, R. H., and Keith, W. L.
[1] Normal forms and center manifold calculation on MACSYMA. In *Applications of Computer Algebra*, ed. R. Pavelle, 309–28. Klumer Academic Publishers, Boston, 1985.

Roels, J., and Louterman, G.
[1] Normalisation des systèms, linèaires canoniques et application au probléme restreint des toris corps. *Celest. Mech.* **3** (1970): 129–40.

Roussarie, R.
[1] On the number of limit cycles which appear by perturbation of separatrix loop of planar vector fields. *Bol. Soc. Bras. Mat.* **17** (1986): 67–101.
[2] A note on finite cyclicity and Hilbert's 16th problem. In *Lecture Notes in Math.*, vol. 1331, 161–8. Springer-Verlag, 1988.
[3] Cyclicré finie des lacets et des points cuspidaux. *Nonlinearity* **2** (1989): 73–117.

Rousseau, C.
[1] Codimension 1 and 2 bifurcations of fixed points of diffeomorphisms and of periodic solutions of vector fields. *Annales des Sciences Math. de Québec* **13** (1989): 55–91.
[2] Universal unfolding of a singularity of a symmetric vector field with 7-jet $C^{\infty}$-equivalent to $y\partial/\partial x + (\pm x^3 \pm x^6 y)\partial/\partial y$. In *Lecture Notes in Math.*, vol. 1455, 334–54. Springer-Verlag, New York, 1989.

Rousseau, C., and Schlomiuk, D.
[1] Generalized Hopf bifurcations and applications to planar quadratic systems. *Ann. Polon. Math.* **49** (1988): 1–16.

Rousseau, C., and Żoładek, H.
[1] Zeros of complete elliptic integrals for 1:2 resonance. *J. Diff. Eq.* **94** (1991): 41–54.

Ruelle, D., and Takens, F.
[1] On the nature of turbulence. *Commun. Math. Phys.* **20** (1971): 167–92.

Sanders, J.
[1] Melnikov's method and averaging. *Celest. Mech.* **28** (1982): 171–81.

Sanders, J. A., and Verhulst, D. F.
[1] *The Theory of Averaging.* Springr-Verlag, New York, 1986.

Schecter, S.
[1] The saddle-node separatrix-loop bifurcation. *SIAM J. Math. Anal.* **18** (1987): 1142–56.
[2] $C^p$ singularity theory and heteroclinic bifurcation with a distinguished parameter, *J. Diff. Eq.* 99 (1992): 306–41.

Schlomiuk, D.
[1] Algebraic integrals, integrability and the problem of center. *Transactions AMS*, **338**, 2 (1993): 799–841.

Sell, G. R.
[1] An optimality condition for approximate inertial manifolds, Turbulence in Fluid Flows: A Dynamical Systems Approach, in *IMA Volumes in Math. and Its Appl.* **55**, eds. G. R. Sell, 165–86, Springer-Verlag, 1993.

Sell, G. R., and You, Y.
[1] Inertial manifolds: The non-self adjoint case, *J. Diff. Eq.* **96** (1992): 203–55.

Shoshitaishvili, A. N.
[1] Bifurcations of topological type of singular points of parameterized vector fields. *Funct. Anal. Appl.* **6**, 2 (1972): 169–70.

Sibirskii, K. S.
[1] On the number of limit cycles in a neighborhood of singular points. *J. Diff. Eq.* **1** (1965): 36–47 (in Russian).

Siegel, C. L.
[1] Über die normalform analytischer differential gleichungen in der nahe einer gleichgewichtslösung. *Math. Phys. kl. IIa* **5** (1952): 21–36.
[2] *Vorlesungen über Himmelsmechanik*. Springer-Verlag, Berlin, 1956.

Sijbrand, J.
[1] Properties of center manifolds. *Trans. AMS* **289** (1985): 431–69.

Silnikov, L. P.
[1] On the generation of a periodic motion from trajectories doubly asymptotic to an equilibrium state of saddle type. *Math. USSR Sb.* **6** (1968): 427–38.
[2] A contribution to the problem of the structure of an extended neighborhood of a structurally stable equilibrium state of saddle-focus type. *Math. USSR Sb.* **10** (1970): 91–102.

Sotomayor, J.
[1] Generic one-parameter families of vector fields on two-dimensional manifolds. *Publ. Math. IHES* **43** (1973): 5–46.
[2] *Curvas Definides Por Equacões diferenciais no Plano*. IMPA, Rio de Janeiro, 1981.

Starzhinskii, V. M.
[1] Normal forms of the fourth-order for nonlinear oscillations. *Mech. of Solids* **7**, 1 (1973): 1–7.
[2] *Applied Methods in the Theory of Nonlinear Oscillations*. Mir Publishers, Moscow, 1980 (Russian original, 1977).

Sternberg, S.
[1] Local construction and a theorem of Poincaré. *Am. J. Math.* **79** (1957): 809–24.
[2] On the structure of local homomorphisms of euclidean $n$-space, I. *Am. J. Math.* **80** (1958): 623–31.
[3] On the structure of local homomorphisms of euclidean $n$-space, II. *Am. J. Math.* **81** (1959): 578–605.
[4] Finite Lie groups and the formal aspects of dynamical systems. *J. Math. Mech.* **10** (1961): 451–74.

Takens, F.
[1] Normal forms for certain singularities of vector fields. *An. Inst. Fourier* **23**, 2 (1973): 163–95.
[2] Forced oscillations and bifurcations: Applications of global analysis I. In *Commun. Math.*, vol. 3. Inst. Rijksuniv. Utrecht, 1974.
[3] Singularities of vector fields. *Publ. Math. IHES* **43** (1974): 47–100.
[4] Applications of global analysis. Symposium, State Univ., Utrecht, 1973.

Temam, R.
[1] Induced trajectories and approximate inertial manifold. *Math. Modelling Numerical Anal.* **23**, 3 (1989): 541–61.
[2] Inertial manifolds and multigrid methods. *SIAM J. Math. Anal.* **21** (1990): 154–78.

Tirapegui, E.
  [1] Normal forms for deterministic and stochastic systems. In *Dynamical Systems, Lecture Notes in Math.*, vol. 1331, eds. A Dold and E. Echmann, 240–50. Springer-Verlag, New York, 1986.
Ushiki, S.
  [1] Normal forms for singularities of vector fields. *Japan J. Appl. Math.* 1 (1984): 1–37.
van Gils, S. A.
  [1] A note on abelian integrals and bifurcation theory. *J. Diff. Eq.* **59** (1985): 437–9.
van Gils, S. A., and Horozov, E.
  [1] Uniqueness of limit cycles in planar vector fields which leave the axes invariant. In Multiparameter Bifurcation Theory, eds. M. Golubitsky and J. Guckenheimer, 117–29. Amer. Math. Soc., Providence, RI, 1986.
van Gils, S. A., and Sanders, J. A.
  [1] On a codimension two bifurcation with two pairs of imaginary eigenvalues, preprint.
van Strien, S.
  [1] Center manifolds and not $C^\infty$. *Math. Z.* **166** (1979): 143–5.
Vanderbauwhede, A.
  [1] Local bifurcation and symmetry. In *Res. Notes in Math.*, vol. 75. Pitman, London, 1982.
  [2] *Center Manifold Theory.* Univ. Ghent, Ghent, 1987.
  [3] Center manifolds, normal forms and elementary bifurcations. In *Dynamics Reported*, Vol. 2, eds. U. Kirchgraber and O. Walther, 89–169. Wiley, New York, 1989.
Vanderbauwhede, A., and Iooss, G.
  [1] Center manifold theory in infinite dimensions, in *Dynamics Reported*, 1 New Series, eds. C.K.R.T. Jones et al., 125–63. Springer-Verlag, 1992.
Vanderbauwhede, A., and van Gils, S. A.
  [1] Center manifolds and contractions on a scale of Banach spaces. *J. Funct. Anal.* **72** (1987): 209–24.
Varadarjan, V. S.
  [1] *Lie Groups, Lie Algebras and Their Representations.* Springer-Verlag, New York, 1984.
Varchenko, A. N.
  [1] Estimate of the number of zeros of abelian integrals depending on parameter and limit cycles. *Funct. Anal. Appl.* **18**, 2 (1984): 98–108.
Villari, G., and Zhang, Z.
  [1] Periodic solutions of a switching dynamical sysem in the plane. *Appl. Anal.* **26**, 3 (1988): 177–98.
Wan, Y.-H.
  [1] Bifurcation into invariant tori at point of resonance. *Arch. Rat. Mech. Anal.* **68** (1978): 345–57.
  [2] On the uniqueness of invariant manifolds. *J. Diff. Eq.* **24** (1977): 268–73.
Wang, D.
  [1] Hopf bifurcation at the nonzero foci in 1:4 resonance. *Acta Math. Sinica, New Series* **6** (1990): 10–17.

[2] An introduction to the normal form theory of ordinary differential equations. *Adv. in Math.* **19**, 1 (1990): 38–71.

[3] Applications of the representation theory of $sl(2, \mathbb{R})$ to normal form theory. *Science in China (Series A)* **33**, 8 (1990): 923–33.

[4] Adjoint operator method in the normal form theory of nonlinear Hamiltonian systems. In *Proceedings of the International Conference on Bifurcation Theory and its Numerical Analysis* (Xian, China, July 1988), eds. K. Li et al., 429–37, Xi'an Jiastong Univ. Press, Xi'an, 1989.

[5] A recursive formula and its application to computatons of normal forms and focal values, in *Dynamical Sysems*, Nankai Series in Pure, Applied Mathematics and Theoretical Physics, vol. 4, eds. S.-T. Liao, T.-R. Ding, and X. Q. Ye, 238–47. World Scientific Publishing, Singapore, 1993.

[6] The critical points of the period function of $x'' - x^2(x - \alpha)(x - 1) = 0$ $(0 \leq \alpha < 1)$. *Nonlinear Analysis, T.M.A.* **11**, 9 (1987): 1029–50.

Wells, J. C.
[1] Invariant manifolds of non-linear operators. *Pacific J. Math.* **62** (1976): 285–93.

Whittaker, E. T.
[1] On the solutions of dynamical problems in terms of trigonometric series. *Proc. London Math. Soc.* **34** (1902): 206–21.

Wiggins, S.
[1] *Global Bifurcations and Chaos: Analytical Methods.* Springer-Verlag, New York, 1988.

[2] *Introduction to Applied Nonlinear Dynamical Systems and Chaos.* Springer-Verlag, New York, 1990.

Williamson, J.
[1] On the algebraic problem concerning the normal forms of linear dynamical systems. *Amer. J. Math.* **58** (1936): 141–63.

Wintner, A.
[1] On the linear conservative dynamic sysems. *Ann. Mat. pura appl. Serie IV* XIII (1935), 105–13.

Xiao, D.
[1] A universal unfolding of a singular vector field with codimension three. Ph.D. thesis, Beijing University, 1991.

Ye, Y.
[1] *Theory of Limit Cycles.* Amer. Math. Soc., Providence, RI, 1986. Monographs, Vol. 66, Trans. Math.

Yi, Y.
[1] Generalized integral manifold theorem, prerint, 1990.

Zehnder, E.
[1] A simple proof of a generalization of a theorem by C. L. Siegel. *Lect. Notes Math.* **597** (1977), 855–66.

Zhang, P., and Cai, S.
[1] The quadratic system with second or third order weak saddle. *Acta Math. Sinica* **30** (1987): 560–5 (in Chinese). English translation.

Zhang, Z.; Ding, T.; Huang, W.; and Dong, Z.
[1] *The Qualitative Theory of Differential Equations.* Science Press, Beijing, 1985 (in Chinese). English Translation.

Zhang, Z., and Li, B.
[1] High order Melnikov functions and the problem of uniformity in global bifurcation, In *Research Report*, vol. 4. Institute of Math. and Dept. of

Math., Peking Univ., 1989; *Ann. Mat. pura appl.*, (IV), CLXI (1992), 181–212.

Żoładek, H.

[1] On the versality of certain family of vector fields on the plane. *Math. USSR Sb.* **48** (1984): 463–92 (in Russian).

[2] Bifurcations of certain family of planar vector fields tangent to axes. *J. Diff. Eq.* **67** (1987): 1–55.

[3] Abelian integrals in unfoldings of codimension 3 singular planar vector field. In *Lecture Notes in Math.*, vol. 1480, 165–224. Springer-Verlag, New York, 1991.

[4] Abelian integrals in non-symmetric perturbation of symmetric Hamiltonian vector field, preprint, Warsaw Univ., 1988.

# Index